Physics and Applications of Quantum Wells and Superlattices

NATO ASI Series

Advanced Science Institutes Series

A series presenting the results of activities sponsored by the NATO Science Committee, which aims at the dissemination of advanced scientific and technological knowledge, with a view to strengthening links between scientific communities.

The series is published by an international board of publishers in conjunction with the NATO Scientific Affairs Division

A	Life Sciences	Plenum Publishing Corporation
B	Physics	New York and London
C	Mathematical and Physical Sciences	Kluwer Academic Publishers
D	Behavioral and Social Sciences	Dordrecht, Boston, and London
E	Applied Sciences	
F	Computer and Systems Sciences	Springer-Verlag
G	Ecological Sciences	Berlin, Heidelberg, New York, London,
H	Cell Biology	Paris, and Tokyo

Recent Volumes in this Series

Volume 165—Relativistic Channeling
edited by Richard A. Carrigan, Jr., and James A. Ellison

Volume 166—Incommensurate Crystals, Liquid Crystals, and Quasi-Crystals
edited by J. F. Scott and N. A. Clark

Volume 167—Time-Dependent Effects in Disordered Materials
edited by Roger Pynn and Tormod Riste

Volume 168—Organic and Inorganic Low-Dimensional Crystalline Materials
edited by Pierre Delhaes and Marc Drillon

Volume 169—Atomic Physics with Positrons
edited by J. W. Humbertson and E. A. G. Armour

Volume 170—Physics and Applications of Quantum Wells and Superlattices
edited by E. E. Mendez and K. von Klitzing

Volume 171—Atomic and Molecular Processes with Short Intense Laser Pulses
edited by André D. Bandrauk

Volume 172—Chemical Physics of Intercalation
edited by A. P. Legrand and S. Flandrois

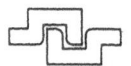

Series B: Physics

Physics and Applications of Quantum Wells and Superlattices

Edited by

E. E. Mendez

IBM Thomas J. Watson Research Center
Yorktown Heights, New York

and

K. von Klitzing

Max Planck Institute for Solid State Research
Stuttgart, Federal Republic of Germany

Plenum Press
New York and London
Published in cooperation with NATO Scientific Affairs Division

Proceedings of a NATO Advanced Study Institute on
Physics and Applications of Quantum Wells and Superlattices,
held April 21–May 1, 1987
in Erice, Sicily, Italy

Library of Congress Cataloging in Publication Data

NATO Advanced Study Institute on Physics and Applications of Quantum Wells
    and Superlattices (1987: Erice, Sicily)
    Physics and applications of quantum wells and superlattices.

    (NATO ASI Series. Series B, Physics; vol. 170)
    "Proceedings of a NATO Advanced Study Institute on Physics and Applica-
tions of Quantum Wells and Superlattices, held April 21–May 1, 1987, in Erice,
Sicily, Italy"—T.p. verso.
    "Published in cooperation with NATO Scientific Affairs Division."
    Includes bibliographical references and index.
    1. Superlattices as materials—Congresses. 2. Quantum wells—Congresses. 3.
Semiconductors—Congresses. 4. Molecular beam epitaxy—Congresses. I.
Mendez, E. E. II. Klitzing, K. von. III. North Atlantic Treaty Organization. Scientific
Affairs Division. IV. Title. V. Series.
QC611.8.S86N38   1987   530.4'1   88-2510
ISBN-13: 978-1-4684-5480-2   e-ISBN-13: 978-1-4684-5478-9
DOI: 10.1007/978-1-4684-5478-9

PREFACE

This book contains the lectures delivered at the NATO Advanced
Study Institute on "Physics and Applications of Quantum Wells and
Superlattices", held in Erice, Italy, on April 21-May 1, 1987. This
course was the fourth one of the International School of Solid-State
Device Research, which is under the auspices of the Ettore Majorana
Center for Scientific Culture.

In the last ten years, we have seen an enormous increase in research in the field of Semiconductor Heterostructures, as evidenced
by the large percentage of papers presented in recent international
conferences on semiconductor physics. Undoubtfully, this expansion
has been made possible by dramatic advances in materials preparation,
mostly by molecular beam epitaxy and organometallic chemical vapor
deposition. The emphasis on epitaxial growth that was prevalent at
the beginning of the decade (thus, the second course of the School,
held in 1983, was devoted to Molecular Beam Epitaxy and
Heterostructures) has given way to a strong interest in new physical
phenomena and new material structures, and to practical applications
that are already emerging from them.

The purpose of this Institute has been to address in an integrated
form the basic physical concepts of semiconductor wells and
superlattices, and their relation to technological applications based
in these novel structures. Although the goal has not been so much
to review the most recent results in the field, as to present the
fundamental ideas and how they can be implemented in new devices, we
hope that the reader will find these lectures also valuable as a re-
source of current research activity in the field.

After an introductory review of the evolution of semiconductor
superlattices since their inception in 1969, the book is divided in
four parts. First, the electronic band structure of superlattices
is discussed in Chapter 2, followed by three chapters describing the
use of molecular beam epitaxy to the growth of heterostructures based

on III-V, II-VI, and IV compounds, respectively. The second part focuses on their transport properties, both parallel and perpendicular to the interface direction. The drastic influence of a strong magnetic field on them -the best example of which is the quantum Hall effect- occupies a significant fraction of this section.

Part three is devoted to optical properties. The characteristics of radiative recombination are studied in Chapter 12, which also considers the principles of quantum-well lasers. The inelastic scattering of light by heterostructures is discussed in the next chapter, followed by a study of far-infrared absorption, and of the effect of a magnetic field on the optical properties. Finally, part four presents applications of quantum wells and superlattices based on either their electrical or optical characteristics, including a discussion of devices that might integrate them both.

We are thankful to A. Gabriele, who provided at the Ettore Majorana Center the right atmosphere for the celebration of a scientific retreat. We thank also L. Esaki, director of the School, for his support during the different phases of the Institute, and to L. L. Chang and K. Ploog, directors of a previous edition of the School, for their advice in the organization of the Course. Finally, we are grateful to the NATO Scientific Affairs Division, to the Italian Ministries of Education and Scientific and Technological Research, to the Sicilian Regional Government, to the U.S. Army European Research Office, to the European and American Physical Societies, and to International Business Machines, whose sponsorhip has made possible this International School.

CONTENTS

INTRODUCTION

A Perspective in Quantum-Structure Development. 1
    L. Esaki

BASIC PROPERTIES AND MATERIALS GROWTH

Electronic States in Semiconductor Heterostructures 21
    G. Bastard

Molecular Beam Epitaxy of Artificially Layered III-V
        Semiconductors on an Atomic Scale 43
    K. Ploog

Molecular Beam Epitaxial Growth and Properties of Hg-
        based Microstructures 71
    J. P. Faurie

Strained Layer Superlattices 101
    E. Kasper

ELECTRICAL PROPERTIES

Electrical Transport in Microstructures 133
    F. Stern

Physics of Resonant Tunneling in Semiconductors 159
    E. E. Mendez

Thermodynamic and Magneto-Optic Investigations of the Landau-
        level Density of States for 2D Electrons 189
    E. Gornik

High Field Magnetotransport: Lectures I and II:
        Analysis of Shubnikov-de Haas Oscillations and
        Parallel Field Magnetotransport 217
    R. J. Nicholas

Physics and Applications of The Quantum Hall Effect 229
    K. von Klitzing

High Field Magnetotransport: Lecture III:
        The Fractional Quantum Hall Effect 249
    R. J. Nicholas, R. G. Clark, A. Usher, J. R. Mallett,
    A. M. Suckling, J. J. Harris, and C. T. Foxon

OPTICAL PROPERTIES

Optical Properties of Quantum Wells 261
   C. Weisbuch

Raman Spectroscopy for the Study of Semiconductor
   Heterostructures and Superlattices 301
   G. Abstreiter

Far-Infrared Spectroscopy in Two-Dimensional Electronic
   Systems 317
   D. Heitmann

Magneto-optical Properties of Heterojunctions, Quantum Wells,
   and Superlattices 347
   J. C. Maan

APPLICATIONS

Band-gap Engineering for New Photonic and Electronic
   Devices 377
   F. Capasso

Hot-Electron Spectroscopy and Transistor Design 393
   M. Kelly

Novel Tunneling Structures: Physics and Device Impli-
   cations 403
   M. J. Kelly, R. A. Davies, N. R. Couch,
   B. Movaghar and T. M. Kerr

Opto-electronics in Semiconductor Quantum Wells
   Structures: Physics and Applications 423
   D. S. Chemla

INDEX 437

A PERSPECTIVE IN QUANTUM-STRUCTURE DEVELOPMENT

L. Esaki

IBM Thomas J. Watson Research Center
Yorktown Heights, New York 10598, U.S.A.

INTRODUCTION

In 1969, research on quantum structures was initiated with a proposal
of an ''engineered'' semiconductor superlattice by Esaki and Tsu (1)
(2). In anticipation of advancement in epitaxy, we envisioned two
types of superlattices with alternating ultrathin layers: doping and
compositional, as shown at the top and bottom of Figure 1, respectively.

The idea occurred to us while examining the feasibility of
structural formation by heteroepitaxy for such ultrathin barriers and
wells to exhibit resonant electron tunneling (3). Figure 2 shows the
calculated transmission coefficient as a function of electron energy
for a double barrier. When the well width L, the barrier width and
height, and the electron effective mass m^* are assumed to be 50 Å,
20Å 0.5eV and one tenth of the free electron mass m_o, respectively,
the bound energies, E_1 and E_2, in the quantum well are derived to be
0.08 and 0.32 eV. As shown in the insert of the figure, if the energy
of incident electrons coincides with such bound energies, the
electrons tunnel through both barriers without attenuation. Such
unity transmissivity at the resonant condition arises from the de-
structive interference of reflected electron waves from inside (R)
with incident waves (I), so that only transmitted waves (T) remain.

The superlattice (SL) was considered a natural extension of
double- and multi-barrier structures where quantum effects are expected to prevail. An important parameter relevant to the observation
of such effects is the phase-coherent distance or, roughly, the

1

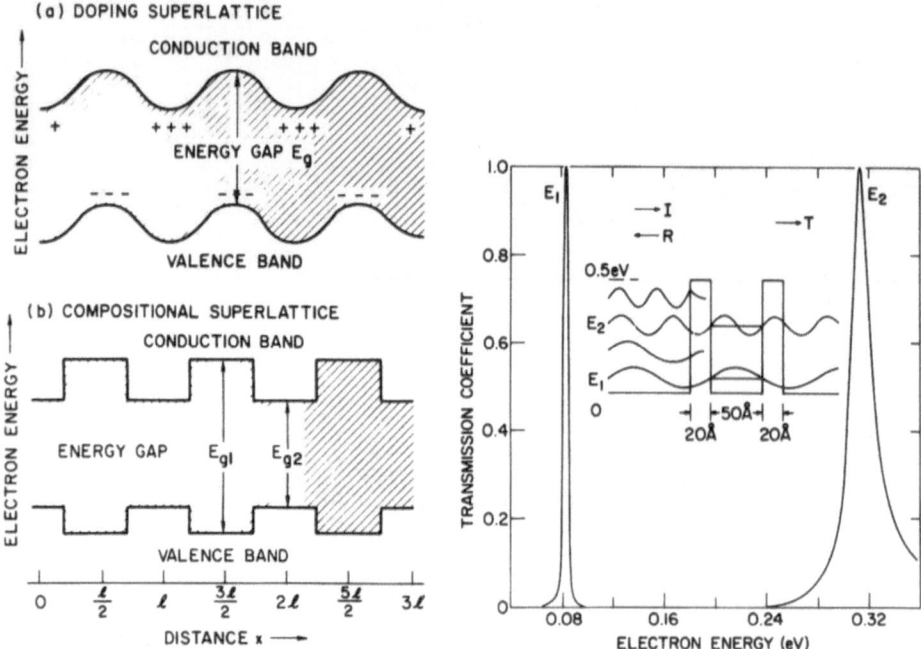

Fig. 1. Spacial variations of the conduction and valence bandedges
in two types of superlattices: doping (top) and compositional.

Fig. 2. Transmission coefficient versus election energy for a double
barrier shown in the insert. At the resonant condition that
the energy of incident elections coincides with one of those
of the bound states, E_1 and E_2, the electrons tunnel through
both barriers without attenuation.

electron inelastic mean free path, which depends heavily on bulk and
interface quality of crystals and also on temperature and values of
the effective mass. As schematically illustrated in Fig. 3, if
characteristic dimensions such as SL periods and well widths are reduced to less than the phase-coherent distance, the entire electron
system will enter a ''mesoscopic'' quantum regime of reduced dimensionality, being placed in the scale between the macroscopic and the
microscopic.

In the early 1970s, we initiated the attempt of the seemingly
formidable task of engineering nanostructures in the search for novel
quantum phenomena (4). It was theoretically shown that the introduction of the SL potential perturbs the band structure of the host
materials, yielding unusual electronic properties of quasi-two-
dimensional character (1) (2). Fig. 4 shows the density of states
$\rho(E)$ for electrons with $m^* = 0.067m_o$ in an SL with a well width of
100Å and the same barrier width, where the first three subbands are
indicated with dashed curves. The figure also includes, for compar-

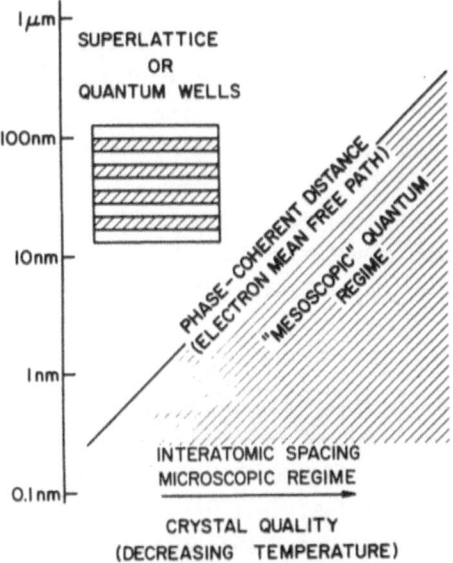

MACROSCOPIC REGIME

Fig. 3. Schematic illustration of a ''mesoscopic'' quantum regime
(hatched) with a superlattice or quantum wells in the insert.

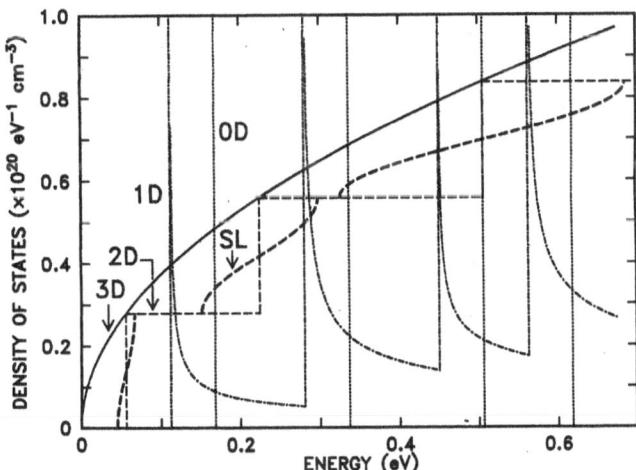

Fig. 4. Comparison of density of states in the three-dimensional (3D)
electron system with those of a superlattice, and the two-
dimensional (2D), one-dimensional (1D), and zero-dimensional
(0D) electron systems.

3

ison, a parabolic curve $E^{\frac{1}{2}}$ for a three-dimensional (3D) system, a steplike density of states for a two-dimensional (2D) system (quantum well), a curve $\sum_{mn}(E-E_m-E_n)^{-\frac{1}{2}}$ for a one-dimensional (1D) system (quantum pipe or wire), and a delta function $\sum\delta(E-E_l-E_m-E_n)$ for a zero-dimensional (0D) system (quantum box or dot) where the quantum unit is taken to be 100Å for all cases and the barrier height is assumed to be infinite in obtaining the quantized energy levels, E_l, E_m and E_n. Notice that the ground state energy increases with decrease in dimensionality if the quantum unit is kept constant. Each quantized energy level in 2D, 1D and 0D is identified with the one, two and three quantum numbers, respectively. The unit for the density of states here is normalized to $eV^{-1}cm^{-3}$ for all the dimensions, although $eV^{-1}cm^{-2}$, $eV^{-1}cm^{-1}$ and eV^{-1} may be commonly used for 2D, 1D and 0D, respectively.

The analysis of the electron dynamics in the SL direction predicted an unusual current-voltage characteristic including a negative differential resistance, and even the occurrence of ''Bloch oscillations.'' The calculated Bloch frequency f is as high as 250 GHz from the equation $f = eFd/h$, for an applied field F and a superlattice period d of $10^3V/cm$ and 100Å, respectively.

Esaki, Chang and Tsu (5) reported an experimental result on a GaAs-GaAsP SL with a period of 200Å synthesized with CVD (chemical vapor deposition) by Blakeslee and Aliotta (6). Although transport measurements failed to show any predicted effect, this system probably constitutes the first strained-layer SL having a lattice mismatch 1.8% between GaAs and $GaAs_{0.5}P_{0.5}$.

Esaki et al. (7) found that an MBE (molecular beam epitaxy)-grown GaAs-GaAlAs SL exhibited a negative resistance in its transport properties, which was, for the first time, interpreted in terms of the above-mentioned SL effect. Although our early efforts focused on transport measurements, Tsu and Esaki (8) calculated optical non-linear response of conduction electrons in an SL medium. Since the first proposal and early attempt, the field of semiconductor SLs and quantum wells (QWs) has proliferated extensively in a cross-disciplinary environment (9).

II. EPITAXY AND SUPERLATTICE GROWTH

Heteroepitaxy is of fundamental interest for the SL and QW growth. Innovations and improvements in growth techniques such as MBE (10), MOCVD (metalorganic chemical vapor deposition) and MOMBE during the last decade have made possible high-quality heterostructures. Such

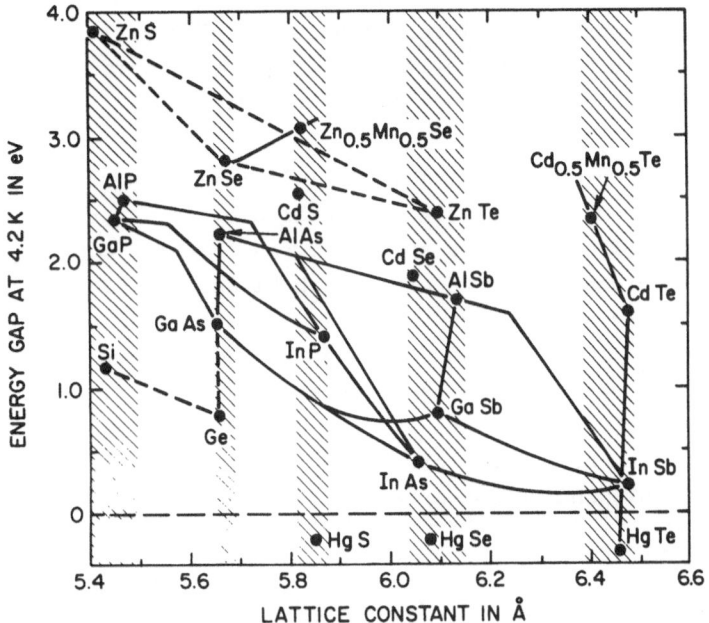

Fig. 5. Plot of energy gaps at 4.2K versus lattice constants.

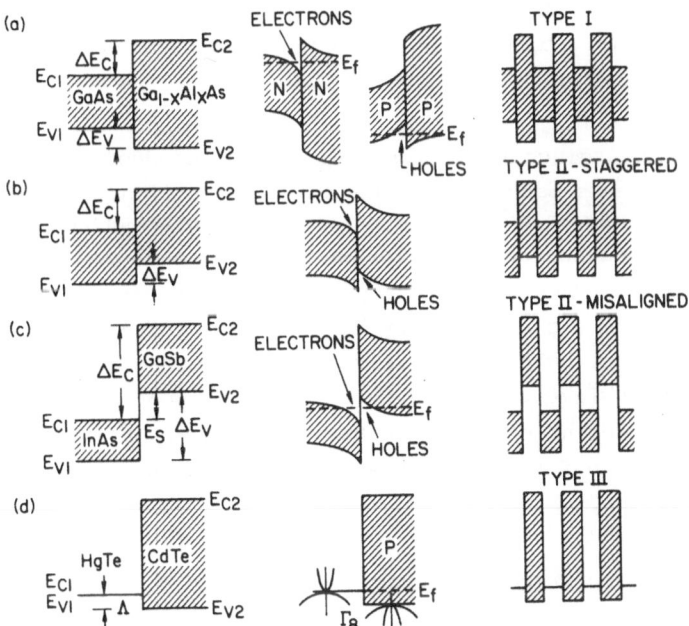

Fig. 6. Discontinuities of bandedge energies at four types of hetero-interfaces: band offsets (left), band bending and carrier confinement (middle), and superlattice (right).

structures possess predesigned potential profiles and impurity distributions with dimensional control close to interatomic spacing and
with virtually defect-free interfaces, particularly, in a lattice-
matched case such as GaAs – $Ga_{1-x}Al_xAs$ .. This great precision has
cleared access to a ''mesoscopic'' quantum regime.

The semiconductor SL structures have been grown with III-V, II-VI
and IV-VI compounds, as well as elemental semiconductors. Fig. 5
shows the plot of energy gaps at 4.2K versus lattice constants for
zinc-blende semiconductors together with Si and Ge. Joining lines
represent ternary alloys except for Si-Ge, GaAs-Ge and InAs-GaSb.
The introduction of II-VI compounds apparently extended the available
range of energy gaps in both the high and the low direction: that
of ZnS is as high as 3.8eV and all the Hg compounds have a negative
energy gap or can be called zero-gap semiconductors. The magnetic
compounds, CdMnTe and ZnMnSe are relative newcomers in the SL and QW
arena.

Semiconductor hetero-interfaces exhibit an abrupt discontinuity
in the local band structure, usually associated with a gradual band-
bending in its neighborhood which reflects space-charge effects.
According to the character of such discontinuity, known hetero-
interfaces can be classified into four kinds: type I, type
II-staggered, type II-misaligned, and type III, as illustrated in
Fig. 6(a)(b)(c)(d): band offsets (left), band bending and carrier
confinement (middle), and SLs (right).

The bandedge discontinuities, ΔE_c for the conduction band and
ΔE_v for the valence band, at the hetero-interfaces obviously command
all properties of QWs and SLs, and thus constitute the most relevant
parameters for device design. For such fundamental parameters, ΔE_c
and ΔE_v , however, the predictive qualities of theoretical models are
not very accurate and precise experimental determination cannot be
done without great care. In this regard, the GaAs-GaAlAs system is
most extensively investigated with both spectroscopic and electrical
measurements. Recent experiments have reduced an early established
value of $\Delta E_c/\Delta E_g$, 85% to somewhat smaller values in the range between
60 and 70 percent.

III. RESONANT TUNNELING AND QUANTUM WELLS

Tsu and Esaki (11) computed the resonant transmission coefficient as
a function of electron energy for multi-barrier structures from the
tunneling point of view, leading to the derivation of the current-
voltage characteristics. The SL band model previously presented,

6

assumed an infinite periodic structure, whereas, in reality, not only
a finite number of periods is prepared with alternating epitaxy, but
also the phase-coherent distance is limited. Thus, this multibarrier
tunneling model provided useful insight into the transport mechanism
and laid the foundation for the following experiment.

Chang, Esaki and Tsu (12) observed resonant tunneling in
double-barriers, and subsequently, Esaki and Chang (13) measured
quantum transport properties for an SL having a tight-binding potential. The current and conductance versus voltage curves for a double
barrier are shown in Fig. 7. The energy diagram is shown in the inset
where resonance is achieved at such applied voltages as to align the
Fermi level of the electrode with the bound states, as shown in cases
(a) and (c). This experiment, together with the quantum transport
measurement, probably constitutes the first observation of man-made
bound states in both single and multiple QWs.

The technological advance in MBE for the last decade resulted
in dramatically-improved resonant-tunneling characteristics, which
renewed interest in such structures, possibly for applications (14).
The observation of resonant tunneling in p-type double barrier
structures (15) revealed fine structure corresponding to each bound

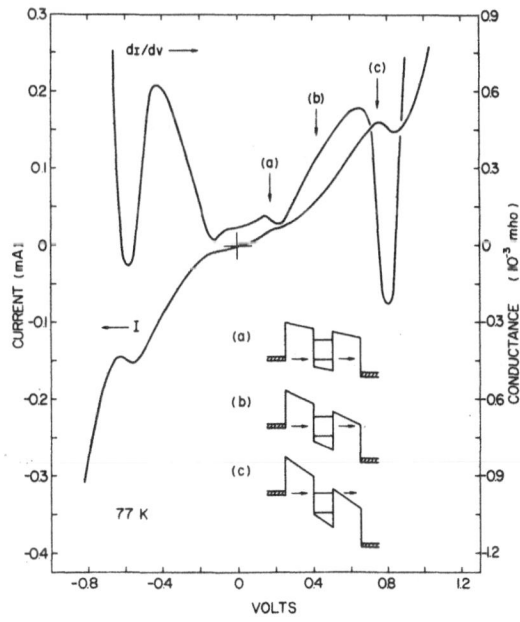

Fig. 7. Current- and conductance-voltage characteristics of a
        double-barrier structure. Conditions at resonance (a) and
        (c), and at off-resonance (b), are indicated by arrows in
        the insert.

state of both heavy and light holes, as shown at the bottom of Fig.
8, confirming, in principle, the calculated tunneling probability at
the top of the figure.

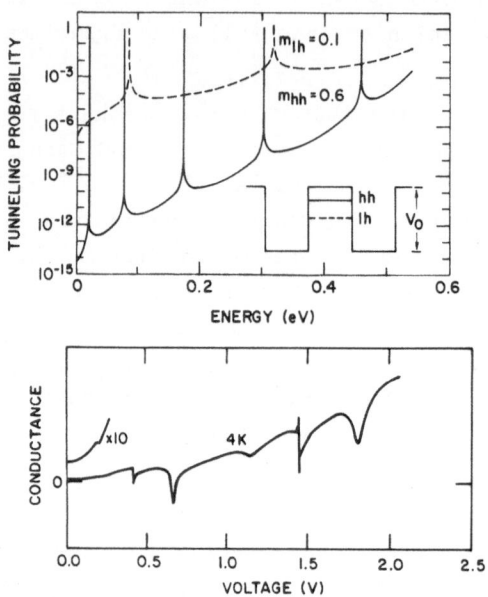

Fig. 8. Tunneling probability versus hole energy for a p-type
         double-barrier structure (top), calcuated for a 50 Å-50
         Å-50 Å structure with a 0.55-eV potential barrier for heavy
         and light hole masses: $0.6m_0$ (continuous line) and $0.1m_0$
         (discontinuous line). Observed conductance-voltage curve
         (bottom).

IV. OPTICAL ABSORPTION, PHOTOCURRENT SPECTROSCOPY,
    PHOTOLUMINESCENCE AND STIMULATED EMISSION.

Optical investigation on the quantum structures during the last decade has revealed the salient features of quantum confinement. Dingle
et al. (16) (17) observed pronounced structure in the optical absorption spectrum, representing bound states in isolated and double
QWs. In low-temperature measurements for such structures, several
exciton peaks, associated with different bound-electron and bound-
hole states, were resolved. The spectra clearly indicate the evolution of resonantly split, discrete states into the lowest subband of
an SL. van der Ziel et al. (18) observed optically pumped laser oscillation from GaAs-GaAlAs QW structures at 15K. Since then, the
application of such structures to semiconductor laser diodes has received considerable attention because of its superior characteristics, such as low threshold current, low temperature dependence,
tunability and directionality (19).

Tsu et al. (20) made photocurrent measurements on GaAs-GaAlAs SLs subject to an electric field perpendicular to the well plane with the use of a semitransparent Schottky contact. The photocurrents as a function of incident photon energy are shown in Fig. 9 for samples of three different configurations, designated A (35Å GaAs - 35Å GaAlAs), B (50Å GaAs - 50Å GaAlAs), and C (110Å GaAs - 110Å GaAlAs). The energy diagram is shown schematically in the upper part of the figure where quantum states created by the periodic potential are labeled E_1 and E_2. These states are essentially discrete if there is a relatively large separation between wells, as is the case in sample C, but broaden into subbands in samples A and B owing to an increase in overlapping of wave functions. Calculated energies and bandwidths for transitions shown in the figure are found to be in satisfactory agreement with the observation. Very recently, Deveaud et al. (21) made transport measurements of 2D carriers in SL minibands by subpicosecond luminescence spectroscopy.

In undoped high-quality GaAs-GaAlAs QWs, the main photoluminescence peak is attributed to the excitonic transition between 2D electrons and heavy holes. Mendez et al. (22) studied the field-induced effect on the photoluminescence in such wells: when the electric field, for the first time in luminescence measurements, was applied perpendicular to the well plane, pronounced field-effects, Stark shifts, were discovered. More recently, Vina et al. (23) studied such shifts on the excitonic coupling as indicated in Figs. 10 and 11. Fig. 10 shows excitation spectrum with the photoluminescence of the heavy-hole exciton, where h_1, $h_1^{(2x)}l_1$, and $l_1^{(2x)}$ denote the ground and excited states of the heavy- and light-hole excitons, respectively, and h_{12a} corresponds to an exciton related to the n=1 conduction and n=2 heavy-hole valence bands. Fig. 11 shows Stark shifts of the light-hole exciton (open circles) and excited states of the heavy-hole exciton (open triangles), exhibiting strong coupling (solid circles) between them. Chemla et al. (24) observed Stark shifts of the excitonic absorption peak, even at room temperature; Miller et al. (25) analyzed such electroabsorption spectra and demonstrated optical bistability, level shifting and modulation.

V. RAMAN SCATTERING

Manuel et al. (26) reported the observation of enhancement in the Raman cross section for photon energies near electronic resonance in GaAs – $Ga_{1-x}Al_xAs$ SLs of a variety of configurations. Both the energy positions and the general shape of the resonant curves agree with theoretical values. Later, it was pointed out that resonant inelastic light scattering yields separate spectra of single particle and col-

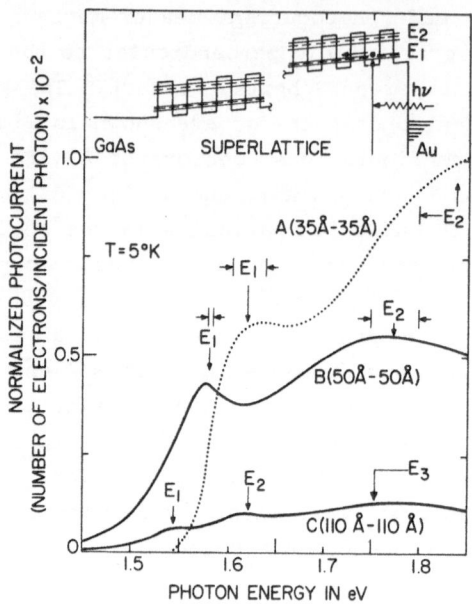

Fig. 9. Photocurrent versus photon energy for three superlattice samples A, B, and C. Calculated energies and bandwidths are indicated. The energy diagram of a Schottky-barrier structure is shown in the upper part.

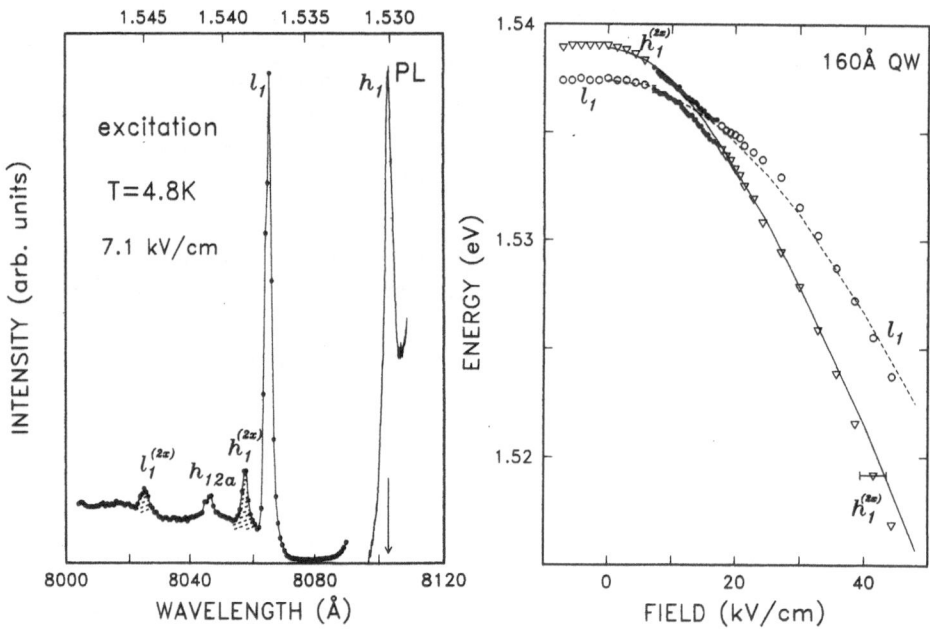

Fig. 10. Excitation spectrum with the photoluminescence of the heavy-hole exciton for a 160-Å quantum well at 7.1 kV/cm, where h_1, $h_1^{(2x)}$, l_1 and $l_1^{(2x)}$ denote the ground and excited states of the heavy- and light-hole excitons, respectively, and h_{12a} corresponds to an exciton related to the n=1 conduction and n=2 heavy-hole valence bands.

Fig. 11. Stark shifts of the light-hole exciton (open circles) and excited states of the heavy-hole exciton (open triangles), exhibiting strong coupling (solid circles) between them.

lective excitations which will lead to the determination of electronic energy levels in QWs as well as Coulomb interactions. Subsequently, this was experimentally confirmed and the technique has now widely been used as a spectroscopic tool to study electronic excitations (27).

Meanwhile, Colvard et al. (28) reported the observation of Raman scattering from folded acoustic longitudinal phonons in a GaAs(13.6Å)-AlAs(11.4Å) SL. The SL periodicity leads to Brillouin zone folding, resulting in the appearance of gaps in the phonon spectrum for wave vectors satisfying the Bragg condition. Recently, Raman scattering revealed confinement of the optical modes into ''phonon quantum well'', leading to further ramification in the field (29).

VI. MODULATION DOPING

In superlattices, in order to prevent usual impurity scattering, it is possible to spatially separate free carriers from their parent impurity atoms by doping impurities in the region of the potential hills. Though this concept was expressed in the original article, Esaki and Tsu, (1), Dingle et al. (30) successfully implemented such

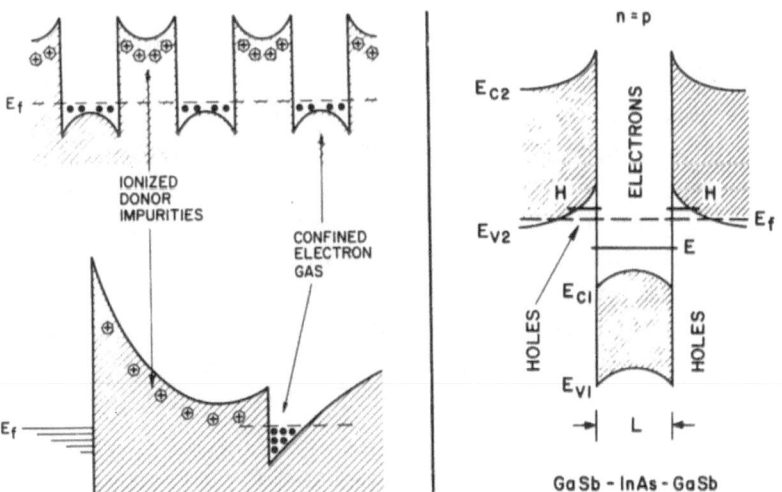

Fig. 12. Modulation doping for a superlattice and a heterojunction with a Schottky junction.

Fig. 13. Energy-band diagram of ideal GaSb-InAs-GaSb quantum wells for electrons and holes.

a concept in modulation-doped GaAs-GaAlAs SLs, as illustrated at the top of Fig.12, achieving unprecedented electron mobilities far exceeding the Brooks-Herring predictions. Subsequently, the similar technique was used to form the high- mobility 2D hole gas. This technique apparently have exerted profound impact on not only the progress of the 2D physics but also device applications such as a high-speed MODFET (modulation-doped field-effect transistor); its band energy diagram is shown at the bottom of Fig. 12 (31).

VII. SHUBNIKOV-DE HAAS OSCILLATIONS, QUANTIZED HALL EFFECT AND FRACTIONAL FILLING

Chang et al. (32) observed Shubnikov-de Haas oscillations associated with the subbands of the nearly 2D character in GaAs-GaAlAs SLs. The oscillatory behaviors agreed well with those predicted from the Fermi surfaces derived from the SL configurations and the electron concentrations.

v. Klitzing et al. (33) demonstrated the interesting proposition that quantized Hall resistance could be used for precision determination of the fine structure constant α, using 2D electrons in the inversion layer of a Si MOSFET. Subsequently, Tsui and Gossard (34) found modulation-doped GaAs-GaAlAs heterostructures desirable for this purpose, primarily because of their high electron mobilities, which led to the determination of α with a great accuracy.

The quantized Hall effect in a 2D electron or hole system is observable at such high magnetic fields and low temperatures as to locate the Fermi level in the localized states between the extended states. Under these conditions, the magnetoresistance ρ_{xx} vanishes and the Hall resistance ρ_{xy} goes through plateaus. This surprising result can be understood by the argument that the localized states do not take part in quantum transport. At the plateaus, the Hall resistance is given by $\rho_{xy} = h/e^2\nu$ where ν is the number of filled Landaù levels; h, Planck's constant; and e, the electronic charge.

Tsui, Stormer, and Gossard (35) discovered a striking phenomenon, an anomalous quantized Hall effect, exhibiting a plateau in ρ_{xy}, and a dip in ρ_{xx} , at a fractional filling factor of 1/3 in the extreme quantum limit at temperatures lower than 4.2K. This discovery has spurred a large number of experimental and theoretical studies.

VIII. InAs-GaSb SYSTEM

Since 1977, the InAs-GaSb system was investigated because of its extraordinary bandedge relationship at the interface, called type II-''misaligned'' in Fig. 6(c). The band calculation for InAs-GaSb SLs (36) revealed a strong dependence of the subband structure on the period: The semiconducting energy gap decreases when increasing the period, becoming zero at 170Å, corresponding to a semiconductor-to-semimetal transition. The electron concentration in SLs was measured as a function of InAs layer thickness, (37); it exhibited a sudden increase of an order-of-magnitude in the neighborhood of 100Å. Such increase indicates the onset of electron transfer from GaSb to InAs which is in good agreement with theoretical prediction. Far-infrared magneto-absorption experiments (38) were performed at 1.6K for semimetallic SLs which confirmed their negative energy-gap.

MBE-grown GaSb-InAs-GaSb quantum wells have been investigated, where the unique bandedge relationship allows the coexistence of electrons and holes across the two interfaces, as shown in Fig.13. Prior to experimental studies, Bastard et al. (39) performed self-consistent calculations for the electronic properties of such QWs, predicting the existence of a semiconductor-to-semimetal transition as a result of electron transfer from GaSb at the threshold thickness of InAs; this is somewhat similar to the mechanism in the InAs-GaSb SLs. Such a transition was confirmed by experiments carried out by Munekata et al.(40), although the electron and hole densities are not the same, probably because of the existence of some extrinsic electronic states. Analysis (41) for such a 2D electron-hole gas elucidated, for the first time, the fact that the quantum Hall effect is determined by the degree of uncompensation of the system. Figure 14 shows magnetoresistance ρ_{xx} and Hall resistance ρ_{xy} for 0.45K, versus magnetic field for a sample with a 200Å InAs well, which has electron and hole densities of 8.4×10^{11} and $3.6 \times 10^{11} cm^{-2}$, and their corresponding mobilities of 1.6×10^5 and $1.4 \times 10^4 cm^2/V.sec$, respectively (42). As noticed in the curves of ρ_{xx} and ρ_{xy}, extra structure exists; there are prominent dips at 7.5, 12.5 and 26 T (shown by arrows) on the Hall plateaux for filling factors, $\nu = 3$, 2 and 1. These filling factors are given by the difference between the electron and hole filling factors, ν_e and ν_h. For instance, $\nu_e = 2$ and $\nu_h = 1$ below the dip at 26T and $\nu_e=1$ and $\nu_h = 0$ above that, although $\nu=1$ for both cases.

IX. OTHER SUPERLATTICES

A doping SL has been pursued for its own interest, for instance, seen in optical properties. If this SL is illuminated, extra electrons

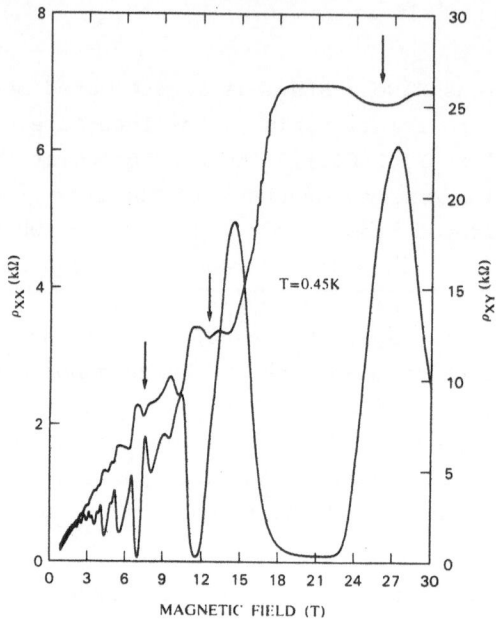

Fig. 14. Magnetoresistance ρ_{xx} and Hall resistance ρ_{xy} versus magnetic field for a GaSb-InAs-GaSb sample with carrier densitites and mobilities:8.4×10^{11} cm^{-2} and 1.6×10^{5} cm^{2}/V.sec for electrons and 3.6×10^{11}cm^{-2} and 1.4×10^{4}cm^{2}/V.sec for holes.

and holes are attracted to minima in the conduction band and to maxima in the valence band so that the amplitude of the periodic potential is reduced by the extra carriers, leading to a crystal which has a variable energy gap (43).

It is certainly desirable to select a pair of materials closely lattice-matched in order to obtain stress- and dislocation-free interfaces. Lattice-mismatched heterostructures, however, can be grown with essentially no misfit dislocations, if the layers are sufficiently thin, because the mismatch is accommodated by uniform lattice strain (44). On the basis of such premise, Osbourn and his co-workers (45) prepared strained-layer SLs from lattice-mismatch pairs, claiming their usefulness for some applications.

Kasper et al. (46) pioneered the MBE growth of Si-SiGe strained-layer SLs, and Manasevit et al. (47) observed unusual mobility enhancement in such structures. The growth of high-quality Si-SiGe SLs (48) attracted much interest in view of possible applications as well as scientific investigations.

Dilute-magnetic SLs such as CdTe-CdMnTe (49) and ZnSe-ZnMnSe are recent additions to the superlattice family, and have already exhibited unique magneto-optical properties.

In 1976, Gossard et al. (50) achieved epitaxial structures with alternate-atomic-layer composition modulation by MBE; these structures were characterized by transmission electron microscope as well as optical measurements. Such short-period SLs of binary compounds apparently serve for interface improvement (10).

Superlattices and quantum wells provide a means to reduce the dimensionality of an electron or hole gas down to 2D. Attempts at further reduction to 1D and 0D have also been made by an ultrafine etching technique (10) (51) or by the combination of confinement by quantum wells and extra structure. With reduction to 1D, one may achieve higher mobilities because of the suppression of scattering (52), and also can enhance exciton binding energies and oscillator strengths in optical properties (53).

X. CONCLUSION

We have witnessed remarkable progress of an interdisciplinary nature on this subject. A variety of ''engineered'' structures exhibited extraordinary transport and optical properties; some of them, such as ultrahigh carrier mobilities, semimetallic coexistence of electrons and holes, and large Stark shifts on the optical properties, may not even exist in any ''natural'' crystal. Thus, this new degree of freedom offered in semiconductor research through advanced material engineering has inspired many ingenious experiments, resulting in observations of not only predicted effects but also totally unknown phenomenon. The growth of papers on the subject for the last decade is indeed phenomenal. For instance, the number of papers and the percentage of those to the total presented at the International Conference on the Physics of Semiconductors (ICPS) being held every other year, increased as shown in Fig. 15. After the incubation period of several years, it really took off, around ten years ago, at an annual growth rate of nearly 70%.

Activities in this new frontier of semiconductor physics, in turn, give immeasurable stimulus to device physics, leading to unprecedented transport and optoelectronic devices or provoking new ideas for applications. Fig. 16 illustrates a correlation diagram in such interdisciplinary research where beneficial cross-fertilizations are prevalent.

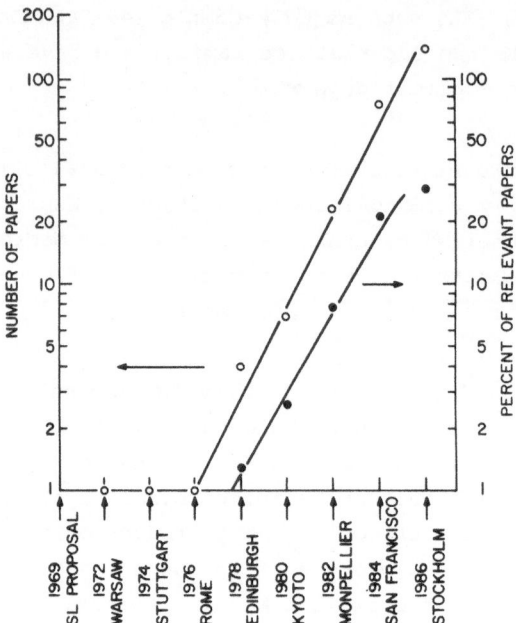

Fig. 15. Growth in the number of papers on SLs and QWs and the percentage of those to the total presented at the International Conference on the Physics of Semiconductors being held every other years.

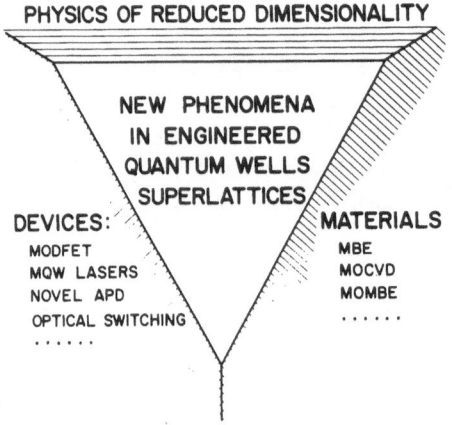

Fig. 16. Correlation diagram in a cross-disciplinary research environment.

I hope this article, which cannot possibly cover every landmark, provides some flavor of the excitement in this field. Finally, I would like to thank the many participants in and out of superlattice research for their contributions and J. M. Hong and E. E. Mendez for their help in preparing Fig. 4, as well as the ARO's partial sponsorship from the initial day of our investigation.

REFERENCES

1. L. Esaki and R. Tsu, ''Superlattice and negative conductivity in semiconductors,'' IBM Research Note RC-2418 (1969).

2. L. Esaki and R.Tsu, ''Superlattice and negative differential conductivity in semiconductors,'' IBM J. Res. Develop. 14: 61 (1970).

3. D. Bohm, ''Quantum Theory:'' (Prentice Hall, Englewood Cliffs, N.J. 1951), p.283.

4. L. Esaki, ''Long journey into tunneling,'' Les Prix Nobel en 1973, Imprimerie Royale, P.A. Norstedt & Soner, Stockholm 1974, p. 66.

5. L. Esaki, L.L. Chang, and R. Tsu, ''A one-dimensional 'superlattice' in semiconductors'' Proceedings of the 12th International Conference on Low Temperature Physics, Kyoto, Japan, 1970 (Keigaku Publishing Co.,Tokyo, Japan), p. 551.

6. A.E. Blakeslee and C.F. Aliotta, ''Man-made superlattice crystals,'' IBM J. Res. Develop. 14: 686 (1970).

7. L. Esaki, L.L. Chang, W.E. Howard, and V.L. Rideout, ''Transport properties of a GaAs-GaAlAs superlattice,'' Proceedings of the 11th International Conference on the Physics of Semiconductors, Warsaw, Poland, 1972, edited by the Polish Academy of Sciences (PWN-Polish Scientific Publishers, Warsaw, Poland), p. 431.

8. R. Tsu and L. Esaki, ''Nonlinear optical response of conduction electrons in a superlattice'' Appl. Phys. Lett. 19: 246 (1971).

9. L. Esaki, ''Semiconductor Superlattices and Quantum Wells'' Proceedings of the 17th International Conference on the Physics of Semiconductors, San Francisco, August, 1984, edited by J.D. Chadi and W.A. Harrison (Springer-Verlag, New York, 1985), p.473; IEEE J. Quantum Electron., QE-22: 1611 (1986).

10. A. C. Gossard, ''Growth of Microstructures by Molecular Beam Epitaxy'' IEEE J. Quantum Electron., QE-22: 1649 (1986)

11. R. Tsu and L. Esaki, ''Tunneling in a finite superlattice,'' Appl. Phys. Lett. 22: 562 (1973).

12. L.L. Chang, L. Esaki, and R. Tsu, ''Resonant tunneling in semiconductor double barriers'' Appl. Phys. Lett. 24: 593 (1974).

13. L. Esaki and L.L. Chang, ''New transport phenomenon in a semiconductor 'superlattice''' Phys. Rev. Lett. 33: 495 (1974).

14. F. Capasso, K. Mohammed and A. Y. Cho, ''Resonant tunneling through double barriers'' IEEE J. Quantum Electron., QE-22: 1853 (1986).

15. E.E. Mendez, W.I. Wang, B. Ricco, and L. Esaki, ''Resonant Tunneling of holes in AlAS-GaAs-AlAS heterostructures'' Appl. Phys. Lett. 47: 415 (1985).

16. R. Dingle, W. Wiegmann, and C.H. Henry, ''Quantum states of confined carriers in very thin $Al_xGa_{1-x}As - GaAs - Al_xGa_{1-x}As$ heterostructures'' Phys. Rev. Lett. 33: 827 (1974).

17. R. Dingle, A.C. Gossard, and W. Wiegmann, ''Direct observation of superlattice formation in a semiconductor heterostructure,'' Phys Rev. Lett. 34: 1327 (1975).

18. J.P. van der Ziel, R. Dingle, R.C. Miller,W. Wiegmann, and W.A. Nordland Jr., ''Laser oscillation from quantum states in very thin $GaAs - Al_{0.2}Ga_{0.8}As$ multilayer structures'' Appl. Phys. Lett. 26: 463 (1975).

19. Y. Arakawa and A. Yariv, ''Quantum well lasers'' IEEE J. Quantum Electron., QE-22: 1887 (1986)

20. R. Tsu, L.L. Chang, G.A. Sai-Halasz, and L. Esaki, ''Effects of quantum states on the photocurrent in a superlattice'' Phys. Rev. Lett. 34: 1509 (1975).

21. B. Deveaud, J. Shah, T. C. Damen, B. Lambert and A. Regreny, ''Bloch transport of electrons and holes in superlattice minibands: direct measurement by subpicosecond luminescence spectroscopy'' <u>Phys. Rev. Lett.</u> 58: 2582 (1987).

22. E.E. Mendez, G Bastard, L.L. Chang, and L. Esaki, ''Effect of an electric field on the luminescence of GaAs quantum wells'' <u>Phys. Rev. B</u> 26: 7101 (1982).

23. L. Vina, R. T. Collins, E. E. Mendez and W. I. Wang, ''Excitonic coupling in GaAs/GaAlAs quantum wells in an electric field'' <u>Phys. Rev. Lett.</u> 58: 832 (1987).

24. D.S. Chemla, T.C. Damen, D.A.B. Miller, A.C. Gossard, and W. Wiegmann, ''Electroabsorption by Stark effect on room-temperature excitons in GaAs/GaAlAs multiple quantum well structures,'' <u>Appl. Phys. Lett.</u> 42: 864 (1983).

25. D.A.B. Miller, J.S. Weiner, and D.S. Chemla, ''Electric-field dependence of linear optical properties in quantum well structures'' <u>IEEE J. Quantum Electron.</u>, QE-22: 1816 (1987).

26. P. Manuel, G.A. Sai-Halasz, L.L. Chang, Chin-An Chang, and L. Esaki, ''Resonant Raman scattering in a semiconductor superlattice,'' <u>Phys. Rev. Lett.</u> 37: 1701 (1976).

27. G. Abstreiter, R. Merlin, and A. Pinczuk, ''Inelastic light scattering by electronic excitations in semiconductor heterostructures'' <u>IEEE J. Quantum Electron.</u>, QE-22: 1771 (1987).

28. C. Colvard, R. Merlin, and M.V. Klein, and A.C. Gossard, ''Observation of folded acoustic phonons in a semiconductor superlattice,'' <u>Phys. Rev. Lett.</u> 45: 298 (1980).

29. M.V. Klein, ''Phonons in semiconductor superlattices'' <u>IEEE J. Quantum Electron.</u>, QE-22: 1760 (1987)

30. R. Dingle, H.L. Stormer, A.C. Gossard, W. Wiegmann, ''Electron mobilities in modulation-doped semiconductor heterojunction superlattices,'' <u>Appl. Phys. Lett.</u> 33: 665 (1978).

31. M. Abe, T. Mimura, K. Nishiuchi, A. Shibatomi and M. Kobayashi, ''Recent advances in ultra-high-speed HEMT technology'' <u>IEEE J. Quantum Electron.</u>, QE-22: 1870 (1986)

32. L. L. Chang, H. Sakaki, C. A. Chang, and L. Esaki, ''Shubnikov-de Haas oscillations in a semiconductor superlattice'' <u>Phys. Rev. Lett.</u> 38: 1489 (1977).

33. K. von Klitzing, G. Doreda, and M. Pepper, ''New method for high-accuracy determination of the fine-structure constant based on quantized hall resistance,'' <u>Phys. Rev. Lett.</u> 45: 494 (1980).

34. D.C. Tsui and A.C. Gossard, ''Resistance standard using quantization of the Hall resistance of $GaAs - Al_xGa_{1-x}As$ heterostructures <u>Appl. Phys. Lett.</u> 38: 550 (1981).

35. D.C. Tsui, H.L. Stormer, and A.C. Gossard, ''Determination of the fine-structure constant using $GaAs - Al_xGa_{1-x}As$ heterostructures,'' <u>Phys. Rev. Lett.</u> 48: 1559 (1982).

36. G.A. Sai-Halasz, R. Tsu, and L. Esaki, ''A new semiconductor superlattice,'' <u>App. Phys. Lett.</u> 30: 651 (1977); G.A. Sai-Halasz, L. Esaki, and W.A. Harrison, ''InAs-GaSb superlattice energy structure and its semiconductor-semimetal transition,'' <u>Phys. Rev. B</u> 18: 2812 (1978).

37. L.L. Chang, N.J. Kawai, G.A. Sai-Halasz, R. Ludeke, and L. Esaki, ''Observation of semiconductor-semimetal transition in InAs-GaSb superlattices'' <u>Appl. Phys. Lett.</u> 35: 939, (1979).

38. Y. Guldner, J.P. Vieren, P. Voisin, M. Voos, L.L. Chang, and L. Esaki, ''Cyclotron resonance and far-infrared magneto-absorption experiments on semimetallic InAs-GaSb superlattices,'' <u>Phys. Rev. Lett.</u> 45: 1719, (1980).

39. G. Bastard, E.E. Mendez, L.L. Chang, L. Esaki, ''Self-consistent calculations in InAs-GaSb heterojunctions,'' <u>J. Vac. Sci. Technol.</u> 21: 531 (1982).

40. H. Munekata, E.E. Mendez, Y. Iye, and L. Esaki, ''Densities and mobilities of coexisting electrons and holes in MBE grown GaSb-InAs-GaSb quantum well,'' Surf. Sci. 174: 449 (1986)

41. E.E. Mendez, L. Esaki, and L.L. Chang, ''Quantum Hall effect in a two-dimensional electron hole gas,'' Phys. Rev. Lett. 55: 2216 (1985).

42. T. P. Smith and H. Munekata, private communication.

43. G.H. Dohler, H. Kunzel, D. Olego, K. Ploog, P. Ruden, H.J. Stolz, and G. Abstreiter, ''Observation of tunable band gap and two-dimensional subbands in a novel GaAs superlattice,'' Phys. Rev. Lett. 47: 864 (1981).

44. J.H. van der Merwe, ''Crystal interfaces,'' J. Appl. Phys. 34: 117 (1963).

45. G.C. Osbourn, R.M. Biefeld and P.L. Gourley, ''A $GaAs_xP_{1-x}$/GaP strained-layer superlattice,'' Appl. Phys. Lett. 41: 172 (1982).

46. E. Kasper, H. J. Herzog and H. Kibbel, ''A one-dimensional SiGe superlattice grown by UHV epitaxy,'' Appl. Phys. 8: 199 (1975).

47. H. M. Manasevit, I. S. Gergis, and A. B. Jones, ''Electron mobility enhancement in epitaxial multilayer $Si - Si_{1-x} Ge_x$ alloy films on (100) Si,'' Appl. Phys. Lett. 41: 464 (1982).

48. J.C. Bean, L.C. Feldman, A.T. Fiory, S. Nakahara, and J.D. Robinson, '' Ge_xSi_{1-x}/Si strained-layer superlattice grown by molecular beam epitaxy,'' J. Vac. Sci. Technol. A2, 436 (1984).

49. A.V. Nurmikko, R.L. Gunshor and L.A. Kolodziejski, ''Optical Properties of CdTe/CdMnTe multiple quantum wells'' IEEE J. Quantum Electron. QE-22: 1785 (1986)

50. A.C. Gossard, P.M. Petroff, W. Weigmann, R. Dingle, and S. Savage, ''Epitaxial structures with alternate-atomic-layer composition modulation,'' Appl. Phys. Lett. 29: 323 (1976).

51. H. Temkin, G.J. Dolan, M.B. Parish, and S.N.G. Chu, ''Low-temperature photoluminescence from InGaAs/InP quantum wires and boxes'' Appl. Phys. Lett. 50: 413 (1987).

52. H. Sakaki, ''Scattering suppression and high-mobility effect of size- quantized electrons in ultrafine semiconductor wire structures'' Jpn. J. Appl. Phys. 19: L735 (1980).

53. Y-C Chang, L. L. Chang and L. Esaki, ''A new one-dimensional quantum well structure'' Appl. Phys. Lett. 47: 1324 (1985).

ELECTRONIC STATES IN SEMICONDUCTOR HETEROSTRUCTURES

G.Bastard

Groupe de Physique des Solides de l'Ecole Normale Supérieure
24 rue Lhomond F-75005 Paris (France)

ABSTRACT

We review some features of the electronic states in semiconductor heterostructures. The emphasis will be placed on the short period superlattices, the in -plane dispersion of the valence subbands, the energy levels in doped heterolayers and in quasi unidimensional semiconductor wires.

I. INTRODUCTION

Since the pioneering work of Esaki and Tsu [1] a considerable amount of results have been obtained as regards the physical and device properties of the semiconductor heterolayers. The present paper is restricted to a brief overview of some features of the electronic states in semiconductor heterolayers. Although there exists a variety of computational schemes [2-4], we concentrate on the envelope function formalism [5-8] which is simple, versatile and accurate enough for most of (but not all) the physical situations met in actual heterostructures. There already exists a large number of reviews devoted to the envelope function formalism. Here, we intend to use this formalism to discuss a number of topics of current interest. For a thorough discussion of the grounds of the envelope function formalism the reader is referred to the existing reviews [9-12].

II. THE ENVELOPE FUNCTION APPROXIMATION : A SUMMARY

Consider two semiconductors A and B which we shall assume to display the same cristallographic structures and to be lattice-matched (although strained layer materials can be described by a suitable modification of the formalism) [13-15]. The two materials have relatively similar band structures. The topmost valence band edges, which occur at the center of the Brillouin zone (Γ point) in III-V and II - VI semiconductors, have no reason to be aligned. In fact, the magnitude of the band offset between the two materials decisively influence the electronic properties of a AB heterolayer (fig.1). One of the main task for physicists is thus to measure these band offsets. This can be achieved through XPS experiments, electrical characterization of isotype heterojunctions or by optical probes.

Suppose then that the valence band offset is known. Then, from ΔE_v, ε_A and ε_B (the direct Γ bandgaps of the two materials) one knows the Γ conduction band offsets as well as all the other offsets (X, L, etc...) . Most of the heterolayer energy states which are relevant for transport or optical properties are relatively close in energy from one of the Γ edge of say, the A material. Thus we need to calculate the heterostructure energy levels in the vicinity of such an edge. The envelope function scheme asserts that the heterolayer wave function Ψ is of the form

$$\Psi(\vec{r}) = \sum_m f_m(\vec{r}) \, u_{m0}(\vec{r}) \tag{1}$$

where the summation over m runs over the $\Gamma_6, \Gamma_7, \Gamma_8$, edges (see fig. 2) which are closely spaced and remote from the other edges of the crystal. In eq. 1 the u_{mo} are assumed to be identical in both kinds of layers. Thus, the differences in the band structure of the A and B materials are only felt by the envelope functions f_m. These are slowly varying on the scale of the hosts' unit cell. The existence of the remote bands is taken into account in each host layer by a second order expansion in $\vec{k}_A$, $\vec{k}_B$ which are the carrier wavevectors in the A and B layers. Note that the projections of $\vec{k}_A$, $\vec{k}_B$ along the growth (z) axis can be either real (propagative states) or imaginary (evanescent states) : the evanescent states are forbidden in bulk materials but relevant in heterolayers which involve layers of finite thicknesses. Inside the Γ_6, Γ_7, Γ_8 subspaces the k.p interaction is exactly taken into account (Kane model) [16]. From these assumptions it follows that the envelope functions f_m are the solutions of a coupled second order 8x8 differential system

$$\sum_{m=1}^{8} \left[(\varepsilon_{l0} + V_l(z) - \varepsilon) \, \delta_{lm} - \frac{i h}{m_0} \, \vec{\nabla}.\vec{P}_{lm} - \frac{h^2}{2} \vec{\nabla} \frac{1}{\mu_{lm}} \, \vec{\nabla} \right] f_m(\vec{r}) = 0 \tag{2}$$

where ε_{l0} is the energy of the lth edge in the A material ; $V_l(z)$ functions which are equal to zero in the A layers and to V_l in the B layer where V_l is the algebraic energy shift of the lth edge when

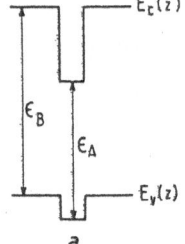
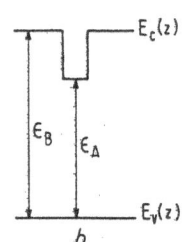
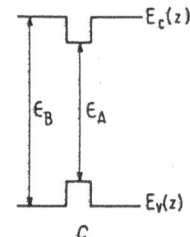
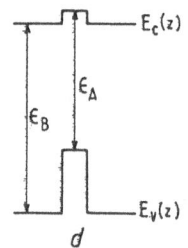

Fig. 1 : Illustration of the part played by different apportionments between the valence and conduction bands of the bandgap energy difference $\varepsilon_B - \varepsilon_A$ of two semiconductors A and B on the electronic states of a BAB rectangular quantum well. The structure c) is the best candidate for low threshold lasing action.

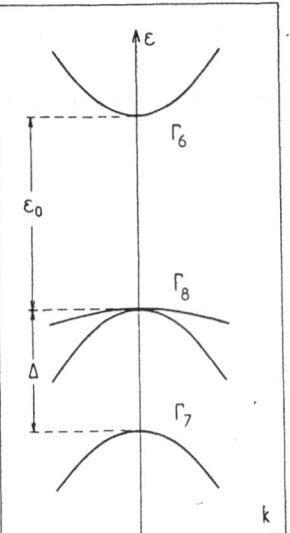

Fig. 2 : Band structure of a direct gap III V compound in the vicinity of the Γ point.

going from the A to the B material. p_{lm} is the interband p matrix elements between the periodic parts of the Bloch functions of the lth and m^{th} edges at the Γ point and μ an effective mass tensor which acccounts for the existence of the remote bands. If a slowly varying band bending potential $- e\phi(z)$ is applied to the heterolayers , one should add the diagonal contributions $- e\phi(z)\delta_{lm}$ to eq.(2)

The boundary conditions fulfilled by the f_m's are

i) f_m is continuous everywhere

ii) $\displaystyle\sum_m \left[-\frac{ih}{m_0} P_{lm}^{(x)} - \frac{\hbar^2}{2} \sum_\alpha \frac{1}{\mu_{lm}^{z\alpha}} \frac{\partial}{\partial x_\alpha} \right] f_m(\vec{r})$ \hfill (3)

The boundary conditions at large z are

$$f_m(\vec{r}) \rightarrow 0 \hfill (4)$$

for bound states in quantum wells and

$$f_m(x, y, z + d) = e^{iqd} f_m(x, y, z) \hfill (5)$$

for superlattices. In eq. (5) d is the superlattice period and q the superlattice wawevector along the growth axis. Without loss of generality one can restrict q to the segment $[- \pi/d, + \pi/d[$.

The perfect heterostructures are translationally invariant in the layer plane. Thus, the envelope functions $f_m(\vec{r})$ should at zero magnetic field be of the form :

$$f_m(\vec{r}) = \chi_m(z) \frac{1}{\sqrt{S}} \exp(i\vec{k}_\perp . \vec{r}_\perp) \hfill (6)$$

Where S is the sample area and $\vec{k}_\perp = (k_x, k_y)$, $\vec{r}_\perp = (x, y)$ are the in-plane projections of the carrier wave and position vectors respectively.

If $\vec{k}_\perp = \vec{0}$ and if the inversion asymmetry splitting of the host materials is neglected, the heterostructure electronic states can be classified according to the z projection of the total angular momentum J_z. $J_z = \pm 3/2$ corresponds to the heavy hole states while $J_z = \pm 1/2$ corresponds to hybrids of Γ_6, Γ_8 (light hole) and Γ_7 states. If, on the other hand, $\vec{k}_\perp \neq \vec{0}$ J_z is no longer a good quantum number and the problem becomes more involved.

The meaning of eq. (2) is that one needs small bulk wavevectors around the Γ edges to build the heterostructure wavefunctions. Such an assumption is valid if the energy level under consideration is closer from one of the Γ edges than from any other bulk extremum. Note that the envelope function scheme is not restricted to the Γ-related extrema. If we want to describe electronic states in the vicinity of one of the bulk X conduction level, an expansion similar to eq. (1) can be used with the u_{lo} replaced by Ψ_{1x} where the Ψ_{1x}'s are the bulk Bloch functions at this X edge in both kinds of layers. To summarize, one should only hybridize bulk states in the vicinity of the same extrema to safely use the envelope function scheme. Thus, this description is reliable when the calculated energy levels belonging to different hosts' extrema are sufficiently energy separated. This is almost always the case for the states of GaAs-Ga$_{0.7}$Al$_{0.3}$As superlattices and quantum wells but not for all the states GaAs-AlAs short period superlattices. When the envelope function calculations predict crossings between levels derived from different extrema one should be suspicious of its validity and switch to more elaborate computational schemes such as pseudopotentials [2] or tight binding calculations [3]. These will lead to a removal of the level crossings. However, one should also inquire wether the differential system given in eq. (2) has included all the relevant k.p details. The envelope function approximation has sometimes been charged of being unable to handle ainticrossings appearing in the Γ_8-related valence subbands. However, a closer inspection would have revealed that the blamed effective envelope hamiltonian was simply incomplete, lacking for the inversion asymmetry splitting contributions (the so called linear k term in the valence band). If properly included they would have explained the anticrossings. In general, we do not expect significant failure of the envelope function framework when calculating the valence-related states. Rather, certain conduction band states of sufficient energies ($\varepsilon - \varepsilon_{\Gamma6} \cong \varepsilon_x - \varepsilon_{\Gamma6}$ or $\varepsilon_L - \varepsilon_{\Gamma6}$) may be poorly described by the envelope function type of calculations. This situation is not without recalling the problem of N defects in GaAs$_{1-x}$P$_x$ alloys.

III. MISCELLANEOUS EXAMPLES

This paragraph provides results of energy level calculations based on eq. (2). The GaAs - Ga$_{1-x}$Al$_x$As heterostructures will be considered for the sake of examples. We first discuss $\vec{k}_\perp = \vec{0}$ energy levels and then the in-plane dispersions of the Γ_8 valence subbands.

III. A. Electronic states in GaAs-Ga$_{0.7}$Al$_{0.3}$As heterolayers ($\vec{k}_\perp = \vec{0}$)

We show in figs (3,4) the Γ_6 - related conduction states of GaAs-Ga$_{0.7}$Al$_{0.3}$As single

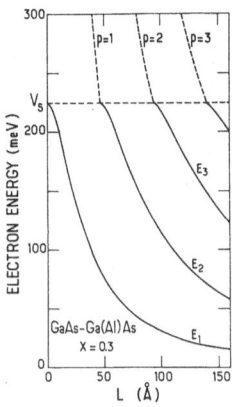

Fig. 3 : Thickness dependence of the Γ_6-related energy levels in GaAs-Ga$_{0.7}$Al$_{0.3}$As single quantum well. Solid line : bound states. Dashed lines : virtual bound states.

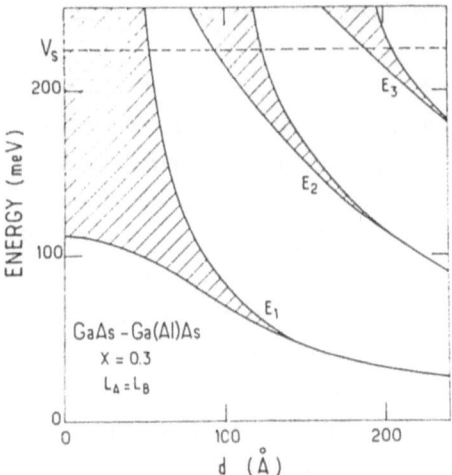

Fig. 4 : the allowed Γ_6-related superlattice states (hatched areas) of GaAs-Ga$_{0.7}$Al$_{0.3}$As superlattices are plotted versus the superlattice period d. Equal layer thicknesses are assumed in the calculations.

quantum wells or superlattices with equal layer thicknesses. The numerical results were obtained by using a conduction band offset of V_s = 224 meV, a band edge effective mass of $0.067m_0$ and by neglecting the remote band effects in eq.(2). Notice that the conduction band non-parabolicity has been included since eq.(2) remains a 6x6 system (the heavy hole states decouple from the light particle states since $\vec{k}_\perp$ is assumed to be zero and the inversion asymmetry splitting has been neglected). The GaAs quantum wells always admit one bound states. The excited states E_n, n>1 pop out of the well for $L<L_n$. They transform into resonances (dashed lines in fig.(3)), or virtual bound states when their energy exceeds the top of the confining barriers. In the superlattices (fig. (4)) the quantum well bound states hybridize to form one-dimensional minibands. Their bandwiths exponentially decay with the Ga$_{0.7}$Al$_{0.3}$As barrier thickness as shown in (fig. (5)).

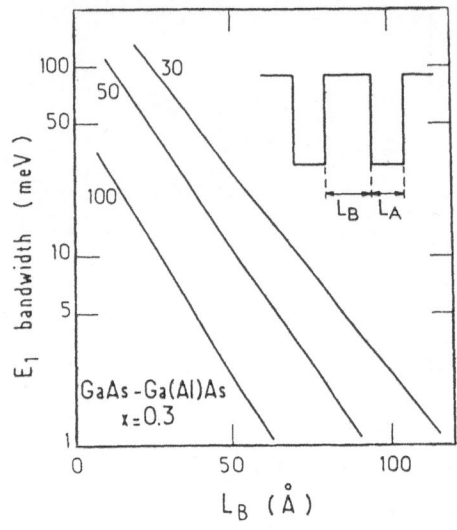

Fig. 5 : The calculated E_1 bandwidth is plotted versus the barrier thickness in GaAs-Ga$_{0.7}$.Al$_{0.3}$As superlattices for three different GaAs slab thicknesses.

III. B. Direct→indirect transitions in GaAs-AlAs short period superlattices

As seen form fig. (4) the lowest lying superlattice state extrapolate when d→0 and $L_A = L_B$ to the mid energy of the GaAs and Ga$_{0.7}$Al$_{0.3}$As Γ_6 edges. More generally, it is easily shown from the inspection of the superlattice dispersion relations derived from eq. (2) that the lowest lying Γ_6-related superlattice state in GaAs-Ga$_{1-y}$Al$_y$As superlattices with thickness ratio $\upsilon = L_{GaAs}/L_{Ga1-yAlyAs}$ extrapolates when d→0 to $V_s(y)/_{1+\upsilon}$ i.e to the value one would naively affix to the conduction bandgap of Ga$_{1-x}$Al$_x$As alloys with x=y/$_{1+\upsilon}$. Moreover, for a superlattice with vanishing period, the superlattice envelope function is uniformly spread out, like in a bulk homogenous material. Thus, one is tempted to identify a short period superlattice to a bulk alloy, except that the carrier in the GaAs-AlAs superlattice will experience no alloy scattering and thus will have a larger electrical mobility than in the alloy.Up to now this mobility improvement remains a tantalizing concept. This is because the growth of short period superlattices is difficult and up to now amounts to the replacement of alloy scattering by interface roughness scattering. In addition, the concept of a uniform spreading of the wavefunction is well founded only in the d=0 limit. Actual short period superlattices are rather grown with d=20-40 Å. Thus, the superlattice wavefunction displays significant spatial modulations depending on wether the carrier stays in GaAs or AlAs layers.This spatial modulation is the larger for the larger effective mass and thus is expected to be more important for X-related states (m*∼ m$_0$) than for Γ_6 related states (m*∼0.1m$_0$) and for Γ_8 valence levels than for Γ_6 conduction ones. Thus, actual short period superlattices are significantly different than true ordered alloys. Nevertheless, the GaAs-AlAs superlattices are very promising from the technological point of view, curing some drawbacks of the Ga$_{1-x}$Al$_x$As alloys. In particular, by doping selectively the GaAs layers of short period GaAs-AlAs superlattices one hopes to a large extent get rid of the DX centers associated with the doping of the G$_{1-x}$Al$_x$As alloys and thus suppress the parasitic effect of persistent photoconductivity.

From the point of view of the energy levels, the short period superlattices and the Ga$_{1-x}$Al$_x$As alloys share many common features. We show in fig. (6) the conduction band states of

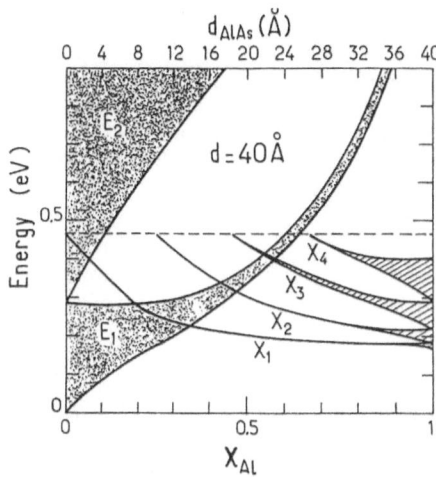

Fig. 6 : Conduction band states (hatched and shaded areas) of short period GaAs-AlAs superlattices d=40 Å. Only the Γ-related and $X_{//}$-related states are shown in the figure.

short period GaAs-AlAs superlattices with a fixed period (d=40 Å) and a varying AlAs thickness. The diagram may also be interpreted in terms of an effective Al content x : $x=d_{AlAs}/d$. We have represented both the Γ- related and X-related states. There are six equivalent X valleys in the host brillouin zone. Let k_0 be one of these X points. The host conduction bands are ellipsoïdal in shape around k_0 :

$$\varepsilon = \varepsilon_{k0} + \frac{\hbar^2}{2m_l} (k_z - k_0)^2 + \frac{\hbar^2}{2m_t} (k_x^2 + k_y^2) \qquad (7)$$

where z lies along the Γk_0 axis. The longitudinal masses are much heavier than the transverse ones : $m_l(GaAs) = 1.98m_0$; $m_t (GaAs) = 0.27\text{-}0.37m_0$; $m_l (AlAs) = 1.56m_0$; $m_t (AlAs) = 0.19m_0$ [17]. The Γ-X energy distances are $\varepsilon_x\text{-}\varepsilon_\Gamma = 0.467$ eV in GaAs and $\varepsilon_{\Gamma6}\text{-}\varepsilon_x = 0.906$ eV in AlAs. The six equivalent X minima are at the same energy in the bulk materials. This degeneracy is lifted in the superlattices due to the existence of a preferential axis : the growth axis. For the superlattices grown along the [001] axis, the two ellipsoïds whose major axis are paralled to [001] provide the lowest lying superlattice states since $m_l \gg m_t$.

The four others remain aquivalent and give rise to excited superlattice states. The only unknown parameters is the relative position of the two structures. Here, we follow Danan et al [18] and take $\varepsilon_{\Gamma6}(AlAs)\text{-}\varepsilon_{\Gamma6}(GaAs) = 1.08$ eV. Thus, the AlAs X point lies at lower energy than the GaAs one (see fig.(7)). For X - related states AlAs is the well-acting layer while GaAs is a potential well for Γ- related states. The valence levels are always preferentially localized in GaAs. Therefore, the GaAs-AlAs short period superlattices may either display a direct Γ-related fondamental bandgap (Ga rich materials) or a Γ−X pseudo-indirect one (Al rich materials). The Γ-X gap is pseudo indirect in the sense that, strictly speaking, it is a direct gap which occurs at the center of the superlattice Brillouin zone but the optical matrix element is fairly weak[18], recalling that the host's states originate from different extrema in the Brillouin zone. In addition, when the equivalent Al content is increased the GaAs-AlAs superlattices undergo a type I→type II transition[18-20]. In Ga-rich superlattices both electrons and holes are mainly localized within the same layer (GaAs) while in Al-rich superlattices the electrons are heavily localized in AlAs and the holes are mainly found in

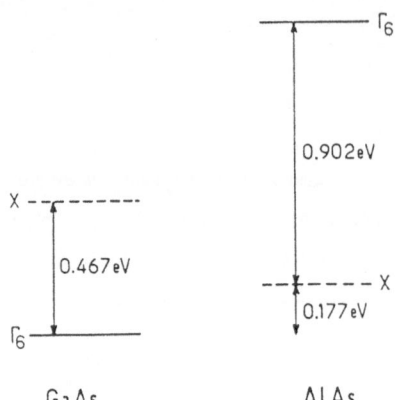

Fig. 7 : Relative alignement of Γ and X states in GaAs and AlAs. After references [17] and [18].

GaAs. The optical responses of type I and type II materials are very different. In type I materials, the photoluminescence and absorption lines are almost coincident, apart from a few meV Stokes shifts due to exciton trapping on interface defects or the occurence of an e-Å recombination line. In type II superlattices the Stokes shift may be very large : the absorption takes place at the Γ–related bandgap and the emission involves the recombination of X-like electron with Γ- like holes. The photoluminescence lines often show phonon replicas like in bulk indirect gap materials. Thus, by studying the optical properties of series of GaAs-AlAs superlattices Danan et al[18] were able to determine the conduction band offset between GaAs and AlAs. Similar results were obtained by studying the pressure dependence of the optical properties[20]. A hydrostatic pressure shifts the Γ and X states differently and one may trigger a type I→type II transition on a given sample.

III. C. In-plane dispersion relation and density of states

The in-plane dispersions of the heterostructures are hardly obtained in closed forms. This is due to the $\vec{k}_\perp$- induced admixture between the heavy hole and light pariticle states. This mixing does not affect too much the Γ_6–related states of wide gap heterostructures : it remains sensible to discard the $1/ \mu_{nm}$ tensor and to deal with the Kane bulk dispersions in each host layer. If one further neglects the band non-parabolicity and assumes a position independent effective mass m* simple results follow. For instance, the density of states associated with the bound states E_n of a single quantum well is written

$$\rho(\varepsilon) = 2 \sum_{n k_\perp} \delta \left[\varepsilon - E_n - \frac{\hbar^2 k_\perp^2}{2m^*} \right] = \frac{m^* S}{\pi \hbar^2} \sum_n Y (\varepsilon - E_n) \qquad (8)$$

where the summation runs over the bound states and Y(x) is the step function. Eq. (8) contrasts with the bulk result :

$$\rho^{bulk}(\varepsilon) = \frac{m^*}{\pi^2 \hbar^2} \sqrt{\frac{2m^* \varepsilon}{\hbar^2}} \qquad (9)$$

in that the quantum well density of states has a much steeper variation at the onset than the bulk one but remains constant above the edge. Correlatively if a phenomenon involves the density of states at a fixed energy, ε_f e.g. the Fermi energy, and if one is capable to vary this energy, there will be no dependence of the phenomenon upon ε_f in the quantum well (unless ε_f crosses E_n) while an ε_f dependence will be evidenced in the bulk case. These considerations apply e.g i) to the band-to-band absorption in the one-electron approximation where plateaus are observed in the absorption coefficient of quantum wells while a continuous increase is found in the bulk case, and ii) to screening effects of an external potential. In the Thomas Fermi approximation (wavevector $<<$ $2k_F$ where k_F is the Fermi wavector) the wavevector dependent dielectric function at T= 0 K is of the form :

$$\varepsilon(q)^{bulk} = 1 + (\frac{q_{TF}}{q})^2 \qquad ; \qquad \varepsilon^{Q.W.}(q) = 1 + \frac{q_{TF}}{q} f_s(q) \qquad (10)$$

where the Electric Quantum Limit has been assumed in the quantum well case and $f_s(q)$ is a form factor which accounts for the finite extension of the E_1 wavefunction along the growth axis. In eq.(10) q^2_{TF} (bulk case) and q_{TF} (quantum well case) are proportional to the respective densities of states.

Thus, q^2_{TF} increases with ε_F in the bulk case which means that the more electrons are present the more effective the screening is. On the other hand q_{TF} stays constant while the two dimensional electron concentration increases on the quantum well. These results imply in particular that a screened coulombic impunity in a quantum well may support one bound state even at high electron concentration [48], while it is known that in the bulk the impurity birding energy vanishes beyond some critical electron concentration.

The in-plane dispersions of the valence subbands are always complicated, even of one restricts the analysis to a parabolic description of the valence states in each host layers and neglects the coupling

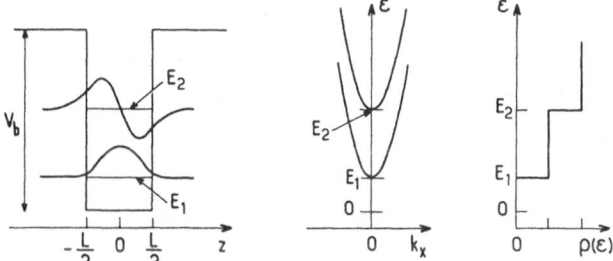

Fig. 8 : A summary of the electronic properties of a
single quantum well . From left to right : band edge profile
and bound states wave functions ; two-dimensional in-plane
dispersions ; density of states. After reference [9].

between Γ_8 and Γ_7 holes. The latter approximation is justified when the energy of the Γ_8- related hole subband remains much smaller than the spin orbit coupling of the host materials. The Γ_8 effective hamiltonian is not a scalar (like the Γ_6 one) but is the sum of a scalar term : αk^2 and of 4x4 operator : $\beta(k.J)^2$ in the simplest approach (isotropic bulk valence bands). Ultimately, these complications stem from the p-like degeneracy of the valence wavefunctions at the top of the valence band oombined with the spin orbit coupling. If one quantizes the total angular momentum along the growth axis the $J_z= \pm 3/2$ components decouple from the $J_z= \pm 1/2$ ones if $k_\perp=0$. The $\pm$ 3/2 components are in a quantum well associated with a series of heavy hole states HH_n while the $\pm$ 1/2 components correspond to the light hole levels. As soon as $\vec{k}_\perp \neq \vec{0}$, the $k_x J_x$, $k_y J_y$ terms become operative and significantly admix the $\vec{k}_{\perp} = \vec{0}$ solutions. Thus, one should resort to numerical diagonalisations to find out the eigenstates. This problem has been the subject of a number of publications [9,10,21-24] which, apart from numerical details, agree in pointing out

i) the marked non parabolicity of the valence subbands

ii) the significant mixing between the heavy and light hole character of the eigenstates

iii) the lifting of the twofold Kramers degeneracy if the heterostructure potential is not centrosymmetric.

The points i) and ii) are illustrated in figs (9,10) where the calculated in-plane valence subbands of GaAs-Ga$_{0.7}$Al$_{0.3}$As single quantum wells are plotted versus $k_{\perp}$ (fig.9) as well as the $k_{\perp}$ dependence of $\sqrt{<J^2_z>}$ (fig.10). In these calculations the linear k terms of the Γ_8 valence bands have been neglected and the axial approximation [10] has been used. In fig.(9) the diagonal approximation, which amounts to neglecting the coupling between the $\pm 3/2$ and $\pm 1/2$ states, gives rise to the dispersion relations shown as dashed lines. In this approximation, the HH$_n$ and LH$_n$ subbands are allowed to cross. The off-diagonal terms replace these crossings by anticrossings, which results in strongly non parabolic dispersions. Note in particular the electron-like behaviour exhibited by the LH$_1$ subband in the vicinity of $k_{\perp}=0$. The anticrossing behaviour means a strong band mixing. In the absence of couplings between HH$_n$ and LH$_n$ subbands $\sqrt{<J^2_z>}$ would be equal to 3/2(1/2) at any $k_{\perp}$ for HH$_n$ (LH$_n$) subbands. One sees in fig.(10) that the actual wavefunctions quickly depart from this non interacting behaviour. In fact, the heavy hole and light hole subbands exchange their character near the anticrossing. Although we keep labelling the valence subbands in the same manner as at $k_{\perp}=0$, i.e. HH$_n$, LH$_m$ the very notion of heavy hole and light hole character is as shown in fig(9,10) very fuzzy. Finally, it should be noticed that the eigenstates of rectangular (or any centrosymmetric) quantum wells are twice degenerate. For heterostructures whose band edge profile or physical property have no center of inversion this twofold degeneracy is lifted at finite $\vec{k}_{\perp}$ and non zero spin orbit coupling. An example of such a degeneracy lifting is shown in fig.(11) which displays the in-plane valence subbands of a rectangular GaAs-Ga$_{0.7}$Al$_{0.3}$As quantum well tilted by an electric field applied along the growth axis.

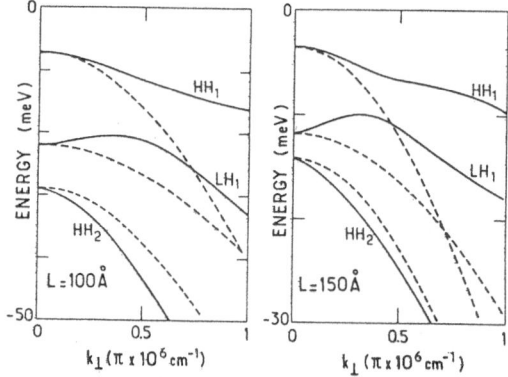

Fig. 9 : In plane dispersion relations of the valence subbands in GaAs-Ga$_{0.7}$Al$_{0.3}$As single quantum wells. L= 100 Å and L = 150 Å. Axial approximation. After reference [9].

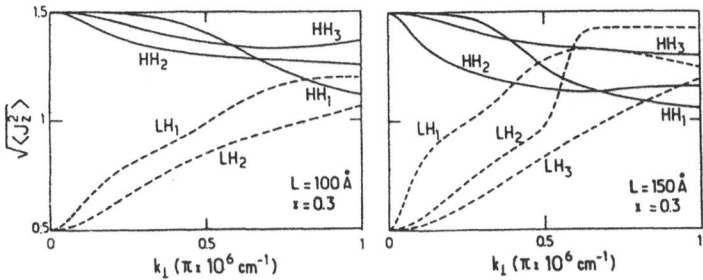

Fig. 10 : The quantity $\sqrt{<J^2_z>}$ is plotted versus the in-plane wavevector $k_\perp$ for the valence subbands of GaAs-Ga$_{0.7}$Al$_{0.3}$As single quantum wells. Axial approximation.

IV. ENERGY LEVELS IN DOPED HETEROLAYERS

If the barrier-acting material of a heterostructure is doped (e.g. with donors) the thermal equilibrium implies that some of the donors loose their electrons which are transferred in the well-acting material (fig.12). This modulation doping technique[25] has led to a tremendous increase of the low temperature mobility of both quasi bi-dimensional electron and hole gases in a variety of heterostructures[26]. This is because the mobile carriers are spatially separated from their parent impurities and thus the ionic scattering is considerably diminished. From the point of view of energy level calculations one has to find the eigenstates of a Schrödinger equation whose potential energy profile is the sum of a fixed contribution (the $V_l(z)$ in eq.(2)) and of contributions due to the free carriers and fixed charges.The simplest scheme is to use the Hartree appoximation which amounts to replacing the true N carrier potential by an average one. This finally leads to simultaneously solving the Schrödinger and Poisson equations :

$$\left[\frac{1}{2} P_z \frac{1}{m(z)} P_z + \frac{\hbar^2 k_\perp^2}{2m(z)} + V_S(z) - e\varphi_{SC}(z) \right] \chi_n(z) = \varepsilon_n \chi_n(z) \tag{11}$$

$$\varphi''_{SC}(z) = \frac{4\pi e}{\kappa} \left[N_{dep}(\vec{r}) + \sum_{j\,occupied} n_j \chi_j^2(z) - N_d^+(z) \right] \tag{12}$$

where we have assumed parabolic host's conduction bands in eq.(11). In eq.(12) κ is the static dielectric constant of the heterostructure (assumed to be position-independent), N_{dep} is the volume concentration of depletion charges, n_j the areal concentration of electrons in the jth subband an $N_d^+(z)$ the net volume concentration of donors located in the barrier. Under most circumstances the details of the self consistent potential in the barrier do not affect the energy levels ε_n very much. On the other hand the background of depletion charges does affect the ε_n by a few meV's and has larger effects on excited subbands than on the ground one [27,28]. There exists a variety of techniques to solve eqs (11,12). In the example chosen below variational calculations [28] performed in the Electric Quantum Limit have been used. Fig.(13) presents the dependence of the ground subband $E_1 (E_1 = \varepsilon_1 (k_\perp = 0))$, upon the areal electron concentration n_e for several GaAs-Ga(Al)As single heterojunctions at T=0 K. A fixed acceptor background has been assumed in the calculations

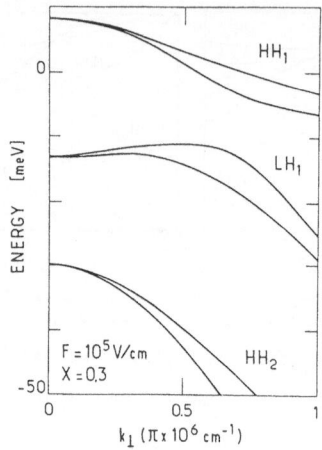

Fig. 11 : Lifting of the Kramers degeneracy of the valence subbands in GaAs-$Ga_{0.7}Al_{0.3}As$ single quantum well (L = 100 Å) tilted by an electric field applied parallel to the growth axis (F=10^5V/cm). After reference [9].

($N_{dep}=10^{14}cm^{-3}$, which corresponds to an equivalent areal concentration $N_{dep}= 4.6x10^{10}$ cm^{-2}). As previously mentionned N_{dep} affects the subband edge E_2 more than E_1. This implies a noticeable dependence of the critical electron concentration N_c above which E_2 becomes populated upon N_{dep}. This point is illustrated in fig.(14). E_2 is calculated to become populated above 7-8x$10^{11}cm^{-2}$ if $N_{dep}= 4.6x10^{10}cm^{-2}$ in $Ga_{0.7}Al_{0.3}As$ single heterojunctions with a residual a p-type doping in the GaAs channel. Notice that in quasi accumulation conditions (n type residual doping) E_2 is more readily populated. This feature may explain the frequent occurence of two populated subbands in n InP-$Ga_{0.47}In_{0.53}As$ channels[29].

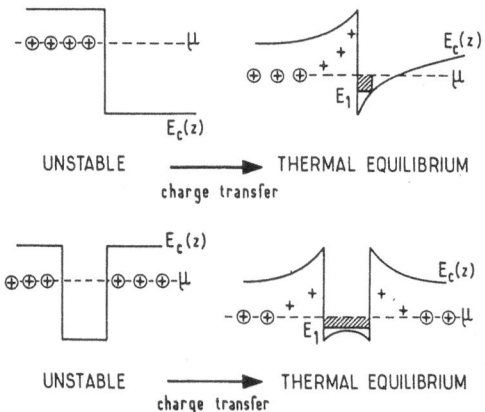

Fig. 12 : Schematic illustration of the modulation-doping process.

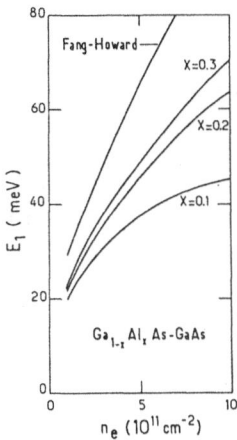

Fig. 13 : The confinement energy of the ground subband E_1 in GaAs-Ga(Al)As single heterojunctions is plotted versus the two dimensional electron concentration n_e. The curve labelled Fang Howard corresponds to an infinite V_b. After reference [12].

Fig. 14 : Critical concentration n_e^c versus 2D concentration of fixed space charges in GaAs channel. Solid line after [28]. Symbols after [27]

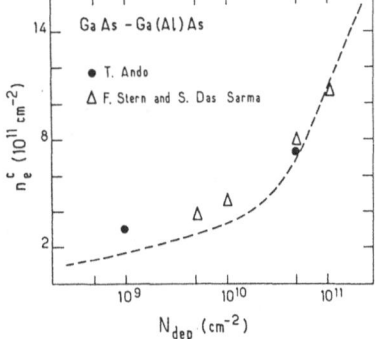

By using the self consistently determined $E_1(n_e)$ one may solve the thermodynamical equations to obtain the amount of transferred charges once the heterostructure design has been specified. Such a charge transfer calculation is presented in fig. (15) for GaAs-Ga$_{0.7}$Al$_{0.3}$As single heterojunctions[12]. The charge transfer is significantly affected by the magnitude of the band offset (V_g for electrons, V_p for holes) which, in turn, allows to determine the offsets by charge transfer measurements. This method has been successfully used by Wang et al[30] in p-type GaAs-Ga(Al)As heterostructure to fit V_p. A valence offset ratio $Q_v = \Delta E_v / \Delta E_g \sim 0.4$ was thus found, in marked disagreement with earlier measurements[31]. We remark in fig. (15) that a conduction band offset ratio $Q_c = 1 - Q_v = 0.6$ also better interprets the electron transfer than the previously accepted value of $Q_c = 0.85$.

Double heterostructures (quantum wells) or multiple quantum wells can also be modulation-doped leading to mobility improvements. However, due impunity segregation near the inverted GaAs-Ga(Al)As interface (binary grown on ternary) the electron mobility hardly exceeds 2×10^5 cm^2/Vs in modulation-doped quantum wells.

Since the quantum wells display size quantization even in the absence of free carriers, one expects to find a cross over in the behaviour of the energy levels versus the quantum well thickness

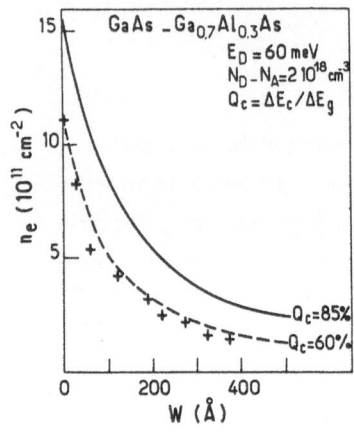

Fig. 15 : Charge transfer in GaAs-Ga$_{0.7}$Al$_{0.3}$As single heterojunctions versus spacer thickness w. The symbols are Hwang et al's data [45].

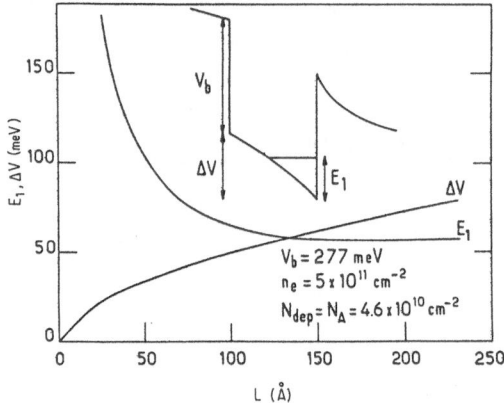

Fig. 16 : Quantum well thickness (L) dependences of the ground electron state (E$_1$) and the conduction band edge drop (ΔV) between the two sides of the well in a one-side doped Ga$_{1-x}$Al$_x$As-GaAs quantum well.

: narrow quantum wells (L 50 Å) are dominated by their intrinsic quantization while thick ones (L 300 Å) have a ground level which is more like those found in two isolated heterojunctions (symmetrical doping) or a single one (one side doped wells). This trend is illustrated in fig.(16).

In n-type doped wells the holes move in the self-consistent band bending potential created by the electrons and have their motions bound by the valence barriers. Thus, conventional optical probes become available to measure the band-to-band transitions. It was found that the quantum well bandgap rigidly shrinks with increasing carrier concentration[32-34], a feature already observed in bulk materials and attributable to exchange and correlation effects. We present in fig. (17) the density dependence of the bandgap renormalization of 150 Å thick GaAs-Ga(Al)As one-side modulation-doped quantum well[34].

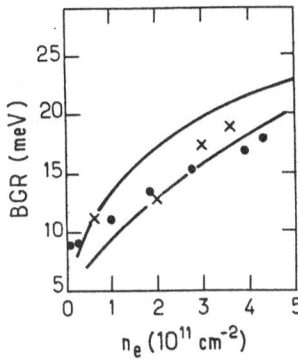

Fig. 17 : Crosses and points : experimental n_s dependence of the bandgap renormalisation in a 150 Å thick GaAs-Ga$_{0.61}$Al$_{0.39}$As single quantum well. The solid lines are a) an interpolation between calculations of reference [46] (upper curve), b) the exchange term [47] $e^2 k_F/\kappa$ where k_F is the Fermi wavector and κ the static dielectric constant (lower curve).

V. ENERGY LEVELS IN QUASI UNI DIMENSIONAL SEMICONDUCTOR HETEROSTRUCTURES

V. I. Energy spectrum

The advances in fine scale lithography makes it possible to realize quasi uni-dimensional heterostructures where the carrier motion is confined in two dimensions while it is free in the third one[35-37]. Tremendous mobility improvements over the quasi bidimensional situation have been predicted to occur if the carriers only occupy the ground one-dimensional subband[38-40]. On the other hand, it has also been predicted that all the quantum states in one dimension are localized by any attractive potential. In this section we shall only discuss the salient features of the energy levels[41-44] in perfect quasi one-dimensional heterostructures and use the envelope function approximation. Let us first notice that the effective Schrödinger equation for a rectangular quantum wire whose potential energy is written

$$V(x,z) = V_b \ Y(z^2 - \frac{L_z^2}{4}) \ Y(x^2 - \frac{L_x^2}{4}) \tag{13}$$

is non separable in x and z if V_b is finite, even if one adopts, as done in the following, the simplest assumption of parabolic host's band with a position independent effective mass. Let us however consider the case where $L_x \gg L_z$ which is the most common one ($L_z \sim 100$ Å, $L_x \sim 1000$ Å). Then it is clear that the z size quantization is much more pronounced than the x one and, as a result, the Schrödinger equation becomes quasi separable. In fact let us write the eigenstates of the hamiltonian as

$$\psi(\vec{r}) = \frac{1}{\sqrt{L_y}} \ \exp(i \, k_y \, y) \sum_n \chi_n(z) \ \alpha_n(z) \tag{14}$$

In eq.(14) the plane wave term takes care of the free motion along the wire axis, $\alpha_n(x)$ are unknown functions and $\chi_n(z)$ are the eigenstates of a rectangular quantum well problem. We write $V(x,z)$ as the sum of a purely z dependent term and of a small $W(x,z)$ contribution and we denote by E_n the eigenvalue associated with χ_n. It is readily found that $\alpha_n(x)$ is the solution of the set of equations.

35

$$\left[T_x + \frac{\hbar^2 ky^2}{2m^*} + E_n + <\chi_n |W(x,z)| \chi_n> \right] \delta_{nm} \alpha_n(x) + \sum_{m \neq n} <\chi_n |W(x,z)| \chi_m> \alpha_m(x) = \varepsilon \, \alpha(x) \quad (15)$$

If one neglects the off diagonal terms in n,m which amounts to assuming a strong size quantization along the z axis the wavefunction factorizes and we have explicitly separated the x and z motions. The eigenenergies are of the form

$$\varepsilon_{n\nu}(ky) = E_n + \varepsilon_\nu + \frac{\hbar^2 ky^2}{2m^*} \qquad (16)$$

where the ν quantum number labels all the possible x motions associated with a given n. We remark that the effective x potential explicitly depends on n. An example of calculated energy levels in rectangular quantum wires is shown in fig.(18) in the decoupled approximation. As expected, each z level carries a set of closely spaced levels related to the x motion. It may happen that a degeneracy ($\varepsilon_{n\nu} = \varepsilon_{m\mu}$) takes place. This degeneracy is removed by the off-diagonal terms of eq.(15) provided that the coupling term does not vanish. In the particular case of fig.(18) the x related levels attached to E_1 can cross those attached to E_2 but should anticross those attached to E_3. This is due to the evenness in z of the rectangular quantum wire potential energy. These anticrossings have been calculated but are so small that they won't be easily visible on figure18. In the actual case of a quantum wire obtained by cutting and etching a one-dimensional quantum well and e.g. regrowing barriers on top of the etched structure, it is likely that the wire will not be rectangular but will display a sharp and a smooth edge (see fig.19). The decoupling procedure will however apply with

$$V(x,z) = V_b \left[Y(-z) + Y(z - L_z) \right] + V_b \; Y(z) \, Y(L_z - z) \, Y \left[\varphi(x,z) > 0 \right] \qquad (17)$$

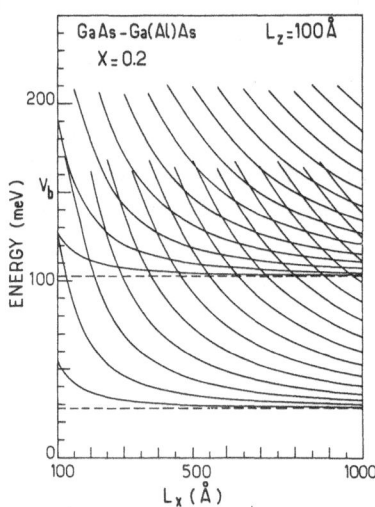

Fig. 18 : Calculated Γ_6-related states in a GaAs-Ga$_{0.8}$Al$_{0.2}$As rectangular wire L_z=100 Å in the decoupled approximation.

where $\varphi(x, z) = 0$, $z > 0$ is the border line between the well-acting and barrier-acting material whose extremum corresponds to $z=L_z$. In the decoupled approximation the effective x dependent potential will be given by

$$V_{eff}(x) = V_b \int_0^{L_x} \chi_n^2(z)\, dz \; Y\left[\varphi(x,z) > 0\right] \qquad (18)$$

which in the case of a simple gaussian shape for the wire cross section $(z=L_z \exp(-x^2/a^2))$ leads for n=1 to

$$V_{eff}(x) = \frac{A^2}{2} V_b \left\{ L_z(1 - e^{-x^2/a^2}) + \frac{1}{k_w} \cos(2k_w L_z e^{-x^2/a^2}) \; \sin\left[2k_w L_z(1 - e^{-x^2/a^2})\right] \right\} \quad (19)$$

where

$$k_w = \sqrt{\frac{2m^*}{h^2} E_1} \qquad ; \qquad A = \chi_1\left(\frac{L_z}{2}\right) \qquad (20)$$

As easily checked $V_{eff}(x=0)$ vanishes while $V_{eff}(x\to 0) \to V_b(1-P_b)$, where P_b is the integrated probability of finding the carrier outside the $[0\,L_z]$ segment while in the E_1 state. One may also imagine an array of parallel and identical wires. If the wires have a rectangular shape and if $L_z <<$ L_x one may use the decoupling procedure to find that the eigenenergies are of the form :

$$\varepsilon_{\nu k_y q_x} = E_n + \frac{\hbar^2 k_y^2}{2m^*} + \varepsilon_\nu (q_x) \qquad (21)$$

where $\varepsilon_\nu (q_x)$ are the eigenvalues of a one dimensional superlattice problem corresponding to the x motion of an electron moving in the effective periodic potential

$$V_{eff}(x) = V_b (1 - P_b) \; Y(x - \frac{L_x}{2}) \; Y(d_x/2 - x)$$
$$V_{eff}(x + d) = V_{eff}(x) \qquad (22)$$
$$V_{eff}(x) = V_{eff}(-x)$$

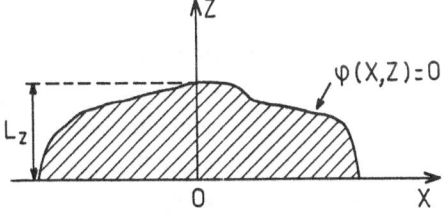

Fig. 19 : Cross section of a quantum wire.

where d_x is the superlattice period along the x axis, q_x the superlattice wavevector that can be restricted to the first Brillouin zone $[-\pi/d_x + \pi/d_x [$ and

$$1 - P_b = \int_0^{L_z} \chi_n^2 (z) \, dz \tag{23}$$

Thus one may calculate the bandwiths, wavefunctions in the same manner as done for one dimensional quantum wells.

The decoupling procedure is well adapted to treat the quasi unidimensional subbands attached to non degenerate extrema in the hosts Brillouin zone. The Γ_8- related valence subbands are more complex. This was shown in section III for quasi bidimensional materials and is even more so if one deals with quasi unidimensional heterolayers. The difficulty arises in both cases from the off diagonal terms of the Γ_8 hamiltonian. Thus, numerical diagonalization should be undertaken to obtain the valence dispersion relations upon k_y. Brum et al [43] first applied the decoupling procedure to the Γ_8 diagonal terms and on the basis spanned by the separable wavefunctions diagonalized both the off diagonal terms of eq.(15) and the off diagonal elements of the Γ_8 matrix. The latters involve p_x and k_y. Thus, they induce band mixing effects ; i.e the eigensolutions are not eigenstates of J_z. In contrast with quasi bi-dimensional heterolayers the band mixing effects do not vanish at the onset of the k_y subbands. This is because k_x is never zero in quantum wires due to the size quantization along the x direction. An example of calculated valence subbands is shown in fig.(20) for a GaAs-Ga$_{0.8}$Al$_{0.2}$As single wire with L_z= 50 Å and L_x= 200 Å. The dashed lines correspond to the valence subband in the wire. An analysis of the corresponding wavefunctions show that, except for the ground subband in the vicinity of k_y= 0, there exists a significant mixing between the $\pm 3/2$ and $\pm 1/2$ states.

V.II. Density of states

The bound z and x motions in a single quantum wire lead to singularities in the density of states. In fact, one readily finds :

$$\rho(\varepsilon) = \frac{L_y}{\pi} \sqrt{\frac{2m^*}{h^2}} \sum_{m,n} \frac{1}{\sqrt{\varepsilon - E_m - \varepsilon_n}} \quad Y(\varepsilon - E_m - \varepsilon_n) \tag{24}$$

This result recalls the density of states of bulk electrons in a quantizing magnetic field. In both cases one deals with carrier motions which are free along one direction but bound in the two others.

The singularities at the onset of quasi unidimensional subbands will be rounded off by imperfections. Nevertheless they will have a profound influence on the carrier transport. When the Fermi level passes through each edge $E_m+\varepsilon_n$, the low temperature carrier mobility will display sharp drops. In fact the mobility should vanish if $\rho(\varepsilon)$ is unbroadened as a result of intersubband elastic scattering. This remark shows the necessity of achieving quantum wires in which the carrier only occupy the ground subband if one wants to benefit of the predicted mobility enhancement.

For an array of weakly coupled parallel wires a tight binding analysis of the superlattice envelope functions along the x direction leads to the dispersion relations :

$$\varepsilon_{nmk_y q_x} = E_n + \varepsilon_m + s_m - 2t_m \cos q_x d_x + \frac{h^2 k_y^2}{2m^*} \tag{25}$$

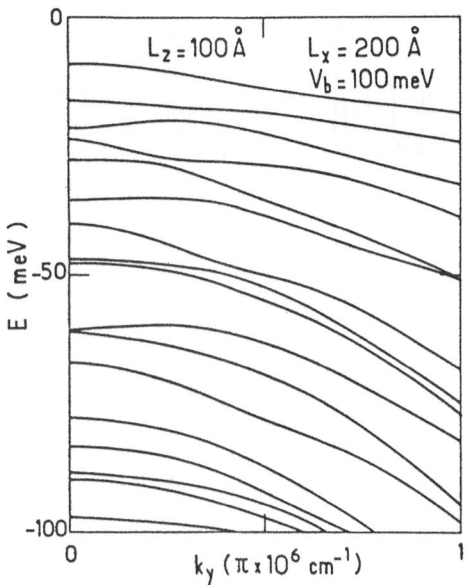

Fig. 20 : Valence dispersion relations of a rectangular
GaAs-Ga$_{0.8}$Al$_{0.2}$As quantum wire with L$_z$=50 Å and
L$_x$=200 Å. After reference [43].

where s_m and t_m are the shifts and transfer integrals respectively. Let us express the energy in the dimensionless form :

$$\varepsilon = E_n + \varepsilon_m + s_m - 2t_m + 2t_m \eta \tag{26}$$

Then, the density of states ρ_{nm} associated with the mth superlattice subband attached to the nth bound state for the z motion is simply.

$$\rho_{nm} = \frac{L_x L_y}{\pi^2 d_x} \sqrt{\frac{m^*}{\hbar^2 t_m}} \int_0^\varphi \frac{dx}{\sqrt{\cos x - \cos \varphi}} \qquad 0 \leq \eta \leq 2 \tag{27}$$

with

$$\cos \varphi = 1 - \eta \tag{28}$$

and

$$\rho_{nm} = \frac{L_x L_y}{\pi^2 d_x} \sqrt{\frac{m^*}{\hbar^2 t_m}} \int_0^\pi \frac{dx}{\sqrt{-1 + \eta + \cos x}} \qquad \eta \geq 2 \tag{29}$$

In eqs (27,29) we have assumed that $t_m > 0$. In the case of negative t_m's, which arises if one hybridizes odd states of the x- dependent quantum wells, eqs(26-29) have to be modified accordingly. Fig. (21) shows a plot of ρ_{nm} versus the dimensionless parameter η. In the vicinity of $\eta = 0$ one recognizes the staircase behaviour of two dimensional subbands while near $\eta = 2$ there is a peak associated with a two dimensional saddle point : near $\eta = 2$, the carrier effective mass is positive for the y motion but negative for the x motion. At large η one recovers the expected $\eta^{-1/2}$ decrease.

39

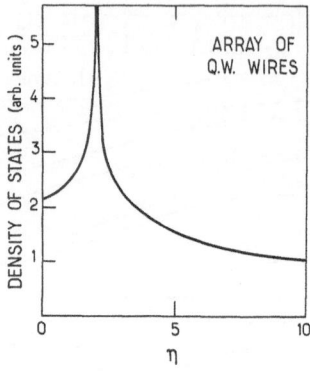

Fig. 21 : Calculated density of states of the lowest subband
of a periodic array of parallel and identical quantum wires.
The singularity occurs at the edge of the Brillouin zone.

Acknowledgment

I am indebted to the organizers of the course "Physics and Applications of Quantum Wells and Superlattices" for the opportunity to participate. I am pleased to thank Drs J.M. Berroir, J.A. Brum, C. Delalande, F. Gerbier, Y. Guldner, M.H. Meynadier, J. Orgonasi, P. Voisin and M. Voos for their active participation to the work reported here and for their constant encouragements. This work has been partly supported by the GRECO "Expérimentations Numériques". The Groupe de Physique des Solides de l'Ecole Normale Supérieure is Laboratoire Associé au CNRS (LA 17).

References

1) L. Esaki and R.Tsu IBM J.Res. Dev. 14, 61 (1970).

2) M. Jaros in "The Physics and Fabrication of Microstructures and Microdevices" edited by M.J Kelly and C. Weisbuch Springer Verlag Berlin (1986).

3) J.N. Schulman and Y.C. Chang Phys Rev B 31, 2056 (1985).

4) Y.C. Chang and J.N. Schulman J.Vac. Sci. Technol 21, 540 (1982).

5) G.Bastard Phys Rev B 24, 5693 (1981) and B 25, 7584 (1982).

6) M. Altarelli Phys Rev B 28, 842 (1983).

7) S. White and L.J. Sham Phys Rev Lett 47, 879 (1981).

8) M.F.H. Schuurmans and G.W.'t Hooft Phys Rev B 31, 8041 (1985).

9) G. Bastard and J.A. Brum IEEE J. of Quant. Elect. Q E 22, 1625 (1986).

10) M. Altarelli in "Heterojunctions and Semiconductor Superlattices" edited by G. Allan, G. Bastar N.Boccara, M. Lannoo and M. Voos. Springer Verlag Berlin (1986).

11) G. Bastard in "Molecular Beam Epitaxy and Heterostructures" edited by L.L. Chang and F Ploog NATO ASI Series E vol. 87. Martinus Nijhoff Dordrecht (1985) p 381.

12) G. Bastard and M. Voos "Wave Mechanics Applied to Semiconductor Heterostructures". L Editions de Physique - Paris - (1987).

13) G.C. Osbourn- J. Vac. Sci. Technol B1(2), 379 (1983).

14) P. Voisin in "Two Dimensional Systems, Heterostructures and Superlattices" edited by G. Bauer, F. Kuchar and H. Heinrich Springer Series in Solid State Sciences $\underline{53}$, 192. Springer Verlag Berlin (1984).

15) J.Y. Marzin ibidem reference 10) p. 161.

16) E.O. Kane J.Phys. Chem. Sol. $\underline{1}$, 249 (1957).

17) "Landölt-Börnstein Numerical Data and Functional Relationships in Science and Technology" edited by O. Madelung Group III, $\underline{17}$, Springer Verlag Berlin (1982).

18) G. Danan, A.M. Jean-Louis, F. Alexandre, B. Etienne, B. Jusserand, G. Le Roux, J.Y. Marzin, F. Mollot, R. Planel, H. Savary and B. Sermage - Phys. Rev B (1987) in press.

19) P. Dawson, B.A. Wilson, C.W. Tu and R.C. Miller Appl. Phys. Lett. $\underline{48}$, 541 (1986).

20) D.J. Wolford, T.F. Kuech, J.A. Bradley, M.A. Gell, D. Ninno and M. Jaros J.Vac. Sci. Technol B $\underline{4}$, 1043 (1986).

21) T.Ando J.Phys. Soc. Japan $\underline{54}$, 1528 (1985).

22) D.A. Broido and L.J. Sham Phys. Rev. B $\underline{31}$, 888 (1985).

23) E. Bangert and G. Landwehr Superl. and Microst. $\underline{1}$, 363 (1985).

24) U. Ekenberg and M. Altarelli Phys. Rev. B $\underline{30}$, 3569 (1984).

25) R. Dingle, H.L. Störmer, A.C. Gossard and W. Wiegmann Appl. Phys. Lett. $\underline{33}$, 665 (1978)

26) See e.g. F. Stern This volume.

27) Exchange and correlation effects have been included in very accurate energy level calculations : see T. Ando, J. Phys Soc Japan $\underline{51}$, 3893, (1982) and F. Stern and S. Das Sarma Phys Rev B $\underline{30}$, 840 (1984) for GaAs-Ga(Al)As single heterojunctions.

28) G. Bastard Surf. Sci. $\underline{142}$, 284 (1984).

29) See e.g. Y. Guldner, J.P. Vieren, M. Voos, F. Delahaye, D. Dominguez, J.P. Hirtz and M. Razeghi Phys Rev B $\underline{33}$, 3990 (1986).

30) W. Wang, E.E. Mendez and F. Stern Appl. Phys. Lett. $\underline{45}$, 639 (1984).

31) R. Dingle in "Festköperprobleme XV" edited by H.J. Queisser. Pergamon Vieweg Braunschweig p. 21 (1975).

32) A. Pinczuk, J.Shah, R.C. Miller, A.C. Gossard and W. Wiegmann Solid State Commun. $\underline{50}$, 735 (1984).

33) G. Trankle, H. Leier, A. Forchel, H. Haug, C. Ell and G. Weimann Phys Rev Lett $\underline{58}$, 419 (1987).

34) C. Delalande, J. Orgonasi, M. H. Meynadier, J.A. Brum, G. Bastard, G. Weimann and W. Schlapp Solid State Commun $\underline{59}$, 613 (1986).

35) See e.g. W. Skocpol, L. D. Jackel, E. L. Hu, R. E. Howard and L. A. Fetter Phys Rev Lett $\underline{49}$, 951 (1982).

36) See e.g. A.B. Fowler, A. Harstein and R.A. Webb Phys. Rev. Lett. $\underline{48}$, 196 (1982).

37) J. Cibert, P. M. Petroff, G. J. Dolan, D. J. Werder, S. J. Pearton, A. C. Gossard and J. H. English Superl. and Microst. (1987) in press.

38) H. Sakaki Jpn. J. Appl. Phys. $\underline{19}$, L 735 (1980).

39) J. Lee and H. N. Spector J. Appl. Phys. $\underline{54}$, 3921 (1983).

40) G. Fishman Phys. Rev. B $\underline{34}$, 2394 (1986).

41) S. E. Laux and F. Stern Appl. Phys. Lett. $\underline{49}$, 91 (1986).

42) W.Y. Lai and S. Das Sarma Phys. Rev. B $\underline{33}$, 8874 (1986).

43) J.A. Brum, G. Bastard, L. L. Chang and L. Esaki   Superl. and Microst. (1987) in press.

44) J.A. Brum, G. Bastard, L. L. Chang and L. Esaki "18th International Conference on the Physic of Semiconductors" Stockholm (1986).

45) J. C. M. Hwang, A. Kastalsky, H.L. Störmer, and V.G. Keramidas - Appl. Phys. Lett. $\underline{44}$ 802 (1984).

46) D.A. Kleinmann and R. C. Miller Phys. Rev. B $\underline{32}$, 2266 (1985).

47) Yia Chung Chang and G .D. Sanders Phys. Rev. B $\underline{32}$, 5521 (1985).

48) J.A. Brum, G. Bastard and C. Guillemot, Phys. Rev. B $\underline{30}$, 905 (1984).

MOLECULAR BEAM EPITAXY OF ARTIFICIALLY LAYERED III-V SEMICONDUCTORS

ON AN ATOMIC SCALE

Klaus Ploog

Max-Planck-Institut für Festkörperforschung
D-7000 Stuttgart-80, FR-Germany

1. INTRODUCTION

In the past decade precisely controlled crystal growth techniques have
emerged, including molecular beam epitaxy (MBE) / 1 / and metalorganic chemical vapour deposition (MO CVD) / 2 / , which have created a variety of new
opportunities for the fabrication of artificially layered III-V semiconductor structures. High-quality superlattices (SL) comprised of thin layers
of, e.g., AlAs and GaAs have been prepared / 3, 4 / with the SL period ranging from a value much larger than the underlying natural periodicity of the
lattice down to the width of a monolayer, where a monolayer is defined as
one layer of cations plus one layer of anions. The interfaces between these
epitaxial layers of different composition are used to confine electrons and/
or holes to two-dimensional (2D) motion (electrical and optical confinement).
Excitons play a more significant role in these quasi-2D systems than in the
corresponding bulk material / 5 / .

In this paper we first briefly review the fundamentals of molecular beam
epitaxy of III-V semiconductors using solid source materials / 6 / . We then
demonstrate the unique capability of MBE to engineer artificially layered
semiconductors on an atomic scale by discussing the optical properties of
all-binary AlAs/GaAs SL with periods ranging from a few tens of nanometer
down to the width of a monolayer.We show that the Kronig-Penney model fails
to describe the observed optical transition energies due to the indirect
nature of the AlAs barrier material. We have, therefore, to pay particular
attention to the influence of the indirect X and/or L minima of the AlAs
barriers on the conduction subband energies even for the larger-period
SL. On the other hand, even ultrathin AlAs barriers only a few monolayers

thick have a significant influence on the excitonic properties of AlAs/GaAs SL. The all-binary AlAs/GaAs short-period superlattices (SPS) with periods below 10 nm are used as substitutes for the ternary $Al_xGa_{1-x}As$ alloy to improve the carrier confinement in GaAs quantum wells (QW). The ultrathin-layer $(AlAs)_m(GaAs)_n$ superlattices (UTLS) with (m, n) = 1, 2, 3 represent a new artificial semiconductor material with novel electronic properties. Finally, we briefly discuss some aspects of MBE using metalorganic (gaseous) source materials. This technique has its merits in the simultaneous generation of controlled phosphorus and arsenic beams for the reproducible growth of mixed phosphide/arsenide compounds such as GaP_yAs_{1-y}, (Ga_xIn_{1-x}) (P_yAs_{1-y}), etc. With the advent of gas-source MBE and MO MBE, also called CBE (chemical beam epitaxy), the gap between the techniques of MBE and low-pressure MO CVD has been closed.

2. FUNDAMENTALS AND CONTROL OF MBE PROCESSES

The unique capability of MBE to create a wide variety of mathematically complex compositional and doping profiles in semiconductors arises from the conceptual simplicity of the growth process / 6 / . The particular merits of MBE are that thin films can be grown with precise control over thickness, alloy composition, and doping level. In Fig. 1 we show schematically the fundamentals for MBE growth of III-V semiconductors. The MBE process consists of a co-evaporation of the constituent elements of the epitaxial layer (Al, Ga, In, P, As, Sb) and dopants (mainly Si for n-type and Be for p-type doping) onto a heated crystalline substrate where they react chemically under ultra-

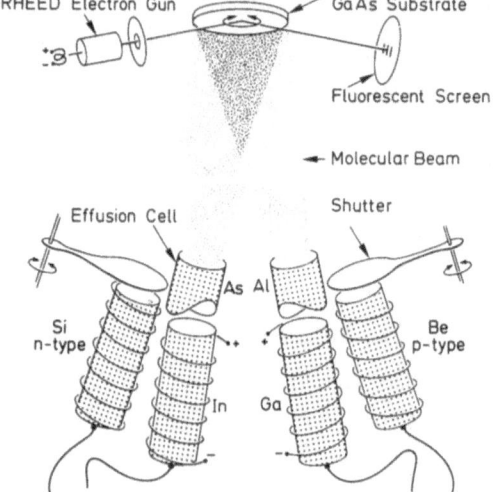

Fig. 1   Schematic illustration of MBE growth process for III-V semiconductors

high vacuum (UHV) conditions. The composition of the layer and its doping level depend on the relative arrival rate of the constituent elements which in turn depend on the evaporation rate of the appropriate sources. Accurately controlled temperatures (to within $\pm$ 0.1° at 1000 °C) have thus a direct, controllable effect upon the growth process. The group-III-elements are always supplied as monomers by evaporation from the respective liquid element and have a unity sticking coefficient over most of the substrate temperature ranges used for film growth (e.g. 500 - 650 °C for GaAs). The group-V-elements, on the other hand, can be supplied as tetramers (P_4, As_4, Sb_4) by sublimation from the respective solid element or as dimers (P_2, As_2, Sb_2) by dissociating the tetrameric molecules in a two-zone furnace. The growth rate of typically 0.5 - 1.5 µm/hr is chosen low enough that migration of the impinging species on the growing surface to the appropriate lattice sites is ensured without incorporating crystalline defects. Simple mechanical shutters in front of the evaporation sources are used to interrupt the beam fluxes to start and stop deposition and doping. Due to the slow growth rate of 1 monolayer/s, changes in composition and doping can thus be abrupt on an atomic scale. The transmission electron (TEM) micrograph of a AlAs/GaAs superlattice displayed in Fig. 2 demonstrates that this independent and accurate control of the individual beam sources allows the precise fabrication of artificially layered semiconductor structures on an atomic scale.

The design of high-quality molecular beam sources fabricated from non-reactive refractory materials is the most important requirement for successful MBE growth. Precise temperature stability ($\pm$ 0.1° at 1000°C) and reproducibility are essential. For accurate kinetic studies relevant to the growth process, (equilibrium) Knudsen effusion cells with small orifice have been used to produce collision-free thermal-energy beams of the constituent elements (Fig. 3a). The analytical formulae used to calculate the angular intensity distribution of molecular beams emerging from Knudsen cells were

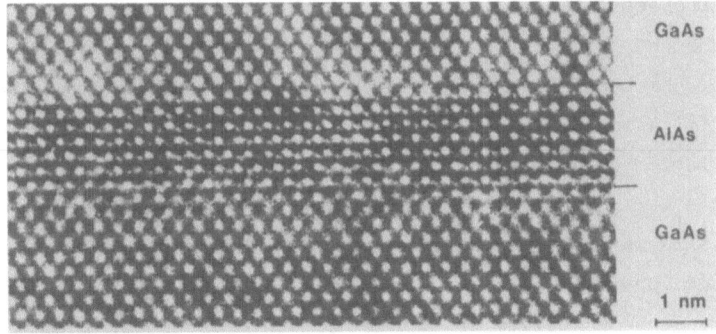

Fig. 2 High-resolution lattice image of a AlAs/GaAs SL using [110] electron beam incidence (the TEM was taken by H. Oppolzer, Siemens AG)

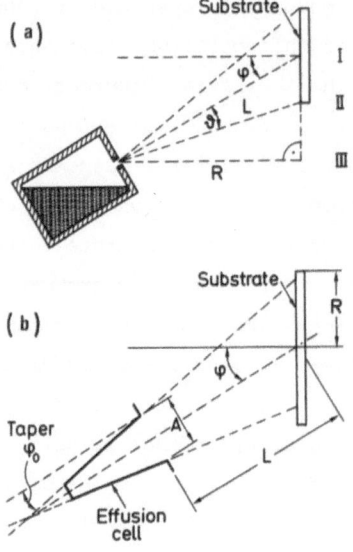

Fig. 3 Schematic illustration of (a) Knudsen effusion cell and (b) nonequilibrium (Langmuir-type) effusion cell used for practical MBE growth

summarized by Herman / 7 / . For practical film growth, nonequilibrium (Langmuir-type) effusion cells of sufficient material capacity (10 - 100 cm^{-3}) for reduced filling cycles are used (Fig. 3b). The crucibles made of PBN without any lid are of either cylindrical or conical (tapered) shape, and they have a length of about 5 - 10 times greater than their largest diameter. Cylindrically shaped effusion cells may yield progressively greater nonuniformity in the films as the charge is depleted, due to some collimating effect. To some extent this variation in uniformity from cell depletion can be balanced by inclining the rotating substrate with respect to the common central axis of the effusion cell assembly. Saito and Shibatomi / 8 / performed a systematic investigation of the dependence of the uniformity of the molecular beams across the substrate on the geometrical relationship between the Langmuir-type effusion cells and the substrate. The authors used conical crucibles having a taper ϕ_o and a diameter 2r of the cell aperture, which are inclined at an angle ϕ to the normal of the continuously rotating substrate. The diameter of the uniform area on the substrate was found to depend primarily on the distance between the substrate and the effusion cell, the taper of the cell wall, and the diameter of the cell aperture. The optimization of these parameters resulted in a reduction of the thickness variation of GaAs and $Al_xGa_{1-x}As$ layers to less than $\pm$ 1% over 3-inch wafers / 9 / . The beam flux emerging from these nonequilibrium effusion cells is monitored intermittantly using a movable ion gauge placed in the substrate position. This procedure between growth runs is mainly used to ensure reproducibility of fluxes, not to measure absolute magnitudes.

In the growth of III-V compounds and alloys, the choice of either dimeric or tetrameric group-V-element species can have a significant influence on the film properties, due to the different surface chemistry involved. There are three methods of producing P_2 and As_2. First, incongruent evaporation of the binary III-V compounds yields dimeric molecules / 10 / . Second, the dimeric species are obtained by thermal decomposition of gaseous PH_3 or AsH_3 / 11, 12 / which, however, produces a large amount of hydrogen in the UHV system. Third, dimers can be produced from the elements by using a two-zone effusion cell, in which a flux of tetramers is formed conventionally and passed through an optically baffled high-temperature stage where complete conversion to P_2 or As_2 occurs above 900 $^\circ$C. The design of two-zone effusion cells has been described by several authors / 13 - 15 / .

Severe problems in the growth of heterojunctions of exact stoichiometry may arise from the existence of transients in the beam flux intensity when the shutter is opened or closed. The temperature of the melt in the crucible is affected by the radiation shielding provided by a closely spaced mechanical beam shutter. Therefore, a flux transient lasting typically 1 - 3 min occurs when the shutter is opened and the cell is establishing a new equilibrium temperature. Several approaches have been proposed to reduce or eliminate these flux transients, including the increase of the distance between cell aperture and shutter to more than 3 cm / 16 / or the application of a two-crucible configuration / 17 / . Elimination of the flux transients is of particular importance for the growth of $Al_xIn_{1-x}As/Ga_xIn_{1-x}As$ heterojunctions lattice-matched to InP substrates.

The stoichiometry of most III-V semiconductors during MBE growth is selfregulating as long as excess group-V-element molecules are impinging on the growing surface. The excess group-V-species do not stick on the heated substrate surface, and their condensation occurs only when group-III-element adatoms are present on the surface / 18, 19 / . The growth rate of the films is essentially determined by the arrival rates of the group-III-elements. A good control of the composition of III-III-V alloys can thus be achieved by supplying excess group-V-species and adjusting the flux densities of the impinging group-III-beams, as long as the substrate temperature is kept below the congruent evaporation limit of the less stable of the constituent binary III-V compounds (e.g. GaAs in the case of Al_xGa_{1-x} As) / 20 / . At higher growth temperatures, however, preferential desorption of the more volatile group-III-element (i.e. Ga from $Al_xGa_{1-x}As$) occurs so that the final film composition is not only determined by the added flux ratios but also by the differences in the desorption rates. To a first approximation we can estimate the loss rate of the group-III-

elements from their vapour pressure data / 21 / . This assumption is reasonable because the vapour pressure of the element over the compounds, i.e.
Ga over GaAs, is similar to the vapour pressure of the element over itself
/ 22 / . The results are summarized in Table 1. The surface of alloys grown
at high temperatures will thus be enriched in the less volatile group-III-
element. As a consequence, we expect a significant loss of In in $Ga_xIn_{1-x}As$
films grown above 550 $^\circ$C and a loss of Ga in $Al_xGa_{1-x}As$ films grown above
650 $^\circ$C, and an intermittant calibration based on measured film composition
is recommended for accurate adjustments of the effusion cell temperatures.
The growth of ternary III-V-V alloy films, like GaP_yAs_{1-y}, by MBE is more
complicated, because even at moderate substrate temperatures, the relative
amounts of the group-V-element incorporated into the growing film are not
simply proportional to their relative arrival rates / 23 / . The factors
controlling this incorporation behaviour are at present not well understood.
It is therefore difficult to obtain a reproducible composition control during MBE growth of ternary III-V-V alloys. For ternary III-III-V alloys, on
the other hand, the simplicity of the MBE process allows composition control
from x = 0 to x = 1 in $Al_xGa_{1-x}As$, $Ga_xIn_{1-x}As$ etc. with a precision of ±0.001
and doping control, both n- and p-type, from the 10^{14} cm^{-3} to the 10^{19} cm^{-3}
range with a precision of a few percent. The accuracy is largely determined
by the care with which the growth rate and doping level was previously calibrated in test layers. Horikoshi et al. / 24 / have recently shown that high
quality GaAs and AlAs can be grown at a substrate temperature as low as 300
$^\circ$C if Ga or Al and As_4 are alternately supplied to the growing surface. In
an arsenic-free environment, which exists for a short period, the surface
migration of Ga and Al is strongly enhanced.

The purity of many MBE grown III-V semiconductors is limited by background acceptor impurities due to residual carbon. Larkins et al. / 25 /
have recently obtained unintentionally doped GaAs with an extremely low
residual acceptor concentration of N_A = 2.4 x 10^{13} cm^{-3}. With a donor concentration of N_D = 1.5 x 10^{14} cm^{-3} they measured a Hall mobility of 1.6 x
10^5 cm^2/Vs at 77 K. These values indicate that the purity of GaAs grown by

TABLE 1 Approximate loss rate of group-III-elements in monolayer per
second estimated from vapour pressure data

Temperature ($^\circ$C)	Al	Ga	In
550	–	–	0.03
600	–	–	0.3
650	–	0.06	1.4
700	–	0.4	8
750	0.05	2	30

MBE now approaches that of GaAs grown by vapour phase epitaxy / 26 / . The most common dopants used during MBE growth are Be for p-type and Si for n-type doping. Be behaves as an almost ideal shallow acceptor in MBE grown III-V semiconductors for doping concentrations up to 2×10^{19} cm^{-3} / 27 / . Each incident Be atom produces one ionized impurity species, providing an acceptor level 29 meV above the valence band edge in GaAs. At doping concentrations above 3×10^{19} cm^{-3}, however, the surface morphology and the luminescence properties degrade / 28 / and the diffusion of Be is enhanced / 29, 30 / , when the samples are grown at substrate temperatures above 550 °C. Lowering of the substrate temperature to 500 °C makes feasible Be acceptor concentrations up to 2×10^{20} cm^{-3} in GaAs with perfect surface morphology and reduced Be diffusion / 31 / .

The group-IV-element Si is primarily incorporated on Ga sites during MBE growth under As-stabilized conditions, yielding n-type material of fairly low compensation. The observed doping level is simply proportional to the dopant arrival rate provided care is taken to reduce the H_2O and CO level during growth. The upper limit of $n = 1 \times 10^{19}$ cm^{-3} for the free-electron concentration in GaAs was originally attributed to the enhanced autocompensation, i.e. the incorporation of Si on As sites and on interstitials / 32, 33 / . Recent investigations / 34 , 35 / indicate, however, that more probably nitrogen evolving from the PBN crucible containing Si and heated to about 1300 °C causes the compensating effect. This effect can be overcome by the use of a very low growth rate of about 0.1 μm/hr / 34 / . Recently, Maguire et al. / 36 / proposed that Si donors in GaAs are mainly compensated by [Si-X] complexes, where X was assigned to Ga vacancy (V_{Ga}), to account for the compensation often found in very heavily doped layers. The possibility of Si migration during MBE growth of $Al_xGa_{1-x}As$ films at high substrate temperatures and/or with high donor concentration has been the subject of controverse discussions, because of its deleterious effects on the properties of selectively doped $Al_xGa_{1-x}As$/GaAs heterostructures (see / 37 / and references therein). Gonzales et al. / 37 / provided some evidence that only at high doping concentrations (> 2×10^{18} cm^{-3}) Si migration might occur in $Al_xGa_{1-x}As$ films as the result of a concentration-dependent diffusion process which is enhanced at high substrate temperatures. Finally, it is important to note that the incorporation of Si atoms on either Ga or As sites during MBE growth depends on the orientation of the GaAs substrate. Wang et al. / 38 / found that in GaAs grown on (111)A, (211)A and (311)A orientations the Si atoms predominantly occupy As sites and act as acceptor, while they occupy Ga sites and act as donors on (001), (111)B, (211)B, (311)B, (511)A, (511)B and higher-index orientations. Based on these results,

Miller / 39 / and Nobuhara et al. / 40 / could grow a series of lateral
p-n junctions on graded steps of a (001) GaAs substrate surface.

In general MBE growth of III-V semiconductors is performed on (001)
oriented substrate slices about 300 - 500 μm thick. The preparation of the
growth face of the substrate from the polishing stage to the in-situ cleaning stage in the MBE system is of crucial importance for epitaxial growth
of ultrathin layers and heterostructures with high purity and crystal perfection and with accurately controlled interfaces on an atomic scale. The
substrate surface should be free of crystallographic defects and clean on
an atomic scale with less than 0.01 monolayer of impurities. Various cleaning methods have been described for GaAs and InP which are the most important substrate materials for deposition of III-V semiconductors / 4, 6,
41 - 43 / . The first step always involves chemical etching, which leaves
the surface covered with some kind of a protective oxide. After insertion
in the MBE system this oxide is removed by heating. This heating must be
carried out in a beam of the group-V-element (arsenic or phosphorus).

Advanced MBE systems consist of three basic UHV building blocks (the
growth chamber, the sample preparation chamber, and the load-lock chamber)
which are separately pumped and interconnected via large diameter channels
and isolation valves. High-quality layered semiconductor structures require
background vacuums in the low 10^{-11} Torr range to avoid incorporation of
impurities into the growing layers. Therefore, extensive LN_2 cryoshrouds
are used around the substrate to achieve locally much lower background
pressures of condensible species. Until recently, most of substrate wafers
were soldered with liquid In (at 160 $^{\circ}$C) to a Mo mounting plate / 41 / .
This practice provides good temperature uniformity due to the excellent
thermal sinking, and it is advantageous for irregularly shaped substrate
slices. However, the increasing demand for production oriented post-growth
processing of large-area GaAs wafers and the availability of large-diameter
(> 2 in.) GaAs substrates has fostered the development of In-free mounting
techniques. Technical details of direct-radiation substrate heaters have
been described by several authors / 44 - 47 / . Immediately after mounting
the Mo plate with the prepared substrate is inserted into the load-lock
chamber. The transfer between the chambers is made by trolleys and magneti-
cally coupled transfer mechanisms. In the growth chamber the Mo mounting
plate holding the substrate wafer is fixed by a bayonet joint to the Mo
heater block which is rigidly attached to a special manipulator. This
manipulator correctly positions the substrate wafer relative to the sources,
heats it to the required temperature, and rotates it azimuthally for optimum
film uniformity.

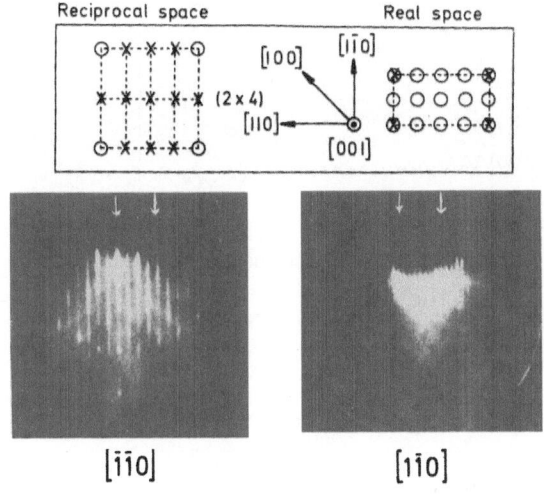

Reciprocal space   Real space

$[100]$ $[\bar{1}\bar{1}0]$

(2×4)

$[110]$

$[001]$

$[\bar{1}10]$ $[\bar{1}\bar{1}0]$

Fig. 4   Real-space and reciprocal space represen-
tation for the (2x4)
surface reconstruction
on (001) GaAs (top).
RHEED patterns of the
(2x4) surface structure
taken at two different
azimuths (bottom).

     The most important method to monitor in-situ surface crystallography
and kinetics during MBE growth is reflection high energy electron diffraction (RHEED) operated at 10 - 50 KeV in the small glancing angle reflection
mode. The diffraction pattern on the fluorescent screen contains information
from the topmost nanometer of the deposited material that can be related
to the topography and structure of the growing surface (Fig. 4). The specific surface reconstruction can be identified and correlated to the surface stoichiometry which is an important growth parameter. In addition,
the temporal intensity oscillations observed in the features of the RHEED
pattern are used to study MBE growth dynamics and the formation of hetero-
interfaces in multilayered structures / 48, 49 / . The periodic intensity
oscillations in the specularly reflected beam of the RHEED pattern shown
in Fig. 5 provides direct evidence that MBE growth occurs predominantly
in a two-dimensional (2D) layer-by-layer growth mode. The period of the
intensity oscillations corresponds exactly to the time required to grow
a monolayer of GaAs (i.e. a complete layer of Ga plus a complete layer of
As), AlAs, or $Al_xGa_{1-x}As$. To a first approximation we can assume that the
oscillation amplitude reaches its maximum when the monolayer is completed
(maximum reflection). Although the fundamental principles underlying the
damping of the amplitude of the oscillations are not completely understood,
the method is now widely used to monitor and to calibrate absolute growth
rates in real time with monolayer resolution.

     The oscillatory nature of the RHEED intensities provides direct real-
time evidence of compositional effects and growth modes during interface
formation. As for the widely used $Al_xGa_{1-x}As$/GaAs heterointerface, the
sequence of layer growth is critical for compositional gradients and crystal perfection, which in turn is important for optimizing 2D transport

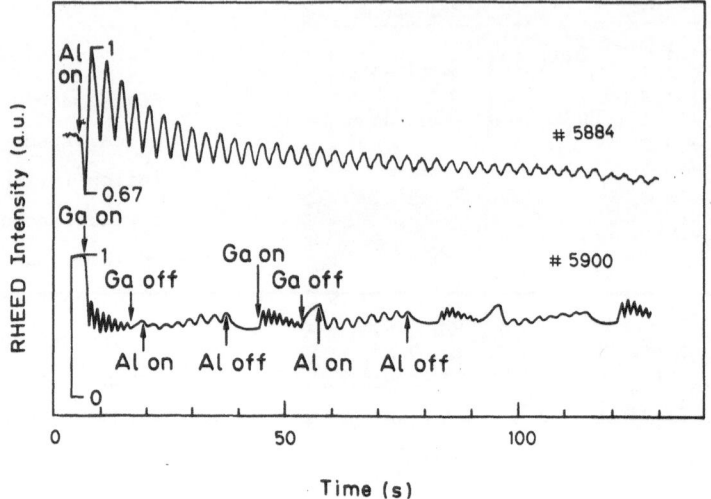

Fig. 5
Fig. 5 RHEED intensity oscillations of the specular beam obtained in
        [100] azimuth on (001) GaAs during growth of a AlAs/GaAs hetero-
        structure.

properties. When the Al flux is switched at the maximum of the intensity
oscillations, the first period for the growth sequence from ternary alloy
to binary compound corresponds neither to the $Al_xGa_{1-x}As$ growth rate nor
to the steady-state GaAs rate, but shows some intermediate value. For the
growth sequence from binary compound to ternary alloy or between the two
binaries an intermediate period does not exist. A possible explanation
for this phenomenon can be found in the relative surface diffusion lengths
of the group-III-elements Al and Ga, which were estimated to be $\lambda_{Al} \cong 3.5$
nm and $\lambda_{Ga} \cong 20$ nm on (001) surfaces under typical MBE growth conditions
/49 / . These differences in cation diffusion rates have striking consequences on the nature of the interface. While a GaAs layer should be covered by smooth terraces of 20 nm average length between monolayer steps,
those on an $Al_xGa_{1-x}As$ layer would be only 3.5 nm apart. The important result of this qualitative estimate is that the $GaAs/Al_xGa_{1-x}As$ interface
is much smoother on an atomic scale than the inverted structure. Direct
experimental evidence for this distinct difference in binary-to-ternary
layer growth sequence is obtained from the high-resolution TEM investigations of Suzuki and Okamoto / 50 / . Their lattice image of a $Al_{0.2}Ga_{0.8}As$
/AlAs superlattice shows clearly that the heterointerface is abrupt to
within one atomic layer only when the ternary alloy is grown on the binary
compound but not for the inverse growth sequence. Since the nature of the
heterointerface is critical for optimising excitonic as well as transport
properties in quantum wells, various attempts have been made to minimize

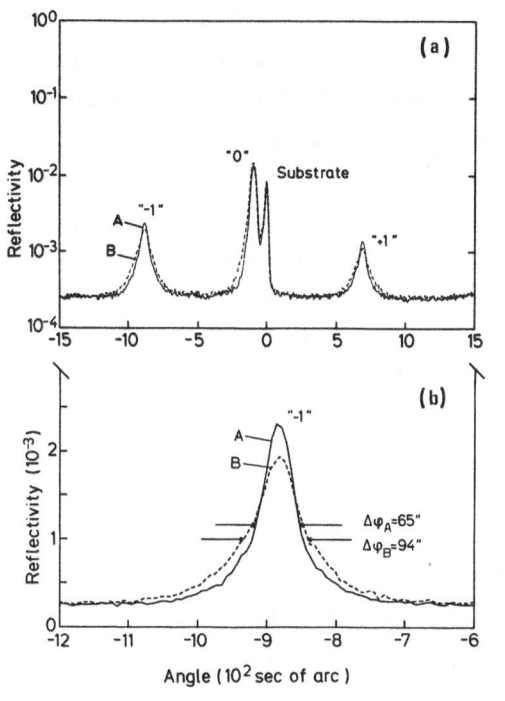

Fig. 6 X-ray diffraction curves
of two AlAs/GaAs super-
lattices grown (A) with
and (B) without growth
interruption at the hetero-
interfaces, recorded with
$CuK\alpha_1$ radiation in the
vicinity of the quasi-
forbidden (002) reflection.

the interface roughness (or disorder) by modified MBE growth conditions.
The most successful modification is probably the method of growth interruption at each interface. Growth interruption allows the small terraces
to relax into larger terraces via diffusion of the surface atoms. This
reduces the step density and thus simultaneously enhances the RHEED specular beam intensity which can be used for real-time monitoring. The time
of closing both the Al and the Ga shutter (while the As shutter is left
open) depends on the actual growth condition. Values ranging from a few
seconds to several minutes have been reported / 51, 52 / .

High-angle X-ray diffraction is a powerful nondestructive technique
for investigation of interface disorder effects in superlattices and multi
quantum well heterostructures, if a detailed analysis of the diffraction
curves is performed / 53, 54 / . As the lattice parameters and scattering
factors are subject to a one-dimensional (1D) modulation in growth direction, the diffraction patterns consist of satellite reflections located
symmetrically around the Bragg reflections, as shown in Fig. 6 for a AlAs/
GaAs superlattice. From the position of the satellite peaks the superlattice periodicity can be deduced. Detailed information about thickness fluctuations of the constituent layers, inhomogeneity of composition, and interface quality can be extracted from the halfwidths and intensities of
the satellite peaks. The existence of interface disorder manifests itself
in an increase of the halfwidths and a decrease of the intensities of the

satellite peaks, shown in Fig. 6. During MBE growth of these two $(AlAs)_{42}$
$(GaAs)_{34}$ superlattices, the adjustment of the shutter motion at the transition from AlAs to GaAs and vice versa was changed in the two growth runs.
Sample A was grown with growth interruption at each AlAs/GaAs and GaAs/AlAs
interface, whereas sample B was grown continuously. While the positions of
all the diffraction peaks of sample A coincide with those of sample B, the
halfwidths of the satellite peaks from sample A are narrower and their reflected intensities are higher. A growth interuption of 10 s was sufficient
to smooth the growing surface which then provides sharp heterointerfaces.
When the heterojunctions are grown continuously, the monolayer roughness of
the growth surface leads to a disorder and thus broadening of the interface.
In X-ray diffraction this broadening manifests itself as a random variation
of the superlattice period of about one lattice constant (~ 5.6 Å) for
sample B. The quantitative evaluation of the interface quality by X-ray
diffraction becomes even more important if the lattice parameters of the
epilayer have to be matched to those of the substrate by appropriate choice
of the layer composition, as for $Al_{0.48}In_{0.52}As/Ga_{0.47}In_{0.53}As$ superlattices
lattice-matched to InP substrates / 55 / .

3. ENGINEERING OF ARTIFICIALLY LAYERED III-V SEMICONDUCTORS
ON AN ATOMIC SCALE

In this section we demonstrate the capability of MBE to engineer artificially layered semiconductors on an atomic scale by discussing the optical properties of all-binary AlAs/GaAs SL with periods ranging from a few
tens of nanometer down to the width of a monolayer. In the AlAs/GaAs UTLS
the concept of artificially layered semiconductor structures is scaled
down to its ultimate physical limit normal to the crystal surface, as the
constituent layers have a spatial extent of less than the lattice constant
of the respective bulk material. We show that the indirect minima of the
AlAs barrier material have a significant influence on the excitonic properties of these superlattices.

3.1 AlAs/GaAs Short-Period Superlattices (SPS)

The research activities on AlAs/GaAs short-period superlattices (SPS)
were initiated by some detrimental structural, electrical, and optical properties of the ternary alloy $Al_xGa_{1-x}As$. First, the interface roughness
for growth of binary GaAs or AlAs on the ternary alloy mentioned before
yields inferior excitonic and transport properties. Second, the electrical
properties of n-type $Al_xGa_{1-x}As$ for $0.2 < x < 0.45$ are controlled by a

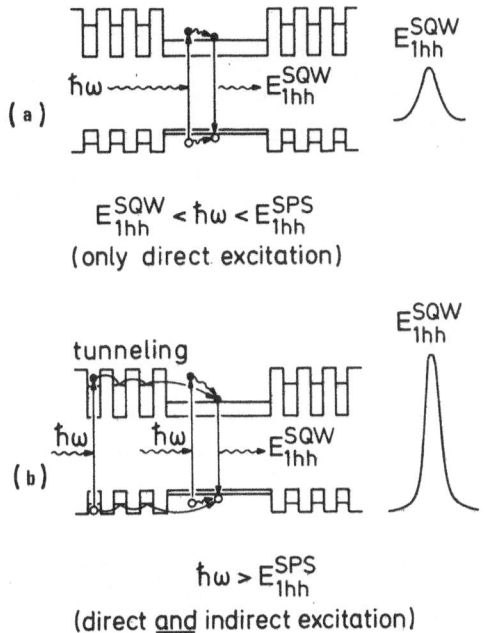

(a)

$E_{1hh}^{SQW} < \hbar\omega < E_{1hh}^{SPS}$
(only direct excitation)

(b)

$\hbar\omega > E_{1hh}^{SPS}$
(direct **and** indirect excitation)

Fig. 7 Schematic real-space energy band diagram of a GaAs quantum well confined by AlAs/GaAs SPS barrier. Also indicated is the process of carrier injection and vertical transport in the SPS. The photoluminescence signal is shown on the right hand side.

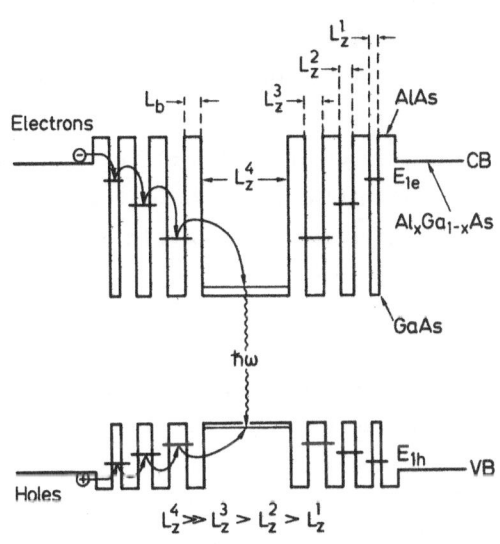

$L_z^4 \gg L_z^3 > L_z^2 > L_z^1$

Fig. 8 Schematic real-space energy band diagram of a GaAs quantum well confined by AlAs/GaAs SPS indicating how a grading of the effective energy gap of the SPS barriers is achieved by gradually changing the widths of the GaAs wells in the SPS.

deep donor ("DX center") in addition to the hydrogen-like shallow donor due to the peculiar band structure of the alloy / 56 - 58 / . Third, the X minimum of the conduction band (CB) becomes the lowest one when x > 0.43 (at 4K), and thus an indirect bandgap of the alloy results / 59 / . The all-binary AlAs/ GaAs short-period and ultrathin-layer superlattices are considered as possible substitutes for the random ternary alloy in advanced device structures. In addition, the electronic properties of these superlattices, which are in the transition region between the extremes of quantum well behaviour for period length > 8 nm (i.e. m, n > 15) and of a possible alloy-like behaviour of monolayer superlattices, are not completely understood.

Confinement layers composed of short-period superlattices $A_m B_n$ with $5 \leq (m, n) \leq 10$ play an important role for highly improved optical properties of GaAs, GaSb, and $Ga_{0.47}In_{0.53}As$ quantum wells / 60 - 62 /. In Fig. 7 we show schematically the real-space energy band diagram of a quantum well confined by SPS barriers. Also indicated in this figure is the process of carrier injection and vertical transport in the SPS towards the quantum well and the process of radiative electron-hole recombination / 63 / . The SPS barriers consist of all-binary AlAs/GaAs for GaAs quantum wells, of all-binary AlSb/GaSb for GaSb quantum wells, and of all-ternary $Al_{0.48}In_{0.52}As$/$Ga_{0.47}In_{0.53}As$ for $Ga_{0.47}In_{0.53}As$ quantum wells lattice matched to InP. The effective barrier height for carrier confinement in the quantum wells is adjusted by the appropriate choice of the layer thickness of the lower-gap material in the SPS. The observed improvement of the optical properties of SPS confined quantum wells is due (i) to a removal of substrate defects by the SPS layer, (ii) to an amelioration of the interface between quantum well and barrier, and (iii) to a modification of the dynamics of photoexcited carriers in the SPS barrier. In particular for GaSb and for $Ga_{0.47}In_{0.53}$As quantum wells we have provided the first direct evidence for intrinsic exciton recombination by application of SPS barriers / 61, 62 / . Detailed studies of the dynamics of photoexcited carriers sinking into SPS confined enlarged GaAs quantum well have clearly demonstrated an efficient vertical transport of electrons and holes through the thin AlAs barriers of the SPS / 63 - 65 / . In addition, the recombination lifetime of photoexcited carriers in GaAs quantum wells is significantly improved by the application of SPS barriers / 66 / .

The improved dynamics of injected carriers and their efficient trapping by the enlarged well make the SPS confined quantum wells very attractive for application in newly designed heterostructure lasers with separate superlattice waveguide and superlattice barriers. In Fig. 8 we show schematically how a grading of the effective band gap can be achieved by gradually changing the width of the GaAs wells in the SPS / 67 / . Recently, graded-index waveguide separate-confinement heterostructure (GRIN-SCH) laser diodes with a very low threshold current density have been fabricated / 68 / , where the graded-index waveguide was constructed by all-binary AlAs/GaAs SPS with gradually changed GaAs layer thicknesses (from 1.8 to 3.3 nm at eight intervals of six periods with a constant AlAs barrier width of 1.9 nm). When n- or p-doped SPS confinement layers are required, this can be achieved by selectively doping the GaAs regions of the SPS / 69 / .

A distinct example for the removal of substrate defects by SPS buffer layers and for the improvement of the interface between quantum well and SPS barrier is given by a modified selectively doped $Al_xGa_{1-x}As/GaAs$ heterostructure with high-mobility 2D electron gas (2DEG) which we have developed recently / 70 / . A 10-period AlAs/GaAs SPS prevents propagation of dislocations from the substrate so that the thickness of the active GaAs layer containing the 2DEG can be reduced to 50 nm. The SPS confined <u>narrow</u> active GaAs layer is of distinct importance for transistor operation, because the electrons cannot escape too far from the 2D channel during pinch-off. This implies a higher transconductance for the high electron mobility transistor (HEMT). In addition, the growth time of the complete heterostructure is reduced to less than 15 min. An additional 15 min for wafer exchange and heat and cool time makes a total of 30 min throughput time per 2-in high-quality heterostructure wafer grown by MBE. We have finally used AlAs/GaAs SPS in one-sided selectively doped $Al_xGa_{1-x}As/AlAs/GaAs$ <u>multi</u> quantum well heterostructures which exhibit an enhanced electron mobility of more than 6×10^5 cm^2/V s at 4K with AlAs spacer as narrow as 4.5 nm / 71 / .

3.2 Effect of Barrier Thickness on Optical Properties of
     AlAs/GaAs Superlattices

When the barrier thickness L_B in $Al_xGa_{1-x}As/GaAs$ superlattices is reduced to below 3 nm, the wavefunctions of the GaAs wells couple through the barriers and subbands of finite width parallel to the layer plane are formed. At this transition from a multi quantum well heterostructure with isolated GaAs wells to a real Esaki-Tsu superlattice the luminescence peak energy should decrease for a constant well width L_Z due to the broadening of the subbands. We have recently found, however, that even for isolated GaAs wells with thick barriers in AlAs/GaAs superlattices the barrier thickness has an unexpected influence on the excitonic transitions / 72 / . In Fig. 9 we show that for constant GaAs well widths of L_Z = 10.2 nm and L_Z = 6.4 nm the excitonic peaks shift to higher energies and the splitting between heavy- (E_{1h}) and light-hole (E_{11}) free excitons becomes larger when the AlAs barrier thickness is reduced from L_B = 16 to L_B = 2nm. The same high-energy shift exists when 3 mole percent Al is added to the well. This phenomenon is in contrast to the expectation from a simple coupling between adjacent wells, and a conclusive explanation has not yet been found.

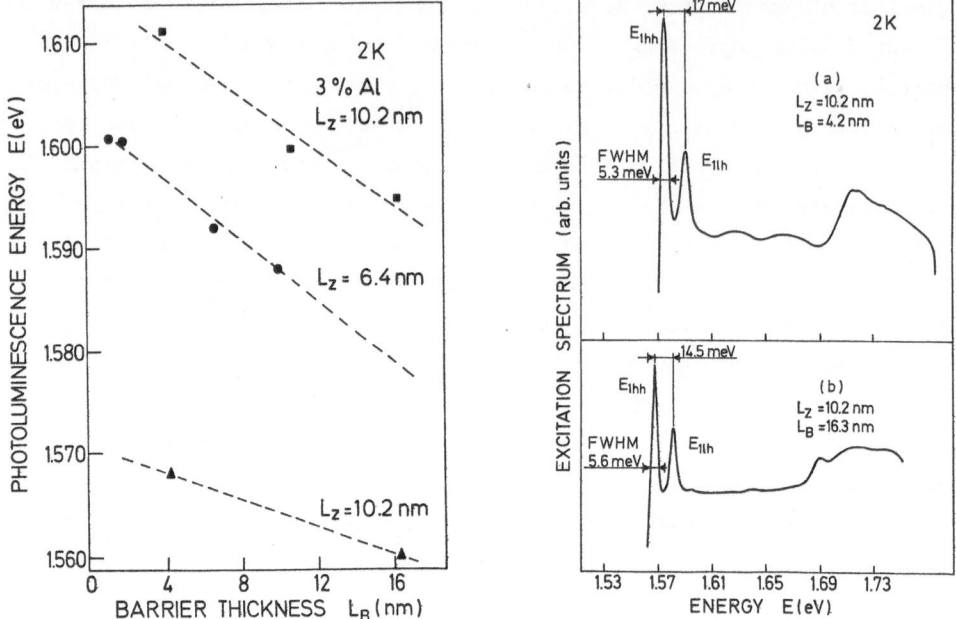

Fig. 9 Variation of luminescence peak-energy (left) and splitting between E_{1h} and E_{11} free-exciton resonances (right) as a function of barrier thickness in AlAs/GaAs superlattices for three series of samples with constant well widths.

For interpretation we have to take into account the complex band structure of GaAs quantum wells arising from (i) the valence band mixing, (ii) the nonparabolicity of the conduction band, and (iii) the indirect nature of the barrier material. For practical application it is important that our results demonstrate the inadequacy of luminescence spectroscopy to determine the well widths of superlattices accurately. For this purpose additional techniques like double-crystal X-ray diffraction are required. We have used this technique to determine the thickness of the constituent layers of the three series of samples precisely, since the knowledge of the actual values is crucial for interpretation of the luminescence data.

We have recently extended our investigation of modified excitonic features in AlAs/GaAs SL particularly to the range of ultrathin AlAs barriers / 73 / . In Fig. 10 we show for 9.2 nm GaAs wells that blue-shifts of 6 meV for the E_{1hh}^{ex} (heavy-hole) and of 8 meV for the E_{11h}^{ex} (light-hole) intersubband transitions occur and that the 1 hh - 1 lh splitting increases from 12.7 meV to 16.4 meV when the barrier thickness L_B is decreased. The faint 2 meV decrease of the transition energy observed for the 1.3 nm AlAs barrier probably indicates the beginning of the well coupling. In addition, we observe a reduction of the oscillator strength (evaluated from the intensities of the E_{1hh}^{ex} and E_{11h}^{ex} peaks) of the 1 lh transition as compared to the 1 hh transitions for SL with thin barriers. In AlAs/GaAs SL with narrower wells of L_z = 6.4 nm, where the confinement effect is larger, the observed blue-shifts (8 meV) and the reduction of the relative oscillator strength are even more pronounced.

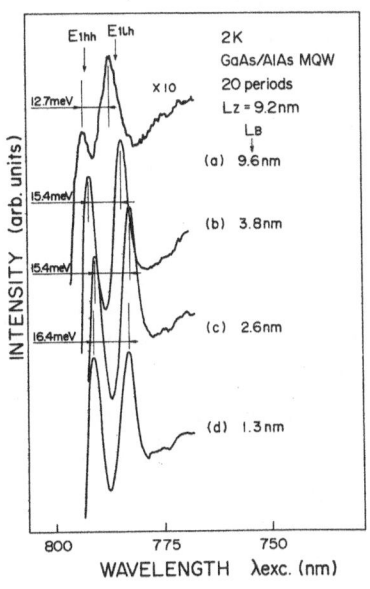

Fig. 10 Photoluminescence excitation spectra of AlAs/GaAs SL with L_z = 9.2 nm wells. The arrows indicate the calculated positions of the lowest heavy-hole (1hh) and light-hole (11h) transitions assuming exciton binding energies of 8 meV for 1hh and 9 meV for 11h.

The important result of our investigation / 73 / is that quantum confinement effects in AlAs/GaAs SL exist down to ultrathin barriers of L_z = 1.3 nm by virtue of the indirect nature of the AlAs conduction band (CB). In addition, the 2D eigen-states which are less confined in the quantum well are more sensitive to a variation of the barrier thickness. We explain these effects by means of the data shown in Fig. 11. In the limit of ultrathin barriers, a strong discrepancy between the measured exciton resonance energies and the calculated values exists (the one-band Kronig-Penney model considering the energies at the T-point was used for these calculations). The variation of the confinement energy as a function of barrier thickness is entirely different for the lowest electron, heavy-hole, and light-hole subband, respectively. In particular for the heavy-hole subband, the energy changes by only 1.5 meV in the range of 10 > L_B > 1 nm. The effective mass of the heavy-hole branch is "heavy" in the GaAs well and also in the AlAs barrier, and therefore the wavefunction is better confined to the well even with ultrathin barriers. The subband energies of the light hole and of the "light" electron, on the other hand,

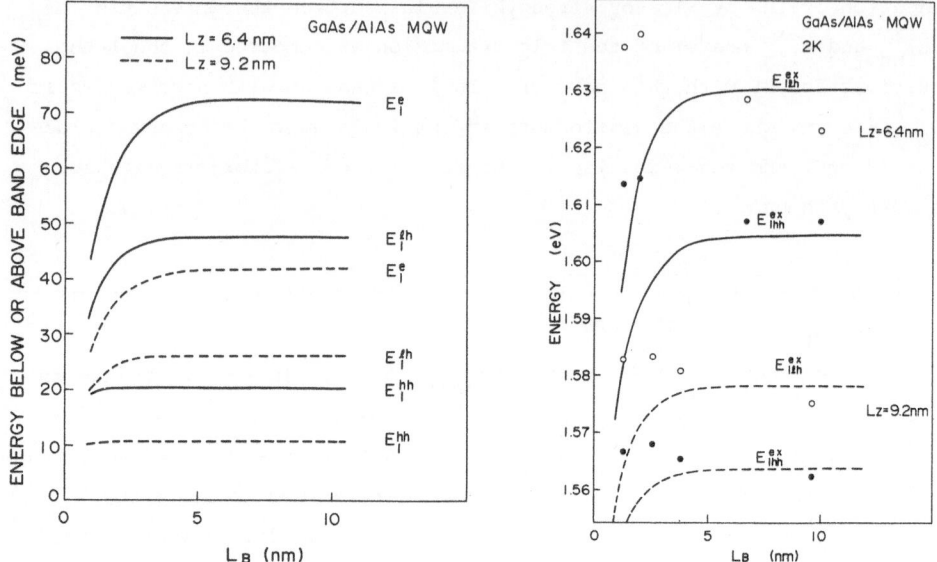

Fig. 11   Comparison of measured and calculated exciton resonance energies
associated with the 1hh and 1lh intersubband transitions for two
sets of AlAs/GaAs SL with L_Z = 6.4 and 9.2 nm (left) and calculated confinement energies of lowest conduction (E_1^e), heavy-hole
(E_1^{hh}), and light-hole (E_1^{lh}) subbands as a function of L_B using
bulk effective mass parameters (right, for details of parameters
see Ref. 73).

depend strongly on the barrier thickness as they are less localized in the
GaAs well. The subbands associated with "light" effective mass parameters
can thus be modified more easily by well coupling and/or band mixing. Taking into account the localized nature of the heavy-hole subband, the observed blue-shifts of 5 and 8 meV for the 1 hh exciton resonances are most probably associated with a modification of the conduction subband. The increasing blue-shift at higher confinement energies in the narrower wells indicates
the importance of band mixing with the upper CB minima at the X and L point
in the AlAs barriers. The decrease of the oscillator strength of the 1lh
transition, on the other hand, is related to the nonparabolicity of the hole
subbands originating from the transverse dispersion / 74 / .

Finally, we have grown a special configuration of AlAs/GaAs SL to allow
the direct observation of photoluminescence (PL) from conduction band states
having both Γ- and X-like character / 75 / . The actual structures consist
of 30-period AlAs/GaAs SL with a few monolayer thick AlAs spikes at the
center of the GaAs wells (see inset of Fig. 12). For the whole set of samples the well and barrier thicknesses are kept constant at values of L_Z =

9 nm and L_B = 5 nm, respectively. Only the thickness l_B of the AlAs spike
at the center of each well is varied from sample to sample, with l_B = 0,
2, 4, 6, and 8 monolayers (ml). Inspection of the low-temperature (2K) PL
spectra in Fig. 12, obtained with the 2.335 eV line of a cw Kr[+] ion laser,
clearly reveals a large shift of the main PL peak to higher energies even
for very narrow (2ml) AlAs spikes. In addition, an increase of the thickness of the spikes results in the appearance of a set of peaks regularly
distributed at intervals of the order of 20 meV, which scarcely evolve in
energy when l_B is further increased up to 8 ml.

The understanding of the observed PL spectra requires a careful description of the band structure of this specific SL configuration, taking
into account the indirect nature of AlAs with the minimum of the CB close
to the X-point. As a consequence of the SL potential along the (001) direction, for the CB at the Γ-point of the SL we have both a well spatially
centered on GaAs for Γ-like states and a well on AlAs for X-like states
(see inset of Fig. 12 at right). The states at these wells will coexist at
the Γ point of the SL. In order to describe all these features properly,
we use a tight-binding Hamiltonian including spin-orbit interaction. A

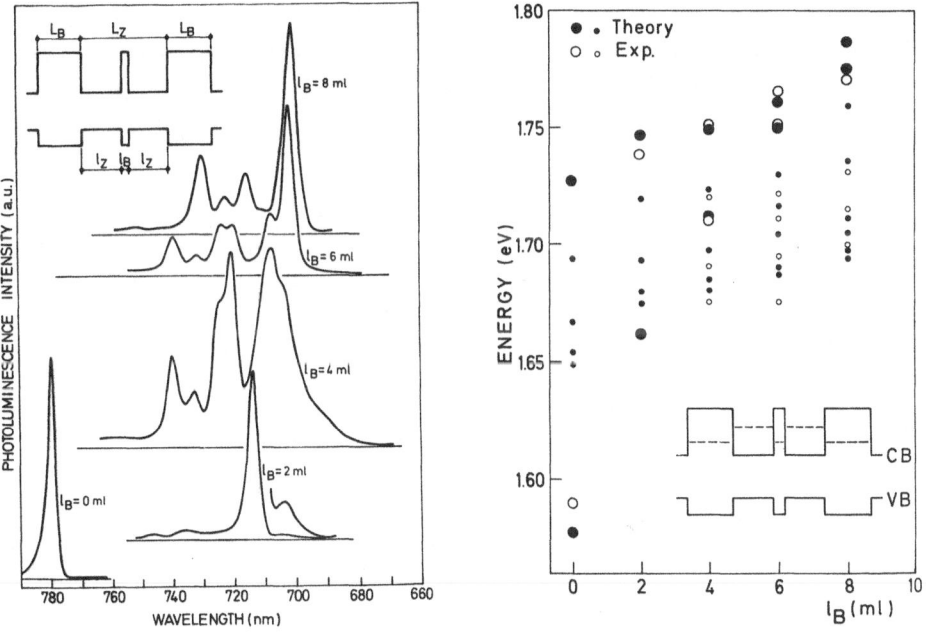

Fig. 12 Low-temperature photoluminescence spectra of AlAs/GaAs SL with
different thickness l_B of AlAs spikes placed at the center of the
GaAs wells (left), and measured (open circles) and calculated
(dots) energies of luminescence transitions in AlAs/GaAs SL with
AlAs spikes of different thickness at the well center (right).
The large circles correspond to the most intense peaks.

simple perturbative approach / 76 / gives us the SL eigen-states both spatially and in terms of the bulk states of the constituent semiconductors. The valence band offset is the only adjustable parameter which has been taken to $\Delta E_V = 0.67$ eV to give the best agreement between theory and experiment. The large open circles in Fig. 12 indicate the energy of the PL peaks that dominate the spectrum at higher temperatures or at high excitation levels, while small open circles represent transitions surviving only under low temperature or low excitation. Large closed circles correspond to calculated transitions which involve CB states concentrated on GaAs. These states give the largest overlap between CB and VB, and are thus responsible for the most intense peaks in the experiment. As far as the other PL peaks are concerned, they are clearly connected with CB states originating from X-like states in AlAs. Now a detailed explanation of the experimental results becomes possible. In the absence of any AlAs spike, the Γ-well is a few tenths of an eV deeper than the X-well, so that the lowest CB state of the SL originates from a GaAs Γ-like state. This state is responsible for the only PL transition to the VB state concentrated on GaAs. The introduction of the AlAs spike implies for the CB the appearance of a barrier for GaAs Γ-like states and a well for AlAs X-like states. So, for a few monolayers spike width, the CB GaAs Γ-like state shifts upwards in energy crossing a set of AlAs X-like states. The main PL peak remains that originating from transitions between states concentrated on GaAs. At the same

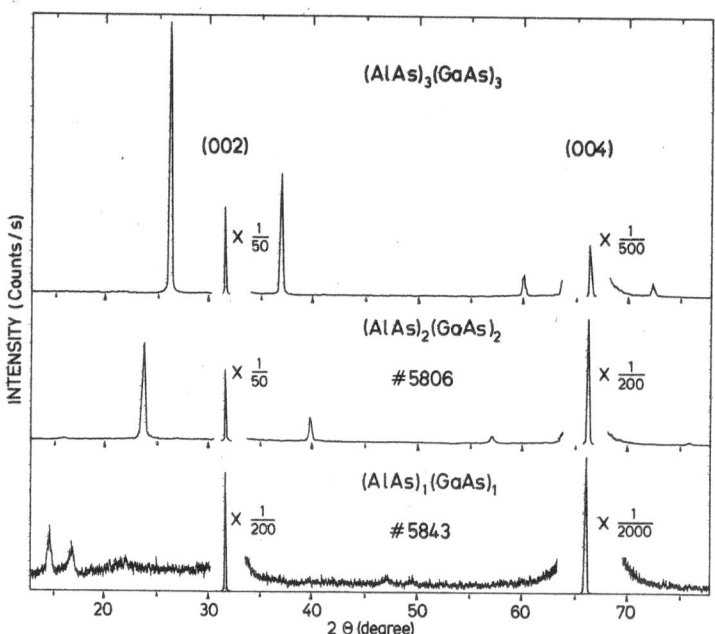

Fig. 13 X-ray diffraction pattern of $(AlAs)_m$ $(GaAs)_m$ superlattices with m = 1, 2, 3 obtained in the vicinity of the (002) and (004) Bragg reflection.

time, electrons transferred to AlAs X-like states cannot thermalize to GaAs Γ-like states of the CB because the latter are higher in energy. Therefore, those electrons in AlAs recombine with holes in GaAs giving a set of less intense peaks in the PL spectra.

3.3 (AlAs)$_m$(GaAs)$_m$ Ultrathin-Layer and Monolayer Superlattices

The investigations of ultrathin-layer (AlAs)$_m$(GaAs)$_m$ superlattices (UTLS) and monolayer superlattices with m from 10 down to 1 were motivated by the possibilities to shift the confined-particle states of the Γ, L and X valleys of the conduction band (and also of Γ of the valence band) to high enough energy to create radiative size-determined "direct-indirect" transitions exceeding in energy the bulk direct-indirect transitions of the ternary alloy Al$_x$Ga$_{1-x}$As at $x_c \cong 0.43$ / 77 / . The recent progress in the control of interface quality using RHEED intensity oscillation and growth interruption has led to the successful growth of (AlAs)$_m$(GaAs)$_m$ with m = 1, 2, 3, and 4 / 78, 79 / . Growth of these superlattices was achieved by monitoring each deposited AlAs and GaAs monolayer from the RHEED oscillation period, interrupting the group-III-element flux at m = 1, 2, 3, or 4 and allowing the RHEED intensity to recover almost to its initial value, and then depositing the next layer. The well-ordered periodic layer-by-layer arrangement of Al and Ga atoms on the appropriate lattice sites in [001]

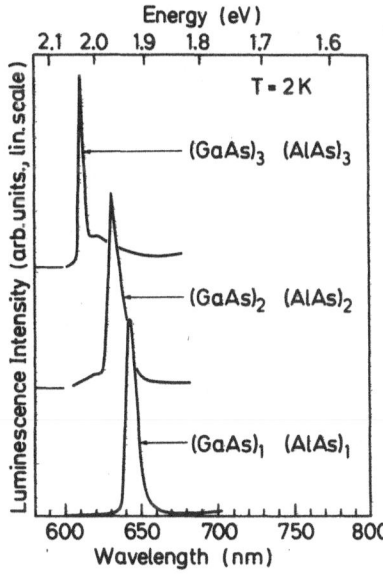

Fig. 14 Low-temperature photoluminescence spectra obtained from (AlAs)$_m$(GaAs)$_m$ superlattices with m = 1, 2, 3.

direction manifests itself in the appearance of distinct satellite peaks around the (002) and (004) reflections of the X-ray diffraction patterns shown in Fig. 13. In these UTLS the electron states have lost their 2D character and become extended throughout the entire superlattice. The mixing of different k-states due to zone folding effects can no longer be neglected, and the electronic structure can therefore not be described by a combination of the two constituent semiconductors in terms of the effective mass theory. The calculations of the energy band structure of UTLS have to be performed by treating the superlattice crystal as a whole / 80, 81 / . In Fig. 14 we show the low-temperature PL spectra obtained from the three prototype $(AlAs)_m (GaAs)_m$ UTLS with m = 1, 2, and 3. The three spectra are dominated by one sharp luminescence line which strongly shifts in energy as a function of period length, i.e. of m. In Table 2 we summarize the luminescence peak energies of the three superlattices. For comparison, we have also included the peak energy obtained from the m = 4 superlattice and the no-phonon (NP) line from the ternary $Al_x Ga_{1-x}As$ alloy. Although all five sample configurations were prepared under similar MBE growth conditions and have nearly the same average composition of $Al_{0.5}Ga_{0.5}As$, distinct differences exist in the peak position and in the line shape of the observed luminescence. The important result is that, except for the ternary alloy, the m = 3 superlattice exhibits the highest PL peak energy of all $(AlAs)_m (GaAs)_m$ UTLS. In particular the luminescence of the monolayer superlattice is shifted by 146 meV to lower energy as compared to the ternary $Al_{0.52}Ga_{0.48}As$ alloy. The observed dependence of the luminescence energy on the period length, i.e. on m, is in contrast to previous results of

TABLE 2   Variation of PL peak energy as a function of period length observed in $(AlAs)_m (GaAs)_m$ UTLS at 2 K. For comparison, also the energy of the no-phonon luminescence detected in the ternary $Al_x Ga_{1-x}As$ alloy is included.

Superlattice configuration	Average Al composition x	Luminescence peak energy (eV)
$(AlAs)_1 (GaAs)_1$	0.51	1.931
$(AlAs)_2 (GaAs)_2$	0.51	1.968
$(AlAs)_3 (GaAs)_3$	0.49	2.033
$(AlAs)_4 (GaAs)_4$	0.49	1.964
$Al_x Ga_{1-x}As$ alloy	0.52	2.077

Ishibashi et al. / 82 / , but it is in good agreement with recent theoretical calculations / 80, 81 / . These results provide definite experimental evidence that the UTLS with m $\leq$ 4 indeed represent a new artificial semiconductor with novel electronic properties. A very intriguing feature of the monolayer superlattice should be the absence of alloy scattering which is dominant in the random ternary alloy / 83 / .

Finally, the question of the origin of the intense PL line in these (AlAs)$_m$(GaAs)$_m$ UTLS arises. To clarify the radiative recombination mechanisms, we have studied the whole series of (AlAs)$_m$(GaAs)$_m$ SL with 70 > m > 1 / 84 / . In this series an intermediate range of SL periods exists between m = 6 and m = 15, which also include the SPS, where the luminescence properties gradually change from the normal quantum well behaviour for m > 15 to a behaviour characteristic for the dominating participation of X-like states. This change of the luminescence properties is clearly seen in the PL excitation spectra depicted in Fig. 15. The m = 15 SL exhibits the expected heavy- and light-hole excitonic features and a Stokes shift of less than 10 meV between E_{1h} luminescence and E_{1h} absorption. The m = 3 UTLS, on the other hand, does not show any feature due to heavy- and light-hole splitting in the excitation spectrum. Instead, we observe a threshold energy E_{th} similar to that of the indirect-gap ternary Al$_{0.52}$Ga$_{0.48}$As alloy. Consequently, the no-phonon excitonic PL transition involves a heavy-hole confined at the Γ-point in the GaAs layer and an electron confined in the (001) X-valley in the adjacent AlAs layer. The

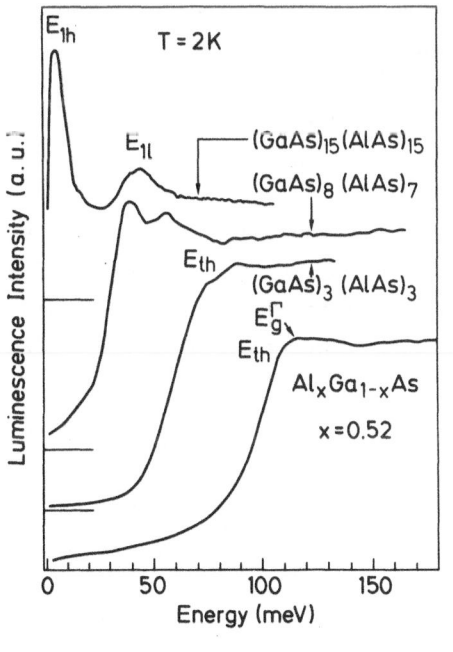

Fig. 15 Photoluminescence excitation spectra of (AlAs)$_m$(GaAs)$_m$ superlattices with 15 > m > 3 and of ternary Al$_x$Ga$_{1-x}$As alloy measured from the detection energy at the main PL peak towards higher energy.

threshold in the absorption process, however, is induced by a Γ-like <u>direct</u> transition process. In the intermediate range of SL periods $6 < m < 15$ the holes are confined in the GaAs while the electrons reside in the indirect-gap AlAs where they are confined to the well formed by the discontinuity of the X-point band edges. The observed energy shift between photoluminescence and absorption threshold thus corresponds directly to the energy separation of the lowest direct (Γ-like) and indirect (X-like) conduction band levels of the respective superlattice configuration.

4. MBE USING GASEOUS SOURCE MATERIALS

The replacement of solid source materials in conventional MBE by gaseous source materials was initiated by the search for a long-lasting arsenic source / 11, 12 / and for a reproducible compositional control during growth of ternary III-V-V alloys based on phosphorus and arsenic / 23 / . Thermal cracking of the hydrides PH_3 and AsH_3 at temperatures around 900 $^{\circ}$C produces the dimers P_2 and As_2 as well as hydrogen. This so-called gas-source MBE, where only the group-V-elements are replaced by their corresponding hydrides, allows epitaxial growth of high-quality $Ga_{0.47}In_{0.53}As/InP$ and $Ga_xIn_{1-x}P_yAs_{1-y}/InP$ heterostructures and superlattices with a high degree of control / 85, 86 / . An extension of this concept was then made by replacement of the group-III-elements by metalorganic compounds, like $Ga(CH_3)$, $Ga(C_2H_5)_3$, $Al(CH_3)_3$, $In(CH_3)_3$ etc. In this so-called metalorganic (MO) MBE / 87 / or chemical beam epitaxy (CBE) / 88 / the metalorganic flows mixed with hydrogen are combined outside the UHV system to form a single beam impinging onto the heated substrate for a good compositional uniformity. On the substrate surface thermal pyrolysis of the metalorganic compounds takes place and in an excess group-V-beam the compound is formed. Also this technique allows reproducible growth of $Ga_{0.47}In_{0.53}As/InP$ and $Ga_xIn_{1-x}P_yAs_{1-y}/InP$ heterojunctions and superlattices of high quality / 87, 88 / .

Gas-source MBE and MO MBE close the gap between the techniques of MBE and low-pressure MO CVD. These two recent developments have advantages for the growth of phosphorus and arsenic containing III-V semiconductors. Although a number of impressive results has been achieved, a detailed investigation of the incorporation behaviour of phosphorus and arsenic in III-V-V-compounds is still lacking. In addition, a thorough comparison of the properties of Al containing heterostructures and superlattices grown by conventional MBE, gas source MBE or MOMBE would reveal the actual state of the art of each of these techniques. Finally, the real challenge for both gas-source MBE and MOMBE will be the replacement of the extremely toxic AsH_3 by suitable safer alkyls of high purity.

ACKNOWLEDGEMENT

This work was sponsored by the Bundesministerium für Forschung und Technologie of the Federal Republic of Germany.

REFERENCES

/ 1 /   For an extensive survey on MBE see: Proc. 4th Int. Conf. Molecular Beam Epitaxy, J. Cryst. Growth 81 (1987)

/ 2 /   For an extensive survey on MOCVD see: Proc. 2nd Int. Conf. Metal-organic Vapour Phase Epitaxy, J. Cryst. Growth 77 (1986)

/ 3 /   A.C. Gossard, Thin Solid Films 57, 3 (1979).

/ 4 /   L.L. Chang and K. Ploog (Eds.):"Molecular Beam Epitaxy and Hetero-structures" (Martinus Mijhoff, Dordrecht, 1985) NATO Adv. Sci. Inst. Ser. E 87 (1985)

/ 5 /   R.C. Miller and D.A. Kleinmann, J. Lumin. 30, 520 (1985)

/ 6 /   E.H.C. Parker (Ed.):"The Technology and Physics of Molecular Beam Epitaxy" (Plenum Press, New York, 1985)

/ 7 /   M.A. Herman, Vacuum 32, 555 (1982)

/ 8 /   J. Saito and A. Shibatomi, Fujitsu Tech. J. 21, 190 (1985)

/ 9 /   J. Saito, T. Igarashi, T. Nakamura, K. Kondo, and A. Shibatomi, J. Cryst. Growth 81, 188 (1987)

/ 10 / C.T. Foxon and B.A. Joyce, In Current Physics in Materials Science, Vol. 7, ed. by E. Kaldis (North-Holland, Amsterdam 1981)

/ 11 / M.B. Panish, J. Electrochem. Soc. 127, 2729 (1980)

/ 12 / A.R. Calawa, Appl. Phys. Lett. 38, 701 (1981)

/ 13 / H. Künzel, J. Knecht, H. Jung, K. Wünstel, and K. Ploog, Appl. Phys. A 28, 167 (1982)

/ 14 / T. Henderson, W. Kopp, R. Fischer, J. Klem, H. Morkoc, L.P. Erickson, and P.W. Palmberg, Rev. Sci. Instrum. 55, 1763 (1984)

/ 15 / D. Huet, M. Lambert, D. Bonnevie, and D.Dufresne, J. Vac. Sci. Technol. B 3, 823 (1985)

/ 16 / T. Mizutani and K. Hirose, Jpn. J. Appl. Phys. 24, L119 (1985)

/ 17 / P.A. Maki, S.C. Palmateer, A.R. Calawa, and B.R. Lee, J. Electrochem. Soc. 132, 2813 (1985)

/ 18 / C.T. Foxon and B.A. Joyce, Surf. Sci. 50, 434 (1975)

/ 19 / C.T. Foxon and B.A. Joyce, Surf. Sci. 64, 293 (1977)

/ 20 / C.T. Foxon and B.A. Joyce, J. Cryst. Growth 44, 75 (1978)

/ 21 / R.E. Honig and D.A. Kramer, RCA Review 30, 285 (1969)

/ 22 / C.T. Foxon, J.A. Harvey, and B.A. Joyce, J, Phys. Chem. Solids 34, 1693 (1973)

/ 23 / C.T. Foxon, B.A. Joyce, and M.T. Norris, J. Cryst. Growth 49, 132
        (1980)

/ 24 / Y. Horikoshi, M. Kawashima, and H. Yamaguchi, Jpn. J. Appl. Phys.
        25, L 868 (1986)

/ 25 / E.C. Larkins, E.S. Hellman, D.G. Schlom, J.S. Harris, M.H. Kim,
        and G.E. Stillman, J. Cryst. Growth 81, 344 (1987)

/ 26 / C.M. Wolfe, G.E. Stillman, and W.T. Lindley, J. Appl. Phys. 41,
        3088 (1970)

/ 27 / M. Ilegems, J. Appl. Phys. 48, 1278 (1977)

/ 28 / N. Duhamel, P. Henoc, F. Alexandre, and E.V.K. Rao, Appl. Phys.
        Lett. 39, 49 (1981)

/ 29 / D.L. Miller and P.M. Asbeck, J. Appl. Phys. 57, 1816 (1985)

/ 30 / Y.C. Pao, T. Hierl, and T. Cooper, J. Appl. Phys. 60, 201 (1986)

/ 31 / J.L. Lievin and F. Alexandre, Electron Lett. 21, 413 (1985)

/ 32 / E. Nottenburg, H.J. Bühlmann, M. Frei, and M. Ilegems, Appl. Phys.
        Lett. 44, 71 (1984)

/ 33 / J.M. Ballingall, B.J. Morris, D.J. Leopold, and D.L. Rode, J. Appl.
        Phys. 59, 3571 (1986)

/ 34 / M. Heiblum, W.I. Wang, L.E. Osterling, and V. Deline, J. Appl. Phys.
        54, 6751 (1983)

/ 35 / R. Sacks and H. Shen, Appl. Phys. Lett. 47, 374 (1985)

/ 36 / J. Maguire, R. Murray, R.C. Newman, R.B. Beal, and J.J. Harris,
        Appl. Phys. Lett. 50, 516 (1987)

/ 37 / L. Gonzales, J.B. Clegg, D. Hilton, J.P. Gowers, C.T. Foxon, and
        B.A. Joyce, Appl. Phys. A 41, 237 (1986)

/ 38 / W.I. Wang, E.E. Mendez, T.S. Kuan, and L. Esaki, Appl. Phys. Lett.
        47, 826 (1985)

/ 39 / D.L. Miller, Appl. Phys. Lett. 47, 1309 (1985)

/ 40 / H. Nobuhara, O Wada, and T. Fujii, Electron. Lett. 23, 35 (1987)

/ 41 / A.Y. Cho and J.R. Arthur, Progr. Solid State Chem. 10, 157 (1975)

/ 42 / G.J. Davies, R. Heckingbottom, H. Ohno, C.E.C. Wood, and A.R.
        Calawa, Appl. Phys. Lett. 37, 290 (1980)

/ 43 / H. Fronius, A. Fischer, and K. Ploog, J. Cryst. Growth 81, 169
        (1987)

/ 44 / K. Oe and Y. Imamura, Jpn. J. Appl. Phys. 24, 779 (1985)

/ 45 / L.P. Erickson, G.L. Carpenter, D.D. Seibel, P.W. Palmberg, P.Pearah,
        W. Kopp, and H. Morkoc, J. Vac. Sci. Technol. B 3, 536 (1985)

/ 46 / D.E. Mars and J.N. Miller, J. Vac. Sci. Technol. B 4, 571 (1986)

/ 47 / A.J. Springthorpe and P. Mandeville, J. Vac. Sci. Technol. B 4,
        853 (1986)

/ 48 / T. Sakamoto, H. Funabashi, K. Ohta, T. Nakagawa, N.J. Kawai, T.
        Kojima, and K. Bando, Superlattices and Microstructures 1, 347 (1985)

/ 49 /   B.A. Joyce, P.J. Dobson., J.H. Neave, K. Woodbridge, J. Zhang, P.K. Larsen, and B. Bölger, Surf. Sci. 168, 423 (1986)

/ 50 /   Y. Suzuki and H. Okamoto, J. Appl. Phys. 58, 3456 (1985)

/ 51 /   M. Tanaka, H. Sakaki, and J. Yoshino, Jpn. J. Appl. Phys. 25, L 155 (1986)

/ 52 /   F. Voillot, A. Madhukar, J.Y. Kim, P. Chen, N.M. Cho, W.C. Tang, and P.G. Newman, Appl. Phys. Lett. 48, 1009 (1986)

/ 53 /   L. Tapfer and K. Ploog, Phys. Rev. B 33, 5565 (1986)

/ 54 /   Y. Kashihara, T. Kase, and J. Harada, Jpn. J. Appl. Phys. 25, 1834 (1986)

/ 55 /   L. Tapfer, W. Stolz, and K. Ploog: Ext. Abstr. 1986 ICSSDM (Jpn. Soc. Appl. Phys., Tokyo, 1986) 603

/ 56 /   D.V. Lang, R.A. Logan, and M. Jaros, Phys. Rev. B 19, 1015 (1979)

/ 57 /   E.F. Schubert and K. Ploog, Phys. Rev. B 30, 7031 (1984)

/ 58 /   T.N. Morgan, Phys. Rev. B 34, 2664 (1986)

/ 59 /   R. Dingle, R.A. Logan, and J.R. Arthur, Inst. Phys. Conf. Ser. 33a, 210 (1977)

/ 60 /   K. Fujiwara, H. Oppolzer, and K. Ploog, Inst. Phys. Conf. Ser. 74, 351 (1985)

/ 61 /   K. Ploog, Y. Ohmori, H. Okamoto, W. Stolz, and J. Wagner, Appl. Phys. Lett. 47, 384 (1985)

/ 62 /   J. Wagner, W. Stolz, J. Knecht, and K. Ploog, Solid State Commun. 57, 781 (1986)

/ 63 /   K. Fujiwara, J.L. de Miguel, and K. Ploog, Jpn. J. Appl. Phys. 24, L 405 (1985)

/ 64 /   B. Deveaud, A. Chomette, B. Lambert, A. Regreny, R. Romestain, and P. Edel, Solid State Commun. 57, 885 (1986)

/ 65 /   A. Nakamura, K. Fujiwara, Y. Tokuda, T. Nakayama, M. Hirai, Phys. Rev. B 34, 9019 (1986)

/ 66 /   K. Fujiwara, A. Nakamura, Y. Tokuda, T. Nakayama, and M. Hirai, Appl. Phys. Lett. 49, 1193 (1986)

/ 67 /   J.J. Coleman, P.D. Dapkus, W.D. Laidig, B.A. Vojak, and H. Holonyak, Appl. Phys. Lett. 38, 63 (1981)

/ 68 /   Y. Tokuda, Y.N. Ohta, K. Fujiwara, and T. Nakayama, J. Appl. Phys. 60, 2729 (1986)

/ 69 /   T. Baba, T. Mizutani, and M. Ogawa, Jpn. J. Appl. Phys. 22, L 627 (1983)

/ 70 /   K. Ploog and A. Fischer, Appl. Phys. Lett. 48, 1392 (1986)

/ 71 /   K. Ploog, H. Fronius, and A. Fischer, Appl. Phys. Lett. 50, (1987)

/ 72 /   J. L. de Miguel, K. Fujiwara, L. Tapfer, and K. Ploog, Appl. Phys. Lett. 47, 836 (1985)

/ 73 / K. Fujiwara, J.L. de Miguel, L. Tapfer, and K. Ploog, Phys. Rev.
        B. to be published (1987)

/ 74 / M. Altarelli, J. Lumin. $\underline{30}$, 472 (1985)

/ 75 / L. Brey, C. Tejedor, J.L. de Miguel, F. Briones, and K. Ploog,
        Proc. 18th Int. Conf. Phys. Semicond., Ed. O. Engström (World
        Scientific Publ., Singapore, 1987) 727

/ 76 / C. Tejedor, J.M. Calleja, F. Meseguer, E.E. Mendez, C.A. Chang,
        and L. Esaki, Phys. Rev. $\underline{B\ 32}$, 5305 (1985)

/ 77 / M.D. Camras, N. Holonyak, K. Hess, J.J. Coleman, R.D. Burnham,
        and D.R. Scifres, Appl. Phys. Lett. $\underline{41}$, 317 (1982)

/ 78 / M. Nakayama, K. Kubota, H. Kato, S. Chika, and N. Sano, Solid
        State Commun. $\underline{53}$, 493 (1985)

/ 79 / T. Isu, D.S. Jian, and K. Ploog, Appl. Phys. $\underline{A\ 43}$, 75   (1987)

/ 80 / T. Nakayama and H. Kamimura, J. Phys. Soc. Jpn. $\underline{54}$, 4726 (1985)

/ 81 / M.A. Gell, D. Ninno, M. Jaros, and D.C. Herbert, Phys. Rev. $\underline{B\ 34}$,
        2416 (1986)

/ 82 / A. Ishibashi, Y. Mori, M. Itabashi, and M. Watanabe, J. Appl. Phys.
        $\underline{58}$, 2691 (1985)

/ 83 / T. Yao, Jpn. J. Appl. Phys. $\underline{22}$, L 680 (1985)

/ 84 / D.S. Jiang, K. Kelting, T. Isu, H.J. Queisser, and K. Ploog, J.
        Appl. Phys. to be published (1987)

/ 85 / For a review on the fundamentals of gas source MBE see: M.B. Panish,
        Prog. Cryst. Growth Charact. $\underline{12}$, 1 (1986)

/ 86 / M.B. Panish, J. Cryst. Growth $\underline{81}$, 249 (1987)

/ 87 / Y. Kawaguchi, H. Asahi, and H. Nagai, Inst. Phys. Conf. Ser. $\underline{79}$,
        79 (1986)

/ 88 / W.T. Tsang, J.Cryst. Growth $\underline{81}$, 261 (1987) and references therein.

MOLECULAR BEAM EPITAXIAL GROWTH AND PROPERTIES OF

Hg-BASED MICROSTRUCTURES

Jean-Pierre Faurie

Department of Physics
University of Illinois at Chicago
P.O. Box 4348, Chicago, IL 60680

ABSTRACT

This review paper reports on growth by Molecular Beam Epitaxy and characterization of $Hg_{1-x}N_xTe-CdTe$ (N = Cd, Mn or Zn) superlattices and $Hg_{1-x}Cd_xTe-HgTe$ heterojunctions with a special attention to the interdiffusion, the valence band offset between HgTe and CdTe and the Type III to Type I transition in these superlattices.

I. INTRODUCTION

HgTe-CdTe superlattices have received a great deal of attention over the last several years as a potential material for far-infrared detectors. Since 1979 when this superlattice (SL) system was first proposed as a new material for application in infrared optoelectronic devices,[1] significant theoretical and experimental attention has been given to the study of this new superlattice system. The interest in HgTe-CdTe SL is due to the fact that it is a new structure involving a II-VI semiconductor and a II-VI semimetal and that it appears to have great potential as a material for infrared detectors.

Most of the studies have focused primarily on the determination of the superlattice bandgap as a function of layer thicknesses and as a function of temperature. Also, the description of the electronic and optical properties at energies close to the fundamental gap has received much attention.[2-4]

The growth of this novel superlattice was first reported in 1982[5] and has subsequently been reported by several other groups.[6-9]

The first theoretical calculations using either the tight binding approximation with spin orbit splitting[1] or the envelope function approximation[2]

71

showed that the bandgap E_g of the SL did vary from 0 to 1.6V demonstrating that it could possibly used as an infrared material. These first calculations assumed that the valence band offset $\Lambda = \Gamma_{8HgTe} - \Gamma_{8CdTe}$ was small or even zero in agreement with the phenomenological common anion rule.[10]

Theoretical calculations predict a narrowing of the SL bandgap E_g compared to the bandgap of the $Hg_{1-x}Cd_xTe$ alloy with the same composition. Also the SL bandgap is predicted to decrease as the thickness of the HgTe layer (d_1) in the superlattice increases. It has also been predicted that in the far infrared the cutoff wavelength of the SL will be easier to control than that of the corresponding alloy since $d\lambda/d(d_1)$ of the SL should be less than $d\lambda/dx$ of the alloy.[3] These three predictions have been confirmed experimentally.[11-14] In the classification proposed for heterointerfaces,[15] the HgTe-CdTe SL appears to belong to a new class of superlattices called Type III. This is due to the inverted band structure (Γ_6 and Γ_8) in the zero gap semiconductor HgTe as compared to those of CdTe, which is a normal semiconductor [Fig. 1]. Thus, the Γ_8 light-hole band in CdTe

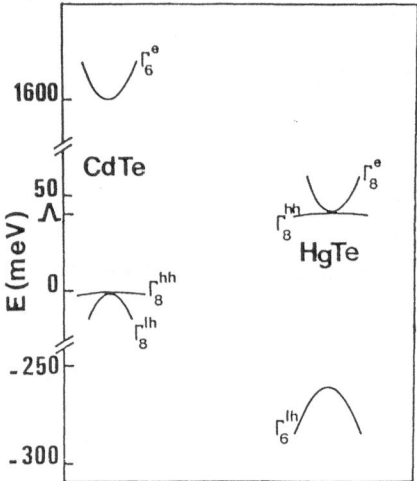

Fig. 1. Band structure of bulk HgTe and CdTe. The lh, hh and e indices refer to light holes, heavy holes and electron respectively.

becomes the conduction band in HgTe. When bulk states made of atomic orbitals of the same symmetry but with effective masses of opposite signs are used, the matching up of bulk states belonging to these bands has, as a consequence, the existence of a quasi-interface state which could contribute significantly to optical and transport properties.[16] Indeed, we have shown that the interface states could be responsible for the high hole mobilities prviously reported and not yet understood.[17,18]

Since then, the theory has been refined, the growth is under better control and more experimental data are available today. Nevertheless, many questions are still to be answered.

In this paper recent developments concerning the growth and characterization of Hg based II-VI superlattices will be presented with a special emphasis on the Type III-Type I transition. In order to investigate the high hole mobility problem $Hg_{1-x}Cd_xTe-CdTe$, $Hg_{1-x}Mn_xTe-CdTe$ and $Hg_{1-x}Zn_xTe-CdTe$ SLs have been grown and characterized by X-ray diffraction and magneto transport experiments.

The thermal stability of the HgTe-CdTe interface has been investigated through temperature-dependent in situ X-ray diffraction measurements and the results on several samples will be reported.

A comparison between the experimental room temperature bandgap (Eg) and the theoretical predictions from the envelope function approximation will be presented. Also a few values of E_g are given at 2K and compared to theoretical predictions. These data have been used to propose an experimental equation relating room temperature cutoff wavelength (λ_c) and HgTe and CdTe layer thicknesses.

The value of the valence band discontinuity Λ between HgTe and CdTe is presently disputed both theoretically and experimentally. XPS measurements carried out on many single heterojunctions grown and characterized in situ will be presented here.

II. GROWTH

HgTe-CdTe superlattices were grown for the first time on a CdTe ($\overline{1}\overline{1}\overline{1}$)Te substrate in a Riber 1000 MBE system. In our laboratory at the University of Illinois, the growth experiments are currently carried out in a Riber 2300 MBE machine using three different effusion cells containing CdTe for the growth of CdTe, Te and Hg for the growth of HgTe. We have shown that on a CdTe substrate, the substrate temperature must be above 180°C in order to grow high-quality superlattice crystals.[5] At this temperature, the condensation coefficient for mercury is close to 10.[-3(19)] This requires a high mercury flux during the growth of HgTe. Nevertheless, the background pressure during the growth is in the high 10^{-7} torr range.

HgTe-CdTe superlattices have also been grown on $Cd_{0.96}Zn_{0.04}Te$($\overline{1}\overline{1}\overline{1}$)Te substrates and on GaAs(100) substrates.[20] On GaAs(100), both (100)-SL//(100)GaAs and (111)SL//(100)GaAs epitaxial relationships have been obtained. The orientation can be controlled by the preheating temperature as previously reported.[21]

For CdTe(111) grown on GaAs(100) we have recently reported that according to selective etching, X-ray photoelectron spectroscopy and electron diffraction investigations, the orientation of the CdTe film is the $(\overline{1}\overline{1}\overline{1})$Te face.[22] We have grown on both CdTe$(\overline{1}\overline{1}\overline{1})$//GaAs(100) and CdTe(100//GaAs(100) substrates and we have seen a difference in the mercury condensation coefficient. This has already been reported for the growth of $Hg_{1-x}Cd_xTe$ films on substrates of different crystallographic orientations.[23] It turns out that growing at 190°C on a (100) orientation requires about 4.4 times more mercury than growing on a $(\overline{1}\overline{1}\overline{1})$Te orientation at the same temperature. But in the (100) orientation no microtwinning due to the formation of antiphase boundaries are observed which makes the growth more easy to control than in the $(\overline{1}\overline{1}\overline{1})$B orientation.

In order to obtain high quality superlattices we use typical growth rates of 3-Ås^{-1} for HgTe and 1Ås^{-1} for CdTe. This represents the best compromise between the low growth rate required for high crystal quality, especially for CdTe which should be grown at a higher temperature than 180°C, and the duration of the growth, which should be as short as possible in order to save mercury and to limit the interdiffusion process which cannot be completely neglected between these interfaces (this will be discussed later).

Compared to the growth of HgTe-CdTe SL that of $Hg_{1-x}Cd_xTe$-CdTe SL presents an additional difficulty since we have to control the ternary alloy $Hg_{1-x}Cd_xTe$ instead of the binary compound HgTe. Furthermore, since our goal is the study of the Type III - Type I transition, the composition (x) of the alloy should be very well controlled. In order to have the necessary flexibility for the composition x, a Cd cell plus a CdTe cell or two CdTe cells are required. The growth of $Hg_{1-x}Cd_xTe$ by MBE has already been discussed in numerous papers[24] and the growth of $Hg_{1-x}Cd_xTe$/CdTe SLs successfully achieved.[18]

$Hg_{1-x}Mn_xTe$-CdTe SLs have been grown with x ranging from 0.02 to 0.12 on CdTe$(\overline{1}\overline{1}\overline{1})$//GaAs(100) substrates using three effusion cells containing Hg, Mn and Te for the growth of the alloy and a CdTe cell for the growth of CdTe.[25]

More recently $Hg_{1-x}Zn_xTe$-CdTe SLs have been grown with x ranging from 0.02 to 0.15 on CdTe$(\overline{1}\overline{1}\overline{1})$//GaAs(100) substrates using three effusion cells containing Hg, Te and ZnTe for the growth of the alloy and a CdTe cell for the growth of CdTe.

The proof that these novel superlattice systems have successfully been grown is attested to by X-ray diffraction, as illustrated in Fig. 2. In addition to the Bragg peak one can see satellite peaks due to the new periodicity. The periods of the SLs were measured from the position of the SL satellite peaks as determined by X-ray. The method for determining the period of a superlattice by X-ray diffraction is commonly used and has been explained elsewhere.[26] The values of the HgTe layer thickness (d_1)

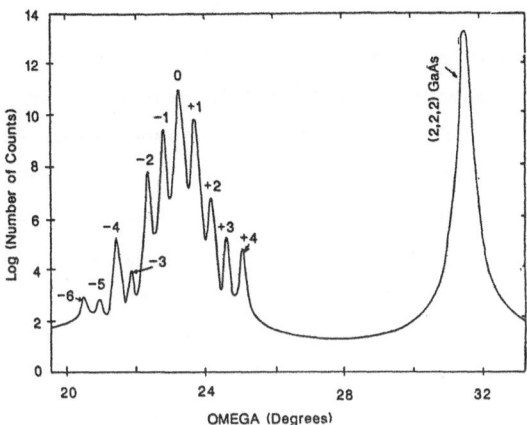

Fig. 2a. Room temperature X-ray diffraction profile about the 222 reflection of a $Hg_{0.92}Cd_{0.08}Te$-CdTe superlattice with a period of 102 Å.

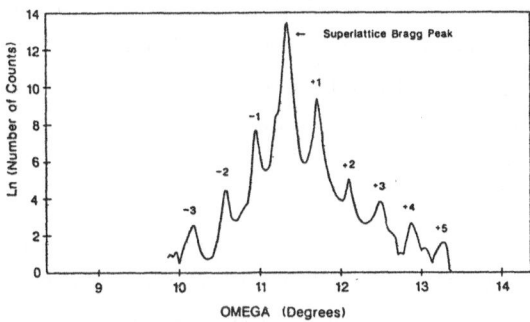

Fig. 2b. Room temperature X-ray diffraction profile about the (111) reflection of a $Hg_{0.87}Mn_{0.13}Te$-CdTe superlattice with a period of 112 Å.

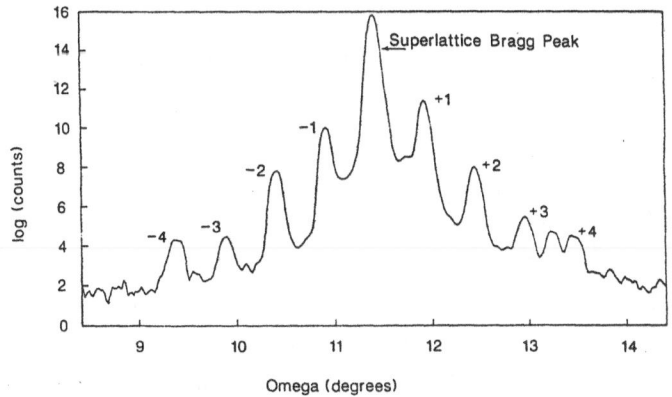

Fig. 2c. Room temperature X-ray diffraction profile about the 111 reflection of a $Hg_{0.99}Zn_{0.01}Te$-CdTe superlattice with a period of 85 Å.

and the CdTe layer thickness (d_2) were then calculated using the Cd and Hg concentrations measured by energy dispersive X-ray analysis (EDAX). In order to prevent the Hg reevaporation from HgTe layers the Hg cell is left open during the growth of CdTe layers. Thus a competition occurs between Hg and Cd. The incorporation of Hg in the CdTe layer depends critically upon several parameters such as the substrate temperature, the crystal orientation and the growth rate. Measurements by secondary ion mass spectroscopy (SIMS), wavelength dispersive spectroscopy (WDS), Raman scattering and EDAX showed that for a thick CdTe film grown under the same condition as we grow our superlattices, there was less than 5% Hg incorporated into the film. Neglecting this small amount of Hg, the ratio of d_2 to the period is just the average Cd composition measured by EDAX. The error in ignoring the Hg in the CdTe and the error in the EDAX measurement itself could lead to errors in d_1 and d_2 of 7 to 8%. Nevertheless, it has been shown recently that more mercury has to be incorporated in the CdTe layers in order to explain the far infrared reflectivity spectra of several super-lattices.[27] Thus the question is: does a thick film represent what fraction of Hg is incorporated in a thin film? Experiments are currently carried out in order to answer this question.

III. INTERDIFFUSION

A very important question for the application of this material to opto-electronic device is the thermal stability of the HgTe-CdTe interface. Because of the lower temperature used in MBE compared to other epitaxial techniques such as LPE, OMCVD, or CSVPE, the diffusion processes are more limited in MBE, but the magnitude of this interdiffusion has not yet been fully determined.

To investigate the extent of this interdiffusion, we have carried out temperature-dependent in situ X-ray diffraction measurements on several HgTe-CdTe samples. The estimated interdiffusion constants D(T) are based on the analysis of the X-ray of the nth satellite intensities as a function of time for given temperatures T.

In $\dfrac{I(t)}{I(t_o)} = -8\left(\dfrac{\pi n}{L}\right)^2 D(t-t_o)$ where $L = d_1 + d_2$ is the periodicity

of the superlattice.[28] The interdiffusion measurements were carried out using several different techniques to hold and heat the sample. These heating methods called respectively radiative or conductive have been described elsewhere.[29] From the slopes of the intensity of the first-order satellite peak versus time we have calculated the interdiffusion coefficients for five different samples annealed at 185°C which is the usual growth temperature. The results reported in table I show a large variation for the diffusion constant calculated at 185°C. D for SL 13 is fifty times higher than for SL 54. We think that part of this difference may be due to the experimental method. Nevertheless a difference of about an order of magnitude is observed for different superlattices measured by the same heating technique. The various values obtained from different

Table 1. Results of in-situ interdiffusion measurements on HgTe–CdTe superlattices grown at 185°C in the (1̄1̄1̄)B orientation

Sample #	# Periods	Period L(Å)	HgTe d_1(Å)	CdTe d_2(Å)	Substrate	Heating method & Environment fordiffusion	Diffusion $D(185°C)$ (cm^2s^{-1})
SL13	250	15	97	60	GaAs	radiative helium	3.0×10^{-18}
SL49	142	97	35	62	CdZnTe	radiative helium	1.8×10^{-18}
SL52	190	97	36	61	GaAs	radiative helium	3.0×10^{-19}
SL54	180	69	44	25	GaAs	conductive mercury	6.3×10^{-20}
SL48	170	94	42	52	CdTe	conductive helium	7.0×10^{-20}

HgTe–CdTe superlattices to this date emphasize that some material specifications should be assessed before attributing too much importance to the direction of the heat flow through the superlattice. These material specifications can all contribute in many ways to the magnitude of the diffusion coefficients. They are (i) the density of the vacancies; (ii) the content and nature of the impurities; (iii) the density and type of dislocations; (iv) the roughness of the heterojunctions; (v) the quality of the superlattice which can be estimated by the number of satellite peaks observed on each side of the control peak; (vi) the growth rate, which might be related to vacancies, impurities and dislocations; and (vii) the nature of the substrate and that of the buffer layer, as well as the lattice mismatch between these two components of the superlattice.

Furthermore our results indicate that interdiffusion is concentration dependent and thus the interpretations of these results will have to be somewhat modified.

It is important to point out that a diffusion constant D(185) in the range of 3×10^{-18} - $3 \times 10^{19} cm^2 s^{-1}$ is consistent with extrapolation of data obtained by a different group working on the interdiffusion in HgTe-CdTe single junction.[30]

Despite this dispersion in the results it turns out that the thickness of the intermixed layer caused by annealing at the growth temperature of 185°C, calculated from the relation $1 = \sqrt{Dt}$, cannot be neglected for thick superlattices required for IR detectors.

Despite these evidences, discussions about interdiffusion during the growth kept on being heard, whether this phenomenon is present or not. In order to answer this question we have grown two thick HgTe-CdTe SLs in the ($\overline{111}$)B orientation on GaAs(100) at 185°C. Table II shows the growth data and the period computed from the 1.476Å X-ray data. Three different wavelengths, 0.709Å, 1.282Å and 1.476Å were used to characterize these two SLs. The absorption of the X-rays is used as a tool to probe various

Table 2. Characteristics of HgTe-CdTe superlattices grown on GaAs(100) in the ($\overline{111}$B orientation at 185°C

SL #	# of period	Period (Å)	Duration of growth	Thickness (μm)
SL93	320	198	6h58	6.34
SL95	420	157	8h18	6.60

depths in the SL. Results are similar for both SLs. It is seen (Fig. 3) that the softest wavelength (1.476Å) produces the cleanest and the best diffraction spectrum, emphasizing that the top of the SL has interfaces much sharper than those near the interface with the buffer layer.

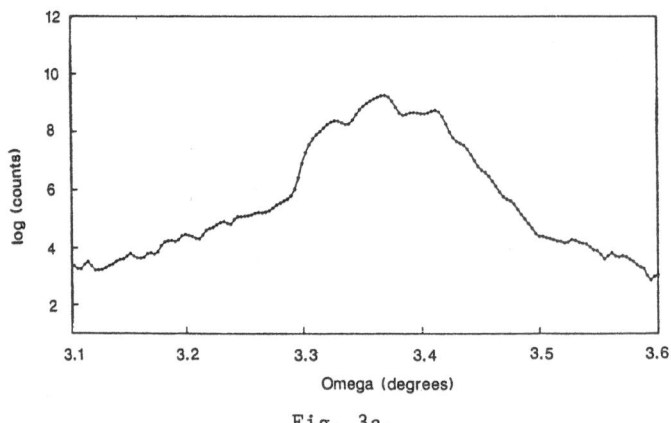

Fig. 3a.

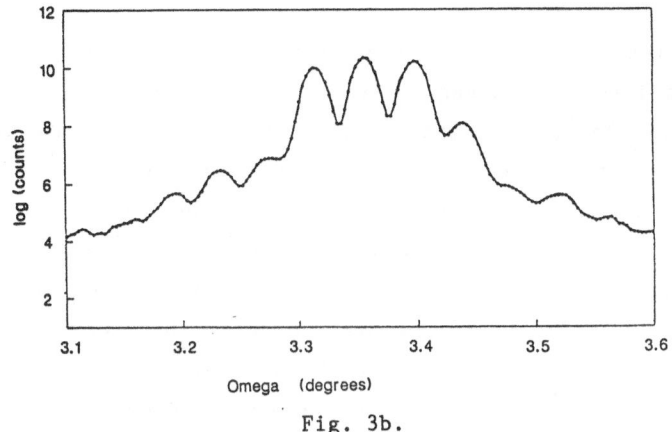

Fig. 3b.

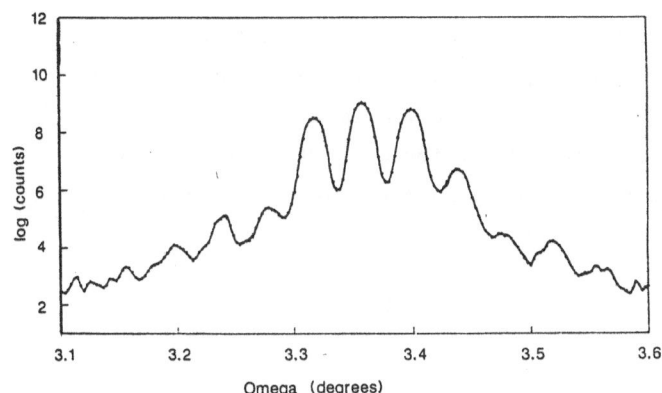

Fig. 3. Room temperature X-ray diffraction profile about the 222 reflection of SL95 X-ray wavelength: (a) 0.709 Å, (b) 1.282 Å, (c) 1.476 Å.

This indeed indicates that significant interdiffusion occurred during the growth of these superlattices in agreement with in situ measurements done on similar superlattices.

In-situ interdiffusion measurements on X-alloyed $Hg_{1-x}X_xTe/CdTe$ superlattices (X = Cd and Mn for this study) reveal that superlattices with x > 0 are more stable than HgTe/CdTe superlattices.[31] These alloyed superlattices have, therefore, a better chemical stability, hence longer device lifetime than the non-alloyed ones, alowing for technological developments to proceed. It is hypothesized that the differences between Fourier components of the diffusion coefficients D(T) are due to Cd- and Mn-substitutions which presumably slow down interstitial and vacancy motions.

IV. SUPERLATTICE BANDGAPS

At low temperature, a precise determination of the SL band gap can be obtained from far-infrared magneto-absorption experiments.[32,33] When a strong magnetic field B is applied perpendicular to the SL layers, the subbands are split into Landau levels. At low temperature (T ~ 4K), the infrared transmission signal, recorded at fixed photon energies as a function of B, presents pronounced minima which correspond to the resonant interband magneto-optical transitions between the valence and conduction Landau levels. The Landau level energies and, therefore, the interband transitions energies, can be calculated in the framework of the envelope function model.[2,34] Good agreement between theory and experiment is obtained for Λ in the range (0-100 meV), taking into account the uncertainties in the sample characteristics. The SL band gap is obtained by extrapolating the energies of the observed transition to B = 0. Figure 4 shows the SL band gap deduced from such experiments in four different samples at 2K. The solid lines in Figure 4 represent the theoretical dependence $E_g(d_1)$ calculated for d_2 = 20, 30 and 50Å using Λ = 40 meV.[35] Experiments and theory are in very satisfying agreement, when Λ is small and positive.

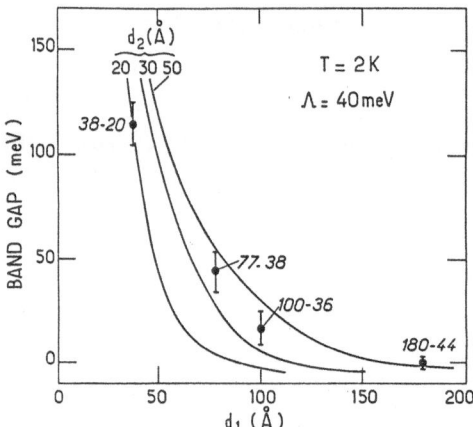

Fig. 4.   Variation of the band gap of different hgTe-CdTe superlattices at 2 K as a function of the HgTe layer thickness (d_1). The experimental data are given by the solid dots; for each sample, the first number corresponds to d_1 and the second one to d_2 which is the CdTe layer thickness. The solid lines are theoretical fits for three values of d_2.

In order to determine the SL cutoff wavelength, infrared transmission spectra were measured between 400 and 5000 cm^{-1} at 300K. The absorption coefficient (α) was calculated versus wavelength and the cutoff wavelength was defined to be the wavelength where α is equal to 1000 cm^{-1}. The absorption coefficient was obtained by taking the negative of the natural logarithm of the transmission spectrum and then dividing by the thickness of the SL. Even though the accuracy of this kind of determination is questionable and the value of 1000 cm^{-1} for α is rather arbitrary, we have found that the values of the cutoff determined in this way are in fairly good agreement with those determined by photoconductivity threshold.[36] The bandgap, in eV, is just 1.24 divided by the cutoff wavelength in µm. We do not mean to imply that this technique gives an absolute measure of the bandgap. Rather, it gives a consistent, first order value. The method is quite reproducible (within 5%) and quite simple. It is also quite useful to determine SL's HgTe layer thickness.

We have previously found that this method is not very accurate for thin superlattices.[11] Nevertheless, these investigations confirm that the bandgap of the SL is less than that of the equivalent alloy and that it decreases as the HgTe layer thickness (d_1) is increased, as illustrated in Table 3.

The theoretical value of the SL bandgap is obtained from the SL band structure at $\vec{K} = 0$, calculated using the envelope function approximation.[2,34] The band structures of HgTe and CdTe near the Γ point are described by the 6 x 6 Kane Hamiltonian taking into account Γ_6 and Γ_8 band edges. The interaction with the higher bands is included up to the second order and is described by the Luttinger parameters γ_1, $\gamma_2 = \gamma_3 = \gamma$ (spherical approximation), and K. In this calculation it is assumed that temperature variation of γ_1, γ and K between 4 and 300 K arises essentially from the variation of the interaction gap ε_0 between the Γ_6 and Γ_8 band edges. For a HgTe-CdTe superlattice, a system of differential equations is established for the multi-components envelope function. The boundary conditions are obtained by writing the continuity of the wave function at the interfaces and by integrating the coupled differential equations across an interface. Taking into account the superlattice periodicity, the dispersion relation of the superlattice is obtained. From this, the superlattice bandgap as a function of the HgTe and CdTe layer thicknesses can be found.[2,32]

Figure 5 presents a comparison of the experimental and theoretical bandgaps. The solid lines correspond to the calculated dependence $E_g(d_1)$ for d_2 = 10, 20, 30, 40 and 100 Å at 300K. An offset Λ = 40 meV[35] between the HgTe and CdTe valence band edges is used in these calculations.

Table 3. Characteristics of HgTe-CdTe superlattice grown in the (111) orientation. The superlattice bandgaps are determined from room temperature infrared transmission.

SL#	HgTe (Å)	CdTe (Å)	SL BANDGAP (meV)	COMPOSITION	ALLOY BANDGAP (meV)
1	40	20	155	0.33	335
2	40	60	225	0.60	712
4	74	36	125	0.33	325
5	97	60	100	0.38	401
6	110	48	90	0.30	295
7	86	50	114	0.37	383
8	81	34	113	0.29	281
9	53	34	167	0.39	413
10	47	30	175	0.39	412
11	83	47	117	0.36	374
14	75	31	116	0.29	281
15	58	47	162	0.45	491
16	58	28	135	0.33	325
17	63	37	145	0.37	385
18	36	61	250	0.63	757
19	50	36	180	0.42	452
22	46	48	200	0.51	581
30	107	91	92	0.46	478
31	66	91	144	0.58	684

There is good agreement if one considers the uncertainties in the HgTe and CdTe parameters used in the theoretical calculation and the uncertainties in the experimental determination of the layer thickness. For the SLs with d_2 between 35 Å and 60 Å, an interesting trend can be seen. For the larger values of d_1, the experimental E_g is larger than the theoretical. For the smaller values of d_1, this is just reversed. This suggests some sort of systematic discrepancy between the experimental data and the theoretical predictions. A similar discrepancy was seen for bandgaps determined by infrared photoluminescence.[37] For this reason, we believe that this discrepancy between theory and experiment is not due to the experimental technique.

This discrepancy suggests that the experimental data may have a different functional form than the theoretically predicted one. To examine this

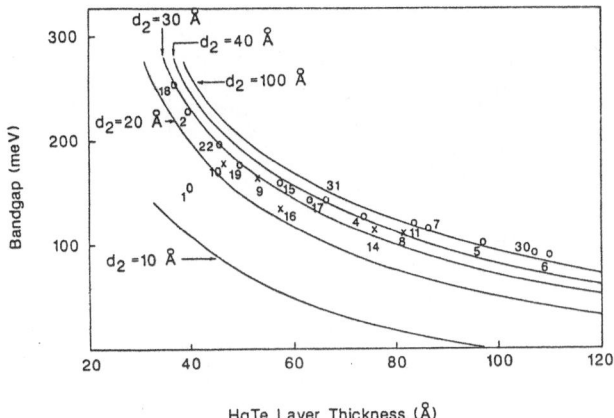

Fig. 5.   Variation of the band gap of different HgTe–CdTe
superlattices at 300 K as a function of the HgTe layer
thickness.   The samples characteristics are listed in
Table II and the experimental data correspond to the
solid circles ($17 \leq d_2 \leq 24$), crosses ($25 \leq d_2 \leq 60$).
The solid lines are theoretical fits for different
values of the CdTe layer thickness (d_2).

problem more closely λ_c has been plotted on Fig. 6 as a function of the
HgTe layer thickness d_1, for SLs having a CdTe layer thickness (d_2) greater
then 40Å, along with the theoretical curve provided by Y. Guldner. Using
the technique described above to determine λ_c, we have measured a value of
13.6μm for λ_c of HgTe. We believe that this is the limit of the technique.

A least squares fit was performed on the data to determine the equation
of the relationship between λ_c and d_1. Both linear and quadratic terms in
d_1 were included in the fit. The coefficient of the quadratic term was
found to be 4 orders of magnitude smaller than the coefficient of the
linear term. The resulting linear relation was:

$$\lambda_c \text{ (μm)} = 0.1184 \ d_1 \text{ (Å)} + 0.78$$

The solid line in Figure 6 corresponds to this equation. If the line is
extrapolated back to $d_1 = 0$, a value of $\lambda_c = 0.78$μm is obtained. The λ_c
of CdTe should be about 0.83μm. This is in reasonable agreement with the
value obtained by extrapolation of our equation. This linear relation
between the cutoff wavelength and the HgTe layer thickness is not predicted
by theory as it can be seen in Fig. 6.

Thus we found a difference between theory and experiment in the functional
form for the relationship between λ_c and d_1. This is true in spite of the

83

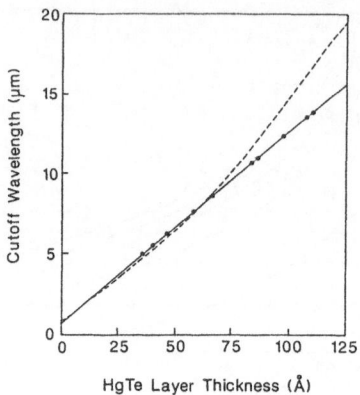

Fig. 6.   Cut-off wavelength of HgTe-CdTe superlattices
at room temperature as a function of the
HgTe layer thickness ($d_2 > 40$ Å).   The
experimental data are given by the circles
and the solid line is the linear fit.   The
dotted line is the theoretical curve calculated
by Y. Guldner.

fact that theoretical predictions and experimental measurements give about
the same value for the superlattice bandgap. The cause for the discrepancy
is not clear at this time.

Infrared photoluminescence of several HgTe-CdTe SLs have been measured
as a function of temperature from liquid helium to 300K.[37] The SLs were
grown on both GaAs and $Cd_{1-x}Zn_xTe$ substrates with HgTe layer thicknesses
ranging from 22Å to 85Å and CdTe layer thicknesses from 18Å to 62Å.
Photoluminescence peak positions were observed over the range from 3μm to
17μm (E - 0.4 to 0.07eV). For SLs with HgTe layers 60Å to 85Å thick it is
found that the photoluminescence4 peak positions as a function of temperature
agree fairly well with the predicted bandgap using a small value (0.04eV)
for Λ.

For superlattices with thinner HgTe layers (22Å to 45Å) the predicted
bandgaps were at higher energy than the photoluminescence peak positions.

It is possible to use the cutoff wavelengths determined by IR transmission to go one step further in finding how λ_c depends upon the layer
thicknesses.[38]   This data can be used to determine an empirical formula
for λ_c as a function of both d_1 and d_2. A least squares linear fit has
been performed on each of the sets of points with a similar value for d_2.
The lines are shown in the Fig. 7. From these fits it is clear that the
intercept is approximately constant and is therefore independent of the
CdTe layer thickness. But the slope of the lines does depend upon d_2.

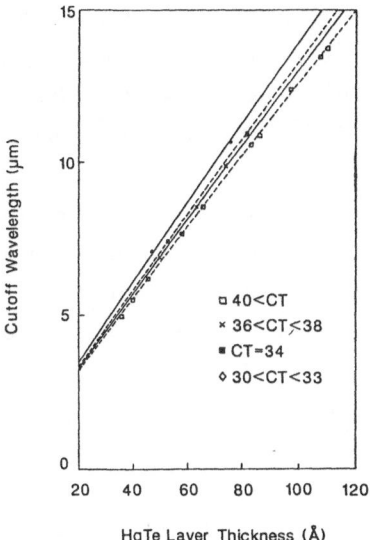

Fig. 7. Variation of the cutoff wavelength of different HgTe-CdTe superlattices at room temperature as a function of the HgTe layer thickness. The data is grouped by CdTe layer thickness. The lines are least squares linear fits of each group of data.

The slopes can then be fit to some functional form. The resulting function of d_1 and d_2 is

$$\lambda_c(\mu m) = [0.3666 \exp(-0.0034d_2^2) + 0.118]\, d_1 + 0.78$$

V. VALENCE-BAND DISCONTINUITY

The band structure of HgTe-CdTe superlattices have been calculated by using the LCAO or the envelope function models which give very similar results. An important parameter, which determines most of the HgTe-CdTe SL's properties, is the valence band discontinuity Λ between HgTe and CdTe. The value of Λ is presently disputed.

Many models have been recently developed to calculate the band discontinuities. For heterojunctions between compounds having the same anion such as tellurium it has been postulated from the phenomenological "common anion rule"[10] that the valence band discontinuity Λ is small i.e.<0.1eV. This prediction is supported by tight binding calculations.[39] But recent theoretical results, based on the role of interface dipoles do not support the common anion rule and predict a much larger value $\Lambda = 0.5$eV.[40] Such a large value has also been calculated recently with a natural lining-up without any dipole contribution: 0.26eV[41] and 0.36eV[42].

The first experimental determination of Λ was obtained from far-infrared magneto-optical experiments at T = 1.6K on a superlattice consisting of 100 periods of HgTe (180Å) - CdTe (44Å). The best agreement between experiment and theory (done in the envelope function approximation) was obtained for Λ = 40 meV.[35]

Since then, additional magneto-absorption experiments have been performed on several other SLs and it has been constantly found that a small positive offset Λ within the limits (0-100meV) provides the best fit.[33]

Resonant Raman Scattering was applied recently to investigate electronic properties of HgTe - CdTe SLs. From these experiments, it has been shown that the Γ_7 holes are confined in the CdTe layers which implies an upper limit of 120 meV for Λ.[43]

As we discussed before, IR photoluminescence measurements agree fairly well with predicted bandgaps using a small value for Λ.

On the other hand, photoemission has been demonstrated to be most valuable for providing direct and microscopic understanding of heterojunction band discontinuities.[44]

Figure 8a illustrates schematically the principle for measuring Λ = ΔE_v at the interface between two semiconductors A and B with XPS. If ΔE_v is small (≤ 0.5 eV), as for the Te-based heterojunctions, a direct investigation of the valence-band edges, E_v^A and E_v^B, is unrealistic. Indirect measurement involving core levels have to be used. By selecting two core levels, E_c^A and E_{cl}^B, well resolved in energy and by measuring their energy difference ΔE_{cl} across the interface, ΔE_v can be directly deduced according to the following relation [see Fig. 8b].[45,46]

$$\Delta E_v(A-B) = \Delta E_{cl}(A-B) + (E_{cl}^A - E_v^A) - (E_{cl}^B - E_v^B) \qquad (1)$$

$E_{cl} - E_v$, the binding-energy (BE) differences between the core level and the top of the valence band for each semiconductor, are determined independently on the bulk semiconductors. All information pertinent to the interface in relation (1) are clearly contained in $\Delta E_{cl}(A-B)$.

Λ was measured recently by XPS and a large value Λ = 0.35 eV was obtained.[45] This value was determined from a unique sample, grown in one chamber and analyzed in another chamber after exposure to air.

Commutativity holds if $\Delta E_V(A/B) = \Delta E_V(B/A)$.

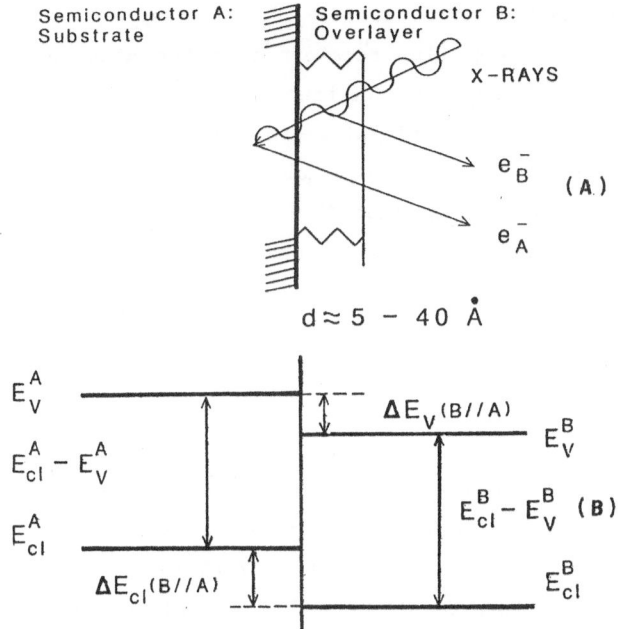

Fig. 8. Principle of determining $\Delta E_V = \Lambda$ with XPS. (a) By irradiating with X-rays semiconductor A covered by overlayer of semiconductor B, XPS spectra of both semiconductors are recorded if overlayer thickness is smaller than electron escape depth. (b) Schematic flat-band energy diagram illustrating relation (1).

In order to clarify whether there is a discrepancy between optical and XPS data and also in order to verify the commutativity rule not obtained for the GaAs/AlAs system we have performed very careful XPS measurements under well controlled conditions.

The investigated interfaces have been grown in situ by molecular-beam epitaxy with a RIBER 2300 system. Epitaxy came out in the (111) orientation with Te-rich face as controlled by reflection high-energy electron diffraction. Linearity test as well as meaningful comparison with theory requires interfaces to be abrupt at the atomic scale. Thus the growth temperature was maintained at 190°C, a temperature known to give no interdiffusion across the interface when growth time is short. Moreover, the interface abruptness is inferred from the exponential attenuation with overlayer thickness of the substrate XPS peak.

The samples were directly transferred to the attached spectrometer, at a pressure of 10^{-10} Torr, without transiting through the air. No contamination occurs, avoiding any cleaning procedure. The XPS spectrometer is a SSX-100 model from Surface Science Laboratories using a monochromatized and focussed Al, Kα excitation line. The overall energy resolution measured on $Au_{4f_{7/2}}$ line at a binding energy of 83.93 eV is 0.7 eV. The core levels selected in this work are the resolved spin-orbit components $Cd_{4d_{5/2}}$, $Hg_{5d_{5/2}}$.

The nearly lattice-matched HgTe–CdTe($\overline{111}$)B heterojunctions have been investigated here in great detail. Figure 9 shows the results for ΔE_{c1}, the binding-energy difference between $E_{Cd4d_{5/2}}$ and $E_{Hg5d_{5/2}}$, at different overlayer thicknesses for the two reverse-growth orders. ΔE_{c1} is found to be independent of the overlayer coverage. Meanwhile the Fermi-level position at the interface is varying by 0.2 eV with coverage. This lack of sensitivity of ΔE_v toward interface Fermi-level position, as previously reported for GaAs–Ge,[47] provides the first hint of linearity, in suggesting a ΔE_v "pinning" by the alignment of some reference levels across the heterojunction. Second, ΔE_{c1} is identical for the two growth orders and equal to 2.696 ± 0.030 eV. The experimental uncertaintly is given by the standard deviation, δ, for the measurements.

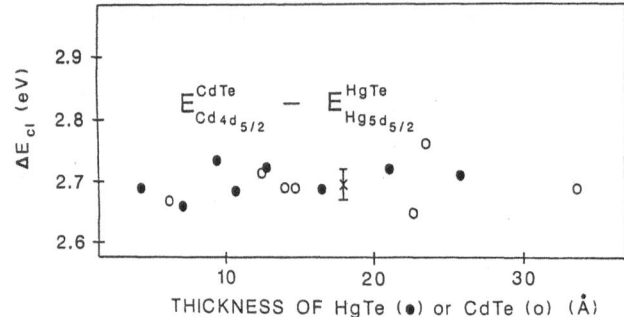

Fig. 9. ΔE_{c1} accross HgTe–CdTe interface as a function of growth order and coverage (open oval, CdTe over HgTe; filled oval HgTe over CdTe).

To obtain $(E_{c1} - E_v)$ used in relation(1), E_v is simply located by linear extrapolation of the valence-band leading edge. This procedure is well justified by the close similarity of the band structure of the tellurides near E_v [see Fig. 10]. It is quite accurate as shown by the δ over numerous

measurements on independent samples: $E^{CdTe}_{Cd4d_{5/2}} - E^{CdTe}_{v} = 10.145 \pm .030$ eV
and $E^{Te}_{Hg5d25/2} - E^{HgTe}_{v} = 7.805 \pm 0.020$ eV.

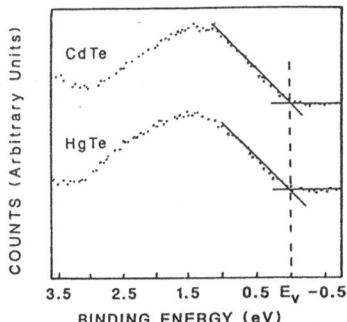

Fig. 10. Linear extrapolation of valence-band leading
edge locates the same characteristic feature
of the bands taken as E_v. The spectra are
shifted to align E_v.

Then $\Delta E_v = \Lambda$ derived for HgTe-CdTe is: 0.36 ± 0.05 eV.[48] This value
compares very closely with ref. 45.   Hence the discrepancy of XPS with
magneto optical data is confirmed but not understood at the present time.
Magneto-optical data at 2K as well as the infrared transmission measurements
that we have performed at 300K cannot be interpreted by using such a large
valence band offset either in the envelope function model or in the LCAO
approach.   In fact, most of the investigated SLs are calculated to be
semimetallic at 4K for $\Lambda = 0.35$eV which is not compatible with the magneto-
optical data.   It should be pointed out that XPS measurements were carried
out at 300K on single and perfectly abrupt heterojunctions whereas magneto-
optical, RRS and IR photoluminescence are performed at low temperature on
multijunction structures where some interdiffusion cannot be completely
ruled out.

In addition, if an appreciable amount of mercury is incorporated during
the growth in the CdTe layers that could change the theoretical calculation
for Eg and hence the fitting parameter which is precisely the valence band
offset.

XPS measurements are currently undertaken in the laboratory on multilayered structures grown under the same conditions than superlattices in
order to shed some light on this discrepancy.

The present center of the theoretical debate on the understanding of
Λ is the role played by dipoles at the interface.   The core level to valence
band maximum binding energy shifts have been measured by XPS for Hg_{1-}

$_x$Cd$_x$Te alloys.[49] Due to small charge transfer and core level shifts, these shifts are mainly due to valence band maximum shifts. We found that the sum of the valence band shift for HgTe and CdTe in Hg$_{1-x}$Cd$_x$Te is constant for the entire range of the alloy composition x and equal to 0.35 eV. This value coincides exactly with the valence band discontinuity measured for HgTe-CdTe heterojunction. Therefore we conclude that there is no need for interface dipole to explain the large valence band offset in agreement with another investigation performed on the alloy Hg$_{0.7}$Cd$_{0.3-}$Te.[50]

VI. TRANSPORT PROPERTIES - TYPE III-TYPE I TRANSITION

One of the most interesting unanswered questions of HgTe-CdTe super-lattices is the mobility enhancement in the p-type structures. Hole mobilities have been reported as high as 30,000 cm^2/V.sec, but all are above 1,000 cm^2/V.sec. Mixing of light and heavy holes has been suggested for the enhancement of the hole mobilities.[17] Several theoretical investigations have been carried out to study this problem. The band structure calculation has been refined using a multi-band tight binding model[51] and the effect of the lattice mismatch between the HgTe and CdTe has been investigated.[51,52] These studies conclude that the light holes should not contribute to the in-plane transport properties.

In order to investigate this interesting problem we have grown related superlattice systems i.e., Hg$_{1-x}$Cd$_x$Te-CdTe, Hg$_{1-x}$Mn$_x$Te-CdTe and Hg$_{1-x}$Zn$_x$Te-CdTe. HgTe-CdTe is called a Type III superlattice because of the inverted band structure of HgTe. In Hg$_{1-x}$Cd$_x$Te-CdTe SL system at T = 77K when x is smaller than 0.14 it is a type III SL. Whereas, when x is larger than 0.14 it is a Type I SL, similar to GaAs-AlGaAs SL, since HgCdTe is now a semiconductor with both electrons and holes confined in the smaller bandgap material [see Fig. 11].

This Type III - Type I transition is also expected to occur in Hg$_{1-x}$Mn$_x$Te-CdTe SLs for x ~ 0.07-0.08 and in Hg$_{1-x}$Zn$_x$Te - CdTe SLs for x ~ 0.10-0.12 (the effect of the strain has not been taken into account).

Near the transition, strain, valence band offset, alloy disorder, native defects, compensation are the same in Type III and Type I superlattices.

The only difference is the existence of interface states in type III SL but not in Type I SL.

In table 4 the Hall characterization of several SL samples is reported. It is interesting to note that if for Hg$_{1-x}$Cd$_x$Te/CdTe and Hg$_{1-x}$Zn$_x$Te/CdTe SL systems p type superlattices have been grown none of the Hg$_{1-x}$Mn$_x$Te-CdTe SLs are p-type. This difference along with the continuous drop of the electron mobility is not currently understood since n and p type Hg$_{1-x}$Mn$_x$Te layers have been grown by MBE with high electron hole mobilities.[53]

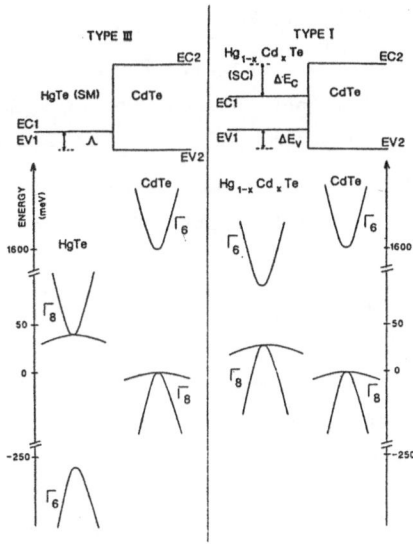

Fig. 11. Band structure of bulk HgTe, CdTe and $Hg_{1-x}Cd_xTe$ illustrating Type III and Type I SL configuration.

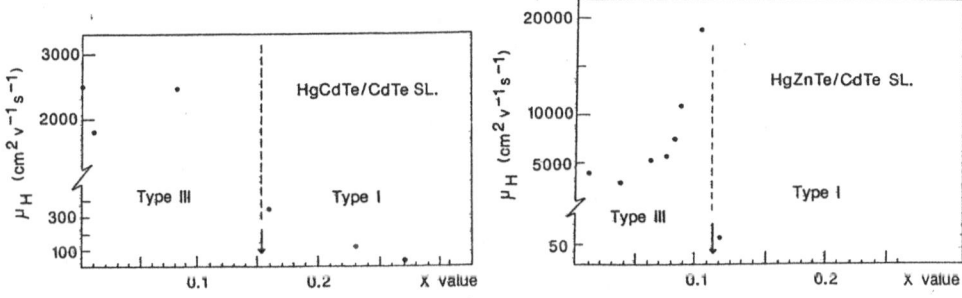

Fig. 12. Hall mobility for holes in $Hg_{1-x}Cd_xTe$-CdTe and $Hg_{1-x}Zn_xTe$-CdTe Type III and Type I SLs.

Table 4 and Fig. 12 show that the hole mobility drops drastically between Type III and Type I. (More $Hg_{1-x}Zn_xTe$-CdTe SLs should be investigated in the Type I region since only one is reported). All these superlattices have been grown in the (111)B orientation on GaAs(100) substrate. It has been previously reported[13] that HgTe-CdTe SLs grown on GaAs exhibit lower p-type mobilities (μ_H – 10^3–10^4cm^{2}V^{-1}s^{-1} range) than those grown on CdTe or CdZnTe (μ_H: 10^4–10^5cm^{2}V^{-1}s^{-1} range). If the same tendency is observed for Hg$_{1-x}$Cd$_x$Te-CdTe SLs it is not the case for Hg$_{1-x}$Zn$_x$Te-CdTe SLs since a hole mobility as high as 2 x 10^4cm^{2}V^{-1}s^{-1} is observed for sample #48688. In this system a uniaxial compressional strain which exists in Hg$_{1-x}$Zn$_x$Te layers could play a role in the hole mobility by pushing up the light hole band.

In comparison hole mobilities in Type I are in the 10^2–10^3cm^{2}V^{-1}s^{-1} range, or even lower, which is the usual range for p type Hg$_{1-x}$N$_x$Te alloys. Thus we are dealing with a hole mobility enhancement in Type III SLs and not a hole mobility decrease in Type I SLs. This strongly suggests that the mobility enhancement is related to the presence of the interface states since it is the only change occuring during the transition. It is not surprising that no sudden change in the electron mobility is observed in n-type Hg$_{1-x}$Mn$_x$Te-CdTe SLs since the interface states involved in the transition have a light hole character and are not supposed to affect the mobility of electrons.

Table 4. Characteristics of Hg$_{1-x}$N$_x$Te-CdTe (N = Cd, Mn or Zn) superlattices grown at 190°C on CdTe(111)/GaAs(100) substrates. The Hall mobilities were measured at 30K except for sample No. 18124 which was measured at 10K. D$_1$ = Hg$_{1-x}$N$_x$Te layer thickness; D$_2$ = CdTe layer thickness; n = numbers of periods; x = Cd, Mn or Zn composition in Hg$_{1-x}$N$_x$Te layers.

SL System	Type	Sample	x	D$_1$ (Å)	D$_2$ (Å)	n	μ_H (cm^{2}V^{-1}s^{-1})
	III	18124	0	70	45	70	p–2.5x10^3
	III	20539	0.01	82	34	120	p–1.8x10^3
Hg$_{1-x}$Cd$_x$Te/	IV	20842	0.08	70	32	100	p–2.5x10^3
CdTe	I	20943	0.16	70	40	100	p–3.5x10^2
	I	18929	0.23	48	22	90	p–1.3x10^2
	I	18728	0.27	69	22	100	p–5 x 10
	III	41880	0.04	168	22	100	n–2.7x10^4
	III	42281	0.05	69	26	105	n–1.5x10^4
Hg$_{1-x}$Mn$_x$Te/	III	4169	0.07	86	14	100	n–5.6x10^3
CdTe	I	32064	0.09	76	40	150	n < 10^2
	I	32266	0.13	66	46	150	n < 10^2

Hg$_{1-x}$Zn$_x$Te/ CdTe						
	III	47685	0.011	56	29	150	p-4.7×10^3
	III	49092	0.038	109	37	150	p-3.6×10^3
	III	48187	0.064			170	p-5.4×10^3
	III	48789	0.077			150	p-6.0×10^3
	III	48991	0.082			150	p-7.6×10^3
	III	47886	0.086			150	p-1.2×10^4
	III	48688	0.103	85	30	150	p-2.0×10^4
	I	46982	0.120			150	p-6×10

Hole mobility enhancement has also been observed recently in p-type HgTe-Hg$_{1-x}$Cd$_x$Te single heterojunctions[54] where Hg$_{1-x}$Cd$_x$Te is a p-type semiconductor. This enhancement was expected since such heterojunction has a Type III interface. The Hall mobility at about 30K of some p-type HgTe-Hg$_{1-x}$Cd$_x$Te heterojunctions and Hg$_{1-x}$Cd$_x$Te alloys are shown in Table 5. The heterojunctions which contain a thin layer of HgTe between the alloy and the CdTe buffer layer exhibit higher mobilities than the epitaxial layers grown without any HgTe layers.

Table 5. Mobilities of some HgTe-Hg$_{1-x}$Cd$_x$Te heterojunctions and Hg$_{1-x}$Cd$_x$Te alloys at about 30 K.

Sample	x	d_{HgTe} (Å)	μ_H(cm^2/V sec)
1	0.20	80	1000
2	0.21	60	1000
3	0.20	70	1200
4	0.28	75	1800
5	0.30	85	1100
6	0.33	70	1200
7	0.20	0	560
8	0.25	0	400
9	0.30	0	400

We have made magneto-transport measurements on both Type III SLs and heterojunctions in magnetic fields up to 22 tesla and temperatures as low as 0.5K. Details have been published elsewhere.[55] The observation of the Shubnikov-de Hass oscillations in HgTe-CdTe SL (sample #SLI) and Hg$_{0.92}$Cd$_{0.08}$Te-CdTe (sample #20842) implies that the hole mobilities are high. We have shown that carriers are in the HgTe or Hg$_{1-x}$Cd$_x$Te layers of the superlattices. Our results determined under identical conditions at 5 tesla are m* = 0.30 and 0.36, respectively, for the two superlattices which are consistent with heavy-hole effective masses in this ternary alloy. Nevertheless, it should be pointed out that the effective mass of holes might be lighter at lower magnetic field where Hall measurements are performed. Indeed, a strong magnetic field dependence has been observed on high hole mobility and from the SL band diagram[33] it can be seen that the curvature of both the heavy hole and interface state bands change much in $\vec{K}$ space.

The Quantized Hall Effect (QHE) has been observed in both Type III p-type superlattices and heterojunctions, confirming the existence of 2D hole gas at the interface.[56] Fig. 13 shows ρ_{xx} and ρ_{xy} for a $HgTe-Hg_{0.72}Cd_{0.28}Te$ heterojunction at 0.5K. The first quantum oscillation for ρ_{xx} is observed at 0.7 tesla. Using the condition $\mu_H B = 10^4$ required for observing quantum oscillation one can deduce that $\mu_H \sim 1.4 \times 10^4 cm^2 V^{-1} s^{-1}$ i.e., one order of magnitude larger than what we are measuring by conventional Hall. This is giving credit to the existence of multi carriers in these structures. The same phenomenon has been observed in $Hg_{0.92}Cd_{0.08}Te - CdTe$ SL. Hall measurement indicates a hole mobility of less than $2 \times 10^3 cm^2 V^{-1} s^{-1}$ at 0.5 tesla where the first quantum oscillation is observed indicating that μ_H should be equal to $2 \times 10^4 cm^2 V^{-1} s^{-1}$ This mixed conduction could explain the apparent difference in hole mobilities between SLs grown directly on CdTe or CdZnTe substrates and those grown on GaAs.

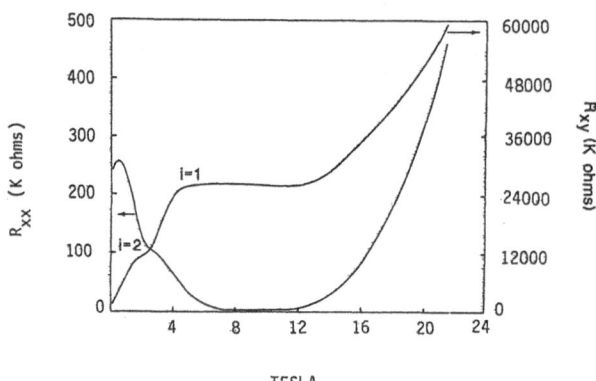

TESLA

Fig. 13. The ρ_{xx} and ρ_{xy} of a $HgTe-Hg_{0.72}Cd_{0.28}Te$ heterojunction at 0.5 K.

The CdTe buffer layer grown on GaAs could be responsible for this difference. Charge transfer at the CdTe–HgCdTe interface involving deep traps in the CdTe buffer layer has been observed recently.[57]

Concerning the QHE reported in Fig. 13 if the Hall resistance at 2.1

tesla corresponds to i = 2 then i = 1 should be at 4.2 tesla and i = 1/3 should be at 12.6 tesla. We observe the i = 1 Quantized Hall Plateau of about 25.820 ohms between 4 tesla and 13 tesla. Our analysis concluded that these features cannot be explained by sample inhomogeneity but could be explained by a magnetic field dependence of the carrier concentration.[56]

VII. CONCLUSION

In this paper we have reported on very recent developments concerning the growth and the characterization of $Hg_{1-x}Cd_xTe-CdTe$ SLs and related Hg based superlattice systems.

These SLs are now currently grown on CdTe, CdZnTe or GaAs substrates. The success of the epitaxial growth on the later substrate represents an important opening due to the high crystal quality of GaAs, its availability in large area and its interest for electronic devices. The only concern with GaAs is its large mismatch with HgTe and CdTe which could generate, even after growth of a buffer layer, some residual strain in the superlattices.

The thermal stability of the HgTe-CdTe interface has been investigated through temperature-dependent in situ X-ray diffraction measurements and, despite the dispersion in the results, it turns out that the interdiffusion cannot be neglected for thick superlattices grown at 185°C. This has been confirmed by probing two thick superlattices grown on GaAs using three different X-ray wavelengths. The softest wavelength produces the cleanest and the best diffraction spectrum emphasizing that the top of the SL has the sharpest interfaces.

A comparison between the experimental room temperature bandgaps and the theoretical predictions from the envelope function approximation has been presented. There is a good agreement when the valence band offset Λ is taken equal to 40 meV. Nevertheless, a systematic discrepancy between the experimental data and the theoretical predictions is observed suggesting that the experimental data may have a different functional form than the theoretically predicted one. We have presented an equation relating λ_c and d_1 for HgTe-CdTe SLs when $d_2 \geq 35$Å. Because the method of determining λ_c is rather simple, this equation makes it possible to determine d_1 quickly and easily as long as d_2 is greater than 35Å. This is very useful to anyone growing HgTe-CdTe superlattices, since in the absence of RHEED oscillations it is difficult to know the layer thicknesses. This linear relation was not predicted by theory. We have proposed an empirical formula for λ_c as a function of both d_1 and d_2

$$\lambda_c(\mu m) = [0.366 \exp (-0.0034 d_2^2) + 0.118]d_1 + 0.78$$

The value of the valence band discontinuity Λ is presently disputed both theoretically and experimentally. The phenomenological "common anion rule" postulates that Λ is small, i.e. < 0.1 eV and this value is supported by magnetooptics, Resonant Raman Scattering, IR photoluminescence and IR transmission experiments. On the other hand, a much larger value of 0.5 eV has been calculated based on the role of interface dipoles. XPS experiments carried out on single heterostructures HgTe/CdTe and CdTe/HgTe agree with a large value of 0.36 eV for Λ.

But XPS carried out on $Hg_{1-x}Cd_xTe$ alloys have shown that there is no need for interface dipole to explain the large valence band offset. The reason of such a discrepancy in the value of Λ is still not clear eventhough several hypothesis are currently under investigation.

The Type III – Type I transition has been investigated in p-type $Hg_{1-x}Cd_xTe$-CdTe and $Hg_{1-x}Zn_xTe$-CdTe SLs. It is reported that the hole mobility is drastically enhanced in Type III superlattices but also in Type III heterojunctions. Such mobility enhancement which is not due to modulation doping as in the $GaAs-Ga_{1-x}Al_xAs$ system is attributed to the presence of interface states in Type III structures. A comparison between Hall data and magneto transport measurement concerning the value of the hole mobility indicates that multi carriers could participate to the transport properties in these structures.

The Quantized Hall Effect has been observed in these Type III heterostructures. The large Hall plateau seen between 4 tesla and 13 tesla in a $HgTe-Hg_{0.72}Cd_{0.28}Te$ heterojunction could be explained by a magnetic field dependence of the carrier concentration.

All these recent investigations confirm once again the specific and fascinating character of these novel microstructures.

ACKNOWLEDGMENT

This work is supported by the Defense Advanced Research Projects Agency under Contract No. MDA 903-85K-0030 and the author would like to acknowledge many participants in the Microphysics Laboratory at the University of Illinois at Chicago. J. Reno, I.K. Sou, X. Chu, S. Sivananthan, P.S. Wijewarnasuriya for their assistance in the growth and characterization of the samples, C. Hsu and Tran Minh Duc for their contribution in XPS analysis, S. Rafol and K.C. Woo for performing the magnetotransport experiment.

REFERENCES

(1) J.N. Schulman and T.C. McGill, Appl. Phys. Lett., vol. 34, 663 (1979).
(2) G. Bastard, Phys. Rev., vol. B25, 7584 (1982).

(3) D.L. Smith, T.C. McGill, and J.N. Schulman, Appl. Phys. Lett., vol. 43, 180 (1983).

(4) Y. Guldner, G. Bastard, and M. Voos, J. Appl. Phys., vol. 57, 1403 (1985).

(5) J.P. Faurie, A. Million, and J. Piaguet, Appl. Phys. Lett., vol. 41, 713 (1982).

(6) J.T. Cheung, J. Bajaj, and M. Khoshnevisan, Proc. Infrared Inform. Symp., Detector Speciality, Boulder, CO, 1983.

(7) P.P. Chow and D. Johnson, J. Vac. Sci. Technol., vol. A3, 67 (1985).

(8) K.A. Harris, S. Hwang, D.K. Blanks, J.W. Cook, Jr., J.F. Schetzina, and N. Otsuka, J. Vac. Sci. Technol., A(4), 2081 (1986).

(9) M.L. Wroge, D.J. Leopold, J.M. Ballingall, D.J. Peterman, B.J. Morris, J.G. Broerman, F.A. Ponle, and G.B. Anderson, J. Vac. Sci. Technol., B4, 1306 (1986).

(10) J.O. McCaldin, T.C. McGill, and C.A. Mead, J. Vac. Sci. Technol 13, 802 (1976).

(11) C.E. Jones, T.N. Casselman, J.P. Faurie, S. Perkowitz, and J.N. Schulman, Appl. Phys. Let., vol. 47, 140 (1985).

(12) S.R. Hetzler, J.P. Baukus, A.T. Hunter, J.P. Faurie, P.P. Chow, and T.C. McGill, Appl. Phys. Lett., vol. 47, 260 (1985).

(13) J.P. Faurie, IEEE Journal of Quantum Electronics 22, 1656 (1986).

(14) J. Reno and J.P Faurie, Appl. Phys. Lett. 49, 409 (1986).

(15) L. Esaki, Proc. 17th Int. Conf. Phys. of Semiconductors, Eds. J.D. Chadi and W. A. Harrison. New York: Springer-Verlag, 473 (1985).

(16) Y.C. Chang, J.N. Schulman, G. Bastard, Y. Guildner, and M. Voos, Phys. Rev. B, vol. 31, 2557 (1985).

(17) J.P. Faurie, M. Boukerche, S. Sivananthan, J. Reno and C. Hsu, Superlattices and Microstructures 1, 237 (1985).

(18) J. Reno, I.K. Sou, P.S. Wijewarnasuriya and J.P. Faurie, Appl. Phys. Lett. 48, 1069 (1986).

(19) J.P. Faurie, A. Million, R. Boch and J.L. Tissot, J. Vac. Sci. Technol. A1, 1593 (1983).

(20) J.P. Faurie, J. Reno and M. Boukerche, J. of Cryst. Growth 72, 11 (1985).

(21) J.P. Faurie, C. Hsu, S. Sivananthan and X. Chu - Surface Science 168, 473 (1986), and references therein.

(22) C. Hsu, X. Chu, S. Sivananthan and J.P. Faurie, Appl. Phys. Lett. 48, 908 (1986).

(23) S. Sivananthan, J. Reno, X. Chu and J.P. Faurie, J. Appl. Phys. 60, 1359 (1986).

(24) J.P. Faurie, M. Boukerche, J. Reno, S. Sivananthan and C. Hsu, J.

Vac. Sci. Technol. A3, 55 (1985) (and references therein).

(25) X. Chu, S. Sivananthan and J.P. Faurie, Appl. Phys. Lett. 50, 597 (1987).

(26) D. DeFontaine, Metallurgical Society Conferences, Vol. 36, Edited by J.B. Cohen and J.E. Hilliard, 51 (1966).

(27) S. Perkowitz, D. Rajavel, I.K. Sou, J. Reno, J.P. Faurie, C.E. Jones, T. Casselman, K.A. Harris, J.W. Cook, Jr., and J.F. Schetzina, Appl. Phys. Lett. 49, 806 (1986).

(28) R.M. Fleming, D.B. McWhan, A.C. Gossard, W. Wiegmann, and R.A. Logan, J. Appl. Phys., vol. 51, 357 (1980).

(29) J.-L. Staudenmann, R.D. Horing, R.D. Knox, J.P. Faurie, J. Reno, I.K. Sou, and D.K. Arch, "In-Situ Interdiffusion Measurements in HgTe-CdTe Superlattices," in "Semiconductor Based Heterostructures: Interfacial Structure and Stability," edited by M.L. Green, J.E.E. Baglin, G.Y. Chin, H.W. Deckman, W. Mayo, and D. Narasinham. A Publication of the Metallurgical Society, Inc., Warrendale, PA 15086, 41-57 (1986).

(30) M.F.S. Tang and D.A. Stevenson, J. Vac. Sci. Technol. A5 (1987) (to be published).

(31) J.L. Staudenmann, R.D. Knox and J.P. Faurie, J. Vac. Sci. Technol. A5 (1987) (to be publlished).

(32) Y. Guldner, Proceedings of the International Winter School, Springer-Verlag, 200 (1984).

(33) J.M. Berroir, Y. Guldner, J.P. Vieren, M. Voos and J.P. Faurie, Phys. Rev. B34, 891 (1986).

(34) M. Altarelli, Phys. Rev. B28, 842 (1983).

(35) Y. Guldner, G. Bastard, J.P. Vieren, M. Voos, J.P. Faurie and A. Million, Phys. Rev. Lett. 51, 907 (1983).

(36) M. DeSouza, M. Boukerche and J.P. Faurie (unpublished results).

(37) J.P. Baukus, A.T. Hunter, J.N. Schulman, and J.P. Faurie, J. Vac. Sci. Technol. A5 (1987) (to be published).

(38) J. Reno and J.P. Faurie (to be published).

(39) W.A. Harrison, Phys. Rev. B24, 5835 (1981).

(40) J. Tersoff, Phys. Rev. Lett. 56, 2755 (1986).

(41) C.G. Van de Walle, J. Vac. Sci. Technol. (to be published).

(42) S. Wei and A. Zunger, J. Vac. Sci. Technol. (to be published).

(43) D.J. Olego, P.M. Raccah and J.P. Faurie, Phys. Rev. Lett. 55, 328 (1985).

(44) For recent review, see G. Margaritondo, Solid-State Electron 29, 123 (1986), and references therein.

(45) S.P. Kowalczyk, J.T. Cheung, E.A. Kraut, and R.W. Grant, Phys. Rev. Lett. 56, 1605 (1986).

(46) The notation (B/A) specifies the growth order as B deposited on A and the notation (A-B) is for whatever the growth order is. In addition, the convention adopted regarding the sign of ΔE_v is the following: $\Delta E_v(A-B) > 0$ corresponds to $E_v^B > E_v^A$ in binding-energy scale.

(47) P. Chiaradia, A.D. Katnani, H.W. Sang, Jr., and R.S. Bauer, Phys. Rev. Lett. 52, 1246 (1984).

(48) Tran Minh Duc, C. Hsu and J.P. Faurie, Phys. Rev. Lett. 58, 1127 (1987).

(49) C. Hsu, Tran Minh Duc and J.P. Faurie (to be published).

(50) C.K. Shih and W.E. Spicer, J. Vac. Sci. Technol (to be published).

(51) G.Y. Wu and T.C. McGill, Apl. Phys. Lett. 47, 634 (1985).

(52) J.N. Schulman and Y.C. Chang, Phys. Rev. B33, 2594 (1986).

(53) X. Chu, S. Sivananthan and J.P. Faurie, Appl. Phys. Lett. 50, 599 (1987).

(54) J.P. Faurie, I.K. Sou, P.S. Wijewarnasuriya, S. Rafol and K.C. Woo, Phys. Rev. B34, 6000 (1986).

(55) K.C. Woo, S. Rafol and J.P. Faurie, Phys. Rev. B34, 5996 (1986).

(56) K.C. Woo, S. Rafol and J.P. Faurie, J.Vac. Sci. Technol. (to be published).

(57) Y. Guldner, G.S. Boebinger, M. Voos, J.P. Vieren and J.P. Faurie, Phys. Rev. B (to be published).

STRAINED LAYER SUPERLATTICES

Erich Kasper

AEG Research Center, Ulm

Sedanstr. 10, 7900 Ulm, F.R.G.

INTRODUCTION

An increasingly important flexibility in the design and fabrication of many types of high performance electronic devices is obtained
by heterostructure or superlattice materials. Properties that cannot
be achieved in bulk materials are provided by quantum size effects or
by the artificial modulation of superlattice structures. Much of the
work on such structures was performed with lattice matched materials
(GaAs/GaAlAs and InP/InGaAsP). Both, scientific and industrial interest
is responsible for the extension of work to lattice mismatched materials.

Scientific Motivation

A much wider range of alloys is allowed by preparation of mismatched material couples with electronic and optical properties not
available in unstrained structures. In crystals with the diamond, zincblende, and wurtzite structures where each atom is surrounded tetrahed-
rally by four other atoms, characteristic radii /1/ can be assigned to
each atom such that the interatomic distance is the sum of the two
radii.

The covalent radii of the low atomic number elements B, C, N are
well below the radii of the other group III, IV, V elements. But the
interatomic distances of the group IV elements and III/V-compounds from
$Z = 13$ to $Z = 51$ are all within $\pm$ 10 %.

Let's consider a thin film B with lattice constant a_B on top of an infinitely thick substrate A with lattice constant a_A. The lattice mismatch ηo between film and substrate is given by $\eta o = (a_B - a_A)/a_A$.

The considerable strain can be used to modify material properties. The band gap and the band offsets are influenced by stress. The stress σ within a thin layer with biaxial strain ε is given by

$$\sigma = \frac{E}{1-\nu}\, \varepsilon \tag{1}$$

Table 1.   Tetrahedral covalent radii (THR) for some group
            III, IV, V elements /1, 2/ (atomic number Z)

Element	B	C	N	Al	Si	P
Z	5	6	7	13	14	15
THR ($\mathring{A}$)	0.88	0.77	0.70	1.26	1.17	1.10

Element	Ga	Ge	As	In	Sn	Sb
Z	31	32	33	49	50	51
THR ($\mathring{A}$)	1.26	1.22	1.19	1.44	1.40	1.38

The following table 2 gives numerical values for σ assuming a strain of ε = 0.01 (1 %)

Table 2:   Stress σ for 1 % strain for some semiconductor
           materials. The numerical values vary with
           surface orientation because of anisotropy of
           the material properties. Values are given for
           (111) and (100) surfaces using elasticity
           data given by /3/.

	(111) surface	(100) surface
GaAs	17.4 kbar	12.4 kbar
GaP	20.5 "	14.9 "
Si	22.9 "	18.0 "
Ge	18.4 "	14.2 "

$$1 \text{ Pa} = 1 \text{ N/m}^2 = 10 \text{ dyn/cm}^2 = 10^{-5} \text{ bar}$$

As shown by table 2 stress values up to more than 100 kbar can be applied with lattice mismatch caused strains of up to 10 %. Strain splitting of degenerate bands can be observed. The latter effect has been realized in removing the valence band degeneracy in p-type InGaAs and observation of carrier transport in the light hole band /4/. The removement of conduction band degeneracy of indirect band gap semi-conductors will be treated in more detail lateron.

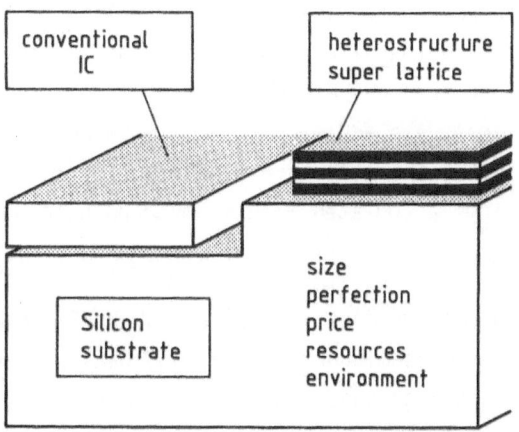

Fig. 1. Material concept for silicon based superlattice integrated circuit (IC)

Industrial Motivation

The manufacturing of todays integrated circuits is dominated by using silicon as a semiconductor material. One can speculate which material system may compete with silicon on a broad manufacturing scale when the progress of silicon technology saturates. In the following sections arguments are given for such a future competing material system consisting of a heterostructure superlattice monolithically integrated with a conventional integrated circuit on top of a silicon substrate (Fig. 1). The superlattice regions form the high performance core and high speed links between conventional parts of the IC.

<u>Choice of substrate material</u>.   Already in existing integrated circuits the device function is confined to a thin surface layer. For bipolar circuits and with increasing tendency also for CMOS-circuits this surface layer is created by epitaxial techniques. The function of the substrate itself is confined to give

- mechanical stability
- growth ordering information for the epitaxial layer
- electrical insulation
- thermal conductance for effective heat removal.

Silicon substrates offer properties superior to other semiconductor materials with regard to wafer size, crystal perfection, thermal conductivity (table 3), handling, and price. The large ressources (silicon is the 2nd most frequent element of earth's crust) and the environmental harmlessness are additional factors favouring a broad usage. But the realization of the structure given in Fig. 1 requires to overcome general problems associated with mismatched heteroepitaxy.

Table 3: Thermal conductivity $\mathcal{H}$ at 300 K /5/. Compared
are the semiconductors Si and GaAs with diamond
(best heat conductor) and the metal Cu

Material	Si	GaAs	diamond	Cu
(W/mK)	145	46	2000	384

Combination of Conventional IC with Superlattice/Heterostructure Devices

Conventional integrated circuits (IC) have obtained an extremely high complexity based on very low defect densities. On the other hand man made semiconductors based on heterostructures/superlattices offer improvement of the device performance itself. With the monolithic integration concept following fig. 1 peripheral and high complexity functions are realized by conventional IC technique whereas high performance cores and interlinks are made with lattice mismatched heterostructure/superlattice devices. A possible fabrication process would start with the conventional IC (except metallization) defined ontop of the silicon substrate at areas given by the chip architecture, then proceed with the deposition (low temperature process) of the superlattice material at the selected areas, and end with the common metallization pattern for conventual IC part and superlattice part of the chip.

The challenge lies in the preparation of materials and in avoiding unacceptable high densities of harmful defects.

104

Nature knows two answers to accomodate lattice mismatched films Fig. 2), namely elastic accomodation by strain and plastic accomodation by misfit dislocations lying in the interface. Nature prefers elastic accomodation for thin films up too critical thickness t_c, whereas above the critical thickness misfit dislocations are generated relaxing the built in strain.

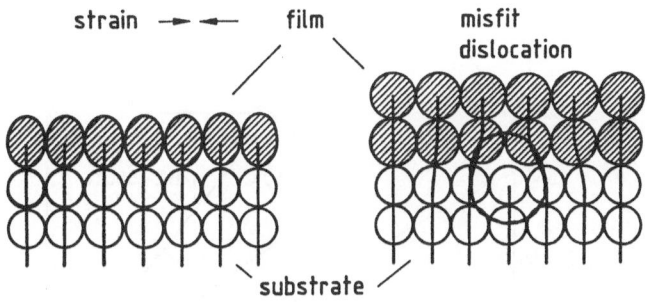

Fig. 2. Mismatch accomodation by strain (left) or misfit dislocations (right)

As a guideline for understanding nature's preference for strained layer epitaxy for a thickness below the critical thickness one can consider thermodynamic equilibrium as was done by v.d. Merwe /6/.

Indeed, his theory predicts qualitatively correct the observed behaviour by minimizing the total energy of the system.

However, the quantitative results of the equilibrium theory differ considerably from the results obtained by molecular beam epitaxy (MBE). This can most clearly be seen by a plot of the critical thickness t_c versus lattice mismatch η_o (Fig. 3). Unfortunately the reader of recent literature may be confused by theory (v.d. Merwe)-curves covering a rather broad range /7, 8/. This is caused by various approximations given by v.d. Merwe and also simply by misprinting. We strongly recommend the use of one data set /9/ which is based on a careful analysis of the approximations valid for SiGe /8/. Most probably kinetic limitations (nucleation and glide) cause the deviations of experiments from theory.

Tetragonal Distortion

Consider a cubic lattice cell which is in plane strained (x, y) by the elastic film stress σ. The in plane strain ε

$$\varepsilon = \varepsilon_x = \varepsilon_y \qquad (2)$$

causes a perpendicular strain ε_z which is given by

$$\varepsilon_z / \varepsilon = -2\nu/(1-\nu) \qquad (3)$$

(ν Poisson's number, M_E modulus of elasticity)

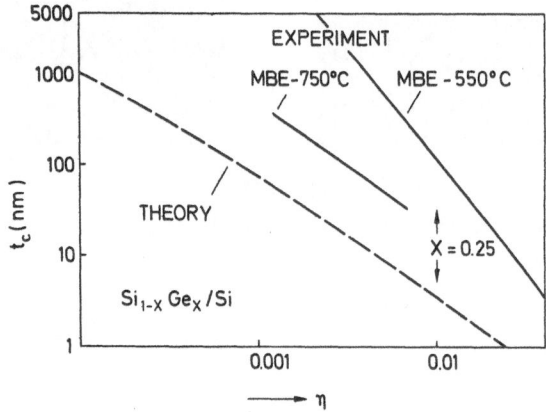

Fig. 3. Critical thickness t_c versus lattice mismatch ηo. Comparison of equilibrium theory with MBE-experiments /11/

The strains distort the cubic lattice cell to a tetragonal cell as can be measured e.g. by X-ray diffractometry (Fig. 4). Both the lattice spacing d as well as the inclination i of lattice planes are changed by the tetragonal distortion. Fig. 4 give examples of the changes in lattice spacing and inclination; Δd and Δi, respectively. The following table gives numerical values for the modulus of elasticity and for Poisson's number derived from anisotropic elasticity theory /3/.

Table 4: Elasticity properties for $\{001\}$ surfaces of semiconductor materials /3/.

	$M_E/(1-\nu)$	ν	
GaAs	1.239	0.312	
GaP	1.488	0.305	M_E in units of
Si	1.805	0.279	10^{11} Pa
Ge	1.420	0.270	($\cong 10^{12}$ dyn/cm^2)

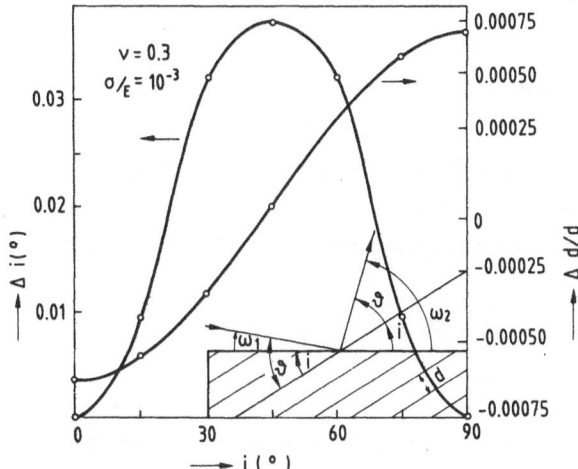

Fig. 4. Tetragonal distortion of the cubic lattice cell by a stress $\sigma = 10^{-3} M_E$ (M_E modulus of elasticity). Relative change of interplanar spacing $\Delta d/d$ and change Δi of inclination versus inclination i of lattice planes. Inset: Scheme of X-ray diffractometric measurement of lattice planes.

The strained crystal stores elastic energy the density E of which is given for an isotropic crystal by

$$E = 2\mu \frac{1+\nu}{1-\nu}\varepsilon^2 \tag{4}$$

with shear modulus μ

$$\mu = \frac{M_E}{2(1+\nu)} \tag{5}$$

The areal energy E_h of an homogeneously strained film of thickness t is then given by

$$E_h = E \times t \qquad (6)$$

Note that a uniform, thin film on a thick substrate is always nearly homogeneously strained in z-axis. Small deviations from homogeneous strain are caused by the strain induced curvature k of the film/ substrate couple. The curvature k (reciprocal of curvature radius R) is given by

$$k = 6 \frac{\epsilon t}{t_s^2} \qquad (7)$$

if $t \ll t_s$ and equal elastic constants of film and substrate (t_s thickness of substrate) are assumed. By the curvature the substrate is also stressed leading to a neutral plane at a distance $t_s/3$ from the backside of the substrate, a stress of the same sign as the film stress at the backside and a reversely signed stress $\sigma_s(o)$ at the substrate surface

$$\sigma_s(o) = -4\sigma \frac{t}{t_s} \qquad (8)$$

The film stress is lowered by the amount of σ_s at the interface, but increases very slightly with distance z from the interface. The variation of strain along the z-axis of a uniform film of 1 μm thickness on top of a 250 μm thick substrate is about 10^{-4} and can therefore be neglected for all practical purposes.

Dislocations

Before treating misfit dislocations itself let's summarize some general properties of dislocations. This chapter can be omitted by readers familiar with dislocation properties. For more details the reader is referred to monographs /10/ about this topic.

Dislocations are line defects the displacement field of which is defined by their Burgers vector b (Fig. 5). The displacement discontinuity, which the Burgers vector measures, remains constant along a given dislocation line. The principle of conservation of the Burgers vector is a general one, and from it follows that a dislocation cannot end inside a crystal. It must end at a surface, form a closed loop or meet other dislocations. A point at which dislocation lines meet is called a node. The conservation of the Burgers vector in this case can be stated in the following way. If the directions of all the dislocation lines are taken to run out from the node, then the sum of the Burgers vectors of all the dislocations is zero. This is the dislocation analogue of Kirchhoff's law of the conservation of electric current in a network of conductors.

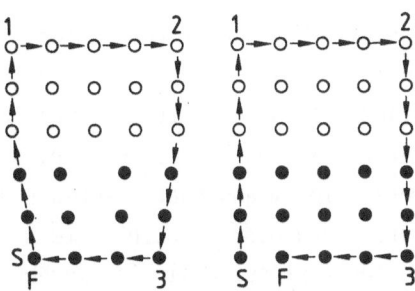

Fig. 5. Burgers circuits in a real crystal (left) and a perfect reference crystal (right). The Burgers vector b defines the displacement field of a dislocation.

In general, a dislocation may lie at any angle to its Burgers vector or may be curved. A length of dislocation which lies normal to its Burgers vector is called edge dislocation, one which lies parallel to its Burgers vector is called a screw dislocation. A dislocation is referred as mixed if it has both edge and screw components.

Dislocation line element and Burgers vector define a slip plane. Dislocation motion in this slip plane is called glide. Dislocation motion outside the slip plane is called climb. When a dislocation glides, there is no problem of acquiring or disposing of extra atoms. Climb can be accomplished by adding or removing atoms. Therefore a dislocation will be able to glide much more quickly than it can climb. So a dislocation loop lying in its own glide plane can expand easily under the action of forces. Loops with Burgers vectors outside the loop plane are called prismatic loops because they could be produced by prismatic punching, in which case its structure corresponds to a penny-shaped disc of extra atoms in the plane of the loop.

Gliding is much the easiest for one type of crystallographic plane so that slip is effectively confined to this plane. For diamond type lattices the slip plane is of the $\{111\}$ type, which contains 4 individual planes. The Burgers vector is of the $1/2 \langle 110 \rangle$ type, which contains 12 individual directions. Frequently the line direction of undisturbed dislocations is also of the $\langle 110 \rangle$ type creating "60°-dislocations" with a 60 ° angle between line direction and Burgers vector, e.g. for the (111) slip plane $[1\bar{1}0]$ and $1/2$ $[01\bar{1}]$ as line direction and Burgers vector, respectively.

A free dislocation will move under the action of an applied stress field. Dislocations with opposite Burgers vector move in opposite directions. The dislocation will bend if it is pinned at a point, e.g. by a node or by an obstacle like an inclusion or by a line segment which is cross slipped. Cross slip means that a screw segment is glided off its original slip plane. Such pinning points are necessary for the operation of multiplication sources. A simple mechanism by which dislocations can multiply was suggested by Frank and Read. Its operation requires only that a dislocation in a slip plane be pinned at two points /10/. The Frank-Read sources and similar source types can generate any number of expanding loops under the action of sufficiently high stress fields.

The crystal lattice is distorted around a dislocation. The rather far reaching strain field is proportional to the length of the Burgers vector b. It fades away with the reciprocal distance r from the dislocation. A considerable amount of strain energy is stored in the elastically distorted region around a dislocation line. The strain energy E_{ds} of a single dislocation is given for an isotropic medium by

$$E_{ds} = \frac{\mu b^2}{4\pi} \quad \ln \left(^R/_{r_i} \right) \tag{9}$$

for a screw dislocation, and by

$$E_{ds} = \frac{\mu b^2}{4\pi(1-\nu)} \quad \ln \left(^R/_{r_i} \right) \tag{10}$$

for an edge dislocation.
(R range of the strain field, r_i inner cut off radius for the dislocation core).

The difficulty in applying these equations is that the strain energy increases without limit as R increases or r_i decreases. Somewhat arbitrary cut off radii for the strain field range and the core have to be chosen to overcome this difficulty. In practice the outer dimension of the crystal or half the distance to the next dislocation are chosen as range R of the dislocation, and values near Burgers vector amount are chosen for the inner cut off radius r_i.

Equilibrium Theory for Misfit Dislocations

On one side the mismatched film can accomodate by strain (tetragonal distortion), on the other side the film can accomodate by formation of a misfit dislocation network at the interface substrate/film. Let's simply consider a (001) surface with a network of misfit dislocations lying in the interface. From the fourfold symmetry of this surface an orthogonal network of misfit dislocations is required, e.g. [110] and [1$\bar{1}$0] directions, respectively. For the threefold (111) surface a network with three different dislocation directions, e.g. [1$\bar{1}$0], [10$\bar{1}$], [01$\bar{1}$] would be required. The perfect misfit dislocation is of the edge type, although edge dislocations require more energy than screw dislocations for their generation (equ. 9, 10). Screw dislocations will rotate the film against the substrate. Consider the effect of a parallel array of edge dislocations a distance p apart from another. Line direction and Burgers vector should lie in the interface as in Fig. 2. The parallel array accomodates a mismatch equal to b/p. The relation between mismatch ηo, film strain $\mathcal{E}$ and dislocation distance p is given by /8/

$$\eta o + \mathcal{E} = b/p \tag{11}$$

(b length of Burgers vector, sign of b is the same as sign of ηo)

A complete accomodation of mismatch by perfect misfit dislocations would require a network with spacing p

$$p = b/\eta o \tag{12}$$

$$\mathcal{E} = 0$$

V.d. Merwe /6/ calculated the energy E_d of a perfect misfit dislocation network and the total energy E_{tot} of a film partly accomodated by strain and partly accomodated by misfit dislocations. The general equation (10) is also valid for a misfit dislocation. For the range R of the strain field the assumption was made that for very thin films R approaches twice the thickness t of the film, whereas for very thick films R approaches half the distance p. For medium values of t a smooth interpolation formula was given

$$R = p/2 \qquad \text{for } t \geq p/2 \tag{13}$$

$$R = 2t/(1 + 4t^2/p^2) \text{ for } t < p/2$$

The inner cut off radius r_i calculated for a Peierls type dislocation core /6/ is given for a (001) interface in the diamond lattice /9/ by

$$r_i = b \ (\pi/2 \sqrt{2} \ \ell \ (1-\nu)) \tag{14}$$

(ℓ base of natural logarithm).

The energy E_d per unit area of an orthogonal network is given by

$$E_d = 2 \ E_{ds}/p \tag{15}$$

The total energy E_{tot} per unit area is given by

$$E_{tot} = E_h + E_d \tag{16}$$

using equ. (4-6, 10, 13-15) for calculating E_h and E_d.

V.d. Merwe assumed that the equilibrium state is given by the minimum of the energy E_{tot} thus neglecting entropy terms.

$$\frac{\partial E_{tot}}{\partial (1/p)} = 0 = \frac{\partial E_h}{\partial \varepsilon} b + 2 \ E_{ds} + \frac{2}{p} \frac{\partial E_{ds}}{\partial (1/p)} \tag{17}$$

Using equ. (6, 10, 13, 14) we obtain

$$\frac{\partial E_h}{\partial \varepsilon} b = 4\mu \ \frac{1+\nu}{1-\nu} b \times t \ \left[(b/p) - \eta_0 \right] \tag{18}$$

$$2 \ E_{ds} = \frac{\mu b^2}{2\pi(1-\nu)} \ \ln (R/r_i)$$

112

$$\frac{2}{p} \frac{\partial E_{ds}}{\partial (1/p)} = - \frac{\mu b^2}{2\pi(1-\nu)} \quad \text{for } t \gg p/2$$

$$= - \frac{\mu b^2}{2(1-\nu)} \frac{8(t/p)^2}{1+4(t/p)^2} \quad \text{for } t < p/2$$

Very thin layers up to a critical thickness t_c grow without misfit dislocations which is called strained layer epitaxy or pseudomorphic or commensurate growth

$$\varepsilon = -\eta_0 \quad \text{for } t < tc \qquad (19)$$
$$p \to \infty$$

The critical thickness t_c is given by /9/

$$t_c \, \eta_0 = \frac{b}{8\pi(1+\nu)} \ln \left\{ \frac{4\sqrt{2}(1-\nu)e\,t_c}{\pi\,b} \right\} \qquad (20)$$

Assuming $b = 0.384$ nm, $\nu = 0.3$ and measuring t_c in nm one obtains the following implicit relation for t_c

$$\eta_0 \, t_c \text{ (nm)} = 1.175 \times 10^{-2} \left\{ 2.19 + \ln t_c \right\} \qquad (21)$$

The film strain ε in films thicker than t_c is relaxed by the formation of the misfit dislocation network.

$$(t\varepsilon) = - \frac{b}{8\pi(1+\nu)} \left[\ln(R/r_i) - \frac{8\left(t/p\right)^2}{1+4(t/p)^2} \right] \qquad (22)$$
$$\cdot \text{ for } t < p/2$$

$$= - \frac{b}{8\pi(1+\nu)} \left[\ln(R/r_i) - 1 \right]$$
$$\text{ for } t \geq p/2$$

Fig. 3 compares the calculated equilibrium theory values with experimental curves for MBE growth at 550 °C and 750 °C /11/. The equilibrium theory allows to understand why thin films up to a certain thickness grow as strained layers but it predicts quantitatively too low values of the critical thickness. The V.d. Merwe formalism gives a rather raw estimate of the energy contribution of the dislocation core. It was shown /8/ that this is unimportant for critical thicknesses of many monolayers. But it should be considered for very thin thicknesses. For this regime numerical calculations are given /12/.

In general the elements of the strained layer superlattice (SLS) contain the thin, mismatched layers which are purely accomodated by strain and stacked to a superlattice, the buffer layer and the substrate (Fig. 6). The mismatch η between the layer is given by the layer materials with lattice constants a_1, a_2

$$\eta = 2 \; \frac{a_2 - a_1}{a_1 + a_2} \tag{23}$$

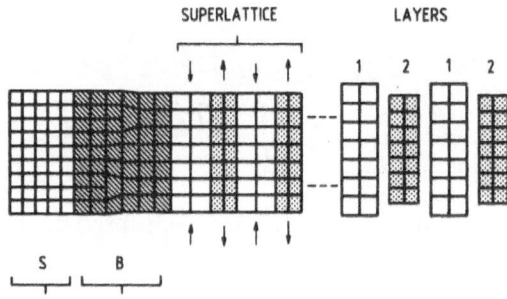

Fig. 6. Elements of a strained layer superlattice (SLS).

The elastic strains ε_1, ε_2 in the superlattice layers are connected with the mismatch η by

$$\eta + \varepsilon_2 - \varepsilon_1 = 0 \tag{24}$$

if no misfit dislocations are in the superlattice interfaces as expected for SLS's. That means that the layer thicknesses t_1, t_2 of the individual superlattice layers meet the requirements for pseudomorphic growth

$$t_1, t_2 < t_c \; (\varepsilon_1, \varepsilon_2) \tag{25}$$

The driving force for introduction of misfit dislocations is the strain energy stored in a layer. For calculation of the critical thicknesses of individual superlattice layers we should insert the strains ε_1, ε_2 instead of ηo for the single layer results (Fig. 3, equ. 21). The superlattice as a whole has the same in-plane lattice constant $a_{\shortparallel}$ (parallel to the interfaces).

$$a_{\shortparallel} = a_1 (1 + \varepsilon_1) = a_2 (1 + \varepsilon_2) \tag{26}$$

The strain jumps at an amount of $(\varepsilon_2 - \varepsilon_1)$ by crossing each super-lattice interface. The strain distribution is determined by the choice of the in plane lattice constant $a_{\shortparallel}$.

$$\varepsilon_1 = \frac{a_{\shortparallel} - a_1}{a_{\shortparallel}} \tag{27}$$

$$\varepsilon_2 = \frac{a_{\shortparallel} - a_2}{a_{\shortparallel}}$$

Strain Symmetrization

The stability of SLS's was investigated for specific material systems as early as 1974, 1975 (19, 20). The general SLS concept was introduced 1982 /21/. In a pioneering work /22/ the concept of a critical thickness t_{cs} of the superlattice as a whole was introduced which should be roughly equal to the critical thickness of a single layer of the same average composition. We will demonstrate that this definition applies only to specific structures. One can give a very general definition /16/ of the stability of a superlattice as a whole against introduction of misfit dislocations. For that we consider the strain energy E_h stored in a SLS. This is simply the sum over all the strain energies of the individual layers (N number of periods with length L)

$$E_{\ell} = 2\mu \frac{1+\nu}{1-\nu} N \sum_{\iota=1}^{2} \varepsilon_\iota^2 t_\iota \tag{28}$$

In equ. (28) equal elastic constants are assumed. It is straight forward also to introduce different elastic properties and anisotropy of the material components. The minimum of strain energy is found by $\partial E_h / \partial \varepsilon_1 = 0$ using equ. (23) as coupling between the ε_i. The result is very simple. Minimum strain energy is obtained, if

$$\varepsilon_1 = \eta \frac{t_2}{L} \tag{29}$$

$$\varepsilon_2 = -\eta \frac{t_1}{L}$$

For the most important case $t_1 = t_2 = L/s$ this condition reduces to

$$\varepsilon_1 = -\varepsilon_2 = \eta/2 \tag{30}$$

That means the SLS is stable if the strain is symmetrized after equ. (30). We use the term strain symmetrization mainly for SLS's with equal layer thickness $t_1 = t_2$, but it means only replacing equ. (30) by equ. (29) without loss of generality if unequally thick layers are used. In this case there is no limiting critical thickness ($t_{cs} \rightarrow \infty$) of the superlattices as a whole. The situation changes if the strain is unsymmetrically distributed. Then the superlattice can gain energy by the introduction of misfit dislocations at its base if the thickness of the superlattice (N x L) raises above a critical superlattice thickness t_{cs}. The critical superlattice thickness will decrease with increasing asymmetry of the strain distribution. The superlattice structure of ref. /16/ was completely unsymmetrical with respect to strain ($\varepsilon_1 = 0$, $\varepsilon_2 = -\eta$). The introduction of misfit dislocations will change $a_{\shortparallel}$, and shift the strain distribution to higher symmetry.

Buffer Layer Design

The buffer layer between superlattice and substrate is grown for several reasons. The main functions of the buffer layer are:

1. Transition from the substrate material to the superlattice material system, eg. from an InP substrat to a InAs/GaAs superlattice via an InGaAs buffer layer, lattice matched to the substrate.

2. Improvement of crystal quality. Especially strained buffer layers or SLS buffers can effectively reduce the density of penetrating substrate dislocations by bending them into the interface plane.

3. Adjustment of superlattice strain distribution by shifting the in plane lattice spacing from the substrate value a_A to the superlattice value $a_{\shortparallel}$. This shift principally involves the generation of misfit dislocations in the buffer layer. The usually used thick (several μm) buffer layer with either linearly or step graded composition suffers from being heavily dislocated which might be an inherent problem for device application. Therefore growth of SLS's without such an strain adjusting buffer layer was performed. High quality growth is only obtained for specific material system where the lattice constant a_A of the substrate /17/ is between the lattice constants a_1, a_2 of the superlattice materials (strain symmetrization by the substrate itself) or for thin unsymmetrically strained systems below the critical superlattice thickness /16, 18/.

We will discuss here the concept of a thin (<0.25 µm), homogeneous buffer layer, the strain adjusting misfit dislocation network of which is mainly confined to the interface between substrate and buffer layer /19/. The in plane lattice spacing $a_{\parallel}$ is controlled (Fig. 7) by the lattice constant a_B of the buffer material and by the strain ε_B in the buffer which depends on mismatch η_0 between substrate and buffer layer and on buffer layer thickness t_B.

$$a_{\parallel} = a_B (1-\varepsilon) \tag{31}$$

The design of the buffer layer is explained here on the example of an $Si_{1-y}Ge_y$ buffer layer ontop of a Si substrate. The lattice constant of the buffer layer is determined by the chemical composition y

$$a_B = a_A (1 + 0.042\ y) \tag{32}$$

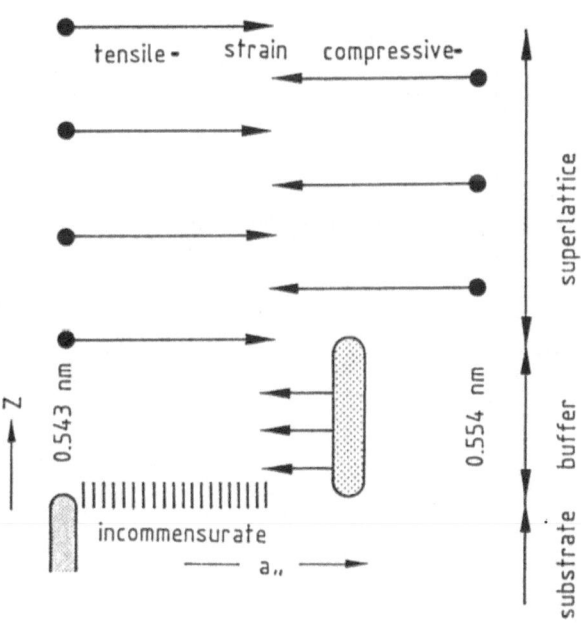

Fig. 7. Scheme of strain symmetrization by a homogeneous, incommensurate buffer layer. Dots: natural lattice spacing. Arrows: Distortion of the in plane lattice spacing $a_{\parallel}$ by strain. Example: $Si_{0.5}Ge_{0.5}$/Si superlattice on Si substrate.

assuming Vegard's law. The lattice mismatch between Ge and Si is 4.2 %.
The strain ε_B in thin SiGe layers on Si substrates was measured by He-
backscattering /20/ (growth temperature: 550 °C). We used for this
calculation the smoothened data set given in /21/. The result is given
in Fig. (8) as in plane spacing $a_{\shortparallel}$ versus buffer layer constant a_B for
various buffer layer thicknesses ranging from 10 nm to 250 nm. One can
consider the couple substrate/buffer layer as a new virtual substrate
which offers the in plane lattice spacing $a_{\shortparallel}$. This would in this example correspond to a SiGe substrate with an effective Ge content y*,
where y* is defined by

$$a_{\shortparallel} = a_A \ (1 + 0.042 \ y^\star) \tag{33}$$

$$y^\star = y + 23.8 \ \varepsilon_B$$

Remember that the effective Ge content y* would be smaller than
the Ge content y of the buffer layer because of the compressive strain
ε_B (negative sign).

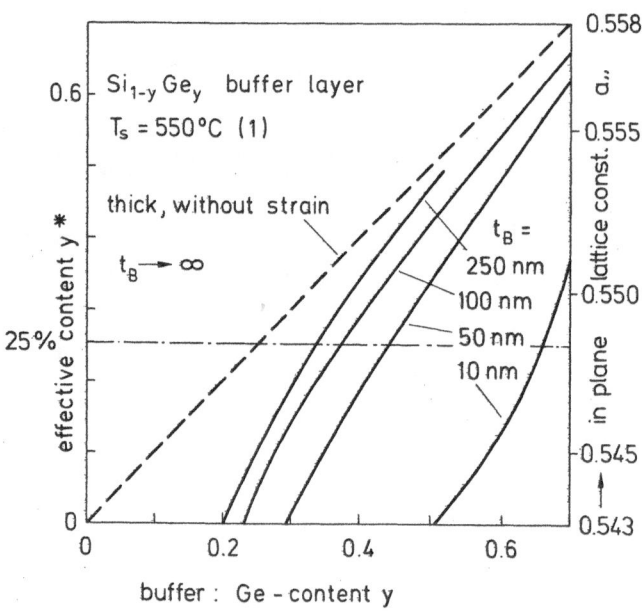

Fig. 8. Design for a $Si_{1-y}Ge_y$ buffer layer on a Si substrate. Effective
Ge content y* of the virtual substrate (Si substrate + buffer)
versus Ge content y of the buffer layer for different buffer
layer thicknesses t_B.

Metastability of the Buffer Layer

The experimentally found critical thicknesses t_c of a mismatched layer are generally higher than that predicted by an equilibrium theory. If one does not assume that the theoretical treatment is based on wrong assumptions one has to consider kinetic reasons for this deviation from equilibrium. These deviations can come from the nucleation and motion of dislocations which are both hampered by the need of an activation energy. Already Matthews /22/ argumented that misfit dislocations which move to the interface by a glide process are not as ideal as perfect misfit dislocations. He considered a substrate dislocation which penetrates the surface (Fig. 9, upper part). The slip plane and the Burgers vector b are inclined to the interface. The dislocation is bended by the film strain and can create a straight misfit dislocation segment if the thickness is raised above the critical thickness. But now only the projection of the edge component of the Burgers vector on the interface is able to accomodate mismatch instead the full Burgers vector as in the case of an ideal edge type misfit dislocation with the interface as slip plane. Let us consider e.g. an (001) surface which is penetrated by a dislocation lying in a $(1\bar{1}1)$ slip plane with a $1/2\,[011]$ Burgers vector b. The misfit dislocation segment extends along the intersection $[110]$ of the slip plane with the interface. The projection of the edge component in $[1\bar{1}0]$ direction amounts to half of the Burgers vector (b/2). For these nonideal misfit dislocation arrangement the relation between strain and misfit dislocation array has to be reformulated.

$$\eta_o + \varepsilon = \beta\, b/\rho \tag{34}$$

$$\beta \leq 1$$

with $\quad \beta b = \vec{b} \cdot [\,\vec{\xi} \times \vec{\iota}\,]$

(b Burgers vector, ξ, ι unit vectors along dislocation line and perpendicular to the interface, respectively).

The weaker accomodation of this dislocation array shifts the predicted critical thickness to higher values. The formal treatment of Matthews uses a force balance instead of energy considerations as v.d. Merwe has done which is proven to be equivalent by the definition of the force on a dislocation. Matthews theory is more close to the experimental results than v.d. Merwe theory especially for medium growth temperatures (e.g. 750 °C MBE results). It should be stated that

it describes a metastable state because principally ideal misfit dislocations can be created by slower process, e.g. climb or glide from
the rim of the sample. The same result is obtained for glide of a dislocation half loop from a nucleation site at the surface (Fig. 9 middle
part) which is a more probable process than the originally suggested
bending of substrate dislocations because of todays low dislocation
density substrates.

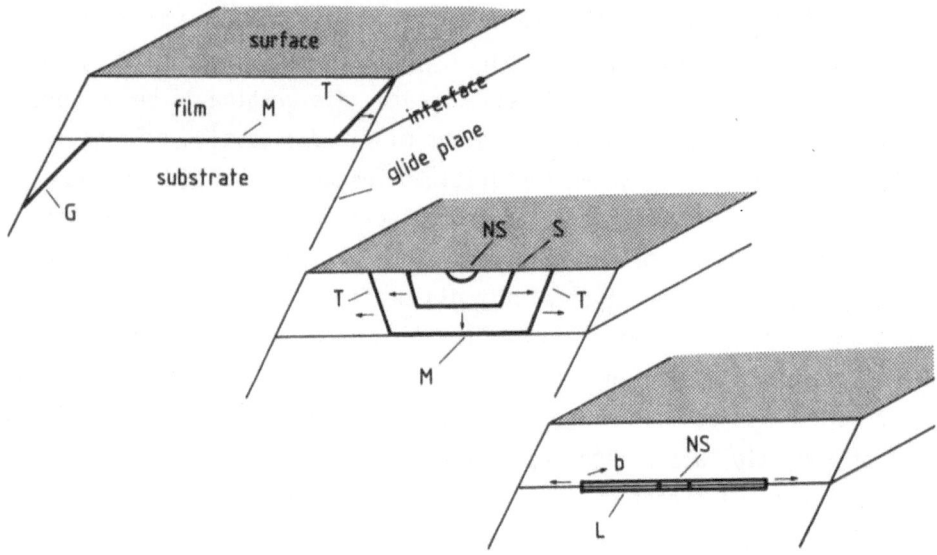

Fig. 9. Basic mechanism for misfit dislocation generation. Upper part:
Bending of a substrate dislocation. Middle: Nucleation (NS) of
a dislocation at the layer surface and movement to the interface. Lower part: Nucleation of a dislocation dipole at the
interface.

The kinetic reason for the increase in critical thickness at low
growth temperatures (550 °C MBE) is not clear. One can speculate that
the barrier for nucleation controls the onset of accomodation by dislocations. TEM images /23/ of films grown at 550 °C are compatible with
the picture of nucleation of dislocation arrays at the interface itself
(Fig. 9 lower part). A fitting curve for the low temperature data /7/
was given (see Fig. 3, 550 °C MBE). In this interesting treatment the
authors compare the elastic energy E_h with what they call areal .energy
density of the dislocation. This may the reader mislead that this fitting curve is obtained by equilibrium considerations. But one can interpret it more as an activation barrier for dislocation nucleation.

GROWTH AND PROPERTIES OF SI/SIGE SUPERLATTICES

The material system silicon/silicon germanium is a model system for strained layer superlattices because of the close relationship in chemistry of Si and Ge. The pure influence of strain can easily be studied in this system with a lattice mismatch ranging from zero to 4.2 % depending on the composition of the completely miscible alloy SiGe. The strain distribution can be adjusted from a symmetrical situation to an unsymmetrical situation with a buffer layer/Si substrate designed after the principles given in the foregoing section. Key issues for the technological realization of the concept of strain adjustment are low growth temperature, control of layer composition and thickness, crystal perfection of the buffer layer.

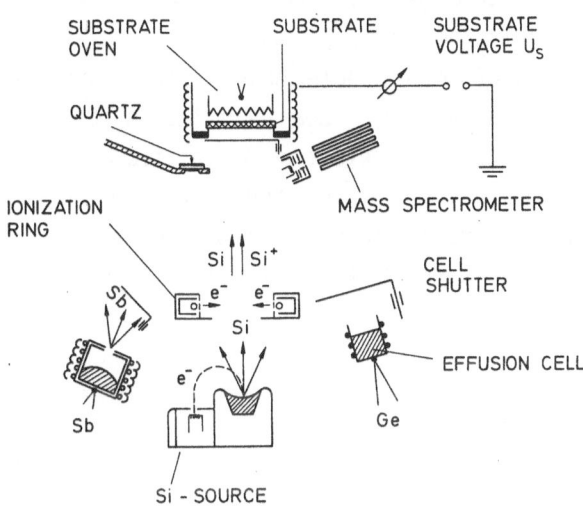

Fig. 10. Scheme of our Si-MBE apparatus /11, 14/ used for growth of SiGe superlattices

Silicon Molecular Beam Epitaxy (Si-MBE)

MBE is ideally suited for meeting the above requirements. For a review of Si-MBE the reader is referred to /21/. The principal arrangement of our MBE apparatus which was used for growth of SiGe superlattices is given in Fig. 10. The following subsystems are shown; (i) material sources (Si, Ge, Sb), (ii) substrate oven, (iii) secondary implantation equipment (ionization ring, substrate voltage), (iv)

in situ monitoring equipment (quartz microbalance, mass spectrometer).
These subsystems are installed inside an ultrahigh vacuum chamber.
We used the method of secondary implantation /24/ for n-type doping
with antimony (Fig. 11). Si ions generated by the electron gun evapora-
tor and/or by an ionizer ring (see Fig. 10) are accelerated toward the
substrate surface by an applied voltage. These Si ions implant the
adsorbed Sb atoms (Fig. 11).

Band Ordering in SiGe/Si

The band gap of the alloy SiGe is smaller than that of silicon.
The ordering of the bands at the interface determines the properties of
two-dimensional carrier gases. The most common case of band ordering is
obtained with the low band gap within the wide band-gap (type I band
ordering). In this case both, electrons and holes would jump from the
wide gap material to the low gap material. For a staggered band ordering (type II) electrons and holes would jump in opposite directions
across the interface. A rough idea about the type of ordering can be
obtained by considering the electron affinities of the materials. The
electron affinity measures the energetic position of the conduction-
band edge against a vacuum level. This simple consideration holds if
both materials are separated. For the system Si/Ge a nearly flat con-

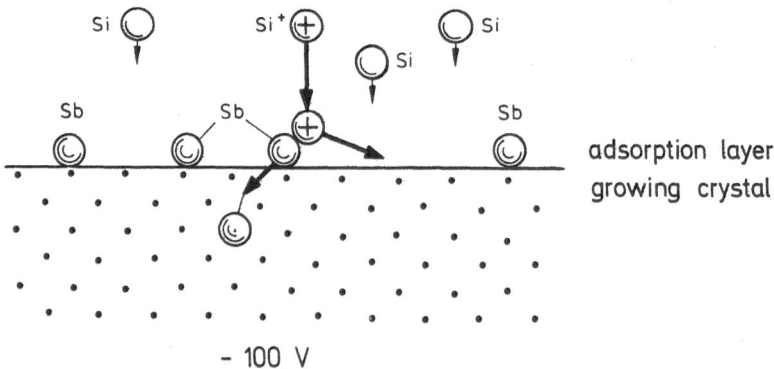

Fig. 11. Doping by secondary implantation (DSI). The adsorbed Sb atoms
        are knocked on by Si-ions accelerated toward the substrate /24/

duction-band ordering (between type I and type II) is proposed. Usually, interface structure, dipole moments, and strain effects contribute
additionally to the band-edge position after facing both materials.

Early experiments with polycrystalline material seemed to confirm
the picture of a flat conduction-band ordering (with very slight type I
character). Also first MBE experiments /25/ of the Bell group with unsymmetrically strained Si/SiGe superlattices demonstrating 2D-hole gas
but no electron gas were in agreement with that flat band picture. In
apparent contrast to that, our group realized a 2D-electron gas in the
wide gap material silicon /26, 27/ with a symmetrically strained Si/
SiGe superlattice. Both groups agree now that the apparent contradic-
tion was caused by the different strain distribution within the superlattices /28, 29/. Fig. 12 shows the calculated conduction-band offset
(type II) as function of the strain distribution /19, 29, 30/ characterized by a substrate with an effective Ge-content. The band ordering
is shifted toward strong type II character as the substrate is made Ge-

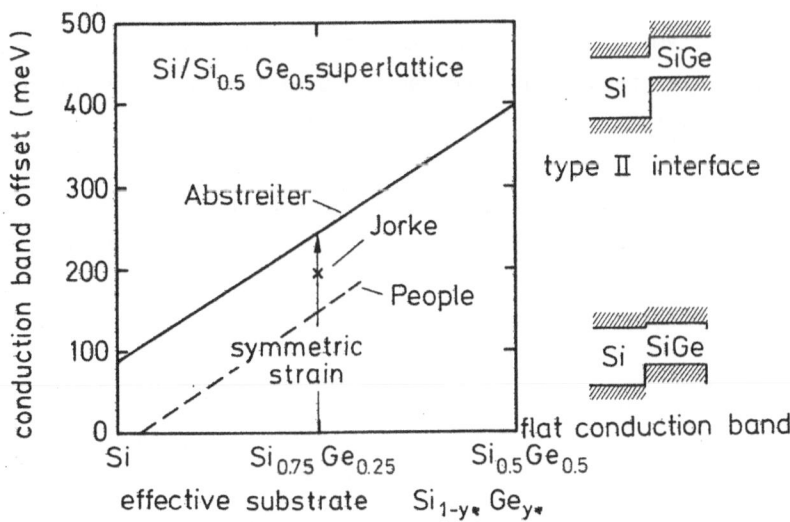

Fig. 12. Conduction band offset for a Si/Si$_{0.5}$Ge$_{0.5}$ superlattice as a
function of the effective substrate which determines the
strain distribution /19, 29-31/

rich. The position of the symmetrically strained superlattice is marked by an arrow suggesting a conduction-band offset of roughly 200 meV for the symmetrically strained superlattice /31/.

The basic mechanism for the shift in conduction band offset was clearly evaluated by G. Abstreiter /29/. The biaxial stress σ in the layers can be described as the sum of a hydrostatic pressure and an uniaxial stress normal to the layers.

$$
\begin{pmatrix} \sigma & & \cdot \\ & \sigma & \cdot \\ \cdot & & 0 \end{pmatrix} = \begin{pmatrix} \sigma & & \cdot \\ & \sigma & \cdot \\ \cdot & & \sigma \end{pmatrix} + \begin{pmatrix} 0 & & \cdot \\ & 0 & \cdot \\ \cdot & & -\sigma \end{pmatrix} \tag{35}
$$

The hydrostatic part basically shifts the conduction and valence band up (compression) or down (tension) depending on the sign of pressure. It causes also small changes in the value of the indirect band gap energies. The conduction band minimum of Si is along the [100] (Δ)-direction and sixfold degenerate. The conduction band ordering remains Si-like up to a Ge content of 85 %. In unstressed Ge-rich alloys the conduction band minimum is at the L-point and consequently fourfold degenerate. The main effect is caused by the uniaxial part of the strain responsible for a splitting of the six-fold degenerate conduction band and the heavy and light hole band. A tensile [100] biaxial stress (= compressive uniaxial part) lowers the two-fold degenerate valley with the large mass normal to the layers and shifts up the light hole band. Fig. 13 (left side) shows the effect of biaxial strain on the band edges for a symmetrical strain situation, tensile in Si, compressive in SiGe. The strain-induced band gaps in SiGe layers, which are grown on different virtual substrates (= substrate + buffer layer) are shown on the right side of Fig. 13. In all cases the energy gap is reduced for the whole concentration region from pure Si to pure Ge. For Si-rich virtual substrates the band structure remains Si-like even for pure Ge.

Modulation Doping

The method of doping by secondary implantation is well suited to grow thin doped layers with abrupt profiles, e.g. in the case of Sb so-called nini structures. When we combine this technique with the growth of strained composition multilayers of Si/SiGe we can study the effect of (n type) modulation doping on a multilayer heterostructure with a periodically varying band gap width.

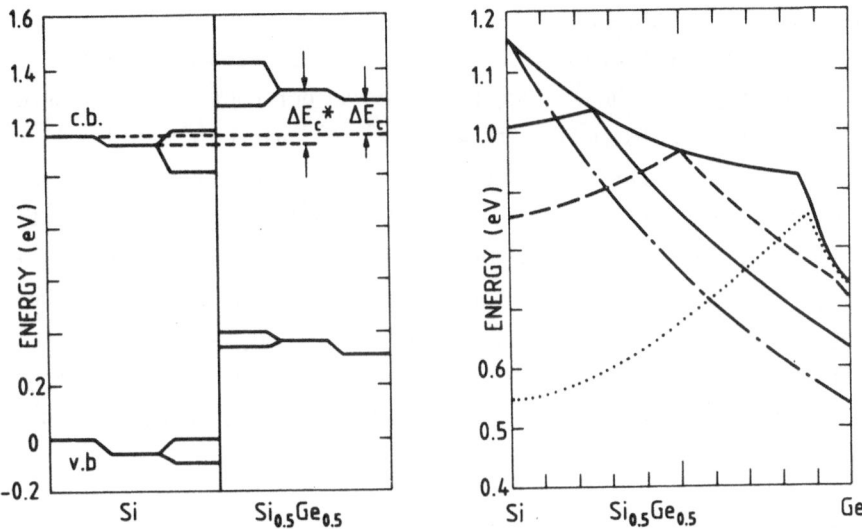

Fig. 13. The effect of strain symmetrization on the band edges of Si/Si$_{0.5}$Ge$_{0.5}$ (left side). Band gap energies in SiGe for different biaxial stress conditions (right side) /29/

In our experiments we applied the same period length L of typically 10 to 20 nm for both, Si/SiGe heterostructure and Sb modulation doping, but we used different phase angles between the two periodic layer structures defined by $\varphi = 2\pi\Delta L/L$ according to Fig. 14. The total

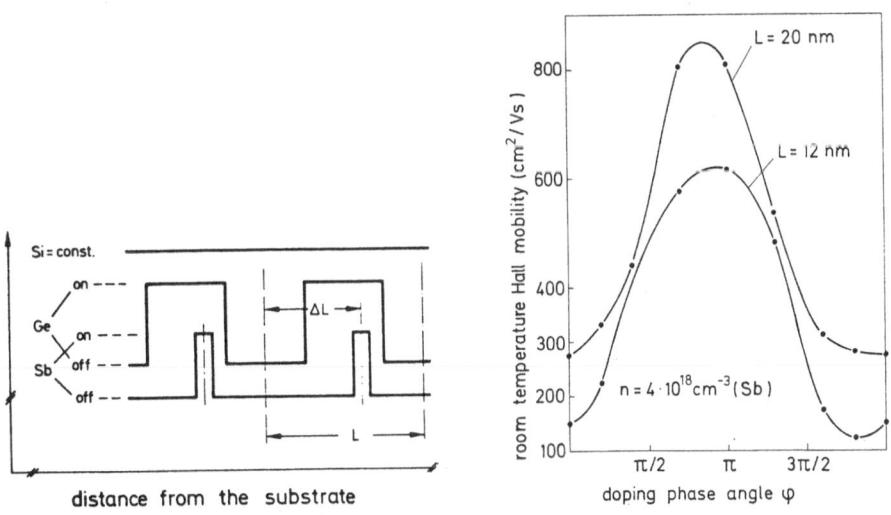

Fig 14. Modulation doped Si/SiGe superlattice with period length L and doping phase angle $\varphi = 2\pi\Delta L/L$ (left side). Room temperature Hall mobility as function of doping phase angle (right side)

number of periods was ten. Mobility and carrier concentration were determined by Hall measurements assuming the Hall factor to unity. The sample geometry was chosen according to van der Pauw. Carrier concentration in all samples amounts to 4×10^{18} cm^{-3} as calculated by the quotient of sheet carrier concentration and the superlattice layer thickness. In Fig. 14 the room temperature Hall mobility as a function of the phase angle φ is shown for symmetrically strained material. The mobility maximum is achieved for $\varphi = \pi$, i.e. in the case where the Sb doping spikes coincide with the center of the $Si_{0.5}Ge_{0.5}$ layers. The temperature dependence of the Hall mobility down to liquid nitrogen temperature is given in Fig. 15. For comparison the corresponding results of equivalently doped bulk Si are added.

The prominent result is a remarkable increase of the electron mobility within the modulation doped strained layer superlattice compared with equally doped Si: Even at room temperature there is a mobility increase of a factor 4 (for L = 12 nm) respective 5.5 (for L = 20 nm, effect of an increased spacer thickness).

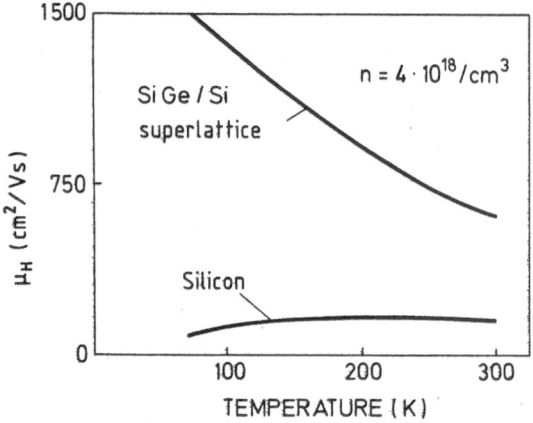

Fig. 15. Hall mobility versus temperature for n-type material with a mean doping level of 4×10^{18}/cm^3. Compared is a modulation doped Si/SiGe superlattice with bulk silicon

Direct confirmation of the existence of a two-dimensional electron gas within the multilayer heterostructure is derived from Shubnikov-deHaas measurements /27/. The oscillation periods, only observed with the magnetic field perpendicular to the layers, lead to a two-dimensional electron density of about 3×10^{12} cm^{-2} per layer period which is in reasonable agreement with the total electron density of 4.8×10^{13}cm^{-2} obtained from Hall measurements of the 10 period structure.

The fact that doping the central region of the $Si_{0.5}Ge_{0.5}$ layers yields the mobility maximum by carrier separation from the ionized parent impurities can only be explained when electron confinement in the Si layers is assumed. This, in turn, suggest a band structure of the Si/SiGe strained multilayer heterostructure as proposed in Fig. 12.

The transport properties of 2-D carrier gases were utilized for the fabrication of MODFET devices. For a review of the first n-channel and p-channel SiGe/Si-MODFET's see ref. 32. A typical n-channel MODFET is shown in Fig. 16. This silicon based heterodevice would well fit into the circuit concept given in Fig. 1. The main elements of this device are the Si substrate, the buffer layer for strain symmetrization, the Si channel for the 2-D electron gas, and the Sb-doping spike in the SiGe-layer delivering the electrons for the channel. A graded layer and a Si overlayer cover the MODFET. By this band gap engineered structure

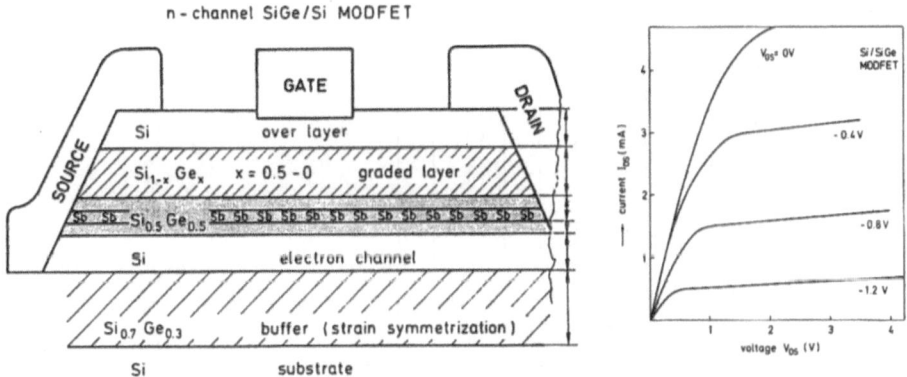

Fig. 16. Sketch of the n-channel Si/SiGe MODFET structure (left side). Current-voltage characteristics (right side)

the escape of electrons to the surface is avoided. Excellent room-temperature characteristics and operation up to microwave frequencies was obtained with the first unoptimized devices of this type /32, 33/ expecting encouraging progress with further development.

Brillouin Zone Folding

The Brillouin zone of a superlattice with period length L shrinks normal to the layers according to

$$-\pi/L \; < \; k_z \; < \; \pi/L \qquad\qquad (36)$$

$$-\pi/a \; < \; k_{x,y} \; < \; \pi/a$$

(a length of lattice cell).

One expects minizones in the momentum space not observed in the host crystal. For a weak superlattice potential the first superlattice Brillouin zone can be constructed by folding the Brillouin zone of the host crystal into the first minizone. Consider, that the derivative (dE/Dk) vanishes at the band edges. Already 1974 it was proposed /34/ that the zone folding effect can transform an indirect band gap to a direct band gap. Fig. 17 shows as example the folding of a host crystal Brillouin zone of a Si-like. band structure by a superlattice period L = 5a. The conduction band with an energy minimum near the X-point [001] is transformed in five mini bands with energy minimum near k_z = 0.

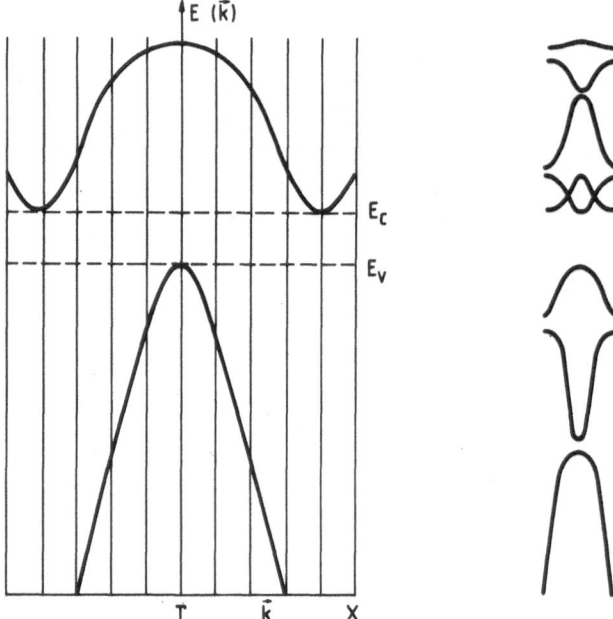

Fig. 17. Brillouin zone folding of a Si-like band structure. Host crystal with superlattice (L = 5a) minizones (left side). Folding of the band structure into the first minizone (right side)

Such an effect would have an tremendeous impact on silicon based opto-electronic devices. First experimental evidence of the zone folding effect in Si/Ge superlattices was announced /35/.

CONCLUSION

Strained layer superlattices will gain increasing importance because of scientific and industrial interest. Driving force for industrial interest is the monolithic integration of lattice mismatched heterostructure/superlattice devices with conventional IC's on a Si substrate.

For the realization of this material concept two basic problems have to be overcome: (i)Stacking of two levels of device structures (conventional IC, superlattice) with completely different layers. (ii) Growth of high quality material on a mismatched substrate. Only the latter problem is addressed in this paper.

Lattice mismatch can be accomodated by strain or misfit dislocations. Up to a critical thickness nature's answer is accomodation by strain. Above the critical thickness accomodation is partly by misfit dislocations and partly by strain. Experimental values of critical thicknesses are larger than the values predicted by equilibrium theory.

Strained layer superlattices with individual layers below the critical thickness are stable for symmetrical strain distribution. Otherwise a critical superlattice thickness would limit the thickness of the whole superlattice itself irrespective of thin individual layers.

Strain symmetrization can be obtained with a thin, homogeneous buffer layer. Design rules for this buffer layer are given.

Si/SiGe is a model system for SLS's. Growth is performed with a Si-MBE system. Strain strongly influences the band ordering switching from a flat conduction band for unsymmetrical strain (Si/SiGe on Si) to a type II interface for symmetrical strain. Mobility enhancement in modulation doped structures was utilized for n- and p-type device operation (MODFET). Zone folding effects are predicted which can strongly influence the optoelectronic properties of indirect gap materials.

REFERENCES

/1/ L. Pauling, The Nature of the Chemical Bond, 3rd Ed., Cornell University Press, 1960

/2/ T. Moriizumi and K. Tokahashi, J. Appl. Phys. 4606 (1972)

/3/ W.A. Brantley, J. Appl. Phys. 44, 534 (1973)

/4/ I.J. Fritz, B.L. Doyle, J.E. Schirber, E.D. Jones, L.R. Dawson and T.J. Drummond, Appl. Phys. Lett. 49, 581 (1986)

/5/ S. Sze, Physics of Semiconductor Devices, 2nd Ed., J. Wiley, New York, 1981

/6/ J.H. v.d. Merwe, Surf. Sci. 31, 198 (1972)

/7/ R. People and J.C. Bean, Appl. Phys. Lett. 47, 332 (1985) Erratum: Appl. Phys. Lett. 49, 229 (1986)

/8/ E. Kasper, H.J. Herzog, Thin Solid Films 44, 357 (1977)

/9/ E. Kasper, H.J. Herzog, H. Dämbkes and G. Abstreiter, Mat. Res. Soc. Symp. Proc., Vol. 56, 347 (1986)

/10/ F.R.N. Nabarro, Theory of Crystal Dislocations, Oxford (1967) J.P. Hirth and J. Lothe, Theory of Dislocations, McGraw-Hill (1968) A. Kelly and G.W. Groves, Crystallography and Crystal Defects, Longman, London (1970)

/11/ E. Kasper, Surf. Sci. 174, 630 (1986)

/12/ B.W. Dodson and P.A. Taylor, Appl. Phys. Lett. 49, 642 (1986)

/13/ J. Matthews and A.E. Blakesle, J. Cryst. Growth 27, 118 (1974)

/14/ E. Kasper, H.J. Herzog and H. Kibbel, Appl. Phys. 8, 199 (1975)

/15/ G.C. Osbourn, J. Appl. Phys. 53, 1985 (1982) and Phys. Rev. B27, 5126 (1983)

/16/ J.C. Bean, 1st Int. Symp. Si-MBE, Electrochem. Soc. Proc. Vol. 85-7, p. 337, ed. by J.C. Bean, Pennington, 1985

/17/ J.Y. Marzin, L. Goldstein, F. Glas and M. Quillec, Surf. Sci. 174, 586 (1986)

/18/ M.Y. Yen, A. Madhukar, B.F. Lewis, R. Fernandez, L. Eng and F.J. Grunthaner, Surf. Sci. 174, 606 (1986)

/19/ E. Kasper, H.J. Herzog, H. Jorke, and G. Abstreiter, Superlattices and Microstructures 3, 141 (1987)

/20/ J.C. Bean, L.C. Feldman, A.T. Fiory, S. Nakakaro, and I.K. Robinson J. Vac. Sci. Technol. A2, 436 (1984)

/21/ J.C. Bean, Silicon Based Heterostructures, in: Silicon Molecular Beam Epitaxy, ed. by E. Kasper and J.C. Bean, CRC Press, Boca Raton, USA (1987)

/22/ Epitaxial Growth, p. 559, ed. by J.W. Matthews, Academic Press, New York (1975)

/23/ H.J. Herzog, pers. communication

/24/ H. Jorke and H. Kibbel, J. Electrochem. Soc. 133, 774 (1986)

/25/ R. People, J.C. Bean, D.V. Lang, A.M. Sergent, H.L. Störmer, K.W. Wecht, R.T. Lynch and K. Baldwin, Appl. Phys. Lett. 45, 1231 (1984)

/26/ H. Jorke and H.J. Herzog, same book as ref. 16, p. 352

/27/ G. Abstreiter, H. Brugger, T. Wolf, H. Jorke and H.J. Herzog, Phys. Rev. Lett. 54, 2441 (1985)

/28/ E. Kasper, H.J. Herzog, H. Dämbkes and Th. Ricker, in: Two-dimensional Systems: Physics and New Devices, ed. by. G. Bauer, F. Kuchar, and H. Heinrich (Springer Series in Solid State Sciences 67), p. 52, Springer, Berlin, 1986

/29/ G. Abstreiter, H. Brugger, T. Wolf, R. Zachai and Ch. Zeller, same book as ref. 28, p. 130

/30/ R. People, and J.C. Bean, Appl. Phys. Lett. 48, 538 (1986)

/31/ H. Jorke, H.J. Herzog, E. Kasper and H. Kibbel, J. Cryst. Growth 81, 440 (1987)

/32/ E. Kasper, and H. Dämbkes, Solid State Devices 1986, p. 93 (Inst. Phys. Conf. Ser. No. 82), ed. by D.F. Moore, I OP Publishing, Bristol (1987)

/33/ H. Dämbkes, H.J. Herzog, H. Jorke, H. Kibbel and E. Kasper. IEEE Trans. ED-33, 633

/34/ U. Gnutzmann and K. Clausecker, Appl. Phys. 3, 9 (1974)

/35/ T.P. Pearsall, J. Bevk, L.C. Feldman, J.M. Bonar, J.P. Mannaerts and A. Ourmazd, Phys. Rev. Lett. 58, 729 (1987)

ELECTRICAL TRANSPORT IN MICROSTRUCTURES

Frank Stern

IBM T.J. Watson Research Center
Yorktown Heights, New York 10598, U.S.A.

INTRODUCTION

These lectures are intended to provide an introduction to carrier
transport in systems with reduced dimensionality. The main application
will be to electrons in silicon inversion layers and GaAs heterojunctions and quantum wells. Knowledge of basic aspects of semiconductor
physics and of transport in solids, as given, for example, in Kittel[1]
or Ziman[2] will be assumed. More detailed discussions of transport,
particularly in semiconductors, can be found in Wilson[3] (which includes
semiconductors in spite of the title), in "big" Ziman,[4] and in
Seeger.[5] General concepts of heterostructure physics are presented in
other lectures in this volume.

The principal aspect of transport to be discussed is ohmic conductance as influenced by impurities, interfaces, phonons, and alloy
scattering. Brief discussions have been added dealing with hot carriers, with device simulation, and with coherence effects in small systems
in the presence of disorder. The presentation will attempt to describe
the basic physical processes and to provide a few references that can
serve as an entry point to the literature. I apologize to those whose
work has been omitted or underrepresented because of time and space
constraints.

Rapid advances in processes for fabricating semiconductor samples
have made possible structures in which carrier motion is restricted in
one (or two or even three) dimensions, with carriers essentially free
to move in the remaining dimensions. In most of what follows, we shall
consider the case in which motion is restricted in a quantum sense, i.e.
we have carriers confined in a "box" of some kind, whose size is less
than or comparable to the Fermi length k_F^{-1}, the reciprocal of the
diameter of the Fermi surface corresponding to the carrier density.

Later we briefly consider cases where the criterion for reduced dimensionality involves a different physical length scale.

ELEMENTARY TRANSPORT CONSIDERATIONS

Although there are many levels from which transport can be approached, we will start with the Boltzmann equation (here simplified to the case of a spatially uniform three-dimensional system)

$$\frac{\partial f(\mathbf{k},t)}{\partial t} - \frac{e\mathbf{E}}{\hbar} \nabla_k f(\mathbf{k},t) = \frac{\partial f(\mathbf{k},t)}{\partial t}\bigg|_{scatt}$$

$$= \frac{V}{(2\pi)^3} \int d\mathbf{k}' \{W(\mathbf{k}' \rightarrow \mathbf{k})f(\mathbf{k}',t)[1 - f(\mathbf{k},t)] - W(\mathbf{k} \rightarrow \mathbf{k}')f(\mathbf{k},t)[1 - f(\mathbf{k}',t)]\} \tag{1}$$

for the carrier distribution function $f(\mathbf{k},t)$ as influenced by external electric fields $\mathbf{E}$ (magnetic fields are ignored in this discussion) and by a scattering rate W which depends on the wave vectors of the initial and final states in the scattering process as well as on temperature and on material parameters such as impurity density. V is the normalization volume.

If all scattering is elastic and for spherical symmetry, the steady-state solution of the Boltzmann equation in the weak-field (i.e. ohmic) regime is

$$f(\mathbf{k}) = f_0 + f_1(\mathbf{k}) = f_0 + e\tau_k \, \mathbf{E} \cdot \mathbf{v} \, \frac{\partial f_0}{\partial E} \tag{2}$$

where f_0 is the Fermi-Dirac distribution and $\mathbf{v} = \nabla_k E$ is equal to $\hbar\mathbf{k}/m$ for spherical and parabolic bands with effective mass m. The current density (including a factor 2 for spin degeneracy) is

$$\mathbf{J} = \frac{-2e}{(2\pi)^3} \int \mathbf{v} \, f(\mathbf{k}) \, d\mathbf{k}, \tag{3}$$

but in the approximation being used here, f can be replaced by f_1 because the equilibrium distribution does not carry any current. The scattering time τ_t that enters in the mobility $\mu = e\tau_t/m$ and in the conductivity $\sigma = n\tau_t e^2/m$, will be given for some specific scattering mechanisms below.

In systems with lower dimensionality the electronic states can be described by subbands within the energy band structure of the bulk material. In the simplest case, in which only one subband has appreciable carrier density, we shall presume that an equivalent Boltzmann equation applies, with the distribution function now referring to a lower dimensional space. It should be noted, however, that the conductance calculated in this way is subject to corrections that modify the transport at very low temperatures. In two dimensions, one finds logarithmic "weak-localization" corrections[6-9] whose magnitude generally does not exceed 10% in the accessible temperature range. In one-dimensional systems these corrections vary as $T^{-1/2}$, and there is a significant temperature range in which Boltzmann equation results will be unreliable.

Early work on transport in two dimensions in the Boltzmann equation regime is that of Stern and Howard,[10] who considered only Coulomb scattering. Work on one-dimensional systems in a similar regime has been carried out by Fishman[11] for the case with only one occupied sub-band and by Das Sarma and Xie[12] for a more general case. Phonon scattering in quasi-one-dimensional systems was treated by Riddoch and Ridley.[13] The one-dimensional case is of interest for a number of reasons, one of them the possibility that the mobility will be high because the only available scattering process is backward scattering, which has a large momentum transfer and therefore a reduced scattering rate.[14]

Most of our discussion applies to electrons, either in silicon inversion layers or in GaAs-based heterostructures, for which there are more data and for which the theory is considerably simpler. A few references to the more limited literature for holes are given after the results for electrons. We consider mainly the conductance of a single electron subband in a single channel. At high carrier densities or high electron temperatures, more than one subband will contribute. The theory for such cases is discussed, for example, by Ando et al.,[15] and the effect was observed in a GaAs-based heterojunction by Störmer et al.[16] In some heterojunctions, there can be conduction both in the GaAs channel and in the $Al_xGa_{1-x}As$ barrier region, a situation which requires care to interpret, as noted by Kane et al.[17] and by Syphers et al.[18]

SCATTERING MECHANISMS

Coulomb scattering

In most cases, the dominant scatterers at low temperatures are charged impurities or defects. The effects of screening by mobile carriers must be included in the scattering cross section. For weak, slowly varying, static potentials, the screened potential ϕ induced by an external charge density ρ_{ext} is given by the solution of

$$\nabla^2 \phi(\mathbf{r}, z) - 2q_s \bar{\phi}(\mathbf{r}) \, g(z) = -\rho_{ext}(\mathbf{r}, z)/\epsilon, \qquad (4)$$

where $g(z)$ is the normalized spatial distribution of the two-dimensional electrons, assumed here to occupy only the lowest subband, and $\bar{\phi}(\mathbf{r}) = \int \phi(\mathbf{r}, z)g(z)dz$. The screening constant q_s is given by[10]

$$q_s = \frac{e^2}{2\epsilon} \frac{dN}{dE_F}. \qquad (5)$$

At low temperatures this reduces to $q_s = g_v me^2/4\pi\epsilon\hbar^2$, where g_v is the valley degeneracy, and at high temperatures (if that is meaningful in a single-subband approximation) $q_s = (e^2 N / 2\epsilon k_B T)$. Note that the screened potential in two dimensions falls off approximately with the third power of the distance,[10] a considerably weaker variation than the exponential decay found in three dimensions.

In most of what follows we shall use the Born approximation to calculate the scattering rate. This approximation, which assumes that the scatterer does not deform the incoming wave substantially, is expected to be valid for the Coulomb scattering rates that dominate the scattering at low temperatures because the dominant scatterers are generally some distance from the center of the electron distribution. That was verified explicitly for silicon inversion layers by Stern and Howard,[10] and is expected to hold for high-mobility GaAs heterostructures as well. If the Born approximation does not apply then the simple treatment of screening used here is also in question.[19]

The detailed evaluation of the scattering rate for electrons in an inversion layer or heterojunction can be found in standard references (see for example, Refs. 10, 15, or 20) and need not be repeated here. The simplest example assumes that the electrons lie in a plane with zero thickness (the extreme two-dimensional limit), that the surrounding material has permittivity ϵ, and that the scatterers are randomly distributed on a plane a distance d from the electron plane. Per singly charged center, the cross section for scattering with wave vector transfer q is then

$$\sigma(\theta) = \frac{me^4}{8\pi\epsilon^2\hbar^3 v} \frac{\exp(-2qd)}{(q + q_s)^2} , \qquad (6)$$

where q_s is the screening parameter defined in (5), $v = \hbar k/m$ is the carrier velocity, and the scattering angle is given by $q = 2k \sin(\theta/2)$. Note that the cross section has the dimensions of a length here, not an area as in three dimensions.

This simple expression, generalized to take into account the permittivities of different layers and the spatial extent of the electron wave function, can be applied to many examples by integrating over the positions of the charged scatterers. For example, it can be used to examine the mobility in GaAs-based heterojunctions. Here a doped $Al_x Ga_{1-x} As$ layer is separated from the GaAs by a nominally undoped spacer layer. Charges transfer from the barrier to the GaAs, and the density of charges in the GaAs channel is determined by the doping density in the $Al_x Ga_{1-x} As$, by the spacer thickness, and by the difference between conduction band offset at the GaAs-$Al_x Ga_{1-x} As$ interface and the binding energy of the donor level in the $Al_x Ga_{1-x} As$.[21]

If the scattering cross section is known, the scattering rate for carriers of a given energy (or wave vector, since we assume isotropic bands here) scattered by N_i scatterers per unit area, with differential cross section $\sigma_i(\theta)$ is given by

$$\tau_{e1}^{-1}(k) = N_i v \int_{-\pi}^{\pi} (1 - \cos\theta)\, \sigma_i(\theta)\, d\theta,\tag{7}$$

where $v = \hbar k/m$. Contributions from various scattering layers simply add.

Taking the charge transfer conditions into account, Stern[21] showed that the highest mobility for a given channel electron density is achieved by maximizing the doping in the $Al_xGa_{1-x}As$ and by keeping the residual density of ionized impurities in the GaAs as low as possible. Calculated values of the reciprocal mobility for one set of parameters are shown in Fig. 1. (Note that the difference between the band offset and the $Al_xGa_{1-x}As$ donor binding energy used in that paper was about 30 meV too large; some reduction in the calculated mobilities would be expected if that difference were reduced. The trends, however, are not affected.) The general features of this figure appear to be supported by experiments on high-quality samples measured in the dark.

Under illumination, the carrier density in GaAs-$Al_xGa_{1-x}As$ heterojunctions usually increases, an effect attributed to lattice relaxation around of donors in $Al_xGa_{1-x}As$ when they are ionized, resulting

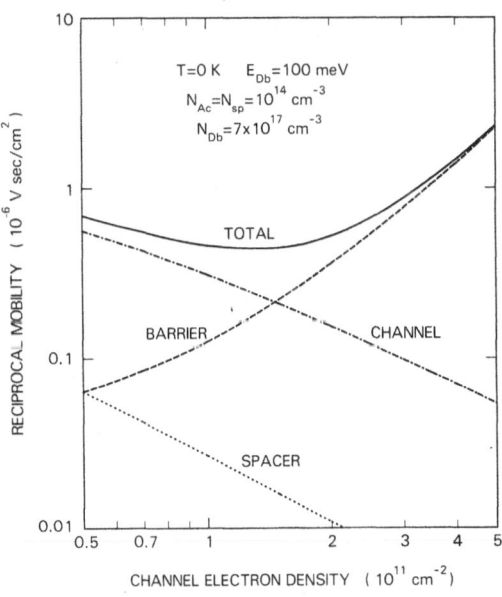

Fig. 1. Calculated "optimum" reciprocal mobility at low temperatures for a GaAs-$Al_xGa_{1-x}As$ heterojunction with background impurity concentration of 10^{14}/cc in the GaAs and in the undoped spacer layer, donor concentration of 7×10^{17} cm^{-3} in the doped portion of the $Al_xGa_{1-x}As$, and a difference of 0.2 eV between the barrier height and the donor binding energy. (After Stern, Ref. 21)

in an energy barrier that impedes electron recombination with the ionized donors at low temperatures. The mobility usually increases upon
illumination, but cannot be calculated from the simple model used here;
screening by mobile carriers in the $Al_xGa_{1-x}As$ must also be included.

We have so far used the static, long-wavelength limit for screening. The results can be generalized to include the effects of wave-
vector-dependent screening, but this is academic in the simplest cases
because for a simple subband in two dimensions the screening is constant
for wave vector transfers up to $q = 2k_F$, the maximum value possible for
scattering in an isotropic system at absolute zero.[22] In three dimensions, on the other hand, screening does depend on q. The wave vector
dependence of static screening in two dimensions is shown in Fig. 2.

While Coulomb scattering is generally thought to be temperature-
independent, it turns out that the singular behavior of screening near
momentum transfer of $2\hbar k_F$ in two-dimensional systems can lead to a
significant temperature dependence at low temperatures. This effect,
an increase in resistance with increasing temperature at temperatures
of order 10 K, is to be be distinguished from the logarithmic "weak
localization" corrections mentioned above, which occur at even lower
temperatures and which lead to an increase in resistance as the temperature decreases. The effect has been observed in silicon inversion
layers by Kawaguchi and Kawaji,[23] Cham and Wheeler,[24] and Dorozhkin and
Dolgopolov.[25] Theoretical treatments have been given by Stern[22] and by

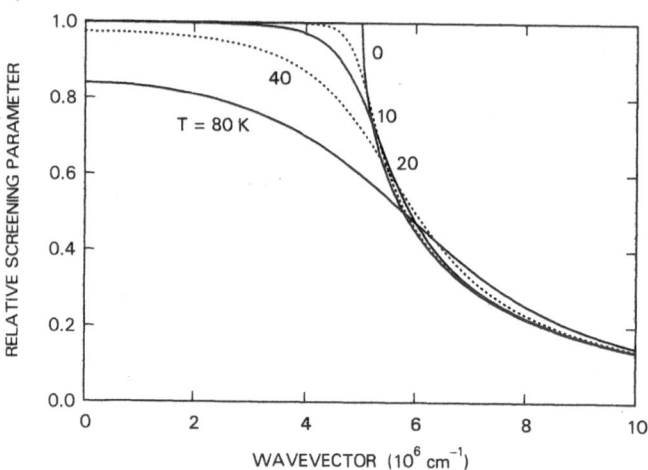

Fig. 2. Wave vector dependence of the screening parameter q_s , normalized to its value at q=0 and T=0, for several temperatures.
The results are for a silicon inversion layer with 2×10^{12}
electrons per cm^2, for which $2k_F = 5 \times 10^6$ cm^{-1}. (After Stern,
Ref. 22)

Gold and Dolgopolov.[26] In GaAs heterostructures other effects appear
to dominate the temperature dependence of mobility, as discussed in the
following section.

Some silicon inversion layer samples are found to have mobilities
above 2 m^2/V s at low temperatures.[27] Those samples, which remain
"metallic" down to electron densities well below 10^{12} cm^{-2}, have mobilities that continue to increase even below 4.2 K,[27,28] consistent with
calculated results for samples with a low density of interface
charges.[26,29]

At very low carrier densities, the conductance in two-dimensional
systems becomes activated, representing a transition from the weak-localization, quasi-metallic regime to the regime of strong localization
and variable-range hopping. Experimental results for the activated
regime in silicon inversion layers have been reviewed by Ando et al.[15]
Theoretical predictions in the framework of the memory function formalism have been made by Gold and Götze[30] for silicon inversion layers
and by Gold for GaAs heterostructures.[31,32] An entirely different
mechanism for activated conduction, magnetic-field-induced freezeout,
has been investigated by Robert et al.[33]

Interface scattering

Deviations of interfaces from planarity can lead to scattering
both in three-dimensional and in two-dimensional systems. The roughness of the Si-SiO_2 interface has been actively studied, and leads to
significant scattering at high electron densities. This subject has
been reviewed by Ando et al.,[15] and recent results have been presented
by Hahn and Henzler[34] and by Goodnick et al.[35]

The scattering rate due to roughness scattering was first derived
by Prange and Nee[36] in connection with magnetic-field-induced states
at the surface of metals. For an inversion layer or a single heterojunction, their result can be simplified to[37]

$$\tau_r^{-1}(k) = \frac{m\Lambda^2\Delta^2 N^{*2}e^4}{\hbar^3} \int_0^\pi \frac{(1 - \cos\theta)\exp(-q^2\Lambda^2/4)}{[\epsilon(q)]^2} \, d\theta, \tag{8}$$

where Δ and Λ are the rms height and the lateral correlation length of
the deviation of the surface from flatness, $q = 2k\sin(\theta/2)$, $N^* = N_d + N_s/2$
is proportional to the average electric field in the inversion layer
(N_d and N_s are the densities of fixed charges and channel electron
charges, respectively, per unit area), and $\epsilon(q)$ includes wave-vector-
dependent screening. Thus roughness scattering increases with the rms
height of the roughness, as one might expect, but also with the electric
field in the channel, and it depends as well on the lateral correlation
length of the roughness. Equation (8) assumes a Gaussian autocorrela-

tion of the roughness, as is customary because of the simple form of the results. However the experiments of Goodnick et al.[35] suggest a correlation falling off more nearly exponentially. The effects of various roughness models were explored by Gold.[38]

While roughness scattering is important in silicon inversion layers at high carrier densities, it was shown by Ando[20] not to be important in GaAs heterojunctions, both because the carrier density is generally lower and because the interfaces can be smoother.

Alloy scattering

The influence of alloy scattering on the mobility of electrons in typical GaAs – $Al_xGa_{1-x}As$ heterostructures[20] is usually weak because the electrons penetrate very little into the alloyed barrier regions. On the other hand, alloy scattering can be significant in structures in which the electron is confined within the alloyed region of the device.

The scattering rate connected with electron motion in a hetero-structure containing a random binary alloy can be written[39]

$$\tau_{alloy}^{-1} = \frac{m\,x(1-x)\,(\delta V)^2 \int \zeta^4(z)dz}{N\hbar^3}\,,$$

(9)

where δV is an effective potential energy difference between the two species in the alloy, x is the alloy fraction, N is the density of primitive cells per unit volume, $\zeta(z)$ is the electron envelope wave function, and the integration is over the alloy region--here the inside of the well. This result assumes that the positions of the two con-stituents are uncorrelated. If correlations do exist, as has been found for at least some $Al_xGa_{1-x}As$ structures,[40] then (9) does not apply. Note that the alloy scattering rate in two dimensions is independent of electron energy.

For the particular case of InP-$In_{0.53}Ga_{0.47}As$-InP quantum wells, Brum and Bastard[39] calculated the alloy scattering vs well thickness. For thicknesses large enough that most of the charge is within the well, the alloy scattering mobility increases approximately linearly with the width of the well (until the second subband is occupied) and is equal to about 5 m^2/Vs for a 10 nm well.

Deformation potential scattering and piezoelectric scattering

Scattering by charged impurities and other defects generally determines the mobility of real devices at very low temperatures, but at higher temperatures scattering by lattice vibrations begins to be felt. Deformation-potential scattering by acoustic-mode phonons in a

140

quantum well of thickness L leads to a scattering rate given in the absence of screening by[41,42]

$$\tau_{DP}^{-1} = \frac{3D^2 k_B Tm}{2\hbar^3 \rho v_L^2 L} \, , \tag{10}$$

where D is the deformation potential (with units of energy), ρ is the mass density, and v_L is the longitudinal sound velocity. Note that both phonon absorption and emission processes contribute to scattering at finite T. Both acoustic-mode and optical-mode scattering are modified by screening effects, as discussed by Price.[43]

In two-dimensional systems with only one subband occupied, the electron wave vectors lie in a two-dimensional space while the dominant phonons are bulk phonons, with wave vectors in three dimensions. Thus an integration over phonon wave vectors in the z direction must be carried out. This leads to a factor $\int \zeta^4 dz$ in the scattering rate, which becomes 3/2L in (10) and corresponds to the factor $1/z_{av}$ in the theory for silicon inversion layers.[44]

In materials without inversion symmetry, acoustic phonons generate electric fields that interact with the carriers. The resulting piezoelectric scattering[41,45] is very anisotropic. Written so that the angle average is equal to the total rate, the unscreened scattering rate is

$$\tau_{piezo}^{-1}(q) = \frac{m(eh_{14})^2}{64\hbar^3} \frac{k_B T}{q} \left(\frac{9\alpha_L}{\rho v_L^2} + \frac{13\alpha_T}{\rho v_T^2} \right) , \tag{11}$$

where h_{14} is the applicable piezoelectric coupling constant (equal to 1.2×10^9 V/m for GaAs), q is the magnitude of the two-dimensional transfer wave vector, and the factors $\alpha_L(q)$ and $\alpha_T(q)$ are equal to 1 for heterolayer thickness small compared to 1/q.

The theory of scattering by acoustic phonons in silicon inversion layers has been reviewed by Ezawa et al.,[46] but the results at intermediate temperatures are masked by the temperature dependence resulting from screened Coulomb scattering, discussed above. In GaAs heterostructures, on the other hand, the phonon scattering is clearly seen in good samples even near 10 K, as shown in the work of Paalanen et al.,[47] Mendez et al.,[48] and Lin et al.[49] Figure 3, from Lin's thesis, shows the temperature dependence of mobility for a series of GaAs heterojunctions. The mobility varies approximately linearly with temperature at low temperatures (the rapid decrease at higer temperatures will be discussed in the following section) and the slope changes from positive to negative as the sample quality improves. This difference is attributed to the usual effect of acoustic phonon scattering in the high-mobility samples, which is masked by an effect associated with the energy dependence of the impurity scattering rate in the low-mobility samples.

Mendez et al.[48] analyzed their data on several high-mobility samples to extract a value for the deformation potential for interaction between the electrons at the GaAs heterojunctions and the acoustic phonons. Their analysis included the piezoelectric scattering and assumed that both deformation potential and piezoelectric scattering are screened. They infer a value of approximately 13.5 eV for the deformation potential constant D. Vinter[50] carried out a similar analysis but used numerical rather than variational envelope wave functions and found D=12 eV. Hirakawa and Sakaki[51] obtained a value 11 ± 1 eV from analysis of electron heating in GaAs heterojunctions.

The low-mobility samples in Fig. 3 have an increasing mobility with increasing temperature. This effect has a much simpler origin than the screening effect discussed above for silicon inversion layers. It is thought to arise simply from the averaging over energies in the expression for the scattering time

$$\tau_t = \frac{\int_0^\infty \tau_k(E) \, E \, (-df_0/dE) \, dE}{\int_0^\infty E \, (-df_0/dE) \, dE} \tag{12}$$

that enters in the conductivity. This leads to a temperature dependence given, to second order in $k_B T/E_F$, by

$$\tau_t(T)/\tau_t(0) = 1 + (\pi^2/6) \, p \, (p + \frac{3}{2}) \, (k_B T/E_F)^2 \tag{13}$$

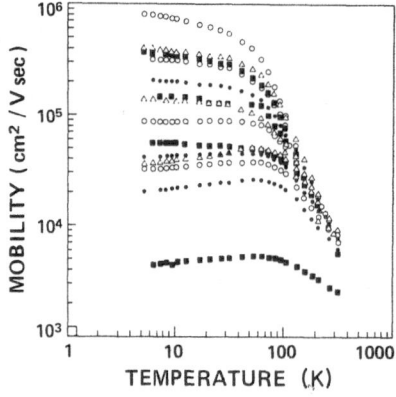

Fig. 3. Electron mobility versus temperature in a series of GaAs-$Al_x Ga_{1-x}As$ heterojunctions. Note especially the change of the initial slope from positive to negative as the sample quality improves. (After the Ph. D. thesis of B. J. F. Lin, Ref. 49)

for a degenerate electron distribution if the scattering time varies
as the p'th power of electron energy and if the Coulomb scattering, here
assumed temperature independent, is much stronger than the phonon
scattering, as will be the case in low-mobility samples. Similar conclusions were obtained by Paalanen et al.,[47] by Lin et al.,[49] and by
Stern and Das Sarma.[52] While the observed behavior agrees approximately
with theoretical expectations, the initial temperature dependence in
the low-mobility samples is approximately linear rather than quadratic
as would be expected from eq. (13).

Hirakawa and Sakaki[53] have given a detailed account of electron
mobility in $Al_xGa_{1-x}As/GaAs$ heterojunctions and have compared theory and
experiment for both the temperature dependence and the carrier concentration dependence of mobility. Electron mobility in quantum wells has
been studied both experimentally and theoretically by Guillemot et
al.[54] for temperatures up to 77 K. They invoke interface scatterers
with densities of order $1-3 \times 10^{10}$ cm^{-2} at the inverted interface of the
quantum well, where GaAs is grown onto $Al_xGa_{1-x}As$, to explain their data.

At very low temperatures, in the so-called Bloch-Grüneisen range,
the acoustic phonon scattering rate changes because of degeneracy combined with energy conservation.[55] The magnitude of the deformation
potential scattering rate is reduced and its temperature dependence
changes from T^1 to T^7 (the exponent is 5, rather than 7, for piezoelectric scattering). This effect is probably masked by other effects
in the mobility but it influences the inelastic phonon scattering rate,
and therefore electron heating effects, at very low temperatures.[51,56]

Optical phonon scattering

At temperatures in the liquid nitrogen range and above, scattering
by optical phonons plays an increasingly important role in limiting the
carrier mobility, especially in GaAs, where the polar character leads
to a strong electron-phonon interaction. Both for silicon inversion
layers and for GaAs heterostructures, the analysis of optical phonon
scattering is complicated because at the elevated temperatures where
this scattering is important one must generally include more than one
subband. Thus the results require not only a careful treatment of the
scattering processes, but also an accurate knowledge of subband energies and wave functions. Ando et al.[15] reviewed optical phonon scattering in silicon inversion layers, where more work is needed.

A number of authors have treated the electron mobility in GaAs
heterostructures at elevated temperatures. Perhaps coincidentally, the
values are similar to those obtained in bulk GaAs. Theoretical treatments have been given by Price,[41,45] Ridley,[42] and Walukiewicz et al.[57]
For experimental results see, for example, the papers of Hirakawa and

Sakaki[53] and Mendez.[58] Perhaps there is a better chance to compare theory and experiment for GaAs than for Si, because of the simpler conduction band structure in GaAs and because it may be possible to study structures--such as thin quantum wells--in which one subband dominates the conductance even at the temperatures where optical phonon scattering becomes important. Here too, however, the contribution of other bands and subbands may need to be considered at temperatures near and above room temperature. Optical phonons play a major role in the energy loss of hot carriers, a subject not considered here.

Hole mobility; Minority carrier mobility

Although it has been shown that holes in GaAs-based heterostructures can have mobilities as high as 3.8×10^5 cm^2/Vs at low temperatures,[59] the literature on hole mobility is sparse. Walukiewicz[60] has given a theoretical treatment, and Mendez[58] has reviewed the literature as of a year ago.

In bipolar transistors, one of the important parameters is the mobility of minority carriers in the base region. When a current is flowing, the drift velocities of minority and majority carriers are oppositely directed, and one expects scattering to have a noticeable effect on the minority carriers,[61] especially minority electrons in GaAs because they have a smaller effective mass than do the holes. Höpfel et al.[62] measured spatially resolved luminescence excited by short pulses of light to infer a negative mobility for minority electrons in GaAs quantum wells at low temperatures and low electric fields.

WARM AND HOT CARRIERS

In this section experimental and theoretical work on carriers excited above ambient temperature by applied electric fields is very briefly noted. Related work is covered in other lectures.

When an electric field is applied to a device carrying a current, the carriers derive power from the field. In the usual treatment of ohmic transport, one assumes that inelastic processes remove this power from the carrier system and transfer it to the lattice, which is assumed to remain in equilibrium. There is an implicit assumption of a large reservoir which can exchange energy with the electronic system without itself being driven out of equilibrium. At sufficiently large applied fields, which at low temperatures can mean surprisingly small values, this assumption fails and the energy applied to the electronic system cannot be removed fast enough, and the carriers get warm or hot. This is a subject widely studied in three dimensions, and now increasingly studied in heterostructures. For an account of recent work, see for example the review article by Hess,[63] the book edited by Reggiani,[64] and the papers in the Proceedings of the Fourth International Conference on Hot Electrons in Semiconductors, published in 1985 in Volume 134 B+C of Physica.

One of the most important properties of hot carriers is the dependence of carrier velocity on electric field. Because a number of dynamic processes are involved, the result can depend on the particular experimental configuration and on the time scale of the measurement, particularly at high fields where intersubband and intervalley processes are involved. Figure 4, from the work of Masselink et al.,[65] shows the electron drift velocity in a GaAs-based heterojunction versus the electric field at 77 K and 300 K, as determined from magnetoresistance measurements in a Corbino-like geometry. These results, although subject to some uncertainty, indicate the qualitative behavior found at low and intermediate electric fields.

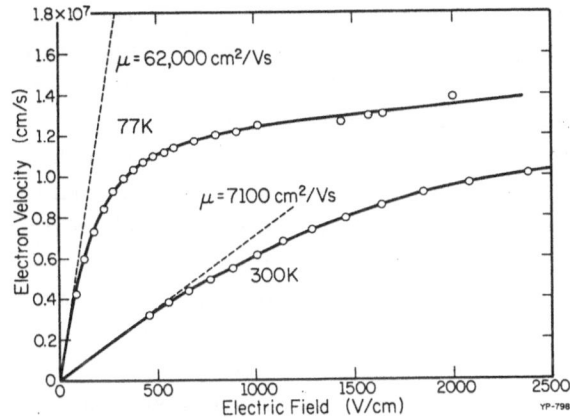

Fig. 4. Velocity-field characteristics of electrons in a GaAs-based heterojunction, as determined from resistance measurements in a Corbino-like geometry. (After Masselink et al., Ref. 65)

When enough power is supplied to the carriers to lead to a hot carrier distribution (T_e above ~ 50 K), it is transferred from the electrons to the lattice primarily via excitation of optical modes, which then decay into other lattice modes of lower phonon energy. If the energy is supplied rapidly enough, the optical modes may themselves be driven out of equilibrium and one speaks of hot phonon effects, which have been discussed in the context of two-dimensional systems by Price.[66] These effects arise primarily in the context of picosecond-pulse optical excitation.

DEVICE SIMULATION

One of the most important applications of transport physics is in the design, implementation, and use of device simulation programs. They range from relatively modest programs that explain the characteristics of a relatively simple device to very large engineering programs which can run to tens of thousands of lines of code, require many man-years to write and maintain, and often have numerous generations of successive refinement and improvement. Many such codes are developed by companies in the semiconductor industry and are proprietary, but others are developed at universities or other research centers, and are in the public domain. Among the more widely known programs are PISCES, developed at Stanford University, and MINIMOS, developed at the Technical University of Vienna.

Most of the large device simulation programs use the so-called drift-diffusion approximation to transport: they solve Poisson's equation and the equations for electron and hole current, but make no provision for changing electron temperature through the device. However, they can deal with complex device configurations involving multiple contacts as well as nonuniform impurity concentrations such as those that might result from diffusions or implantations. Solutions are generally obtained in two space dimensions but extensions to three space dimensions are becoming feasible. The calculations can require substantial computer resources, both memory and computing time, and are used in device design. Because of the simplicity of the underlying physics, these programs are often calibrated to the device or device family being studied, generally by supplying them with empirical expressions for the carrier mobilities as functions of electric field, impurity concentration, etc., and for the parameters describing carrier recombination and--if very high fields are involved--carrier multiplication.

An interesting question that arises in the drift-diffusion equation is how to include carrier heating effects in the mobility that enters in these equations: Does it depend on the local electric field? On the gradient of the local quasi-Fermi level? This question has a history that goes back at least 20 years and has recently been revisited by Higman and Hess.[67]

The device equations are solved numerically on a grid taylored (either manually or automatically) to the needs of the device being considered, with grid points close together where quantities of interest are varying rapidly and farther apart where the variation is gradual. Both finite difference and finite element methods are widely used. Even a simple device may have three unknowns (electrostatic potential, electron density, and hole density) at each of, say, 1000 grid points for a two-dimensional mesh, giving 3000 unknowns in all. There are at

least two general approaches in solving these equations. One approach
tackles the full set of 3000 simultaneous equations at once. The other
freezes the values of two of the unknowns, solves for the third unkown
at each grid point, and then iterates through the remaining sets of
equations until covergence is obtained--which is not always assured.
The problems, both in generating the mesh and in solving the equations,
become more difficult for three-dimensional device models.

A description of this field beyond the cursory description given
here is outside the scope of the present lectures. Interested readers
can consult, for example, the book by Selberherr[68] or the Proceedings
of the biennial NASECODE conferences[69] for entry into the literature
of this important application of transport physics.

Because of the severe physical approximations underlying the
drift-diffusion equations, there has been considerable effort to find
practical alternatives for situations where those approximations fail.
Examples of such cases are: devices in which some electrons get very
hot and can be transferred over potential barriers to other parts of
the device (where they can be trapped and can produce gradual deterioration of device properties); very short devices, where electrons do
not have many collisions during their transit through the active part
of the device; and many GaAs-based devices, whose band structure leads
to tunneling and other effects not dealt with by the simple models.

There have been two main routes for coping with these more complicated situations. One, sometimes called the "hydrodynamic" method
or the higher moments method, introduces an additional moment of the
velocity from the Boltzmann equation to capture additional information
about the carriers. For information about this method, consult, for
example, the papers of Blotekjaer,[70] Cook and Frey,[71] and Rudan and
Odeh.[72]

A more powerful method, increasingly used in recent years, is to
solve the full Boltzmann equation by the Monte Carlo method, which
follows a single carrier or an ensemble of carriers through a series
of ballistic flights interrupted by collisions, all chosen in a statistical way to simulate the system being studied. This method, which
can in principle approach the solution of the Boltzmann equation arbitrarily closely if enough steps are taken, requires substantial computer time and requires special techniques to deal with the relatively
few very hot electrons that are often of special interest. Descriptions
of Monte Carlo methods have been given by Price[73] and by Jacoboni and
Reggiani.[74]

Monte Carlo methods have been widely applied to calculate electron
motion in two-dimensional systems, and in recent years have included
the subband structure connected with carrier confinement. An example

of such work is the paper of Yokoyama and Hess,[75] who found the quantum
level structure of a GaAs-$Al_xGa_{1-x}As$ heterojunction at 77 K and 300 K,
evaluated matrix elements needed for transport calculations, and then
carried out a Monte Carlo calculation of the response of the system
versus time and under steady-state conditions. Perhaps surprisingly,
the velocity-field characteristics in the heterojunction, illustrated
in Fig. 5, are quite similar to those found for bulk GaAs by Ruch and
Fawcett.[76]

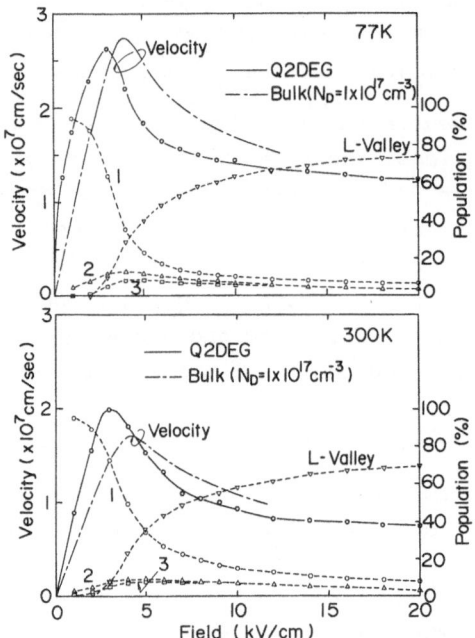

Fig. 5. Calculated steady-state electron velocity and populations of
subbands and of the L valley in a GaAs – $Al_xGa_{1-x}As$ heterojunction at 77 K and 300 K versus electric field. The dot-dash
curve is the corresponding velocity curve for bulk GaAs, from
the work of Ruch and Fawcett, Ref. 76. (After Yokoyama and
Hess, Ref. 75)

An application of the Monte Carlo method to a perhaps less well-
known system is the work of Fischetti et al.[77] on hot electron motion
in SiO_2.

A number of authors have attempted to combine features of Monte
Carlo calculations with those of the more traditional simulation
schemes, using each method for the part of the problem for which it
offers particular advantages. Nguyen et al.[78] use a regional approach:
they physically partition the device and use different methods in different regions.

This very brief discussion of device simulation has been included to call attention to a field that is increasingly active--partly because of the many new device configurations being designed and built and partly because the availability of rapidly growing computer resources makes it possible to attempt simulations of devices and phenomena that were heretofore out of reach--but usually does not come to the attention of physicists.

COHERENCE EFFECTS

The conventional theory of transport in semiconductors assumes that there are many scatterers within an appropriate physical region, and that one need only specify their average density. The theory takes an average over an ensemble of the positions of the scatterers, and it is presumed that all real samples with the same density of scatterers will give the same physical response. At some size scale, however, such averaging is no longer valid and the measured response will depend on the details of the arrangement of scatterers in a particular sample. This size scale, sometimes called mesoscopic, has received increased attention in recent years.

In this section, I will give a few examples and a small number of references. The literature is already vast and the subject might be considered outside the scope of conventional semiconductor transport. Nevertheless, the experimental and theoretical results obtained in the last ten years clearly show the limits of conventional tranport theory. As devices continue to shrink in size and as the possibility of device operation at lower temperatures grows, it is important to recognize the influence of phenomena, such as "universal" conductance fluctuations and Aharonov-Bohm[79] effect, which I have grouped under the heading "coherence effects."

Scattering events that change the direction of current flow need not necessarily destroy the coherence of the electron wave function. There are several length scales affecting loss of coherence, but the principal one is the distance a carrier diffuses between phase-breaking collisions. In the simplest case, these are inelastic collisions associated with phonon scattering or carrier-carrier scattering, and the relevant length scale in transport processes is the inelastic diffusion length

$$L_{in} = (D\tau_{in})^{1/2}. \qquad (14)$$

Samples smaller than this in one or more dimensions have reduced dimensionality in a transport sense.[80] Since the inelastic scattering time generally increases at low temperatures, the inelastic diffusion length can be of order microns or larger, and many samples exhibit

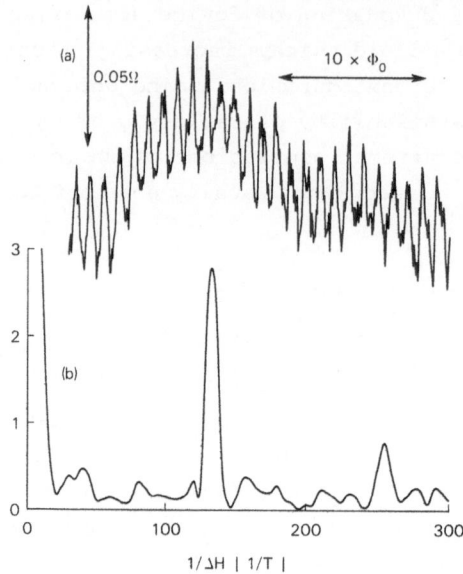

Fig. 6. Magnetic field dependence of resistance in a metal ring (upper
curve), and its power spectrum versus reciprocal field (lower
curve). The large peak at 130 T^{-1} is the Aharonov-Bohm oscillation. (After Webb et al., Ref. 81)

consequences of reduced dimensionality at sufficiently low temperatures.

Two consequences of quantum coherence have recently received considerable experimental and theoretical attention. The first deals with
the oscillations of conductance--most pronounced in rings whose width
is small compared to their diameter--as the magnetic field perpendicular to the plane of the ring is varied. The periodic variation of
resistance illustrated in Fig. 6 is associated with a change in the
phase of electronic wave functions proportional to the flux inside the
ring, and would be present even in cases where there is no magnetic
field acting on the electrons themselves, according to the theory of
Aharonov and Bohm.[79] The effect was found first in metal rings[81] but
is expected in semiconducting rings as well. In high-mobility heterostructures, where even the elastic mean free path can become comparable
to the sample size, the effect can be even more pronounced.[82]

The second class of effects resulting from coherent scattering is
the variation in conductance observed in high-resistance samples as
some control parameter, such as the gate voltage in a gated device, is
changed to change the scattering path seen by electrons near the Fermi
level. Three kinds of effects have been observed: so-called "universal" conductance fluctuations in quasi-metallic samples,[83-88] whose
variation is of order e^2/h provided the sample is long compared to the

elastic mean free path but not long compared to the inelastic diffusion length; large fluctuations in the conductance or resistance at low temperatures for samples in the "hopping conduction" regime;[89-91] and occasional large resistance fluctuations at very low temperatures that are attributed to resonant tunneling.[92,93] Although detailed studies of these fluctuations have been carried out only in the last few years, there has been evidence of structure in the conductance of two-dimensional devices for more than 20 years.[94]

This is only a glimpse of the physics of coherent transport in small devices. Extensive experimental and theoretical work has been done in the United States, the Soviet Union, Israel, and elsewhere, with a rapid series of advances that are beyond the scope of these lectures but represent a fine example of the interaction of experiment and theory in unraveling a series of puzzles. The subject has recently been reviewed by Washburn and Webb,[95] and continues to develop. The importance of including the effects of contact leads in the analysis of experiments has been emphasized by Büttiker.[96]

Other considerations, which could perhaps be subsumed under the name quantum transport, also influence the analysis of the properties of small structures and are expected to play an increasing role in the physics of small devices. They are beyond the scope of these lectures. One aspect of this subject has been examined by Mahan.[97]

I am indebted to Michael Artaki, Markus Büttiker, Carlo Jacoboni, Steve Laux, Barry Lin, Ted Masselink, and Sean Washburn for providing figures and references and for helpful comments and suggestions and to Peter Price for a critical reading of the manuscript.

REFERENCES

1. C. Kittel, "Introduction to Solid State Physics," 5th ed., Wiley, New York (1976).
2. J. M. Ziman, "Principles of the Theory of Solids," 2nd ed., Cambridge (1972).
3. A. H. Wilson, "The Theory of Metals", 2nd ed., Cambridge (1953).
4. J. M. Ziman, "Electrons and Phonons," Oxford (1960).
5. K. Seeger, "Semiconductor Physics," Springer, Wien (1973).
6. S. Hikami, A. I. Larkin, and Y. Nagaoka, Spin-orbit interaction and magnetoresistance in the two dimensional random system, Prog. Theor. Phys. 63:707 (1980).
7. H. Fukuyama, Theory of weakly localized regime of the Anderson localization in two dimensions, Surf. Sci. 113:489 (1982).
8. P. A. Lee and T. V. Ramakrishnan, Disordered electronic systems, Rev. Mod. Phys. 57:287 (1985).

9. S. Kawaji, Weak localization and negative magnetoresistance in semiconductor two-dimensional systems, <u>Surf. Sci</u>. 170:682 (1986).

10. F. Stern and W. E. Howard, Properties of semiconductor surface inversion layers in the electric quantum limit, <u>Phys. Rev</u>. 163:816 (1967).

11. G. Fishman, Mobility in a quasi-one-dimensional semiconductor: An analytical approach, <u>Phys. Rev. B</u> 34:2394 (1986).

12. S. Das Sarma and X. C. Xie, Calculated transport properties of ultrasubmicron quasi-one-dimensional inversion lines, <u>Phys. Rev. B</u> (1987).

13. F. A. Riddoch and B. K. Ridley, Phonon scattering of electrons in quasi-one-dimensional and quasi-two-dimensional quantum wells, <u>Surf. Sci</u>. 142:260 (1984).

14. H. Sakaki, Scattering suppression and high-mobility effect of size-quantized electrons in ultrafine semiconductor wire structures, <u>Jpn. J. Appl. Phys</u>. 19:L735 (1980).

15. T. Ando, A. B. Fowler, and F. Stern, Electronic properties of two-dimensional systems, <u>Rev. Mod. Phys</u>. 54:437 (1982).

16. H. L. Störmer, A. C. Gossard, and W. Wiegmann, Observation of intersubband scattering in a 2-dimensional electron system, <u>Solid State Commun</u>. 41:707 (1982).

17. M. J. Kane, N. Apsley, D. A. Anderson, L. L. Taylor, and T. Kerr, Parallel conduction in $GaAs/Al_xGa_{1-x}As$ modulation doped heterojunctions, <u>J. Phys. C</u> 18:5629 (1985).

18. D. A. Syphers, K. P. Martin, and R. J. Higgins, Determination of transport coefficients in high mobility heterostructure systems in the presence of parallel conduction, <u>Appl. Phys. Lett</u>. 49:534 (1986).

19. F. Stern, Friedel phase-shift sum rule for semiconductors, <u>Phys. Rev</u>. 158:697 (1967).

20. T. Ando, Self-consistent results for a $GaAs/Al_xGa_{1-x}As$ heterojunction. II. Low temperature mobility, <u>J. Phys. Soc. Japan</u> 51:3900 (1982).

21. F. Stern, Doping considerations for heterojunctions, <u>Appl. Phys. Lett</u>. 43:974 (1983).

22. F. Stern, Calculated temperature dependence of mobility in silicon inversion layers, <u>Phys. Rev. Lett</u>. 44:1469 (1980).

23. Y. Kawaguchi and S. Kawaji, Lattice scattering mobility of n-inversion layers in Si(100) at low temperatures, <u>Surf. Sci</u>. 98:211 (1980).

24. K. M. Cham and R. G. Wheeler, Temperature-dependent resistivities in silicon inversion layers at low temperatures, <u>Phys. Rev. Lett</u>. 44:1472 (1980).

25. S. I. Dorozhkin and V. T. Dolgopolov, Conductivity increase of a 2D electron gas with decreasing temperature in Si(100) metal-insulator-semiconductor structures, Pis'ma Zh. Eksp. Teor. Fiz. 40:245 (1984) [JETP Lett. 40:1019 (1984)].

26. A. Gold and V. T. Dolgopolov, Temperature dependence of the conductivity for the two-dimensional electron gas: Analytical results for low temperatures, Phys. Rev. B 33:1076 (1986).

27. V. M. Pudalov, S. G. Semenchinskii, and V. S. Edel'man, Oscillations of the chemical potential and the energy spectrum of electrons in the inversion layer at a silicon surface in a magnetic field, Zh. Eksp. Teor. Fiz. 89:1870 (1985) [Sov. Phys. JETP 62:1079 (1985)].

28. R. P. Smith and P. J. Stiles, Temperature dependence of the conductivity of a silicon inversion layer at low temperatures, Solid State Commun. 58:511 (1986).

29. F. Stern and S. Das Sarma, Self-consistent treatment of screening and Coulomb scattering in silicon inversion layers at low temperatures, Solid-State Electron. 28:211 (1985) (extended abstract).

30. A. Gold and W. Götze, Localization and screening anomalies in two-dimensional systems, Phys. Rev. B 33:2495 (1986).

31. A. Gold, Transport and cyclotron resonance theory for GaAs-AlGaAs heterostructures, Z. Phys. B 63:1 (1986).

32. A. Gold, Metal insulator transition due to surface roughness scattering in a quantum well, Solid State Commun. 60:531 (1986).

33. J. L. Robert, A. Raymond, L. Konczewicz, C. Bousquet, W. Zawadski, F. Alexandre, I. M. Masson, J. P. Andre, and P. M. Frijlink, Magnetic-field-induced metal-nonmetal transition in GaAs-$Ga_{1-x}Al_xAs$ heterostructures, Phys. Rev. B 33:5935 (1986).

34. P. O. Hahn and M. Henzler, The $Si-SiO_2$ interface: Correlation of atomic structure and electrical properties, J. Vac. Sci. Technol. A 2:574 (1984).

35. S. M. Goodnick, D. K. Ferry, C. W. Wilmsen, Z. Liliental, D. Fathy, and O. L. Krivanek, Surface roughness at the $Si(100)-SiO_2$ interface, Phys. Rev. B 32:8171 (1985).

36. R. E. Prange and T. W. Nee, Quantum spectroscopy of the low-field oscillations in the surface impedance, Phys. Rev. 168:779 (1968).

37. Y. Matsumoto and Y. Uemura, Scattering mechanism and low temperature mobility of MOS inversion layers, in "Proceedings of the Second International Conference on Solid Surfaces, Kyoto," Jpn. J. Appl. Phys. Suppl. 2, Pt. 2, p. 367 (1974).

38. A. Gold, Conductivity, plasmon, and cyclotron-resonance anomalies in Si(100) metal-oxide-semiconductor systems, Phys. Rev. B 32:4014 (1985).

39. J. A. Brum and G. Bastard, Self-consistent calculations of charge transfer and alloy scattering-limited mobility in $InP-Ga_{1-x}In_xAs_yP_{1-y}$ single quantum wells, Solid State Commun. 53:727 (1985). See also P. K. Basu and B. R. Nag, Estimation of alloy scattering potential in ternaries from the study of two-dimensional electron electron transport, Appl. Phys. Lett. 43:689 (1983).

40. T. S. Kuan, T. F. Kuech, W. I. Wang, and E. L. Wilkie, Long-range order in $Al_xGa_{1-x}As$, Phys. Rev. Lett. 54:201 (1985).

41. P. J. Price, Two-dimensional electron transport in semiconductor layers. I. Phonon scattering, Ann. Phys. (N.Y.) 133:217 (1981).

42. B. K. Ridley, The electron-phonon interaction in quasi-two-dimensional quantum-well structures, J. Phys. C 15:5899 (1982).

43. P. J. Price, Two-dimensional electron transport in semiconductor layers. II. Screening, J. Vac. Sci. Technol. 19:599 (1981).

44. S. Kawaji, The two-dimensional lattice scattering mobility in a semiconductor inversion layer, J. Phys. Soc. Japan 27:906 (1969).

45. P. J. Price, Electron transport in polar heterolayers, Surf. Sci. 113:199 (1982).

46. H. Ezawa, S. Kawaji, and K. Nakamura, Surfons and the electron mobility in silicon inversion layers, Jpn. J. Appl. Phys. 13:126 (1974); 14:921(E) (1975).

47. M. A. Paalanen, D. C. Tsui, A. C. Gossard, and J. C. M. Hwang, Temperature dependence of electron mobility in $GaAs-Al_xGa_{1-x}As$ heterostructures from 1 to 10 K, Phys. Rev. B 29:6003 (1984).

48. E. E. Mendez, P. J. Price, and M. Heiblum, Temperature dependence of the electron mobility in GaAs-GaAlAs heterostructures, Appl. Phys. Lett. 45:294 (1984).

49. B. J. F. Lin, D. C. Tsui, and G. Weimann, Mobility transition in the two-dimensional electron gas in GaAs-AlGaAs heterostructures, Solid State Commun. 56:287 (1985); B. J. F. Lin, Ph. D. Thesis, Electrical Engineering Department, Princeton University (1985).

50. B. Vinter, Low-temperature phonon-limited electron mobility in modulation-doped heterostructures, Phys. Rev. B 33:5904 (1986); Maximum finite-temperature mobility in heterostructures: Influence of screening of electron-acoustic phonon interactions, Surf. Sci. 170:445 (1986).

51. K. Hirakawa and H. Sakaki, Energy relaxation of two-dimensional electrons and the deformation potential constant in selectively doped AlGaAs/GaAs heterojunctions, Appl. Phys. Lett. 49:889 (1986).

52. F. Stern and S. Das Sarma, Self-consistent treatment of screening and Coulomb scattering in silicon inversion layers at low temperatures, Solid-State Electron. 28:211 (1985) (abstract).

53. K. Hirakawa and H. Sakaki, Mobility of the two-dimensional electron gas at selectively doped n-type $Al_xGa_{1-x}As/GaAs$ heterojunctions with controlled electron concentrations, Phys. Rev. B 33:8291 (1986).

54. C. Guillemot, M. Baudet, M. Gauneau, A. Regreny, and J. C. Portal, Temperature dependence of electron mobility in $GaAs-Ga_{1-x}Al_xAs$ modulation-doped quantum wells, Phys. Rev. B. 35:2799 (1987).

55. P. J. Price, Heterolayer mobility in the Bloch-Grüneisen range, Solid State Commun. 51:607 (1984).

56. P. J. Price, Hot electrons in a GaAs heterolayer at low temperature, J. Appl. Phys. 53:6863 (1982).

57. W. Walukiewicz, H. E. Ruda, J. Lagowski, and H. C. Gatos, Electron mobility in modulation-doped heterostructures, Phys. Rev. B. 30:4571 (1984).

58. E. E. Mendez, Electronic mobility in semiconductor heterostructures, IEEE J. Quant. Electron. QE-22:1720 (1986).

59. W. I. Wang, E. E. Mendez, Y. Iye, B. Lee, M. H. Kim, and G. E. Stillman, High mobility two-dimensional hole gas in an $Al_{0.26}Ga_{0.74}As/GaAs$ heterojunction, J. Appl. Phys. 60:1834 (1986).

60. W. Walukiewicz, Hole-scattering mechanisms in modulation-doped heterostructures, J. Appl. Phys. 59:3577 (1986).

61. T. P. McLean and E. G. S. Paige, A theory of the effects of carrier-carrier scattering on mobility in semiconductors, J. Phys. Chem. Solids 16:220 (1960).

62. R. A. Höpfel, J. Shah, P. A. Wolff, and A. C. Gossard, Negative absolute mobility of minority electrons in GaAs quantum wells, Phys. Rev. Lett. 56:2736 (1986).

63. K. Hess, Aspects of high-field transport in semiconductor heterolayers and semiconductor devices, Adv. Electronics Electron Phys. 59:239 (1982).

64. L. Reggiani, ed., "Hot-Electron Transport in Semiconductors," Springer, Berlin (1985).

65. W. T. Masselink, T. S. Henderson, J. Klem, W. F. Kopp, and H. Morkoc, The dependence of 77 K electron velocity-field characteristics on low-field mobility in AlGaAs-GaAs modulation-doped structures, IEEE Trans. Electron Dev. ED-33:639 (1986).

66. P. J. Price, Hot phonon effects in heterolayers, Physica 134B:164 (1985).

67. J. Higman and K. Hess, Comment on the use of the electron temperature concept for nonlinear transport problems in semiconductor p-n junctions, Solid-State Electron. 29:915 (1986).

68. S. Selberherr, "Analysis and Simulation of Semiconductor Devices," Springer, Wien (1984).

69. J. J. H. Miller, ed., "NASECODE IV: Proceedings of the Fourth International Conference on the Numerical Analysis of Semiconductor Devices and Integrated Circuits," Boole Press, Dublin (1985).

70. K. Blotekjaer, Transport equations for electrons in two-valley semiconductors, IEEE Trans. Electron Dev. ED-17:38 (1970).

71. R. K. Cook and J. Frey, Two-dimensional numerical simulation of energy transport effects in Si and GaAs MESFETs, IEEE Trans. Electron Dev. ED-29:970 (1982).

72. M. Rudan and F. Odeh, Multidimensional discretization scheme for the hydrodynamic model of semiconductor devices, COMPEL 5:149 (1986).

73. P. J. Price, Monte Carlo calculation of electron transport in solids, in "Semiconductors and Semimetals," ed. by R. K. Willardson and A. C. Beer, Academic Press, New York (1979), Vol. 14, p. 249.

74. C. Jacoboni and L. Reggiani, The Monte Carlo method for the solution of charge transport in semiconductors with applications to covalent materials, Rev. Mod. Phys. 55:645 (1983).

75. K. Yokoyama and K. Hess, Monte Carlo study of electronic transport in $Al_{1-x}Ga_xAs/GaAs$ single-well heterostuctures, Phys. Rev. B 33:5595 (1986).

76. J. G. Ruch and W. Fawcett, Temperature dependence of the transport properties of gallium arsenide determined by a Monte Carlo method, J. Appl. Phys. 41:3843 (1970).

77. M. V. Fischetti, D. J. DiMaria, L. Dori, J. Batey, E. Tierney, and J. Stasiak, Ballistic electron transport in thin silicon dioxide films, Phys. Rev. B 35:4404 (1987).

78. P. T. Nguyen, D. H. Navon, and T. W. Tang, Boundary conditions in regional Monte Carlo device analysis, IEEE Trans. Electron Dev. ED-32:783 (1985).

79. Y. Aharonov and D. Bohm, Significance of electromagnetic potentials in quantum theory, Phys. Rev. 115:485 (1959).

80. Another scale length, the thermal length $L_T = (hD/k_BT)^{1/2}$, also enters under some conditions. See, for example, A. D. Stone and Y. Imry, Periodicity of the Aharonov-Bohm effect in normal-metal rings, Phys. Rev. Lett. 56:189 (1986).

81. R. A. Webb, S. Washburn, C. P. Umbach, and R. B. Laibowitz, Observation of h/e Aharonov-Bohm oscillations in normal-metal rings, Phys. Rev. Lett. 54:2696 (1985).

82. S. Datta and S. Bandyopadhyay, Aharonov-Bohm effect in semiconductor microstructures, Phys. Rev. Lett. 58:717 (1987).

83. P. A. Lee and A. D. Stone, Universal conductance fluctuations in metals, Phys. Rev. Lett. 55:1622 (1985).

84. C. P. Umbach, S. Washburn, R. B. Laibowitz, and R. A. Webb, Magnetoresistance of small, quasi-one-dimensional, normal-metal rings and lines, Phys. Rev. B 30:4048 (1984).

85. J. C. Licini, D. J. Bishop, M. A. Kastner, and J. Melngailis, Aperiodic magnetoresistance oscillations in narrow inversion layers in Si, Phys. Rev. Lett. 55:2987 (1985).

86. S. B. Kaplan and A. Hartstein, Universal conductance fluctuations in narrow Si accumulation layers, Phys. Rev. Lett. 56:2403 (1986).

87. W. J. Skocpol, P. M. Mankiewich, R. E. Howard, L. D. Jackel, D. M. Tennant, and A. D. Stone, Universal conductance fluctuations in silicon inversion-layer nanostructures, Phys. Rev. Lett. 56:2865 (1986).

88. A. D. Benoit, S. Washburn, C. P. Umbach, R. B. Laibowitz, and R. A. Webb, Asymmetry in the magnetoconductance of metal wires and loops, Phys. Rev. Lett. 57:1765 (1986).

89. A. B. Fowler, A. Hartstein, and R. A. Webb, Conductance in restricted-dimensionality accumulation layers, Phys. Rev. Lett 48:196 (1982).

90. P. A. Lee, Variable-range hopping in finite one-dimensional wires, Phys. Rev. Lett. 53:2042 (1984).

91. J. A. McInnes and P. N. Butcher, Numerical calculations of variable-range hopping in one-dimensional MOSFETs, J. Phys. C 18:L921 (1985).

92. M. Ya. Azbel, A. Hartstein, and D. P. DiVincenzo, T dependence of the conductance in quasi one-dimensional systems, Phys. Rev. Lett. 52:1641 (1984); M. Ya. Azbel and D. P. DiVincenzo, Finite-temperature conductance in one dimension, Phys. Rev. B 30:6877 (1984).

93. A. B. Fowler, G. L. Timp, J. J. Wainer, and R. A. Webb, Observation of resonant tunneling in silicon inversion layers, Phys. Rev. Lett. 57:138 (1986).

94. See, for example, the papers cited in F. Stern, Recent progress in electronic properties of quasi-two-dimensional systems, Surf. Sci. 58:333 (1976), or p. 520 of Ref. 15.

95. S. Washburn and R. A. Webb, Aharonov-Bohm effect in normal metal quantum coherence and transport, Adv. Phys. 35:375 (1986).

96. M. Büttiker, Four-terminal phase-coherent conductance, Phys. Rev. Lett. 57:1761 (1986); Voltage fluctuations in small conductors, Phys. Rev. B 35:4123 (1987).

97. G. D. Mahan, Quantum transport equation for electric and magnetic fields, Phys. Repts. 145:251 (1987).

PHYSICS OF RESONANT TUNNELING IN SEMICONDUCTORS

E. E. Mendez

IBM Thomas J. Watson Research Center
Yorktown Heights, New York 10598, U.S.A.

INTRODUCTION

The concept of tunneling through a potential barrier lies at the core of quantum mechanics, and its experimental observation is a manifestation of the wave-like behavior of matter. Since the early days of quantum mechanics, tunneling models have been used to explain fundamental experiments such as the ionization of hydrogen by an electric field and the emission of alpha particles by heavy nuclei. The idea of tunneling was also incorporated very soon into solid-state physics, and, thus, was used in 1928 by Fowler and Nordheim to describe field emission from metals, and by Zener in 1934 to account for internal field emission in semiconductors.

Closer to us in time, Esaki's diode relies on the concept of tunneling, and so does Giaever's work in metal-insulator-metal junctions. For their work, they shared the 1973 Nobel prize in physics with Josephson, who predicted phase coherence between two superconductors separated by a thin tunnel barrier. Last year, inventions based on tunneling were recognized again, with the award of the Nobel prize for the electron microscope (Ruska) and the scanning tunneling microscope (Binnig and Röhrer).

The subject of tunneling in solids has been thoroughly reviewed by Duke[1]. In his 1969 book he stated: "A new type of quantum size effect can occur in metal-insulator-metal-insulator-metal junctions when the intermediate metal becomes atomically thin... (Similar effects) would be predicted if the intermediate 'metal' were a semiconductor or semimetal. These effects ... have not been observed."

This "new quantum-size effect" -resonant tunneling- was indeed first observed in 1974 by Chang, Esaki, and Tsu[2], in a double-barrier semiconductor heterostructure, and is the main subject of this set of lectures.

I review first those aspects of the physics of tunneling through a single potential barrier that are relevant to resonant tunneling. Then, I use a matrix-transfer formalism to deal with the transmission probability through a potential configuration of arbitrary shape, and apply it to a few simple cases. The concept of resonant tunneling follows naturally from the formalism, and it is then applied to rectangular double-barriers under an electric field. The time involved in the resonant tunneling process is discussed in detail, in the WKB approximation, as well as the accumulation of charge that takes place between the barriers.

The second part of the lectures is devoted to some topics of current interest in resonant tunneling in semiconductors. I review resonant tunneling via Landau levels, when a strong magnetic field is applied parallel to the tunnel current, I consider the existence of confined states in the potential barrier, through which resonant tunneling can proceed, and I examine the question of which barrier controls tunneling when dealing with an indirect-gap semiconductor. As a final point, I contrast resonant tunneling with sequential tunneling, both of which can give rise to negative-resistance effects.

TUNNELING CURRENT

When talking about electronic tunneling in a semiconductor heterojunction the first question that arises is: what is the physical origin of the barrier potential? The electrons in both materials move in a screened potential due to the ion cores, while interacting with each other via the Coulomb force. The Hamiltonian of the system is complicated, and therefore it is frequently simplified by using one-electron states that take into account the periodic component of the electron-ion potential. The charge distribution near the junction is introduced as a slowly-varying potential determined by Poisson's equation.

In what follows we will make further simplifications:

1. The wavefunctions are expanded in terms of a single band on either side of the junction.

2. Schrödinger's equation is separated into two components, parallel and perpendicular to the junction plane.

3. The eigenstates of interest have energies sufficiently near those of critical points in the energy band structure on both sides of the interface.

4. The total energy, E, and the momentum parallel to the interface, $k_{\parallel}$, are conserved (specular tunneling).

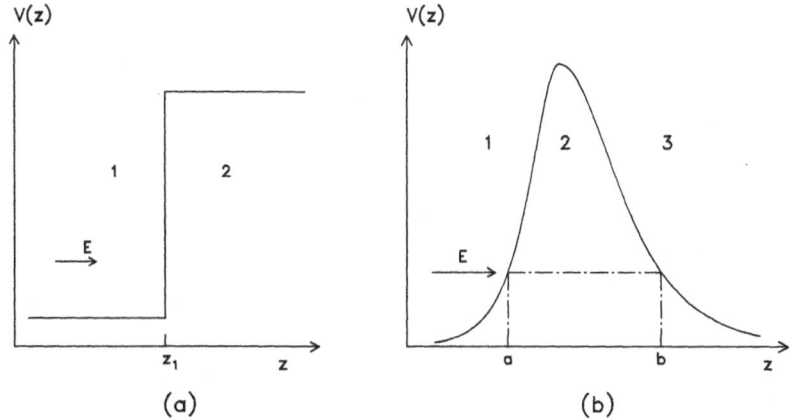

Fig. 1. (a) Electronic potential energy at an interface between two materials. The motion of an electron in material 1 is restricted along the z direction because of a potential barrier at the junction.

(b) Generic potential profile for a heterostructure of three materials, in which material 2 serves as a potential barrier to electronic motion along the z direction. As the thickness of the barrier is finite, there is a non-zero probability that an electron with energy E can tunnel through it.

Since there is no electrostatic potential parallel to the interface, we need to consider only a one-dimensional Schrödinger equation. In this limit (free-electron approximation), the problem is reduced to solving a one-dimensional Schrödinger's equation of the form,

$$\left[-\frac{\hbar^2}{2m} \frac{d^2}{dz^2} + V(z) - E \right] \psi_E(z) = 0 \tag{1}$$

V(z) is the electrostatic potential near the interface, and ψ_E is an envelope function subject, at any interface (see Fig.1a), to the following boundary conditions that guarantee current conservation:

$$\psi_E(z_1^-) = \psi_E(z_1^+) \tag{2}$$

161

$$\frac{1}{m_1} \frac{d\psi_E}{dz} \bigg|_{z_1^-} = \frac{1}{m_2} \frac{d\psi_E}{dz} \bigg|_{z_1^+}$$ [3]

Very frequently, the interest resides in calculating the tunneling current through a potential barrier of finite width and arbitrary shape, like the one shown in Fig.1b. It is easy to demonstrate that the current per unit area, J, can be written as

$$J = \frac{e}{4\pi^3 \hbar} \int dk_z d^2 k_\parallel f(E) T(E_z) \left(\frac{\partial E}{\partial k_z} \right)$$ [4]

where $f(E)$ is the Fermi-Dirac distribution, and $T(E_z)$ is the tunneling probability, defined as the ratio between the incident and transmitted probability currents.

If an external bias V is applied to the barrier, the net current flowing through it is the difference between the current from left to right and that from right to left. Thus,

$$J = \frac{e}{4\pi^3 \hbar} \int dE_z d^2 k_\parallel [f(E) - f(E + eV)] T(E_z)$$ [5]

obtained after taking into account energy conservation and after substracting [7] from [6],

$$J_{LR} = \frac{e}{4\pi^3 \hbar} \int dE_z d^2 k_\parallel f_L(E) [1 - f_R(E)] T(E_z)$$ [6]

$$J_{RL} = \frac{e}{4\pi^3 \hbar} \int dE_z d^2 k_\parallel [1 - f_L(E)] f_R(E) T(E_z)$$ [7]

Further physical insight can be obtained by considering the zero-temperature limit, and realizing that in the free-electron approximation it is

$$d^2 k_\parallel = \frac{2\pi m}{\hbar^2} dE_\parallel$$ [8]

After some simple algebra the tunneling current can be written as,

$$J = \frac{e}{2\pi^2 \hbar^3} \left[eV \int_0^{E_F - eV} dE_z T(E_z) + \int_{E_F - eV}^{E_F} dE_z (E_F - E_z) T(E_z) \right] \quad \text{if} \quad eV \le E_F$$
$$J = \frac{em}{2\pi^2 \hbar^3} \int_0^{E_F} dE_z (E_F - E_z) T(E_z) \quad \text{if} \quad eV \ge E_F$$ [9]

Equation [9] can be readily evaluated as long as the tunneling probability through the barrier is known.

TUNNELING PROBABILITY

The tunneling probability is determined by solving Schrödinger
equation. Unfortunately, analytical solutions can be obtained only
for a few barrier profiles. The simplest one is the rectangular barrier (Fig.2a), that appears in almost every textbook in quantum
mechanics[3], and which can serve as a first approximation to the potential profile of a thin, undoped, semiconductor, clad between thick
layers of a heavily-doped semiconductor of larger energy gap.

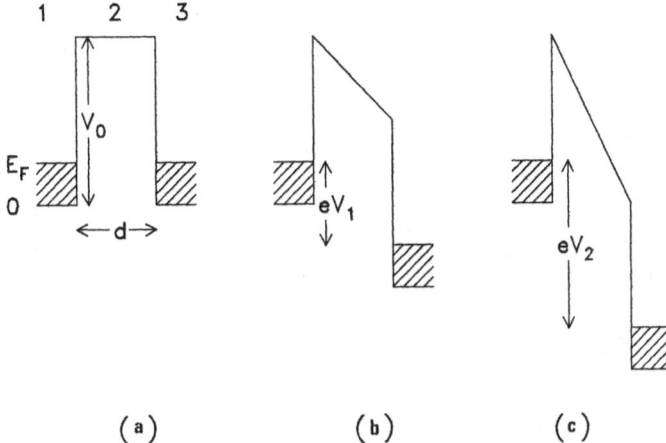

Fig. 2. Rectangular-potential model, (a), used to describe the effect
of an insulator, 2, between two metals, 1 and 3. When a
negative bias is applied to 1, electrons, with energies up
to the Fermi energy E_F, can tunnel through the barrier. For
small voltages, (b), the barrier becomes trapezoidal, but
at high bias (c), it becomes triangular. In this case, we
talk of Fowler-Nordheim tunneling.

In the presence of an external voltage V_1 between the doped regions (acting as electrodes), the barrier becomes trapezoidal
(Fig.2b), and even triangular, if the bias exceeds the barrier height
V_0 (Fig.2c). In this last case the effective width of the barrier
decreases linearly with V, and we normally refer to this situation
as Fowler-Nordheim tunneling. As in the case of the rectangular
barrier, in regions 1 and 3 the solution of [1] is expressed in terms
of plane waves. In region 2, ψ_E can be written as a linear combination
of the Airy functions, Ai(z) and Bi(z). By imposing the boundary
conditions [2] and [3] at the two interfaces, the relative amplitudes
of the incident and transmitted waves are determined, from which a
transmission probability follows[4].

Figure 3a shows the current-voltage (I-V) characteristics for a rectangular barrier 0.2V high and 100Å wide, assuming T=0K, and a Fermi energy for the electrodes of 54 meV. The electron mass in the barrier was taken to be $0.2m_0$ (m_0 is the free-electron mass). At the

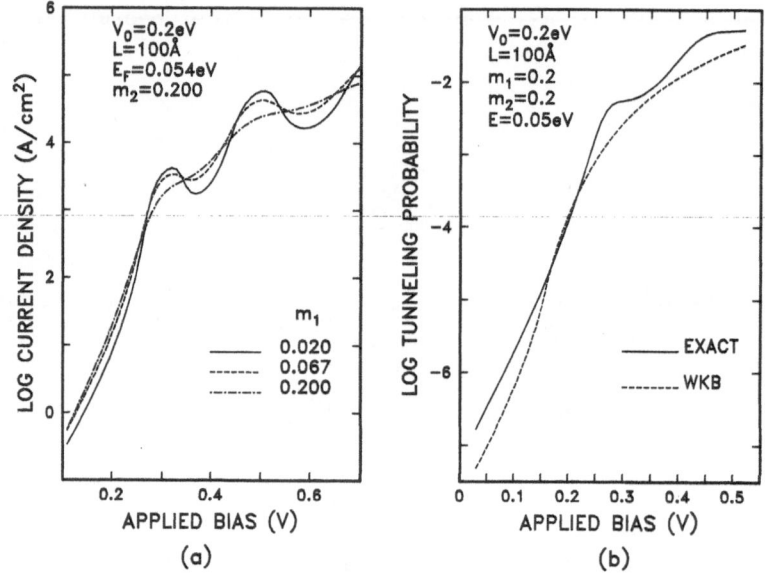

Fig. 3. (a) Tunneling current through a rectangular barrier like the one of Fig.2a, as a function of bias. The thickness of the barrier is 100Å, its width 0.2eV, and the electronic effective mass in the barrier is $0.2m_0$. The electron mass at the metals is varied between $0.02m_0$ and $0.2m_0$. The temperature is taken to be 0K, and the Fermi energy is 0.054eV. For voltages larger than the barrier height the current shows oscillations due to constructive interference of the incident and reflected wavefunctions in the insulator-metal interface (junction 2-3 in Fig.2b).

(b) Comparison of an exact calculation of the tunneling probability through a potential barrier under an external bias, with an approximate result obtained using the WKB method. The barrier parameters are the same as in (a), and the energy of an incident electron, of mass $0.2m_0$, is 0.05eV.

electrodes, various mass values were used, between $0.2m_0$ and $0.02m_0$. In addition to an expected drastic increase of the current with increasing bias, oscillatory structure is observed at ~0.3V and ~0.5V.

These features reflect an enhanced tunneling probability for certain voltages and are a consequence of constructive interference of the incident and reflected waves in region 2. As can be seen in Fig.3a, the oscillations are most pronounced when the difference between the electronic mass at the barrier and at the electrodes is the largest. This interference phenomenon is frequently called resonant Fowler-Nordheim tunneling and has been observed in metal-oxide-semiconductor heterostructures[5] and in GaAs-Ga$_{1-x}$Al$_x$As-GaAs capacitors[6].

THE WKB METHOD

In many cases of interest Eq.[1] cannot be solved analitically. Although, nowadays, computers make it easy to obtain numerical solutions, physical insight is gained frequently by using approximation methods that yield expressions for the tunneling probability. The most widely used is the Wentzel-Kramers-Brillouin (WKB) semiclassical approximation. Its applicability is limited to those situations in which the change of potential energy within a de Broglie wavelength is small compared with the kinetic energy. The method, explained in detail in many quantum-mechanics books[3], can be applied to the penetration through a barrier, provided that the energy of the particle is small compared to the height of the barrier. Although the approximation is not strictly valid for those points in which the kinetic energy is zero (turning points), formulas can be derived connecting a region where E >V with another one where E<V. After some algebra[3], it is possible to arrive to the following expression for the tunneling probability through a potential barrier like the one in Fig.1b:

$$T \sim \exp\left[-2 \int_a^b k(z)dz \right] \qquad [10]$$

where

$$k(z) = \sqrt{\frac{2m}{\hbar^2} [V(z) - E]} \qquad [11]$$

This expression is easily applied to the case of Fowler-Nordheim tunneling (Fig.2c), for which we obtain

$$T(E) \sim \exp\left[-\frac{4}{3} \left(\frac{2m}{\hbar^2} \right)^{1/2} \frac{d}{eV} (V_0 - E)^{3/2} \right] \qquad [12]$$

It is instructive to compare this approximate result with the exact result for a trapezoidal barrier, as done in Fig.3b. We see

that the WKB method gives a reasonable agreement with an exact calculation, except for the oscillatory behavior observable in the latter. As we have discussed above, this results from interference
between the forward and reflected wave inside the barrier. As the
WKB approximation is a semiclassical one, in which the reflections
at the interfaces are absent, those resonant effects are missing.

THE TRANSFER-MATRIX METHOD

As we know, when the potential V(z) is constant in a given region
the general solution of Eq.[1] has the form

$$\psi_e(z) = A \exp[ikz] + B \exp[-ikz] \qquad [13]$$

with

$$\frac{\hbar^2 k^2}{2m} = E - V \qquad [14]$$

When E-V >0, k is real and the wave functions are plane waves. When
E-V <0, k is imaginary and the wave functions are growing and decaying
waves. Thus, the overall wavefunction for a step profile like the
one in Fig.1 has a plane-wave form for $z_1<0$ and is an exponentially-
decaying wave when $z_1>0$. The boundary conditions [2] and [3] determine
the coefficients A and B, that can be described by a 2 x 2 matrix R,
such that

$$\begin{pmatrix} A_1 \\ B_1 \end{pmatrix} = R \begin{pmatrix} A_2 \\ B_2 \end{pmatrix} \qquad [15]$$

where R is

$$R = \frac{1}{2k_1 m_2} \begin{pmatrix} (k_1 m_2 + k_2 m_1) \exp[i(k_2 - k_1)z_1] & (k_1 m_2 - k_2 m_1) \exp[-i(k_2 + k_1)z_1] \\ (k_1 m_2 - k_2 m_1) \exp[i(k_2 + k_1)z_1] & (k_1 m_2 + k_2 m_1) \exp[-i(k_2 - k_1)z_1] \end{pmatrix}$$
$$[16]$$

This method has been used by Kane[7] for the case of more than one
interface, assuming that the mass of the particle is constant. In
general, if the potential profile consists of n regions, characterized by the potential values V_i and the masses m_i (i=1,2,...n), separated by n-1 interfaces at positions z_i (i=1...n-1), then

$$\begin{pmatrix} A_1 \\ B_1 \end{pmatrix} = (R_1 R_2 \ldots R_{n-1}) \begin{pmatrix} A_n \\ B_n \end{pmatrix} \qquad [17]$$

The elements of R_i are

$$R_i \big|_{11} = \left(\frac{1}{2} + \frac{k_{i+1} m_i}{2 k_i m_{i+1}} \right) \exp[i(k_{i+1} - k_i)z_i] \qquad [18a]$$

$$R_i \big|_{12} = \left(\frac{1}{2} - \frac{k_{i+1}m_i}{2k_i m_{i+1}} \right) \exp\left[-i(k_{i+1} + k_i)z_i \right] \qquad [18b]$$

$$R_i \big|_{21} = \left(\frac{1}{2} - \frac{k_{i+1}m_i}{2k_i m_{i+1}} \right) \exp\left[i(k_{i+1} + k_i)z_i \right] \qquad [18c]$$

$$R_i \big|_{22} = \left(\frac{1}{2} + \frac{k_{i+1}m_i}{2k_i m_{i+1}} \right) \exp\left[-i(k_{i+1} - k_i)z_i \right] \qquad [18d]$$

If an electron is incident from the left (region 1) only a transmitted wave will appear in region n, and therefore $B_n = 0$. The transmission probability current is then given by

$$T = \frac{k_1 m_n}{k_n m_1} \frac{|A_n|^2}{|A_1|^2} \qquad [19]$$

This method yields an analytical expression[7] for the tunneling probability through a double-barrier profile like the one shown on Fig.4a (inset). Under certain conditions, a particle incident on the left can appear on the right without attenuation. This situation, called resonant tunneling, corresponds to a constructive interference between the two plane waves coexisting in the region between the barriers (quantum well). From this point of view this phenomenon is similar to resonant Fowler-Nordheim tunneling. As we will see later, there is a basic difference, however, resulting from charge accumulation in the case of resonant tunneling.

The tunneling probability through such a profile for a particle of mass $0.067m_0$ is illustrated on Fig.4a. The height of the barriers was taken to be 0.3 eV, their widths were 50Å and their separation was 60Å. As observed in the figure, for certain energies below the barrier height the particle can tunnel unattenuated. These energies correspond precisely to the eigen-energies of the quantum well; this is understandable, since the solutions of Schrödinger's equation for the isolated well are standing waves. We note in Fig.4a that for incident energies above the barrier, the transmission probability reaches one only for certain values, when interference is constructive. These "quasi-levels" above the barrier are frequently called virtual or resonant levels and resemble the Fowler-Nordheim case discussed before.

When the widths of the two barriers are different, the tunneling probability -although showing maxima for incident energies corresponding to the bound and virtual states- does not reach unity. As pointed out by Ricco and Azbel[8], this fact may be relevant in the case

of double-barrier heterostructures that are symmetric in the absence
of any external field, since under bias the symmetry is destroyed.

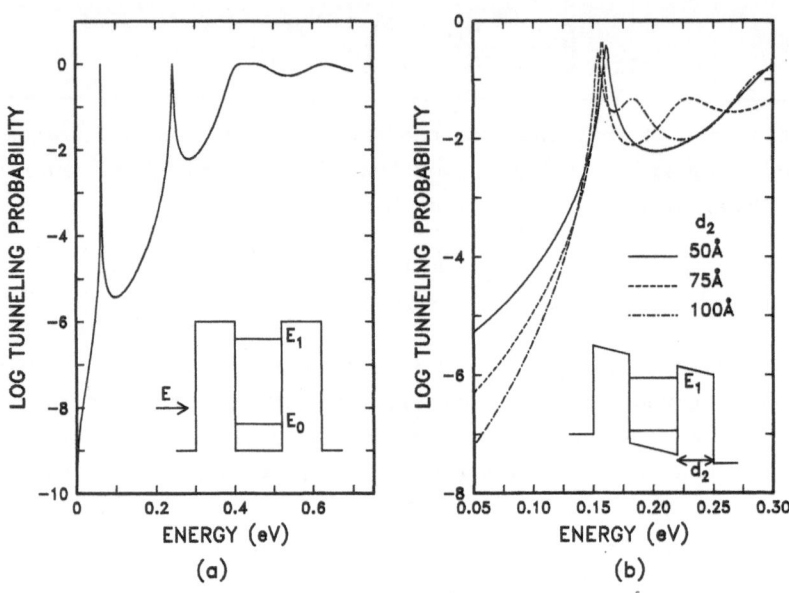

Fig. 4. (a) Probability of tunneling through a double rectangular
barrier as a function of energy. The heights of the barriers
are 0.3eV, their widths 50Å, and their separation is 60Å.
The mass of the particle is taken to be $0.1m_0$ in the barrier
and $0.067m_0$ outside of it. The sharp peaks in the transmission probability, below 0.3eV, correspond to resonant
tunneling through the quasi-bound states in the quantum well
formed between the barriers. Broader transmission maxima,
above 0.3eV, are associated with tunneling via virtual states
in the well.

(b) Tunneling probability through a double-barrier structure,
subject to an electric field of 1×10^5 V/cm. The width of the
left barrier is 50Å, while that of the right barrier is
varied between 50Å and 100Å. The peak at ~ 0.16 eV corresponds to resonant tunneling through the first excited state,
E_1, of the quantum well. The optimum transmission is obtained when the width of the right barrier is ~75Å, because
then the degree of "effective barrier symmetry" for tunneling
is maximum.

Although the transfer-matrix method was developed for rectangular
barriers[7], it can be generalized to profiles of arbitrary shape, by

dividing it into steps of infinitesimal width. This procedure is
illustrated on Fig.5a for the case of a barrier of height V_0 under a
bias V. The exact tunneling probability as a function of incident
energy is shown on Fig.5b, along with approximate results obtained
by dividing the actual profile into either two or five steps. As is
apparent, the method converges quite rapidly to the exact result and,
thus, for five steps it is almost undistinguishable from an exact
treatment.

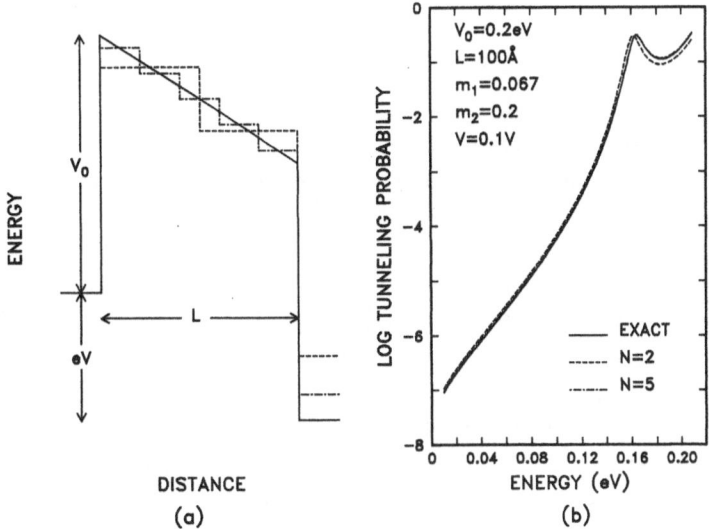

DISTANCE

(a)

ENERGY (eV)

(b)

Fig. 5. (a) Rectangular potential barrier of height V_0 under an applied
         bias V (continuous line) and its succesive approximations
         by either two- or five-step barriers.
         (b) Comparison of the exact tunneling probability versus in-
         cident energy through the barrier of (a), with approximate
         results obtained by the transfer-matrix method (see text),
         for either two- or five-step barriers. The effective mass
         of the particle is $0.2m_0$ in the barrier, and $0.067m_0$ outside.

This method can be used to calculate the tunneling probability
through a double-barrier potential like the one of Fig.4a (inset),
when a finite bias is applied to it. The result, for a selected range
of energies, is shown on Fig.4b, for an electric field of 10^5V/cm.
The tunneling probability peaks at 0.158eV, corresponding to the E_1
resonance. In contrast with the zero-field case (rectangular symmetrical barriers) the probability does not reach one, because of the
asymmetry introduced by the electric field. The thickness of the

169

second barrier can be varied to optimize tunneling, as illustrated on Fig.4b, where we show results for barriers ranging from 50Å to 100Å. Although an optimum value is achieved for an ~75Å width, the maximum is still below unity.

TRANSIT TIME

We have seen that the tunneling probability through a double-barrier structure is maximum whenever the energy of the incident particle coincides with an eigenenergy of the quantum-mechanical system. When the barriers are infinitely wide those eigenenergies correspond to the bound states of the quantum well. For finite-width barriers we can only speak of quasi-bound states, since a particle localized in the well at t=0 can tunnel out of it. The lifetime associated with any of these states is related to the tunneling probability near resonance, and, by the uncertainty principle, to the energy width of the resonance.

We can visualize the resonant tunneling process starting with a wavepacket incident on the left of a barrier structure like the one on Fig.4a. (inset). If the packet's energy distribution is peaked at one of the eigenenergies, the waves are trapped in the well, reflecting back and forth between the barriers in such a phase as to continually reinforce themselves, and leaking out very slowly. Thus, resonant tunneling is a slow process. The time τ it takes for a particle to leak out of the well can be estimated qualitatively from the tunneling probability and the number of times the particle hits the barrier per second. Then,

$$\frac{1}{\tau} = \frac{\sqrt{\frac{2E}{m}}}{2w} T(E)$$ [20]

It can be shown that this expression indeed corresponds to the lifetime of the state, in the WKB approximation[3], with T(E) given by [10]. It follows also from the definition of τ and of the width of the resonance ΔE, that $\tau\Delta E = \hbar$, which, in agreement with uncertainty's principle, means that to make a quasi-bound state, we must use a wave packet of width $\Delta E \sim \hbar/\tau$.

Let us estimate this transit time for a specific case. Consider two rectangular barriers 0.3eV high and 30Å wide, separated from each other by 50Å. For a particle of mass $0.067m_0$, the first resonance occurs when its energy is 0.089eV; the transit time in the WKB approximation is, using [20], 5.6×10^{-13} sec. A better estimate can be

obtained from the width of the transmission probability near resonance, which, when calculated exactly, yields 2.6 x 10^{-13} sec. Because
of the exponential nature of the tunneling probability off resonance,
a small change of the barrier width affects drastically the transit
time. For example, increasing the width from 30Å to 50Å increases
the lifetime in the well by an order of magnitude. Similar increases
take place with either larger barrier heights or heavier tunneling
masses.

Although the WKB method gives a longer time than the exact result,
the order of magnitude of both is the same, confirming the suitability
of that approximation to estimate relevant quantities involved in
tunneling. This conclusion is in sharp contrast to a recent work[9] that
dismisses the WKB approximation as inappropriate for estimation of
tunneling times.

CHARGE ACCUMULATION

The process of resonant tunneling requires a wave-function
build-up in the well, which for charged particles leads to an accumulation of electric charge. The charge density is given by

$$Q = \tau J \qquad [21]$$

where J is the tunneling current density. Although τ may vary by
several orders of magnitude for different barrier parameters (width,
height, and effective mass), the values of Q are quite insensitive
to them, since the variations of τ are compensated by an opposite
change of the tunneling current.

Before giving an estimate of the charge accumulated in the well
let us calculate the current density under some simplifying assumptions. We consider T=0K and voltages such that eV≥E$_F$, so that we can
use [9]. We write the tunneling probability T(E) near resonance as,

$$T(E) = \frac{T_0}{1 + \left(\dfrac{E - (E_0 - eV)}{\Delta E} \right)^2} \qquad [22]$$

where T$_0$ is of the order of unity and V is the voltage applied to the
center of the quantum well relative to the left electrode. E$_0$ is the
energy of the quantum state in the well, which we assume to remain
unchanged after the application of the bias (except for the rigid
shift included in [22]). If we now use the definition of the δ function

$$\delta(x - a) = \lim_{n \to \infty} \frac{n}{\pi} \frac{1}{1 + n^2(x - a)^2} \qquad [23]$$

we immediately get

$$J \sim \frac{em\Delta E}{2\pi\hbar^3} T_0(E_F - E_0 + eV) \qquad 0 \leq E_0 - eV \leq E_F \qquad [24a]$$

$$J = 0 \qquad\qquad\qquad \text{elsewhere} \qquad\qquad [24b]$$

We see that in this simple model the current has a voltage threshold and then increases linearly with bias, reaching a maximum when $E_0 = eV$, that is, when the bias is such that the quantum state coincides with the bottom of the conduction band of the left electrode. The peak value is

$$J_{peak} \sim \frac{em\Delta E}{2\pi\hbar^3} T_0 E_F. \qquad\qquad [25]$$

Consequently, the charge accumulation in the well increases with voltage, and reaches a maximum value of

$$Q_{max} \sim \frac{em}{2\pi\hbar^2} E_F T_0 \qquad\qquad [26]$$

which, as anticipated, is independent of any structural parameters. Since T_0 is approximately one at resonance, the above expression tells us that the maximum charge accumulated in the well is of the order of the two-dimensional (2D) density of states times the Fermi energy. This result is in agreement with the charge stored in a 2D system filled up to an energy E_F.

RESONANT TUNNELING IN SEMICONDUCTOR HETEROSTRUCTURES

Up to now we have been discussing the phenomena of tunneling and resonant tunneling in general terms, without specifying any material structures where those processes can take place. In the remainder, we will focus on semiconductor heterostructures made out of two materials that have a significant discontinuity (e.g. larger than 0.1eV) between their conduction-band edges. Although the examples we will use refer to III-V compounds -in particular to $Ga_{1-x}Al_xAs$ ($0 \leq x \leq 1$)- the concepts are quite general and in principle are equally applicable to other material systems.

Let us consider an undoped $Ga_{1-x}Al_xAs$-GaAs-$Ga_{1-x}Al_xAs$ heterostructure, where the typical width of the individual layers is in the range 30Å-100Å, clad between thick layers of n^+-GaAs. A profile of the potential energy along the direction perpendicular to the material interfaces, sketched on Fig.6, shows that $Ga_{1-x}Al_xAs$ acts as a barrier for the motion of electrons from one reservoir to another (the n⁺ regions).

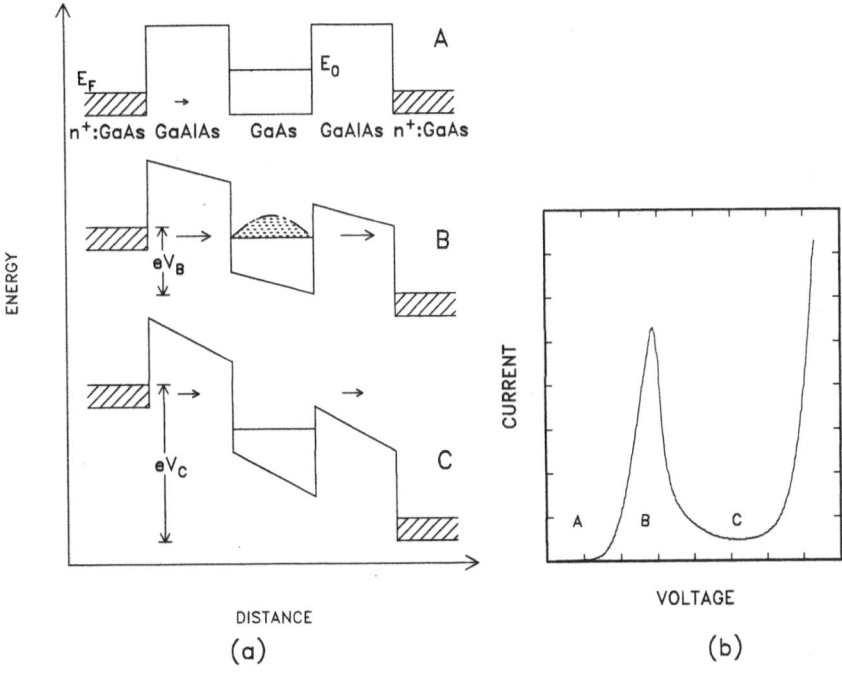

Fig. 6. (a) Rectangular-barrier model for a GaAlAs-GaAs-GaAlAs heterostructure clad between heavily-doped GaAs electrodes. When the bias applied to he structures is such that the energy of the quantum state E_0 is within the Fermi energy of one of the electrodes, resonant tunneling occurs and the tunneling current increases by several orders of magnitude. When E_0 falls below the conduction-band edge of the electrode the current drops to zero because of conservation of momentum parallel to the interfaces.

(b) The current-voltage characteristics shows a strong negative resistance that can be exploited in high-frequency devices.

Under a small, positive bias, V_A, applied to the electrode on the right, the tunneling probability is small, and so is the current between the two electrodes. When the voltage V_B is such that the energy of the quantum state E_0 is in the range of the energies of available electrons on the left electrode, resonant tunneling occurs and the current increases dramatically. As we have seen above, the process continues until E_0 falls below the bottom of the conduction band (region C), at which point resonant tunneling is not possible because momentum parallel to the interface would not be conserved, and the current drops drastically. At very high bias, one of the barriers effectively disappears and the current increases again by Fowler-Nordheim tunneling through the other.

The first observation of resonant tunneling was done by Chang, Esaki, and Tsu, in 1974, on heterostructures grown by molecular beam epitaxy[2]. Their I-V characteristics showed peaks at voltages near the quasibound states of the potential well, although the current did not decrease to values near zero. Since then, the quality of epitaxial layers made by this technique (and more recently also by metalorganic chemical vapor deposition) has improved dramatically and the peak-to-valley ratios of the I-V characteristics have increased by an order of magnitude[10]. It is the intrinsic negative resistance associated with resonant tunneling that has drawn interest for this kind of heterostructures as possible high-frequency devices, some of which are discussed elsewhere in this Course.

The number of negative-resistance features on the I-V characteristics depends only on the width L of the quantum well, as illustrated on Fig.7. The figure shows the characteristics of three heterostructures whose barriers are in all cases 100Å thick, but the width of the well, ranges from 40Å to 60Å. The uppermost curve reflects the existence of only one quasi-bound state in the 40Å well, while the lowest one proves the existence of two localized states for 60Å. From the voltage at which the negative resistance occurs it is in principle possible to find out the energy of the quantum states E_0 and E_1, and, consequently, to get information on effective masses and barrier heights.

If the two barriers are identical, the energy of a quantum state is, to very first approximation, half of the total voltage at which its resonance occurs. In practice, the situation is much more complicated, since part of the applied voltage drops at the electrodes (that have either accumulation or depletion regions) and since charge may be trapped in the barriers, that distorts the potential profile. Moreover, the charge accumulation in the well that accompanies resonant tunneling produces an additional, and significant, deviation of the ideal linear profile. Only when all these factors are taken into account it is possible to calculate realistic current-voltage characteristics that agree quantitatively with experimental results[11].

Although so far, most of the work on resonant tunneling has concentrated on electrons, holes have also been found to tunnel resonantly in $Ga_{1-x}Al_xAs-GaAs-Ga_{1-x}Al_xAs$, sandwiched between

p⁺-electrodes[12]. The reported I-V characteristics show at moderate
temperature (T≤250K) a set of negative-resistance features associated
with resonant tunneling of light holes. The resonant contribution of
the heavy holes is much smaller, because of the narrower resonance
in T(E) (see Eq.[25]), so that their presence can only be detected at
low temperature, once the thermionic current disappeared.

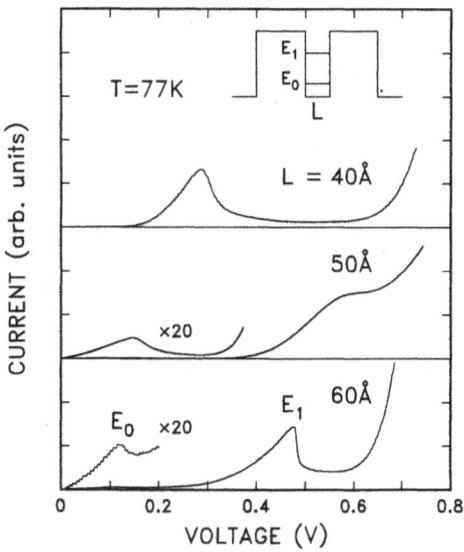

Fig. 7. Experimental current-voltage characteristics for three
    double-barrier $Ga_{0.60}Al_{0.40}As$-GaAs-$Ga_{0.60}Al_{0.40}As$
    heterostructures that differ only on the width, L, of the
    quantum well, ranging from 40Å to 60Å. The width of the
    barriers was in all cases 100Å. For a 40Å well only the
    ground state is bound, while for 60Å two levels are confined
    in the well, as illustrated by the presence of either one
    of two negative-resistance features in the I-V character-
    istics. The characteristic for the 50Å well is intermediate
    between the other two, with the excited state marginally
    bound under an external bias.

    In addition to resonances associated to quasi-bound states, the
I-V characteristics of heterostructures like the ones of Fig.7 might
have, at high bias, features indicative of resonant tunneling through
virtual states above the barriers. As we will discuss later, high-
energy states have been observed but its origin has been attributed
to a different potential profile[13]. Only very recently, unambiguous

observation of virtual states has been claimed, in $In_{0.53}Ga_{0.47}As-InP$ double-barrier structures[14].

RESONANT TUNNELING VIA LANDAU LEVELS

As discussed above, resonant tunneling occurs in double-barrier structures whenever the energy of a quasi-bound state of the quantum well is between the Fermi level and the conduction-band edge of the electrode that supplies the electrons. The width of the resulting triangular I-V characteristic is, therefore, of the order of the Fermi energy. This is a consequence of the constant density of states in the well, parallel to the interfaces. The presence of a strong magnetic field parallel to the tunneling direction has a deep effect on that density of states and consequently on the resonant tunneling process.

The magnetic field quantizes the electronic motion perpendicular to the tunneling direction, in both the electrodes and the quantum well. Their energy spectra are then given by

$$E_N = (N + \frac{1}{2}) \hbar\omega_c \qquad [27]$$

where N (N=0,1,2,...) is called the Landau-level index, and $\hbar\omega_c$ is the cyclotron energy ($\omega_c = eB/m$), which in principle can be different for the various layers. The situation is illustrated on Fig.8, where we sketch the dispersion relation of the left electrode and, for representative voltages, the one-dimensional density of states in the well.

The conservation of momentum parallel to the interface requires, in the presence of a magnetic field, the conservation of level index N. In other words, electrons that in the electrode are in the i-th level, can tunnel only through the i-th level in the well. The threshold voltage for resonant tunneling becomes now

$$eV_{th} < E_0 - E_F + \frac{\hbar\omega_c^{2D}}{2} \qquad [28]$$

where ω_c^{2D} is the cyclotron frequency in the well.

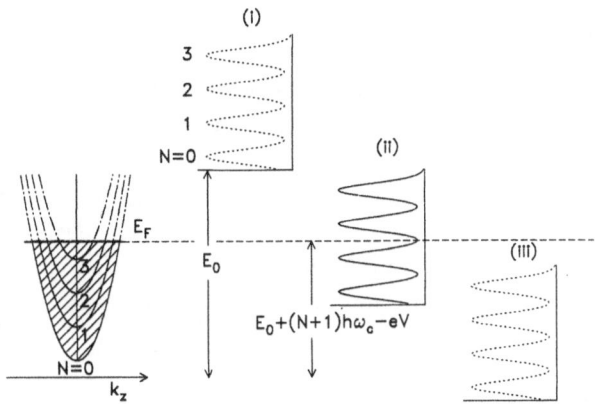

Fig. 8. Dispersion of the electrode subbands in the presence of a
magnetic field in the z direction, and density of states of
the quantum well for representative voltages, for a
heterostructure like those of Fig.7. (i) corresponds to V=0;
(ii) to the case when resonant tunneling via Landau levels
occurs, and (iii) to larger voltages, when momentum conservation prohibits tunneling. (After Ref. 15.)

Beyond the threshold voltage the current increases fast ini-
tially, but, since the one-dimensional density of states is sharply
peaked, a higher voltage does not increase further the current, which
remains constant until the state peaked at N=1 coincides with the
electrode Fermi energy. Then, more electrons can tunnel via this new
state and the current increases again. The process continues until
ultimately the Landau levels in the well pass through the band-edge
of the electrode and tunneling cannot be sustained. When the effective electron mass at the electrode is the same as in the well, the
shift of the N=0 level with magnetic field is the same on both regions, and the voltage for negative resistance is independent of
field. However, when the masses are different, this voltage shifts
by an amount $\hbar(\omega_c^{2D} - \omega_c^{3D})/2$, where ω_c^{3D} is the cyclotron frequency in the
electrode.

Ideally, the tunneling current in a strong magnetic field should
be characterized by a series of steep rises followed by plateaus, and
a final sharp drop to zero. In practice, these features are expected
to be smoothed by Landau-level broadening that results from scattering, thickness fluctuations, etc. In Fig.9 we show experimental

values of the conductance for a GaAlAs-GaAs-GaAlAs heterostructure at various magnetic fields[15]. The zero-field curve reflects the quasi-triangular shape of the I-V characteristic. When the field increases beyond ~3T, oscillations are apparent at voltages below the negative-conductance region. At higher fields, their amplitudes increase and they shift to higher bias until they merge with the principal minimum, whose position remains unchanged. Finally, at fields above 20T all the oscillations have disappeared and the only significant effect is a change of the lineshape relative to the zero-field conductance.

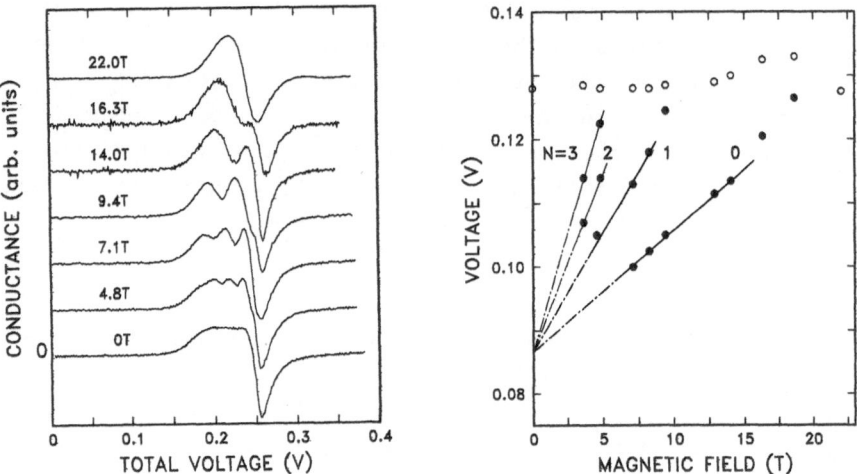

Fig. 9.   Conductance vs total voltage applied for a 100Å-40Å-100Å Ga$_{0.60}$Al$_{0.40}$As-GaAs-Ga$_{0.60}$Al$_{0.40}$As heterostructure, for various magnetic fields. The voltage drop between an electrode and the center of the quantum well is half of the total voltage. The traces shown were taken at 0.55K. (After Ref. 15.)

Fig. 10. Position of the negative conductance minimum (open circles) in Fig.9, and of the field-induced minima (closed circles) as a function of magnetic field. The quantum index N of each series is included. The voltages plotted correspond to half of the total voltage between the two electrodes. The solid lines correspond to a least squares fit to Eq.[29] in the text. (After Ref. 15.)

For a given magnetic field, the onset of each oscillation corresponds to the alignment of a Landau level in the well with the electrode Fermi energy. The conductance dips represent regions between Landau levels, where the one-dimensional density of states is minimal. Thus, we see on Fig.9 that at 14T the conductance almost reaches zero at ~0.2V, as qualitatively predicted above. However, at lower fields the dips are smaller because of level broadening. The fact that the position of the negative-conductance minimum does not change significantly up to the highest field indicates that the 2D and 3D effective masses are the same. Indeed, we can determine directly the 2D mass from an analysis of the voltage of the oscillation, as in Fig.10. The minima, represented by circles, lie on straight lines, each of them corresponding to a given magnetic level. From a fit of the data to the expression

$$eV_{min} = E_0 - E_F + (N + 1) \hbar\omega_c^{2D}, \qquad N = 0,1,2, \ldots \qquad [29]$$

we get directly the effective mass and the Fermi energy at the electrode, since E_0 is known from the voltage of negative-conductance. For the example of Fig.9 a value m=(0.063±0.002)m_0 was obtained, in good agreement with the effective mass of bulk GaAs, 0.067m_0. The small discrepancy may be due, as pointed by Goldman et al.[16], to the fact that in the analysis of Fig.10 the voltage drop in the electrodes was ignored. Although we have discussed only the conductance oscillations as a function of bias for a fixed magnetic field, similar effects are obtained keeping the voltage fixed and sweeping the field.

Magnetotunneling experiments have also been done for holes; however, no conductance oscillations were observed, even up to 22T. The only noticeable effects were shifts of the negative-conductance structures, suggesting drastic differences in the 2D and 3D effective masses. Since both light and heavy holes contribute to resonant tunneling at low temperature, a change of their character during the tunneling process may be responsible for the shifts observed.

TUNNELING THROUGH INDIRECT-GAP BARRIERS

Until now we have assumed that the total wavefunctions can be expanded in terms of a single band on either side of a heterojunction, and that the eigenstates of interest have energies close to the same critical point in the energy-band structure on both sides of the interface. In particular, for the examples discussed above we have been referring to states near the Γ point, in both GaAs and $Ga_{1-x}Al_xAs$. While this assumption may be correct for small x, when other critical points in the conduction band have much larger energy than Γ, for x ≤0.40 it may fail since then the energy of the X point E_X is comparable in energy to that of Γ, E_Γ. For x<0.45 $E_X < E_\Gamma$ and $Ga_{1-x}Al_xAs$ becomes an indirect-bandgap semiconductor.

The question that arises then, is the following: A Γ electron that encounters a potential barrier whose minimum height is determined by the X point (above which is Γ), what barrier does it "feel", the X or the Γ barrier? The situation is sketched on Fig.11 for a double-barrier heterostructure but the question is equally meaningful for a single barrier. This problem has been considered experimentally by doing tunneling measurements on $Ga_{0.60}Al_{0.40}As$-GaAs-$Ga_{0.60}Al_{0.40}As$ structures subject to hydrostatic pressure[17]. At atmospheric, or low, pressure the X point of $Ga_{1-x}Al_xAs$ is higher in energy than the corresponding Γ point. However, at a critical pressure p_c, that depends on x, a X-Γ crossover occurs and at even higher pressure $Ga_{1-x}Al_xAs$ becomes an indirect-gap material. For $Ga_{0.60}Al_{0.40}As$ p_c is ~ 5kbar.

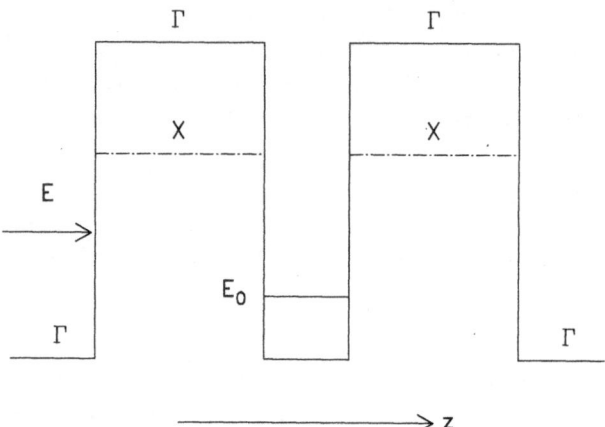

Fig. 11. Potential profile for a $Ga_{1-x}Al_xAs$-GaAs-$Ga_{1-x}Al_xAs$
heterostructure for x such that $Ga_{1-x}Al_xAs$ is an indirect-gap
material. The direct-indirect transition takes place at
atmospheric pressure for x~ 0.45. For a given x (x≤0.45),
the transition can be induced by applying hydrostatic
pressure to the heterostructure. The effect of 1kbar of
pressure on the relative motion of the Γ and X points is
roughly equivalent to an increase of x by 0.01.

Figure 12 shows the logarithm of the tunneling current for a given voltage, as a function of external pressure. At low pressure the current increases slowly with p, but above ~4kbar the slope is much larger. Since the pressure coefficient of the Γ barrier is ~1meV/kbar and that of the X barrier is ~12meV/kbar, the data of Fig.12 show that above the critical pressure the X barrier controls tunneling. For the structures of Fig.12, with layer thicknesses of 100Å-60Å-100Å, at 0.55V there is effectively only one barrier for

180

electronic motion, and tunneling through it takes place in the
Fowler-Nordheim regime.

Using Eq.[12] we easily obtain

$$\frac{d(\ln I)}{dp} = -0.685 \frac{m^{1/2}V_B^{3/2}d}{V} \left[\frac{1}{2} d \frac{(\ln m)}{dp} + \frac{3}{2} \frac{d(\ln V_B)}{dp} \right] \qquad |30|$$

where the mass is in units of the free-electron mass, the effective
height of the barrier V_B is in eV, its width d is in angstroms, and
the voltage drop accross it is in volts. From Eq.[30] and the data
of Fig.12 we obtain the effective mass for tunneling, for pressures

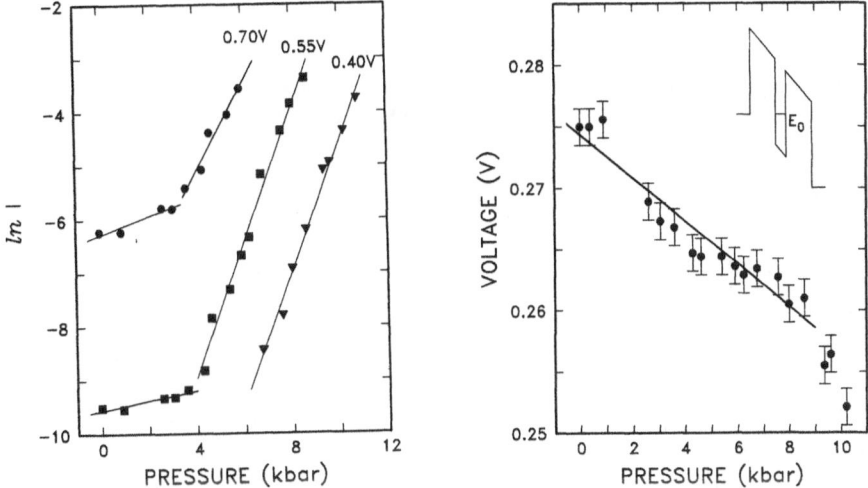

Fig. 12. Tunneling current vs hydrostatic pressure, at constant bias,
for the same heterostructure as in Fig.9. For biases above
0.5V the tunnel current is non-resonant and shows an abrupt
change with pressure at ~4kbar. This change is due to the
direct-indirect transition of the GaAlAs barrier. For
V≤0.4V, the current is dominated at low pressures by resonant tunneling, and its value does not change appreciably
with pressure. (After Ref. 17.)

Fig. 13. Voltage of minimum conductance vs pressure, for the sample
of Fig. 12. The energy of the quantum state in the GaAs well
is approximately half this voltage. A least-squares fit
to the data (·) up to 8.5 kbar gives a slope of -
1.76 mV kbar⁻¹. A larger decrease above 9 kbar results from
the dominant presence of non-resonant current. The inset
sketches the potential profile that leads to negative resistance. (After Ref. 17.)

above 4kbar. The deduced value, m ~0.18, is significantly larger than the effective mass at the Γ point, but is in reasonable agreement with the mass value at X. This confirms once more that tunneling proceeds through the lowest potential barrier.

Those pressure experiments have also provided information about another important question: What potential profile determines the energy of the quantum state E_0? Below the critical pressure a profile derived fully from the Γ point seems appropriate, but the situation may be entirely different at higher pressures, when a considerable amount of band mixing may occur.

The pressure dependence of the voltage at which resonant tunneling through E_0 is maximum is plotted on Fig.13, for the same heterostructure described above. This voltage decreases monotonically with pressure, but, in contrast to the results of Fig.12, no change in slope is observed above 4kbar, implying that the quantum-state energies are determined by a profile that preserves symmetry, in this case that of Γ.

RESONANT TUNNELING VIA X-POINT STATES

In the previous section we have considered the effect in the tunneling probability of the $Ga_{1-x}Al_xAs$ X point, once it becomes the bottom of the conduction band. However, nothing was said of the position of the X point in GaAs, except that is significantly higher than Γ. A sketch of the X-point energy as a function of distance, shown in the inset of Fig.14 for the extreme case of AlAs-GaAs-AlAs, reveals that in such a profile the roles of well and barrier are reversed relative to a Γ-profile. At X the electronic states are localized in AlAs wells, and confined by GaAs barriers, as confirmed by luminescence experiments[18].

Recently, it has been demonstrated that electrons that have a Γ character in the GaAs electrode, can tunnel resonantly via X states[13,19]. The I-V characteristics seen on Fig.14 correspond to a AlAs (50Å)-GaAs (20Å)-AlAs (50Å) heterostructure, in which the ground-state of the system has an X character. The tunneling current shows a weak feature at ~0.16V and then a plateau at ~0.34V, corresponding to resonant tunneling via E_X^x and E_X^y, which are degenerate in the absence of an external bias. At very high bias, submerged in a strong background of non-resonant tunneling current, there is a faint structure (see Fig.14, inset (b)) associated with resonant tunneling via the only quasi-bound Γ state.

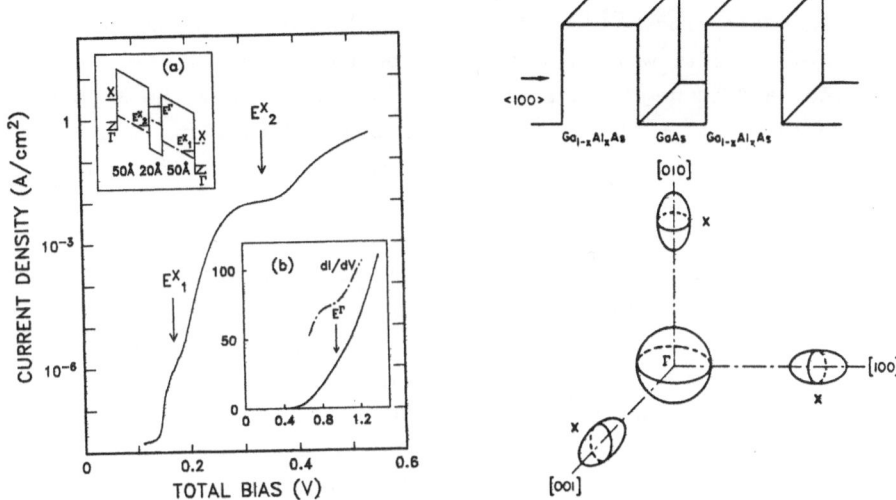

Fig. 14. Tunneling current for a 50Å-20Å-50Å AlAs-GaAs-AlAs structure, as a function of voltage. The X- and Γ-profiles are
sketched in insert (a), with their lowest-energy quantum
states. The I-V characteristic shows features associated
with the first two X-profile states. At high bias, the
current (see insert (b)) exhibits a weak feature related
to resonant tunneling via the first Γ-profile state. (After
Ref. 19.)

Fig. 15. Potential profile (top) of a heterostructure grown along the
<100> direction, and the constant-energy surfaces (bottom)
relevant to tunneling through the X-point barrier. The
Fermi sphere of the GaAs electrode is at the Γ point. The
three ellipsoids represent constant-energy surfaces at the
equivalent X points of $Ga_{1-x}Al_xAs$. Electrons from the
electrode can tunnel through $Ga_{1-x}Al_xAs$ via states near the
X point, conserving momentum parallel to the interfaces.
Specular tunneling occurs only via the ellipsoid along the
[100] direction; the effective mass for tunneling is then
of the order of m_0. Inelastic tunneling takes place through
the other ellipsoids, with a lighter mass, ~$0.2m_0$.

Since X is lower in energy for AlAs than for GaAs, it follows
that a single AlAs film between GaAs electrodes could provide electronic confinement. The I-V characteristics of such heterolayers have
indeed shown features that can be interpreted as resonant tunneling.

The results are not conclusive, however. As the electrons collected in the GaAs electrode have a Γ character, one could think in terms of a Γ→X→Γ profile, which might give rise to resonant Fowler-Nordheim tunneling, and to negative resistance. Although only the X-point state would exist in the absence of an external electric field, the measurement of the current at a given bias cannot distinguish one effect from the other.

One of the basic assumptions so far has been that both the total energy and the momentum parallel to the interfaces are conserved in a tunneling process. Since Γ is at the center of the Brillouin zone and X is at the edge, along the (001) direction, how valid is that assumption in the case of X-point tunneling? The answer is that for tunneling along the (001) axis -which is the case for all the examples discussed here- parallel momentum can be conserved even when going from Γ to X.

Figure 15 can help to clarify this point. The Fermi sphere of the GaAs electrode is centered around the Γ point, on the lower part of the figure. In GaAlAs there are three ellipsoids, each centered around one of the three equivalent X points. For two of them $k_\parallel = k_X$, which is much larger than the Fermi wavevector, k_F, and therefore parallel momentum cannot be conserved. On the other hand, for the third ellipsoid, with its long axis along the tunneling direction, $k_\parallel$ can be smaller than k_F, conserving momentum. It follows also that the effective mass for tunneling via this ellipsoid is heavy, of the order of the free-electron mass for $Ga_{1-x}Al_xAs$.

If the momentum conservation law is relaxed because of electronic scattering (due to impurities, phonons, surface roughness, etc.) then tunneling can proceed via the other ellipsoids, through which the tunneling probability is much larger because their effective mass along the tunneling direction is lighter, ~ $0.2m_0$. The results mentioned above on the effect of pressure on the I-V characteristics with indirect-gap barriers, yielding a tunneling mass of $0.18m_0$, indicate that indeed scattering processes are dominant. A similar conclusion can be drawn from the experiments of Hase et al.[20] and Solomon et al.[21].

RESONANT TUNNELING VERSUS SEQUENTIAL TUNNELING

The focus of these lectures has been resonant tunneling on double-barrier semiconductor heterostructures, a phenomenon -as we have seen- that assumes phase coherence and leads to negative-resistance effects that are measured experimentally. In real systems scattering is unavoidable and therefore the question arises, up to what point is it valid to talk of resonant tunneling?

As shown by Stone and Lee[22], the effect of inelastic scattering on tunneling resonances is to decrease the peak transmission. The extreme case occurs when phase coherence is completely lost. Then we should talk about sequential tunneling through the two barriers, rather than resonant tunneling. Luryi has lucidly discussed this point[23], demonstrating that for negative resistance to exist only a quantum well is needed, and, therefore, the second barrier can be infinitely wide[24].

The material parameter that determines which of the two mechanisms dominate is the barrier width. We have seen that resonant tunneling implies charge accumulation in the well and a long traversal time τ. If this time is much shorter than the scattering time in the well, resonant tunneling may be dominant. In the opposite case, tunneling is sequential. Since τ increases exponentially with the barrier width (see eqs. [12] and [20]) only for thin barriers we can properly talk of resonant tunneling. However, the concepts illustrated in these lectures are equally applicable to both mechanisms. The only basic difference between them is that the quantity T_0, introduced in Eq.[22] to represent the tunneling probability near "resonance", will be of the order of unity in one limit (resonant tunneling) and of the order of the tunneling probability through a single barrier[25] in the other (sequential tunneling).

Information about T_0 can be obtained from a comparison of the measured current densities and those predicted theoretically. For example, for the heterostructures described in Ref.17, with 100Å barriers and a 40Å well, the current density at resonance is $\sim 10^{-3}$ A/cm². Using a WKB expression (trapezoidal barrier) for the transmission probability that enters in Eq.[20] the lifetime τ is estimated to be $\sim 4 \times 10^{-9}$s, and, therefore, the width of the resonance is $\sim 2 \times 10^{-7}$ eV. The ideal current density for specular resonant tunneling would then be (see Eq.[21]) $J_{Res} \sim 30$A/cm² , which is much larger than the experimental value, indicating that, indeed, scattering is very important. This is understandable, since the electron spends about 4 ns in the well before leaking out of it. On the other hand, the scattering time of undoped GaAs at 77K is at most 10ps, and, thus, an electron suffers many collisions during the tunneling process. For a purely sequential tunneling event T_0 would be approximately the tunneling probability through a 100Å barrier, $\sim 3 \times 10^{-6}$. This value is an order of magnitude smaller than the ratio of experimental to theoretical current densities, and therefore suggests that tunneling is not fully sequential and that a small degree of coherence is preserved.

CONCLUSION

The main objective of these lectures has been to give a basic understanding of the physical concepts involved in resonant tunneling in semiconductors. Thus, the path has proceeded from the simplest idea, of tunneling through a potential barrier, to complicated profiles that cannot be described in terms of a potential configuration derived from a single critical point of the Brillouin zone. By going from the simple to the complex, we have moved from areas that are reasonably well understood to regions that are still under active research. In the spirit of a tutorial lecture, the emphasis has been more in choosing selective examples that illustrate a physical concept, rather than in reviewing the specialized literature. However, a certain bias in selecting those topics will not be denied.

A second goal has been to provide numerical methods to evaluate tunneling probabilities and current densities, and to get simple analytical expressions to estimate quantities that can be measured in the laboratory. By comparing different methods and approximations I have tried to give a feeling for their usefulness and range of validity. Most of these tools are not original, and can be found scattered in the literature in one way or another. But, if as Wigner stated, "each generation of physicists has to rediscover Quantum Mechanics by themselves", I will be happy if I have saved somebody part of the time I spent in "discovering" it myself.

REFERENCES

1. C. B. Duke, "Tunneling in Solids," Academic Press, New York (1969).

2. L. L. Chang, L. Esaki, and R. Tsu, Resonant tunneling in semiconductor double barriers, Appl. Phys. Lett., 24:593 (1974).

3. See, e.g., D. Bohm, "Quantum Theory," Prentice-Hall, New York (1951).

4. K. H. Gundlach, Zur Berechnung des Tunnelstroms durch eine Trapezförmige Potentialstufe, Solid State Electron. 9:949 (1966).

5. J. Maserjian, Tunneling in thin MOS structures, J. Vac. Sci. Technol., 11:996 (1974).

6. T. W. Hickmott, P. Solomon, R. Fischer, and H. Morkoc, Resonant Fowler-Nordheim tunneling in n⁻GaAs-undoped $Al_xGa_{1-x}As$ -n⁺GaAs capacitors, Appl. Phys. Lett., 44:90 (1984).

7. E. O. Kane, Basic concepts of tunneling, in: "Tunneling phenomena in solids," E. Burstein and S. Lundqvist, ed., Plenum, New York (1969).

8. B. Ricco, and M. Ya. Azbel, Physics of resonant tunneling. The one-dimensional double-barrier case, Phys. Rev. B, 29:1970 (1984).

9. T. C. L. G. Sollner, E. R. Brown, W. D. Goodhue, and H. Q. Le, Observation of millimeter-wave oscillations from resonant tunneling diodes and some theoretical considerations of ultimate frequency limits, Appl. Phys. Lett., 50:332 (1987).

10. T. C. L. G. Sollner, W. D. Goodhue, P. E. Tannenwald, C. D. Parker, and D. D. Peck, Resonant tunneling through quantum wells at frequencies up to 2.5 THz, Appl. Phys. Lett., 43:588 (1983).

11. H. Ohnishi, T. Inata, S. Muto, N. Yokoyama, and A. Shibatomi, Self-consistent analysis of resonant tunneling current, Appl. Phys. Lett., 49:1248 (1986).

12. E. E. Mendez, W. I. Wang, B. Ricco, and L. Esaki, Resonant tunneling of holes in AlAs-GaAs-AlAs heterostructures, Appl. Phys. Lett., 47:415 (1985).

13. E. E. Mendez, E. Calleja, C. E. T. Goncalves da Silva, L. L. Chang, and W. I. Wang, Observation by resonant tunneling of high-energy states in GaAs-$Ga_{1-x}Al_xAs$ quantum wells, Phys. Rev. B, 33:7368 (1986).

14. T. H. H. Vuong, D. C. Tsui, and W. T. Tsang, Tunneling in $In_{0.53}Ga_{0.47}As$-InP double-barrier structures, Appl. Phys. Lett. 50:212 (1987).

15. E. E. Mendez, L. Esaki, and W. I. Wang, Resonant magnetotunneling in GaAlAs-GaAs-GaAlAs heterostructures, Phys. Rev. B, 33:2893 (1986).

16. V. J. Goldman, D. C. Tsui, and J. E. Cunningham, Resonant tunneling in magnetic field: evidence for space-charge build-up, Phys. Rev. B, 35:xxxx (1987).

17. E. E. Mendez, E. Calleja, and W. I. Wang, Tunneling through indirect-gap semiconductor barriers, Phys. Rev. B, 34:6026 (1986).

18. E. Finkman, M. D. Sturge, and M. C. Tamargo, X-point excitons in AlAs/GaAs superlattices, Appl. Phys. Lett., 49:1299 (1986).

19. E. E. Mendez, W. I. Wang, E. Calleja, and C. E. T. Goncalves da Silva, Resonant tunneling via X-point states in AlAs-GaAs-AlAs heterostructures, Appl. Phys. Lett., 50:1263 (1987).

20. I. Hase, H. Kawai, K. Kaneko, and N. Watanabe, Current-voltage characteristics through GaAs/AlGaAs/GaAs heterobarriers grown by metalorganic chemical vapor deposition, J. Appl. Phys., 59:3792 (1986).

21. P. M. Solomon, S. L. Wright, and C. Lanza, Perpendicular transport across (Al,Ga)As and the Γ to X transition, Superlattices and Microstructures, 2:521 (1986).

22. A. D. Stone, and P. A. Lee, Effect of inelastic processes on resonant tunneling in one dimension, Phys. Rev. Lett., 54:1196 (1985).

23. S. Luryi, Frequency limit of double-barrier resonant-tunneling oscillators, Appl. Phys. Lett., 47:490 (1985).

24. H. Morkoc, J. Chen, U. K. Reddy, T. Henderson, and S. Luryi, Observation of a negative differential resistance due to tunneling through a single barrier into a quantum well, Appl. Phys. Lett., 49:70 (1986).

25. M. Büttiker, Role of quantum coherence in series resistors, Phys. Rev. B, 33:3020 (1986).

THERMODYNAMIC AND MAGNETO-OPTIC INVESTIGATIONS OF THE

LANDAU LEVEL DENSITY OF STATES FOR 2D ELECTRONS

E. Gornik

Institut für Experimentalphysik
University of Innsbruck
A-6020 Innsbruck, Austria

1. INTRODUCTION

In a 2-dimensional electron system (2DES) a magnetic field perpendicular to the plane of electrical confinement leads to full quantization of the electron motion. The energy spectrum consists of sharp Landau levels (LL) separated by the cyclotron energy $\hbar\omega_c$.

In a real system the LL are broadened due to scattering by impurities, phonons or other scattering mechanisms. In the simplest approximation the levels are described by a level width Γ. In the case of high magnetic fields, where $\hbar\omega_c \gg \Gamma$, real gaps appear between the LL. This leads to an oscillatory structure of practically all physical quantities as a function of the magnetic field.

The most fundamental quantity underlying all these physical properties of the system is the form of the density of states $D(E)$. Experimental investigations on the $D(E)$ of 2D electrons in GaAs have first been performed with thermodynamic techniques: From specific heat measurements[1] a Gaussian density of state on a flat background was determined. From the analysis of the magnetization[2] an increasing Gaussian level width with magnetic field was found. Temperature dependent resistivity[3] measurements reveal also a magnetic field dependent Gaussian density with a flat density of states between LL. Similar results were obtained from capacitance experiments[4].

A complete theoretical description of D(E) has not been performed until now. Diagrammatic[5,6] as well as path integral[7,8,9] techniques have been used to calculate D(E). In the self-consistent Born approximation[6,10] a semi-elliptic form of D(E) without background is obtained. Using a path integral technique within lowest order cumulant expansion Gerhardts[8] obtained a pure Gaussian D(E) for long range potentials. Exact results have been obtained by Wegner[8] and Brezin et al.[9] for some restricted types of short range scattering distributions for the lowest LL. For a white noise potential a Gaussian D(E) is obtained. A Poisson distribution of scatterers (with nonzero higher order correlations) yields a peak density of states at the center of the LL and a weakly decaying D(E) towards the next LL. In a very recent paper Gudmunsson and Gerhardts[11,12] introduced a statistical model for a spatially inhomogeneous 2D-electron gas, which simulates the effect of Poisson's equation and some essential properties of self-consistent screening. The model yields an effective background density of states between LL with a few % of inhomogeneity.

It is the aim of this paper to give an overview of the present understanding of the observed form of D(E). The results from thermodynamic techniques will be critically compared with conductivity and capacitance techniques. Additional information on D(E) is obtained from cyclotron resonance studies[13]. A correlation between the cyclotron linewidth and the backgrond density of states is found. For integer filling factors the linewidth is inversely proportional to the square root of the zero field mobility. The LL density of states influences also relaxation phenomena. From cyclotron resonance saturation experiments an oscillation in the inter LL lifetime with filling factor is found.

2. SPECIFIC HEAT

A direct method to determine D(E) is the measurement of the electronic specific heat given by

$$C_{el} = (\frac{\partial U}{\partial T})_{B,V} = \frac{d}{dT} \int_0^\infty E\ D(E)\ f(E,E_F)\ dE \tag{1}$$

where $f(E, E_F)$ is the Fermi distribution function. An externally induced temperature change leads to a reordering of the electrons. The heat capacity is proportional to $D(E)$ at the Fermi energy, sampling localized and delocalized states.

The first calculation of the specific heat in 2D systems was performed by Zawadzki and Lassnig[14]. They assumed a Gaussian density of states $D(E) \approx e^{-E^2/2\Gamma_G^2}$ independent of the magnetic field. Two contributions to the specific heat are found: intra- and inter-Landau level contributions. Results for a level width $\Gamma_G = 0.25$ meV are shown in Fig. 1 for two temperatures. The intra LL contributions lead to an scillartory behavior with a vanishing specific heat at integer filling factors. The filling factor is defined as $v = n_s \cdot 2\pi l^2$ (neglecting spin splitting) with n_s the electron concentration and $l = \sqrt{\hbar/eB}$ the cyclotron radius.

The inter-LL contributions appear as sharp spikes at the position of integer v-values. These spikes are only present at low magnetic fields and "high" temperatures where kT is comparable with the LL splitting. At the lower temperature (dashed curve) the inter-LL peaks have disappeared. The intra-LL contributions for a given filling factor depend on kT/Γ_G. A maximum is found for $kT/\Gamma_G \approx 0.2$, which means that the specific heat is not sensitive to Γ_G in this range.

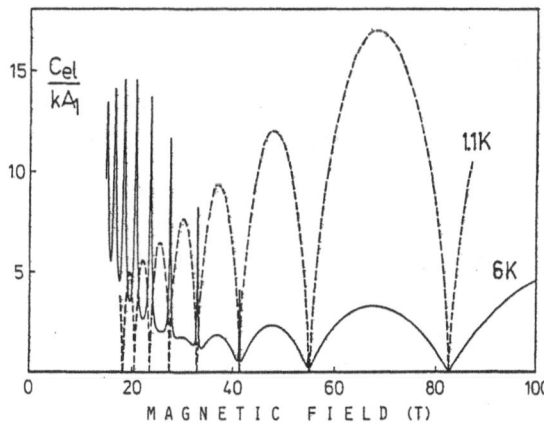

Fig. 1. Specific heat of 2D electrons versus magnetic field for two temperatures after Ref. 14. Full curve: $T = 6$ K; dashed curve: $T = 1.1$ K, $n_s = 8.0 \times 10^{11}$cm^{-2}, $\Gamma_G = 0.25$ meV, $A_1 = 1/\pi \cdot (eB/\hbar)$.

A heat pulse technique was applied to determine the electronic specific heat: In this technique a short heat pulse heats the sample adiabatically. The thermal time constant of the system has to be considerably longer than the heating pulse. The thermal isolation was achieved with thin 5 to 10 μm diameter superconducting wires. The change in sample temperature was measured with a Au:Ge film, a thin Ni-Cr film served as a heater. A detailed description of the experimental set-up is given in Ref.[15].

The experiments were performed on two different multilayer materials. In this paper results on a sample consisting of 94 layers of 220 A GaAs and 500 A GaAlAs will be discussed. The mobility at 4.2 K was 80.000 cm^2/Vs, and n_s = (7.7 ± 0.3) x 10^{11}cm^{-2}. The total sample thickness was 20 μm.

Fig. 2 shows the observed temperature change expressed as curves ΔR versus the magnetic field for three temperatures as obtained from averaging over 10 runs. The applied heat pulse raised the sample temperature by the values indicated as ΔT. The dashed curves ΔR_F show the backgrond dc-resistance variation of the detector film on an extended scale. Oscillations of the sample temperature are clearly observed. The temperature changes are most pronounced for integer υ-values which represent the number of fully occupied LL. The υ-values are determined from the oscillatory conductivity measured on the same sample before being thinned down, shown as dotted curve ρ_{xy}.

The size of the ΔR signal is proportional to the rise in sample temperature. The data show that the sample temperature is higher for all integer υ-values at T = 2 K and up to υ = 4 for the higher temperature. At lower fields (υ > 4) peaks with opposite sign are observed for T = 4.2 and 5.0 K. From a qualitative comparison with the calculations we can identify the behavior at higher fields as due to intra LL contribution and the peaks at lower fields and higher temperature as due to inter LL contribution (see Fig. 1).

From the experimental data the form of the density of states can be determined by comparing the observed temperature change with calculations of ΔT. The main influence on ΔT comes from the form of the electronic specific heat $C_{el}(T,B,\Gamma)$ which is calculated using different types of model density of states. In Fig. 2 the T = 2 K data are compared with a) a Gaussian density of states

192

$$D_G(E) = (\pi e^2)^{-1} \sum_n (2/\pi)^{\frac{1}{2}} \Gamma_G^{-1} \exp[E - \lambda_n)/2\Gamma_G^2]^2 \qquad (2)$$

and b) a Gaussian with background

$$D_{GB}(E) = D_G E)(1 - x) + (\pi l^2)^{-1}(\frac{x}{\hbar\omega_c}) \; \theta(E) \qquad (3)$$

with $\lambda_n = \hbar\omega_c(n + 1/2)$ and x is the percentage of background states.

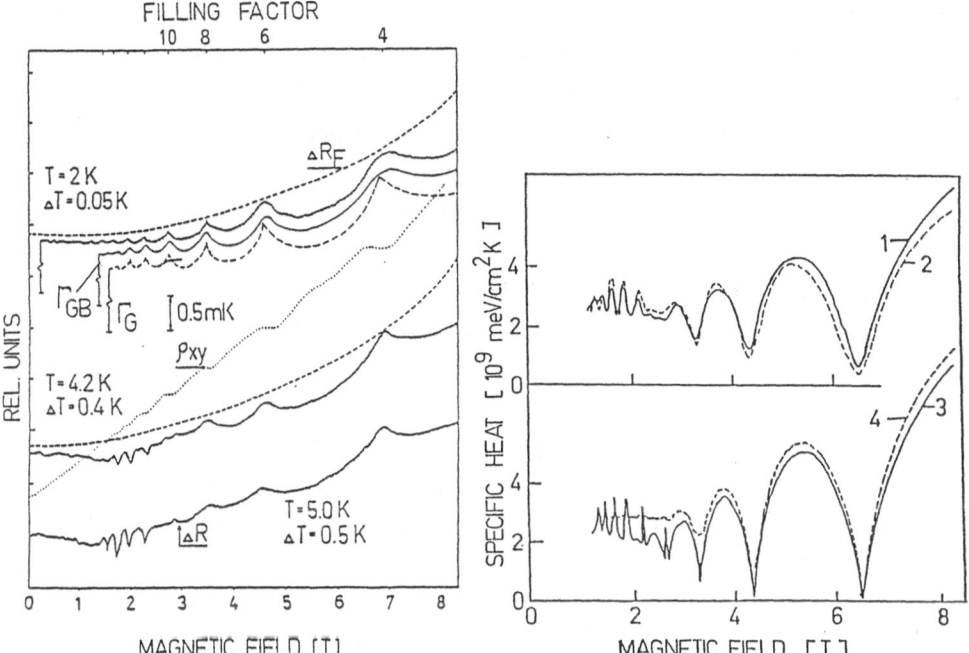

Fig. 2. Temperature change measured with the Au:Ge film versus magnetic field (curves ΔR), for a heat pulse rising the sample temperature by ΔT (ΔR_F change in d.c. detector resistance). The resistivity ρ_{xy} is also plotted for T = 2 K. Theoretical calculations for a pure Gaussian levelwidth (Γ_G) and a Gaussian with background (Γ_{GB}) are also shown for 2 K.

Fig. 3. Comparison of calculated and experimental C_{el} for a sample with 94 layers (described in the text) vs. magnetic field at 4.2 K: Curve 1: experiment; curve 2: $\Gamma_G = 0.6$ meV·$\sqrt{B}$, x = 0.2; curve 3: $\Gamma_G = 0.75$ meV, x = 0; curve 4: $\Gamma_G = 1.5$ meV, x = 0.

The levelwidth Γ is defined as total width at half maximum which correlates to Γ_G as: $\Gamma = 2.4 \times \Gamma_G$. The resulting temperature change (including the behavior of the background) is shown in Fig. 2 by the full (denoted GB) and dotted (denoted G) curves. It is clearly evident that the curve GB fits the data considerably better than curve G. The flat part of the ΔT curve at $T = 2$ K is the direct evidence for the existence of a flat background density of states. The pure Gaussian $D_G(E)$ results in a sharp spike-like change in ΔT even if we increase Γ_G considerably. It has been shown previously, that a Lorentzian density of states does not fit the data[1].

The next question is whether Γ_G is constant with B or not. Other experimental techniques give evidence for a magnetic field dependent level width[2,3,4]. The influence of different level widths on C_{el} is shown in Fig. 3. The experimental result (curve 1 for $T = 4.2$ K) is compared first with a magnetic field independent Gaussian width $\Gamma_G = 0.75$ meV (curve 3) and $\Gamma_G = 1.5$ meV (curve 4). It is evident that inter LL peaks are very sensitive to Γ_G, while intra LL peaks at high field are not. As a consequence the data are consistent with a magnetic field dependent Γ_G. The best linewidth fit to the data is achieved at $B \approx 2$ T where inter LL contributions are dominant, giving $\Gamma_G = 0.75$ meV. A good fit over the whole magnetic field range can be achieved with $\Gamma_{GB} = 0.6$ meV $\times \sqrt{B}$ (B in T) and a background of $x = 0.2$ (curve 2). The $\sqrt{B}$ dependence was taken as used in Ref. /2/. If we try to fit the sample with 172 double layers described in Ref. /1/ with a magnetic field dependent width we obtain $\Gamma_{GB} = 0.9$ meV $\times \sqrt{B}$ and $x = 0.30$.

Summarizing the specific heat data we can state that there is clear evidence for a Gaussian density of states at the positions of the LL, sitting on a flat background. The analysis of the experiments is consistent with a $\sqrt{B}$ dependent Gaussian width. However, due to the rather weak sensitivity of C_{el} to Γ_G at higher magnetic fields, this result relies more on other experimental techniques.

3. MAGNETIZATION

Another thermodynamic technique to determine D(E) is the measurement of the magnetization of the 2D electron gas. The magnetization is given by

$$M = -\left(\frac{\partial F}{\partial B}\right)_{n_s,T} \qquad\qquad \text{with}$$

$$F = n_s \cdot E_F - \frac{1}{k_B T} \int_{-\infty}^{\infty} D(E)\, \ln(1 + e^{-(E - E_F)/k_B T})\, dE \qquad (4)$$

where F is the free energy. At absolute zero, both the Fermi level and the magnetization exhibit a saw-tooth oscillation periodic in the inverse field, with discontinuities at integer filling factors. The magnetization oscillations are of constant amplitude $M_o = n_s A \mu_B^*$, where A is the sample area and μ_B^* the effective Bohr magneton. The first calculation of M for a Gaussian density of states was performed by Zawadzki and Lassnig[16]. A comparison of the calculated shape of the magnetization oscillation with the behavior of the Fermi level for a multilayer sample (sample with 172 double layers) is shown in Fig. 4 for a pure Gaussian Γ_G and a Gaussian including background Γ_{GB} level width. It is clearly evident that a pure Gaussian level width will always show a steplike behavior at high magnetic fields. The main effect of the background is to flatten the steps. With increasing level width the amplitudes of the oscillations are strongly reduced.

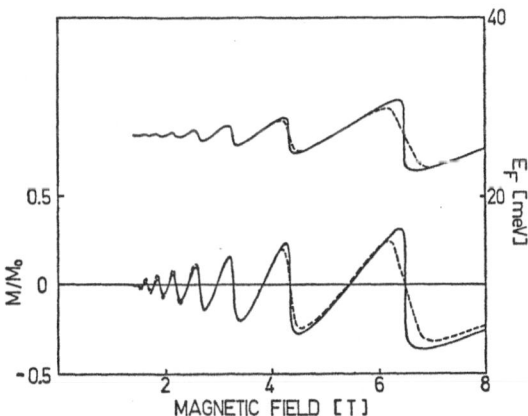

Fig. 4. Normalized calculated magnetization M/M_o and Fermi level for a sample with 172 double layers of GaAs (200 A) and GaAlAs (200 A) with $n_s = 6 \times 10^{11} \text{cm}^{-2}$, $\mu_{4.2} = 40\,000$ cm^2/Vs vs. magnetic field at 1.5 K. Full curve: $\Gamma_G = 1.04$ meV, x = 0.25; dashed curve: $\Gamma_G = 1.04$ meV, x = 0.

Successful measurements of the magnetization were performed by Eisenstein et al.[2]. An extremely sensitive capacitive technique was developed to measure the torque of the sample due to the change in the magnetic moment[17]. In this paper we only will discuss the results: Fig. 5 shows the maximum amplitudes of the de- Haas-van-Alphen oscillations normalized by the ideal amplitude M_o versus magnetic field. The solid lines represent calculations with a magnetic field independent Gaussian $D_G(E)$ with $\Gamma_G = 1$ meV and 2 meV. The data are plotted for two samples: A GaAs/GaAlAs multilayer (50 periods, $\mu_{4.2K} = 8.0 \times 10^4 cm^2/Vs$, $n_s = 5.4 \times 10^{11} cm^{-2}$) and a single layer heterostructure ($\mu_{4.2K} = 2.85 \times 10^5 cm^2/Vs$ and $n_s = 2.7 \times 10^{11} cm^{-2}$). It is evident that a constant Γ_G cannot account for the observed behavior. The best fit for the multilayer sample is obtained with $\Gamma_G = 1$ meV$\cdot\sqrt{B(T)}$ shown as dashed curve. No background density of states is assumed.

The data can also be fitted with a reduced $\Gamma_{GB} = 0.8$ meV$\cdot\sqrt{B(T)}$ and a background density of x = 20 % shown as curve denoted Γ_{GB}. The fit is quite good up to 4 T but deviates for higher fields. The single heterolayer sample has a somewhat smaller width. The data are fitted with x = 10 % and $\Gamma_{GB} = 0.65\cdot\sqrt{B(T)}$. However, the fit with background is not as good as without background. Unfortunately the method has only

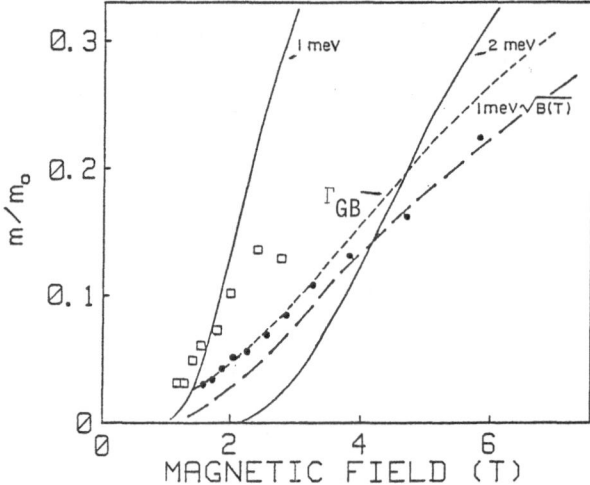

Fig. 5. Normalized dHvA oscillation amplitude vs. magnetic field after Ref. 2; multilayer: full circles, single layer: square. The curves are calculated with Γ_G-values indicated and $\Gamma_{GB} = 0.8$ meV$\cdot\sqrt{B(T)}$, x = 0.2.

been applied in a rather narrow magnetic field range. It should also be noted that this technique is most sensitive when the LL filling is of integer values. Under this condition the maxima in amplitude appear.

4. MAGNETOCAPACITANCE

The capacitance of a metal-insulator-semiconductor structure depends on the thickness of the insulator, on the density of states in the semiconductor and on material parameters. Capacitance measurements seem to be a straight-forward method to obtain direct information on $D(E)$. A variation of the magnetic field results in oscillations of the magnetocapacitance measurements which directly reflect changes in $D(E)$. In a first attempt Smith et al.[18] tried to deduce $D(E)$ from magnetocapacitance measurements in GaAs/GaAlAs heterostructures. They analysed mainly the minima of the oscillations at 1.3 K; however, the method failed for fields higher 1.6 T due to strong contributions of non-capacitive signals. Recently Mosser et al.[4] have investigated the magnetocapacitance in the same system up to fields of 8 T analysing only the maxima, thus extracting informaton on $D(E)$ for half filled LL.

Fig. 6 shows the capacitance data at different temperatures (full curves) for a sample with n_s = 2.25 x $10^{11} cm^{-2}$ and μ = 220.000 cm^2/Vs. Strong oscillations are clearly observed with minima at integer filling factors. From a critical analysis of the experimental situation only the maxima of the capacitance can be used to determine $D(E)$, since only for half filled LL the capacitance is well defined and the signal remains mainly capacitive.

The experimental results are compared with calculations using a Gaussian density of states with background $D_{GB}(E)$. The fit shown in Fig. 6 assumes Γ_G = 0.3 x $\sqrt{B(T)}$ meV and x = 0.13 (corresponding to D(background) = 3.9 x $10^9 cm^{-2} meV^{-1}$. At all investigated temperatures the calculated maxima are in good agreement with the data for fields up to 5 T. However, the fit is not as good for a filling factor of ν = 1. The analysis of a sample with nearly the same density and a mobility of μ = 800.000 cm^2/Vs yields a Gaussian width of Γ_G = 0.25 x $\sqrt{B(T)}$ meV and a reduced background of x = 0.07. It is stated in the paper[4] that a constant background is even not necessary to fit the data. The obtained Gaussian width Γ_G on the other hand does not seem to depend critically on the mobility. The result of the capacitive technique is a weakly mobility dependent Γ_G for half filled LL.

Most recently the capacitive technique was also applied to GaAs-heterostructures by Hickmott et al.[19] and Smith et al.[20]. An oscillating density of states is revealed with residual density between LL. Even an indication for density of states oscillations in the fractional Quantum Hall situation is found. However, in the case of integer filling factors, the data have to be treated with care since the signal is not purely capacitive.

5. TEMPERATURE DEPENDENCE OF ρ_{xx}

For a 2D electron system the resistivity ρ_{xx} exhibits strong oscillation with magnetic field (Shubnikov-de Haas oscillation). Minima of ρ_{xx} occur at integer filling factors which are strongly temperature dependent. As the Fermi level moves towards the center of a LL ,the temperature dependence of ρ_{xx} becomes weaker. The analysis of this effect was used by Weiss et al.[21] to determine the density of states between LL:

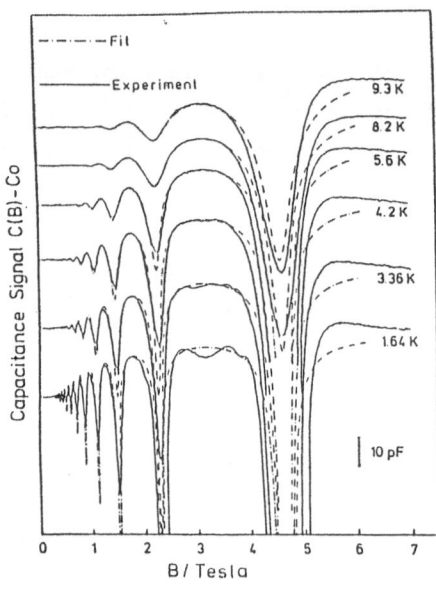

Fig. 6. Measured magnetocapacitance of a sample with $n_s= 2.25 \times 10^{11} cm^{-2}$ and $\mu_{4.2K} = 220\ 000\ cm^2/Vs$. The fit was performed with a Gaussian width $\Gamma_G = 0.3 \cdot \sqrt{B(T)}$meV and x = 0.13 (after Ref. 4).

The analysis assumes that nearly all states are localized except states in the center of LL. An example of the temperature dependence of ρ_{xx} for different magnetic fields is shown in Fig. 7. By fitting the activated behavior of ρ_{xx} the density of states for integer filling factors is determined according to

$$\rho_{xx} \approx \rho_o(T) \sum_n \exp(-(E_n - E_F)/kT) \tag{5}$$

where $\rho_o(T)$ is a temperature dependent prefactor. A Gaussian density of states with background D_{GB} is included in the calculation of the Fermi-level: $n_s = \int D_{GB}(E)f(E,E_F)dE$. In addition the level width Γ_G is assumed to be $\sqrt{B}$ dependent. The main result of this technique is a quite accurate determination of the background density of states, which varies significantly with sample mobility. The Gaussian width Γ_G is rather weakly sample dependent which might be due to the assumption of a very narrow range of extended states in the center of LL.

6. DENSITY OF STATES FROM RADIATIVE RECOMBINATION SPECTRA

Very recently a new technique to measure directly the 2D-density of states was reported by Kukushkin and Timofeev [22]. It is shown that radiative spectra due to the recombination of 2D-electrons with holes at acceptor levels in Si(100) MOS structures reveal directly D(E). The

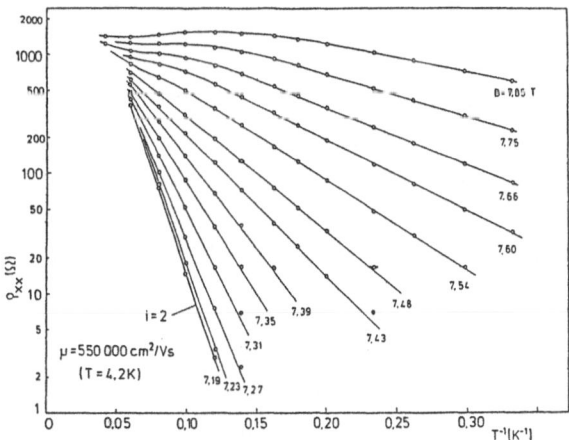

Fig. 7. Temperature dependence of the resistivity ρ_{xx} versus inverse temperature at different magnetic fields close to a filling factor $v = 2$ (after Ref. 19).

recombination is indirect in Si and thus its probability is not energy dependent. Since the recombination process is a free electron to bound acceptor transition, the resolution is determined by the rather narrow acceptor width of 0.6 meV.

An example of observed spectra is shown in Fig. 8 for 0 T and 7T. The LL structure with a strong overlap is observed. In the zero field case the constant $D(E)$ is clearly demonstrated. The insert shows the principle of the emission process. The advantage of this technique is that all fully and partly occupied LL are observed at the same time. Thus a narrowing of the LL width for half filled LL is seen together with quite broad completely filled levels. Maxima appear when the LL are filled and minima for partly filled levels. The oscillation of Γ with the filling factor is thus unambiguously demonstrated in this paper[22]. This technique has up to now only been applied to Si. The basic behavior

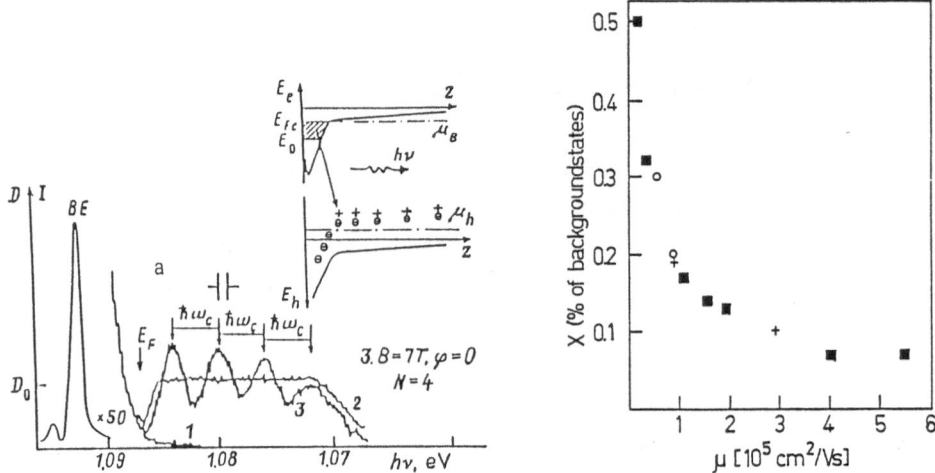

Fig. 8. The inset is a band diagram of the recombination of 2D electrons with injected holes. The intense line in the emission spectrum (BE), with a peak at $\hbar\omega = 1.0928$ eV corresponds to a volume emission of excitons bound by Boron atoms. Spectrum 1 is the long-wave tail of line BE, magnified by a factor of 50 with $n_s = 0$. Curves 2 and 3 show the emission spectra of 2D electrons found for T = 1.6 K, $n_s = 2.7 \times 10^{12} cm^{-2}$ and B = 0 (spectrum 2) or B = 7 T (spectrum 3).

Fig. 9. Summary of determined background density of states in % of the B = 0 density. (■) from temperature dependent ρ_{xx}; (○) from specific heat; (+) from magnetization.

of Γ that maxima appear when the LL are filled and minima for partly filled levels is expected to be the same in Si and GaAs. However, the scattering mechanisms are quite different since Si is covered with an amorphous oxide while the GaAs structures are made by molecular beam epitaxy. A summary of the determined background values from the activated resistivity measurements together with the results from specific heat and magnetization is given in Fig. 9. The x-values are calculated from Ref. 21 by the ratio of the midpoint density to the density at B = 0 given by $D(E)_{B=0} = 28 \times 10^9 cm^{-2}meV^{-1}$. The results from all three techniques are in good agreement. It is evident that the background density of states increases very steeply for very low mobility samples while it decreases slowly for high mobilities. A hint for the origin of the background is given through the analysis of an electron irradiated sample[21] with a mobility of 14 000 cm^2/Vs showing that 50 % of all the states were contained within the background density. Annealing of the sample results in a significant reduction of the background and an increase in mobility. A direct correlation between the created point defects through electron irradiation and the background is not possible since the mobility change is even larger.

An analysis of the temperature dependence of ρ_{xx} and ρ_{xy} for integer filling factors was performed by Pudalov and Semenchinsky[23] in Si-MOSFETs. A considerable amount of density of states assigned to localized state was found in the gap between LL giving also a similar density of state as found in GaAs.

From the results of the various techniques we can conclude that there exist Gaussian peaks sitting on a flat background density of states. However, the width of the Gaussian peaks cannot be extracted from the ρ_{xx} data to the same accuracy as the background. The specific heat data give quite accurate values for Γ_G only at low magnetic fields while the magnetization data rather at higher fields. Therefore there is still need for additional techniques to get information about the width of the Gaussian peaks. In addition, it should be mentioned that the ρ_{xx} and specific heat data give accurate information only for integer filling factors. About the form of D(E) for half filled levels we have little information. The magnetization is sensitive rather for a non-integer filling factor. This gives a hint for a dependence of the background density on filling factor since the magnetization data can be fitted better without background.

7. CYCLOTRON RESONANCE SPECTROSCOPY

In cyclotron resonance (CR) the observed linewidth of the transmission spectrum is determined by the broadening of both the initial and final LL. The linewidth contains information on the individual level width Γ but not in a trivial way. In the techniques described in the previous chapters the pure LL width was extracted from experiments which were not influenced by scattering processes or transitions between different LL; in the analysis the level width was always assumed to be the same for all LL.

While there is consistent information on the background, the value of the Gaussian width Γ_G and its dependence on the zero field mobility has not been derived yet in a satisfactory way. One technique which should give information on changes of Γ_G is the analysis of the CR linewidth. It is clear that this method will only give good values of Γ_G after a careful theoretical fit of the data. However, tendencies of Γ_G as a function of certain external parameters for a situation where no significant physical quantity changes can be determined.

Previous experimental[24-27] and theoretical[6,10] investigations have revealed a filling factor dependent CR linewidth due to screening. Maxima of the linewidth occur at integer values of the filling factor and minima in between. In the following two experimental techniques will be described which give information on the form of D(E): First results from cyclotron emission – the inverse effect of CR absorption – performed on several GaAs/GaAlAs heterojunction samples will be shown. These data will mainly depend on D(E) in the upperexcited level. Second results from CR-transmission experiments will be analysed for a filling factor of $v = 2$ and correlated to the zero magnetic field mobility.

7.1. Cyclotron emission

Cyclotron emission is generated by heating up the electron gas by electric field pulses. As a result a nonequilibrium carrier distribution with average carrier temperatures above the lattice temperature is obtained. Radiative transitions between neighbouring LL in the vicinity of the Fermi level occur. The emitted radiation is measured and analysed with a narrow-band and magnetic field tunable GaAs detector[13,28]. The detector has a narrowband peak sensitivity at 4.4 meV ($35cm^{-1}$) at B = 0. In the magnetic field this line splits into 3 narrow lines (resolution 0.25 cm^{-1}).

The emission technique is applied to several GaAs/GaAlAs heterostructures. Figures 10a – 10c show emission spectra as obtained with a GaAs-photoconductive detector at B = 0. In the experiment the emission signal is monitored as a function of the 2DEG magnetic field for various electric fields. The most prominent feature is that the emission spectra differ considerably from the transmission spectra, obtained on identical samples[28].

This effect is extremely pronounced for sample 1 (Fig. 10a, sample data given in figure captions): For very low electric fields the emission spectrum consists of a broad line, which is well below the bulk cyclotron position (corresponding to a higher energy). With increasing electric field which is equivalent to increasing the carrier temperature this peak becomes smaller and a line at the position of the transmission resonance grows until it dominates the spectrum at high fields. The dashed line separates the pure CR-line from the "other" structure. The 2D character of the structure was tested by tilting the magnetic field. For sample 2 (Fig. 10b) the peak at lower magnetic field is not that strong. Sample 2 has a higher mobility and a narrower linewidth. A shoulder on the low magnetic field side in the emission spectrum has

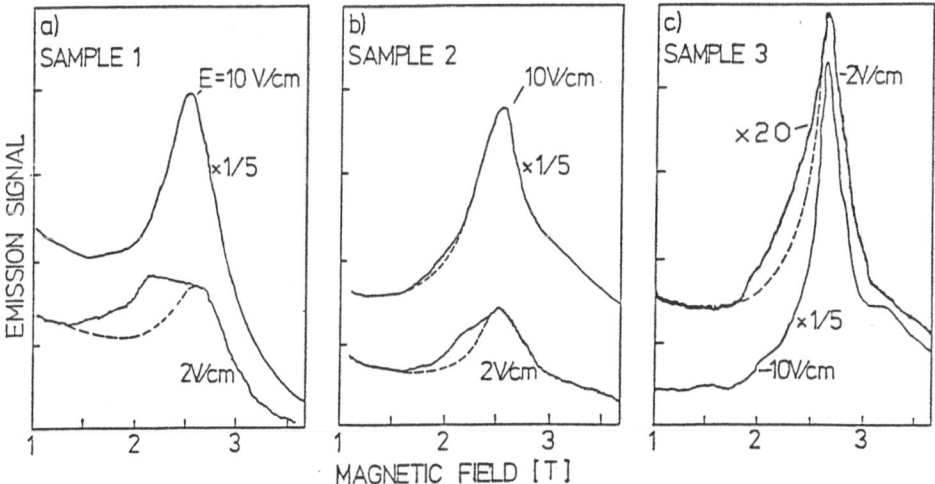

Fig. 10: Cyclotron emission signal detected with a GaAs detector at 0 T
(see Fig. 10) as a function of magnetic field for
sample 1 (n_s = 4.6 x 10^{11}cm^{-2}, μ = 1.3 x 10^5cm^2/Vs) at 4.2 K
sample 2 (n_s = 2.6 x 10^{11}cm^{-2}, μ = 2.4 x 10^5cm^2/Vs) at 4.2 K
sample 3 (n_s = 2.2 x 10^{11}cm^{-2}, μ = 5.0 x 10^5cm^2/Vs) at 4.2 K.

been found in all investigated samples. A correlation of the emission data with the d.c. mobility is found. The higher the mobility the less pronounced the additional peak becomes (Fig. 10a). This feature is not restricted to the 3 T magnetic field range. We have found this behavior in the whole investigated magnetic field ranges up to 8 T.

As the additional peak is only observed in emission it must be correlated to the special excitation technique used: The electric field excites carriers to the next higher LL. The carriers first thermalize within the upper LL and occupy only states at the lower edge. If localized states exist below the edge, the electrons will get trapped in these states. As the number of excited electrons is very small at low electric fields, the recombination radiation will mainly be due to transitions out of these localized states. The emission frequency will be shifted to higher energy due to the binding potential similar to the bulk case, where a so called impurity shifted CR line is observed. With increasing electric field the binding potentials are washed out and the pure CR line from extended states is observed.

The integral emission intensity of the additional structure decreases linearly with increasing mobility. That means that the impurities which are responsible for the mobility also influence the emission signal. This conclusion can be drawn since the normalized cyclotron emission signal (main peak) is practically independent of the sample properties if plotted versus the input power per electron ($P_e = e\mu E^2$).

The spectral analysis is performed with a GaAs-detector in a magnetic field of 3.0 T. Fig. 11 shows emission spectra of sample 1 and 2 (same as in Fig. 10) as a function of the sample magnetic field. The spectra consist of a main CR line and a shoulder on the lower magnetic field side for all three frequencies. For comparison an emission spectrum from a high purity bulk GaAs sample is also shown indicating a "pure" single CR-spectrum and no evidence for impurity related emission. The number of donors in the bulk sample is $8 \times 10^{13} \text{cm}^{-3}$. 2D samples with mobilities higher $2 \times 10^6 \text{cm}^2/\text{Vs}$ also show single line spectra indicating impurity levels comparable with the bulk sample. From this comparison and a calculation of the total impurity related emission intensity we estimate a number of several 10^8 impurities per cm^{-2} for the highest mobility sample. Samples with a mobility around $4 \times 10^5 \text{cm}^2/\text{Vs}$ have 10^9 impurities close to the 2D gas which are responsible for the observed impurity line.

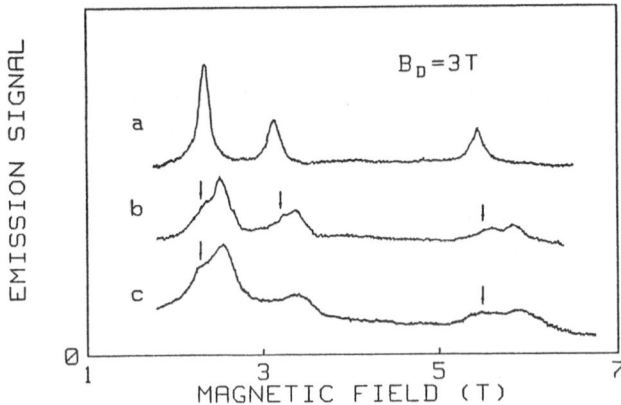

Fig. 11. Emission spectra as obtained from a GaAs detector at 3.0 T
first line at 32.5 cm^{-1}, second at 43.5 cm^{-1}, third at 75 cm^{-1})
for bulk GaAs (a), sample 2 (b) and sample 3 (c). The impurity
line is indicated by arrows.

To assign binding potentials to the observed "impurity lines" in
Fig. 11 we use a simple oscillator model where the observed frequency
ω_{obs} is given by

$$\hbar\omega_{obs} = \sqrt{(\hbar\omega_c)^2 + (\hbar\omega_B)^2} \qquad (6)$$

with $\hbar\omega_c$ the CR frequency and $\hbar\omega_B$ the binding energy. The observed
energy is defined by the detector magnetic field; the spectra show that
the dominant impurity feature does not shift for different samples but
only decreases. An analysis with the above formula gives binding
potentials of 2.0 meV at 2.3 T, 2.7 meV at 4.7 T and 3.0 meV at 6 T. The
binding potentials are increasing with magnetic field.

The derived binding potentials allow us to determine the position
of the impurities. It has been shown[29-32] that the impurity binding
potentials depend strongly on the position in respect to the 2D gas. The
impurities behind the spacer cannot be responsible for the observed
potentials[32], since they are found to be well below 1 meV in the
investigated magnetic field range for comparable samples. The rather
defined structure of the impurity line indicates that only impurities at
defined positions can be responsible for the observed spectra. A
distinction between impurities at the interface between GaAs and GaAlAs
and donors (compensating) in the GaAs can be made through the expected
binding energies. At the interface the binding energy is considerably

smaller than in the center of the 2D channel. The compensating impurities in the GaAs are randomly distributed and a rather broad impurity peak with binding energies comparable or higher than in the bulk are expected. The bulk binding energies are well above 5 meV. Therefore we conclude that impurities at the interface are responsible for the observed phenomena. These impurities are most likely Si-impurities which have diffused from the high doped region in the GaAlAs to the GaAs interface which acts as a diffusion stop.

These results give evidence for an accumulation of donor impurities at the interface of GaAlAs/GaAs which are correlated with the sample mobility. An impurity shifted cyclotron emission line can be observed for samples with mobilities up to $1 \times 10^6 cm^2/Vs$. For higher mobilities we can state that the number of impurities is in the $10^8 cm^{-2}$ range. The defined structure in the emission spectra gives evidence for an impurity band which exists below every LL. A defined structure was found for several magnetic fields (filling factors) in samples with mobilities up to a few $10^5 cm^2/Vs$. This conclusion is in agreement with recent work by Raymond et al.[33].

7.2. Cyclotron resonance transmission

CR transmission experiments with the use of far infrared lasers were performed by several groups[24-27] which all found a systematic oscillation of the linewidth with filling factor. Pronounced maxima in the observed linewidth were found by Englert et al.[24] at filling factors of 2 and 4 and minima in between. This behavior was found only in low density samples (n $\langle$ $1.5 \times 10^{11} cm^{-2}$) and was absent for higher densities (n $\rangle$ $2.5 \times 10^{11} cm^{-2}$). Weak oscillations in the linewidth were also found by Gornik et al.[28], while rather large oscillations were observed by Rikken et al.[27]. In a recent paper Heitmann et al.[34] have shown large oscillations of the linewidth with the filling factor up to $v = 5$ for 2D electrons in InAs.

Here results from a recent systematic study of the linewidth as a function of mobility[35] are presented. Fig. 12 shows a plot of the measured linewidth as a function of filling factor for 3 different samples at a temperature of 4.2 K. All samples show a clearly defined maximum of the linewidth for $v = 2$. A systematic behavior of the linewidth with the sample mobility is evident, while the linewidth for a

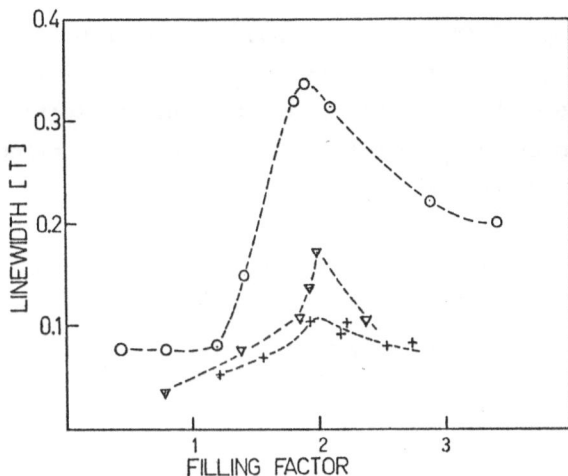

Fig. 12. Measured cyclotron resonance transmission linewidth as a
function of filling factor for 3 different samples:

$n_s = 1.2 \times 10^{11} cm^{-2}$, $\mu = 4.5 \times 10^5 cm^2/Vs$,

$n_s = 1.2 \times 10^{11} cm^{-2}$, $\mu = 1.2 \times 10^5 cm^2/Vs$,

$n_s = 2.3 \times 10^{11} cm^{-2}$, $\mu = 1.0 \times 10^6 cm^2/Vs$.

noninteger filling factor seems not to be correlated clearly to the
mobility. There is also clear evidence for a temperature dependence of
the linewidth but we only want to compare here data at one temperature.

To get information on D(E) from these data we have to make the
assumption that the CR linewidth at even filling factors is directly
correlated to the Gaussian level width Γ_G. This correlation is
demonstrated in Fig. 14 where the CR linewidth at $\nu = 2$ is plotted as a
function of mobility (*). Several samples with different mobilities and
densities varying only between 1.2×10^{11} and $2.4 \times 10^{11} cm^{-2}$ were used.
The most conclusive results are from samples with the same density and
considerably different mobilities. The interpretation of the data is
straight forward. The zero field mobility and thus the scattering time
τ_{DC} is inversely proportional to the number of scatterers[36,37]. That
means a plot as a function of mobility is equivalent to a plot versus
the reciprocal number of impurities N_i. On the other hand the main
process which determines the LL broadening and thus the CR linewidth is
the virtual double scattering at the impurities. Thus in the above
situation the linewidth is proportional to $\sqrt{N_i}$ as demonstrated in Fig.
13. The same behavior has been found in bulk GaAs previously[13]. The
level width predicted by the theory for short range scatterers (point

scatterers)[5] $\Gamma_{SR} = (2/\pi \cdot (\hbar^2 e^2 B / m^{\times 2} \mu))^{1/2}$ is also plotted over μ in Fig. 13 for a magnetic field of 5 T. The obtained values lie systematically a factor of 2 above the experimental values. This reflects the long-range nature of the impurity potentials, which are less effective in the dynamic (local) CR process than in the d.c.-response.

For partly filled LL intra LL screening strongly reduces the levelwidth which explains the drastic drop of the linewidth close to the even values. For $\nu \blacktriangleleft 1$ extremely narrow lines are observed, approaching

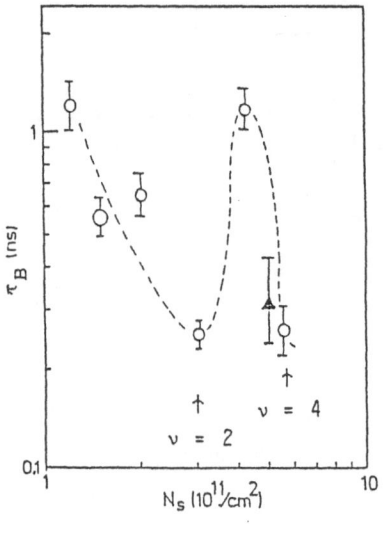

Fig. 13. LL lifetimes from CR-saturation measurements as a function of n_s (o). Photoluminescence data after Ref. 43 are also included ($\triangle$). The dashed curve indicates the tentency.

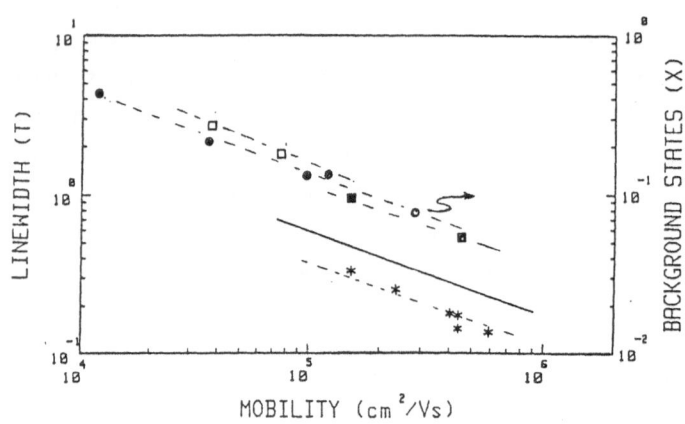

Fig. 14. Plot of LL-width (in T) at half maximum as a function of zero field mobility from different experimental techniques: CR-absorption linewidth (X), from specific heat and magnetization ($\square$) and from cyclotron emission ($\blacksquare$). In addition the percentage of background states ($\bullet$) (Fig. 9) is also plotted over mobility. The dashed curves indicate tendencies. The full curve shows the linewidth due to short range scatterers after Ref. 5.

the limiting values due to "saturation" effects[38,39]. The saturation effects are also the reason that the main features in the linewidth are masked for samples with densities higher $2.5 \times 10^{11} cm^{-2}$ and mobilities higher than 500 000 cm^2/Vs. Furthermore the superposition of individual CR transitions, which differ due to nonparabolicity, can alter the behavior. It could even lead to maxima in linewidth at half fillings.

8. ENERGY RELAXATION BETWEEN LANDAU LEVELS

A property which is also keyed to the density of states is the energy relaxation time τ_B between LL. A direct way to obtain τ_B is the measurement of the incoherent saturation of the CR transmission. This technique has been previously applied to bulk n-type InSb[41] and GaAs[42]. It was found that a combination of electron-electron scattering and optical phonon emission governs the energy relaxation. Decreasing lifetimes with increasing electron concentration were achieved.

This technique was applied to GaAs/GaAlAs heterostructures by Helm et al.[43]. A high power CW optically pumped FIR laser at 118 μm was used to perform the experiments. At powers of about 1 W/cm^2 the transmission starts to get reduced for all investigated samples. In the analysis of the data a rate equation for 3 LL and a constant relaxation time τ_B was used.

A systematic increase of τ_B with decreasing filling factor below $v = 2$ is found. However, well pronounced minima in τ_B are observed for $v = 2$ and $v = 4$, while a maximum appears for $v = 3$. Since the spin splitting for low magnetic field (B < 8 T) is rather small uneven filling factors represent half filled levels, while only even filling factors correspond to filled levels.

Data on relaxation rates between LL were also reported by Ryan et al.[43] and Hollering et al.[44] using psec time-resolved photoluminescence techniques. Ryan et al.[43] find relaxation times in the order of 0.3 ns between the first excited and lowest LL for MQW samples with $n_s = 5 \times 10^{11} cm^{-2}$ (This value is indicated in Fig. 13). The relaxation times in magnetic fields of 7 and 8 T are longer than without magnetic field. On the other hand Hollering et al.[44] find a considerably shorter time for fields of 20 T as compared to the zero magnetic field case.

The CR-saturation data indicate that the energy relaxation is more effective for filled levels where the density of states is also large between LL. For narrow LL (half filling) the lifetime is increased considerably. However, a quantitative analysis of the data is rather difficult since the LL width will not only influence the relaxation time but also the excitation process. It is well known that saturation cannot be achieved in an equidistant Landau ladder. This effect has been observed by Schlesinger et al.[45] for resonant fields between 2 and 3 T. Relaxation times in the order of 10 psec were observed which probably directly reveal the optical phonon relaxation time as final limiting process.

In the presented situation for high mobility samples it can be assumed that the LL width is still narrow enough to allow an analysis with a 3 level system. The fourth LL is already very close to the optical phonon energy which induces a strong polaron shift of the level.

Two mechanisms can be responsible for the energy relaxation: a) acoustic deformation potential scattering and b) electron-electron scattering. Both processes will critically depend on the density of states which oscillates. The systematic increase of τ_B with decreasing density below $v = 2$ .favors an interpretation via the electron-electron scattering process. However, more detailed investigations are necessary to get clear evidence for the mechanism.

CONCLUSIONS

A consistent picture of the density of states can be drawn by summarizing the results from all experimental techniques. Fig. 14 shows a plot of the percentage of background states from Fig. 9 on a double logarithmic scale as a function of mobility. In addition the CR-linewidth at a filling factor of $v = 2$ and the total half width Γ from the specific heat and magnetization data at $v = 4$ are plotted on the same scale.

This plot shows the same slope for all derived quantities. This is strong evidence that the background has the same origin as the CR-linewidth, which is governed by ionized impurity scattering. A direct correlation of the Gaussian width Γ with the CR-linewidth is difficult since CR probes potential fluctuations in the range of the cyclotron radius, while specific heat and $\rho_{xx}(T)$ sample all ranges of impurity potentials. This is clearly evident when we compare the absolute values of the CR-linewidth with Γ from the specific heat data. The Γ-values are larger by a factor of 5 indicating a considerable amount of large range fluctuations in all samples for even filling factors. Another experiment which gives evidence for an extended density of states at the tails of LL was cyclotron emission. If we add to the central LL peak the impurity peak and assign to the total density of state an average width we obtain values which are below the results from specific heat. The difference between these data and the specific heat data can only be due to fluctuations which are induced by the impurities beyond the spacer. Only these impurities will induce very small binding energies (not observable in emission) and thus extremely long range fluctuations.

For half filled LL screening is effective and only short range impurity potentials contribute to the level width. It is evident from Fig. 12 that the CR-linewidth is quite narrow and only weakly dependent on the sample mobility. In a most recent paper Weiss and v.Klitzing[47] derived the thermodynamic density of states for half filled LL from capacitance experiments. It is found that the level width can be fitted without background and is not far away from the predictions of the self-consistent Born approximation for short range scatterers[38]. The expected short range levelwidth for a magnetic field of 5 T is shown in Fig. 14 as full curve. The upper dashed curve and the full curve thus shows the limiting cases of the oscillation of the thermodynamic level width between filled and half filled LL.

A consistent empirical model explaining all the observed phenomena has been developed by Gerhardts and Gudmundsson[11,12]. Impurity fluctuations will lead to a certain amount of local density fluctuations. Local density fluctuations of a few % will be strongly enhanced in a magnetic field for even filling factors. In this case potential fluctuations of the size of the LL splitting are possible over long ranges. The CR-linewidth is considerably smaller than the thermodynamic level width. An analysis with this model assigns the observed thermodynamic width to fluctuations of the LL edges in respect to the Fermi level. Values of average long range fluctuations in the order of several meV are determined for the lowest mobility samples and to less than 1 meV for high mobility samples at even filling factors. In this model the background is only present for filled levels. For half filled levels the long range potentials are screened and the density of states becomes narrow without a background.

ACKNOWLEDGEMENT

This work is partly supportet by the Stiftung Volkswagenwerk and the Jubiläumsfonds der Österreichischen Nationalbank.

REFERENCES

1. E. Gornik, R. Lassnig, G. Strasser, H.L. Stoermer, A.C. Gossard, W. Wiegmann, Specific heat of two-dimensional electrons in GaAs-GaAlAs multilayers, Phys.Rev.Lett. 54:1820 (1985).

2. J.P. Eisenstein, H.L. Stoermer, V. Narayanamurti, A.Y. Cho, A.C. Gossard, and C.W. Tu, Density of states and de Haas-van Alphen effect in two-dimensional electron systems, Phys.Rev.Lett., 55:875 (1985)

3. E. Stahl, D. Weiss, G. Weimann, K. v.Klitzing, K. Ploog, Density of states of a 2D electron gas in a strong magnetic field, J.Phys.C, 18:L783 (1985)

4. V. Mosser, D. Weiss, K. v.Klitzing, K. Ploog, G. Weimann, Density of states of GaAs-AlGaAs heterostructures dediced from temperature dependent magneto-capacitance measurements, Solid State Commun., 58:5 (1986).

5. T. Ando, Y. Uemura, Theory of quantum transport in a 2D electron system under magnetic fields, Characteristics of level broadening and transport under strong fields, J.Phys.Soc.Japan, 36:959 (1974).

6. R. Lassnig, E. Gornik, Calculation of the cyclotron resonance linewidth in GaAs-AlGaAs heterostructures, Solid State Commun.,47: 959 (1983)

7. R.R. Gerhardts, Path-integral approach to the 2D magneto conductivity problem, Z.Phys.B, 21:275 (1975) and Surf.Sci., 58:227 (1976).

8. F. Wegner, Exact density of states for lowest Landau level in white noise potential superfield representation for interacting systems, Z.Phys.B, 51:279 (1983).

9. E. Brezin, D.I. Gross, C. Itzykson, Density of states in the presence of a strong magnetic field and random impurities, Nuclear Phys., B235:24 (1984).

10. T. Ando, Y. Murayama, Landau-level broadening in GaAs/AlGaAs heterojunctions, J.Phys.Soc.Japan, 53:693 (1985).

11. R.R. Gerhardts, V. Gudmundsson, Statistical model for inhomogeneities in a two-dimensional electron gas implying a background density of states between Landau levels, Phys.Rev. B, 34:2999 (1986).

12. V. Gudmundsson, R.R. Gerhardts, Interpretation of experiments implying density of states between Landau levels of a 2DEG by a statistical model for inhomogeneities, Phys.Rev. B, to be published (preprint).

13. E. Gornik, Far infrared light emitters and detectors, Physica 127B:95 (1984).

14. W. Zawadzki and R. Lassnig, Specific heat and magneto-thermal oscillations of two-dimensional electron gas in a magnetic field, Solid State Commun., 56:537 (1984).

15. E. Gornik, R. Lassnig, G. Strasser, H.L. Stoermer, A.C. Gossard, Landau level density of states through specific heat in GaAs/GaAlAs multilayers, Surf.Sci. 170:277 (1986).

16. W. Zawadzki, R. Lassnig, Magnetization, specific heat, magneto-thermal effect and thermoelectric power of two-dimensional electron gas in a quantizing magnetic field, Surf.Sci. 142:225 (1984).

17. J.P. Eisenstein, High precision torsional magnetometer: Application to 2D electron systems, Appl.Phys.Lett., 46:695 (1985).

18. T.P. Smith, B.B. Goldberg, P.J. Stiles, M. Heiblum, Direct measurement of the density of states of a two-dimensional electron gas, Phys.Rev.B, 32:2696 (1985).

19. D. Weiss, K. v.Klitzing, V. Mosser, Density of states of landau levels from activated transport and capacitance experiments, Springer Series in Solid States Sciences, 67:204 (1986).

20. T.W. Hickmott, Fractional quantization in ac conductance of $Al_xGa_{1-x}As$ capacitors Phys.Rev.Lett., 57:751 (1986).

21. T.P. Smith, W.I. Wang, P.J. Stiles, The two-dimensional density of states at fractional filling factors, Springer Series in Solid State Sciences, 71:173 (1987).

22. I.V. Kukushkin, V.B. Timofeev, Direct determination of the state density of 2D electrons in a transverse magnetic field, JETP Lett., 43:499 (1986).

23. V.M. Pudalov, S.G. Semenchinsky, Density of states for a two-dimensional electron Landau band in Si Mosfet, Solid State Commun., 55:593 (1985).

24. Th. Englert, J.C. Maan, Ch. Uihlein, D.C. Tsui, A.C. Gossard, Oscillations of the cyclotron resonance linewidth with Landau level filling factor in $GaAs/Al_xGa_{1-x}As$ heterostructures, Physica, 117B & 118B:631 (1983).

25. W. Seidenbusch, G. Lindemann, R. Lassnig, J. Edlinger, E. Gornik, Cyclotron resonance studies of screening and polaron effects in GaAs-AlGaAs heterostructures, Surf.Sci., 142:375 (1984).

26. R. Lassnig, W. Seidenbusch, E. Gornik, G. Weimann, Landau level width and cyclotron resonance in 2D systems, in: "Proc. of the 18th Int. Conf. on The Physics of Semiconductors, Stockholm", O. Engström, ed., World Scientific, Singapore (1987).

27. G.L. Rikken, H.P. Wyder, G. Weimann, W. Schlapp, R.E. Horstman, J. Wolter, Anomalous cyclotron resonance linewidth in hetero-junctions displaying the fractional quantum Hall effect, <u>Surf.Sci.</u>, 170 (1986).

28. E. Gornik, W. Seidenbusch, R. Lassnig, H.L. Stoermer, A.C. Gossard, W. Wiegmann, FIR investigations of GaAs/GaAlAs heterostructures, <u>Springer Series in Solid State Sciences</u>, 53:60 (1984).

29. B. Bastard, Hydrogenic impurity states in quantum well structures, <u>Phys.Rev.B</u>, 24:4714 (1981).

30. R.L. Greene, K.K. Bajaj, Energy levels of hydrogenic impurity states in $GaAs-Ga_{1-x}Al_xAs$ quantum well structures, <u>Solid State Commun.</u>, 45:825 (1983).

31. N.C. Jarosik, B.D. McCombe, B.V. Shanabrook, I. Comas, I. Ralston, and G. Wicks, Binding of shallow donor impurities in quantum well structures, <u>Phys.Rev.Lett.</u>, 54:1283 (1985).

32. J.L. Robert, A. Raymond, L. Konczewicz, C. Bousequet, W. Zawadzki, F. Alexandre, I.M. Masson, J.P. Andre, P.M. Frijlink, Magneto-impurities and quantum wells, <u>Phys.Rev.B</u>, 33:5935 (1986).

33. A. Raymond, J.P. Andre, The zero resistance state in GaAs-GaAlAs-heterojunctions: evidence of a nearest-neighbour hopping process, <u>Phys.Rev.</u>, to be published.

34. D. Heitmann, M. Ziesmann, L.L. Chang, Cyclotron resonance oscillations in InAs quantum wells, <u>Phys.Rev. B</u>, 34:7463 (1986).

35. R. Lassnig, W. Seidenbusch, E. Gornik, G. Weimann, Landau level width and cyclotron resonance in 2D systems, in: "Proc. of the 18th Int. Conf. on The Physics of Semiconductors, Stockholm", O. Engström, ed., World Scientific, Singapore (1987), p.593.

36. F. Stern and W.E. Howard, Properties of semiconductor surface inversion layers in the electric quantum limit, <u>Phys.Rev.</u>, 163:816 (1967).

37. W. Walukiewicz, H.E. Ruda, J. Lagowski, and H.C. Gatos, Electron modulation-doped heterostructures, <u>Phys.Rev.B</u>, 30:4571 (1984).

38. F. Thiele, W. Hansen, M. Horst, J.P. Kotthaus, J.C. Maan, U. Merkt, K. Ploog, G. Weimann, and A.D. Wieck, Cyclotron resonance in n-GaAs/GaAlAs heterojunctions, <u>Springer Series in Solid State Sci.</u>, 71:252 (1987).

39. W. Seidenbusch, E. Gornik, G. Weimann, Cyclotron resonance linewidth oscillations in the integer and fractional quantum hall regimes, Phys. Rev. B, to be published.

40. E. Gornik, T.Y. Chang, T.J. Bridges, V.T. Nguyen, J.D. Mc Gee, W. Müller, Landau level electron lifetimes in n-InSb, Phys.Rev. Lett., 40:1151 (1978).

41. G.R. Allan, A. Black, C.R. Pidgeon, E. Gornik, W. Seidenbusch, P. Colter, Impurity and Landau-level electron lifetimes in n-type GaAs, Phys.Rev. B, 31:3560 (1985).

42. M. Helm, E. Gornik, A. Black, G.R. Allan, C.R. Pidgeon, K.Mitchell, Hot electron Landau level lifetime in GaAs/GaAlAs heterostructures, Physica, 134B:323 (1985).

43. J.F. Ryan, Time-resolved photoluminescence for quantum well semi-conductor heterostructures, Physica, 134B:403 (1985).

44. R.W. Hollering, T.T. Berendschot, H.J. Bluyssen, H.A. Reinen, P. Wyder, Energy relaxation of lower dimensional hot carriers studied with picosecond photoluminescence, in: "Proc. of the 18th Int. Conf. on The Physics of Semiconductors, Stockholm", O. Engström, ed., World Scientific, Singapore (1987), p. 1323.

45. G.A. Rodriguez, R.M. Hart, A.J. Sievers, F. Keilmann, Z. Schlesinger, S. Wright, W.I. Wang, Intensity dependent CR in a GaAs/GaAlAs 2D electron gas, Appl.Phys.Lett., 49:458 (1986).

HIGH FIELD MAGNETOTRANSPORT: LECTURES I AND II: ANALYSIS OF SHUBNIKOV de

HAAS OSCILLATIONS AND PARALLEL FIELD MAGNETOTRANSPORT

R.J. Nicholas

Clarendon Laboratory
Parks Road
Oxford

INTRODUCTION

The study of high field magnetotransport, and in particular the Shubnikov-de Haas effect, is one of the most useful and direct ways of characterising semiconductor heterostructures. At low temperatures a two dimensional gas of carriers bound in a heterostructure acts like a metal with a small Fermi energy, typically of order 10-100 meV. The magnetic field causes a quantisation of the free carrier states into a ladder of Landau levels. Changing the magnetic field sweeps the levels through the Fermi energy causing the familiar oscillations in the magnetoresistance, known as the Shubnikov-de Haas effect. This phenomenon was first used by Fowler et al,[1] in the first demonstration of the existence of a two-dimensional gas of electrons bound at the surface of a (100) silicon MOSFET. The periodicity of the oscillations is directly proportional to the carrier concentration bound in the layer, and is independent of the number of layers which may be present. Since the cyclotron motion induced by the field is in the plane perpendicular to its direction, this means that a two-dimensional system is sensitive only to the component of field parallel to the surface normal, so that rotation of the sample relative to the field can be used to provide a very simple and direct proof of the two-dimensional nature of any system under investigation.

Using the Landau gauge ($A = B(0,x,0)$) we may write the Hamiltonian for the carriers in a 2-D system as

$$H = 1/2m^* (\underline{p} - e\underline{A})^2 + V(z) \tag{1}$$

where $V(z)$ is the confining potential which leads to 2-D behaviour. P_y commutes with this Hamiltonian, and $V(z)$ leads to a quantum number defining the subband, and a confinement energy E_z. Thus we have

$$H = p^2/2m^* - e\hbar B/m^* xky + e^2B^2x^2/2m^* + E_z \tag{2}$$

and by substituting $x' = x - \hbar ky/m^*\omega_c$, this leads to

$$H = p^2/2m^* + m^*/2 \, \omega_c^2 x'^2 - \hbar^2ky^2/2m^* + E_z \tag{3}$$

and

$$E = (n+\tfrac{1}{2})\hbar\omega_c + E_z \tag{4}$$

The corresponding wavefunctions are:

$$\psi = e^{ik_yY} \times H_n(x-x_o) \times \phi(z) \tag{5}$$

where $H_n(x-x_o)$ is the simple Harmonic oscillator wavefunction and $\phi(z)$ comes from the z quantisation. The degeneracy of the levels comes from the constraint that the centre of motion x_o must lie within the sample. Thus for a sample of width L_x (and length L_y) we have

$$0 < \hbar k_y/m^*\omega_c < L_x \tag{6}$$

giving a range of $k_y = m^*\omega_c L_x/\hbar$, and leading to a total number of states per level of n,

$$n = \Delta k_y/2\pi/L_y = m^*\omega_c/h\, L_x\, L_y \tag{7}$$

$$= eB/h \text{ per unit area}$$

The resulting final density of states for a 2-D system is shown in fig. 1. On the left (fig. 1a) is shown the formation of Landau levels in the absence of scattering, giving simply a set of delta functions. Figs. 1b and 1c include a broadening of the levels due to scattering of the carriers by the presence of impurities and imperfections in the surface or confining potential. In fig. 1b the density of carriers in the system is such that one level is only half filled, and in this case the system will act as a metallic conductor. In fig. 1c a few more carriers have been added (or the magnetic field reduced), so that the level becomes completely filled, and we have an insulator. The conductivity will thus oscillate strongly as a function of carrier concentration, or as a function of magnetic field, since the degeneracy of each level is proportional to magnetic field, as shown above. This occurs whenever

$$n_e = N\, 2eB_N/h = NB_N \times 0.484 \times 10^{11}/cm^2.T, \tag{8}$$

where N is an integer, and we have included an additional factor of two for spin degeneracy. At these points the Hall voltage displays the quantised Hall effect first discovered by von Klitzing et al,[2]. At high fields or low temperatures spin splitting of the oscillations can be seen very easily, and the periodicity will halve. A common way to define the state of the system at high fields is in terms of the occupancy factor v, which counts the number of filled levels (counting spin states separately) and is defined by

$$v = n_e/n = h\, n_e/eB \tag{9}$$

When we come to make experimental measurements of Shubnikov-de Haas oscillations, then it is also important to consider the geometry of the sample studied. The most useful case is a long bar shape, with separate potential probes to look at both resistive and Hall fields. The resistivity ρ_{xx} and ρ_{xy} is related to the conductivity σ_{xx} and σ_{xy} via the relations

$$\rho_{xx} = E_x/I_x = \sigma_{xx}/\sigma_{xx}^2 + \sigma_{xy}^2 \,, \tag{10}$$

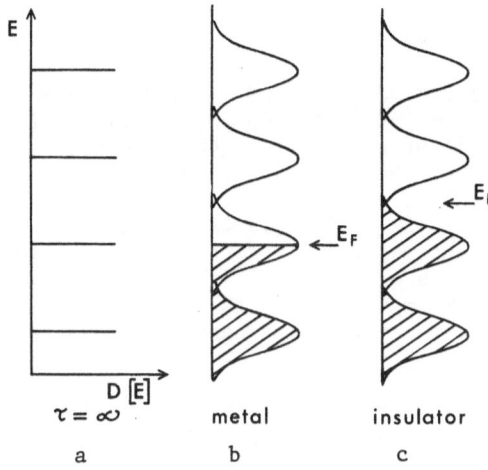

Fig. 1. A schematic view of the density
of states in a two-dimensional system.
Figs. 1b and 1c include the effects of
uncertainty broadening of the Landau
levels.

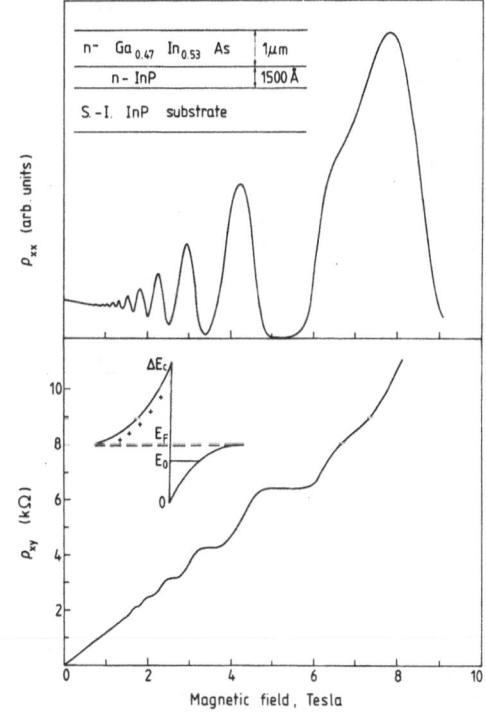

Fig. 2. Typical experimental recordings
of Shubnikov-de Haas oscillations
in ρ_{xx}, and quantum Hall steps in
ρ_{xy}, for a Ga.47In.53As-InP heterojunction at 1.6K.

$$\rho_{xy} = E_y/I_x = \sigma_{xy}/\sigma_{xx}^2 + \sigma_{xy}^2 \,. \qquad (11)$$

At high magnetic fields, $E_y/E_x = \sigma_{xy}/\sigma_{xx} = \mu B \gg 1$, so we may write $\rho_{xx} \sim \sigma_{xx}/\sigma_{xy}^2$ and $\rho_{xy} \sim 1/\sigma_{xy} = R_H B$, where R_H is the Hall coefficient. The result that the resistivity is proportional to the conductivity is due to the very important influence of the Hall field, and the way in which resistivity and conductivity have been defined, for zero Hall current and zero Hall field respectively. If one is more interested in the regions in which both the conductivity and resistivity go towards zero, for completely filled Landau levels, then it is better to use the "Corbino" geometry in which resistance is measured from the inside to the outside of an annulus. By symmetry the Hall field must be zero, and hence the measured resistance is proportional to $1/\sigma_{xx}$.

Some typical experimental recordings of the resistivity and Hall voltage are shown in fig. 2, which show the Shubnikov-de Haas oscillations and the quantised Hall effect. This type of measurement has now become a standard way of characterising 2-D systems. Typical information which can be derived from such measurements is

i) The periodicity, which gives the 2-D carrier concentration independent of geometrical conditions.
ii) The temperature dependence of the amplitude of the oscillations at low fields can be used to deduce the effective mass; the dependence of the amplitude of the oscillations upon electric field can then be used to measure the electron temperature as a function of electric field.
iii) The magnetic field dependence of the amplitude can be used to deduce the scattering time,[3].
iv) When more than one quantised subband is occupied, then the oscillations may be analysed to give the relative populations in the different subbands. This will be discussed in more detail below.
v) The magnitude of the spin splitting of the Landau levels is determined by the total magnetic field, whereas the Landau level separation comes from the perpendicular component of magnetic field. By rotating the sample relative to the magnetic field direction it is thus possible to alter the relative magnitudes of the spin and Landau splittings, leading to level coincidences which then allow one to make measurements of the electron g-factor,[4].

MULTIPLE SUB-BAND OCCUPANCY

The condition necessary for the establishment of two-dimensional behaviour is that all carriers present in the system occupy the same quantum state for motion in the third dimension. In typical semiconductor lattices and heterojunctions, it is quite often the case that more than one electric subband is occupied; however, it is still very useful to consider the system as essentially two-dimensional. If, for example the Fermi energy is sufficiently high that two subbands are populated, then the application of a magnetic field will cause two separate ladders of Landau levels to be populated, but the two ladders will be offset relative to each other by the energy separation between the subbands. At low magnetic fields, where the Landau levels are only weakly resolved, the population of each subband will remain approximately constant and two independent series of Shubnikov-de Haas oscillations will be observed, each determined by eq. (8) with the appropriate concentration n_i of the ith subband. At higher fields, once the levels are completely resolved, then the field positions of the conductivity minima will be determined by the total carrier concentration. In this second case, however, the amplitudes of the oscillations will become somewhat irregular, since they will depend

upon the relative positions of Landau levels arising from different subbands. A further complicating factor will be that this will cause a
redistribution of population between the two subbands, and since the
potential causing the subband quantisation is self-consistently determined,
the separation between the subbands will itself change as a function of
magnetic field.

An example of this behaviour is shown in fig. 3 for a heterojunction
of $Ga_{0.47}In_{0.53}As/Al_{0.48}In_{0.52}As$,[5]. In this case two electric subbands are
occupied, with carrier concentrations of 0.9 and 6.0 x 10^{11} cm^{-2},
respectively. At low fields it is fairly clear that there are two independent series of oscillations present, with NB products of 1.9 and 12.3 T,
corresponding to the two subbands acting independently. At higher fields
the oscillations become much more irregular, but there is a very pronounced
minimum in the resistivity at 14.2 T. This corresponds to the complete
occupation of a single Landau level by the total electron population, and
so the upper subband has been completely depopulated by the magnetic field.
What is happening is that the Fermi energy is a very strong function of
magnetic field, once the Landau levels are well resolved. The upper
section shows schematically the motion of the Fermi energy as a function of
magnetic field, calculated on the assumption of infinitely sharp Landau
levels. The two fans of lines originating from E=0 and E=26 meV, show the
Landau levels of the first two electric subbands, including spin splitting.
The motion of the Fermi energy is shown by the bold line, and is calculated
from the known electron concentration and the degeneracy of each Landau
level. At very high fields all of the electrons can be put into the
lowest level, as the field falls, there comes a critical value at which this
level is completely filled, whereupon the Fermi energy must jump up to the
next available level, irrespective of from which subband it originates. A
more pronounced example of this behaviour is shown in fig. 4, for a
$Ga_{0.47}In_{0.53}As/InP$ heterojunction with three populated subbands at lower
temperatures,[6]. The Shubnikov-de Haas oscillations are now very well
defined, leading to clear zeros in the conductivity, but still with irregular maxima. The Hall voltage also shows irregular, but quantised Hall
steps. Due to the mixing up of the spin states and levels from the
different subbands, unexpected filling factors give rise to strong, wide
plateaus, e.g. for $\nu = 5$.

Heterojunctions and surface accumulation layers formed on narrow band
gap semiconductors lead to the most extreme examples of multiple subband
occupancy, particularly for the case of accumulation layers. In
$Hg_{0.8}Cd_{0.2}Te$ M.I.S. structures up to five occupied subbands have been
detected,[7] which leads to very complex behaviour for the Shubnikov-de Haas
oscillations. This is shown in fig. 5, for three different surface
accumulation layer concentrations. In order to illustrate this more clearly
it is necessary to enhance the strength of the oscillatory structure. This
is usually done by the use of derivative techniques, taking either the
first or second derivatives of the resistivity with respect to magnetic
field, or by taking derivatives with respect to the carrier concentration
for gated structures. This is illustrated in fig. 6, where the second
derivative of resistivity is shown, measured experimentally by using high
frequency pass filters, with a gain proportional to frequency squared,
which generates the second derivative for sinusoidal oscillations.
Different magnetic field sweep rates (or filter characteristics) can then
be used to optimise different regions of the oscillations. The shape of
the oscillations is rather complex, so that in order to extract the
periodicities it is necessary to use Fourier analysis. Fig. 7 shows both
the direct oscillations taken this time with a gate voltage derivative,[8]
and the resulting Fourier spectrum. The oscillations are periodic in
1/B, so that the analysis must be done using a 1/B sampling. For a single

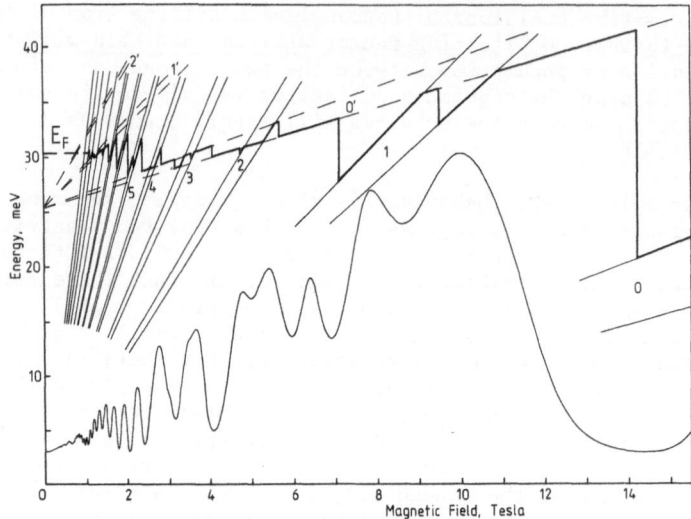

Fig. 3. Shubnikov-de Haas oscillations (lower curve) in
a Ga.$_{47}$In.$_{53}$As/Al.$_{48}$In.$_{52}$As heterojunction at
1.6K. The upper fan diagram shows the Landau
levels originating from the two occupied subbands
and the bold line shows the motion of the Fermi
level as a function of magnetic field,[5].

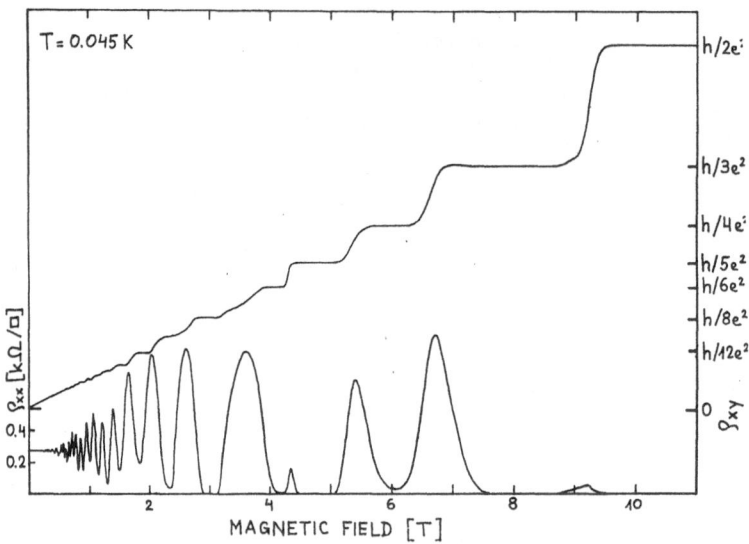

Fig. 4. Shubnikov-de Haas and quantum Hall steps at 45mK in
a Ga.$_{47}$In.$_{52}$As-InP heterojunction with two occupied
subbands,[6].

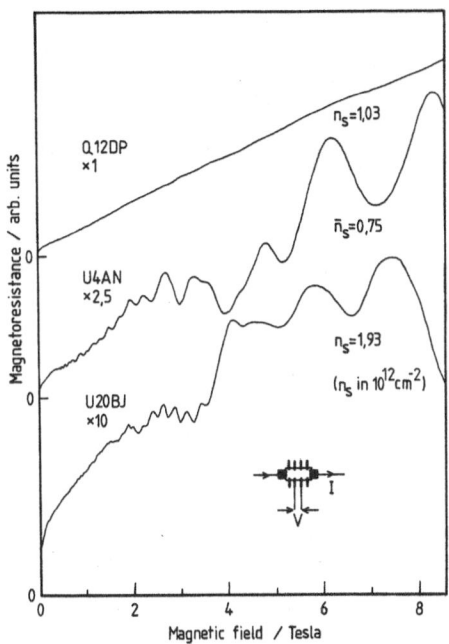

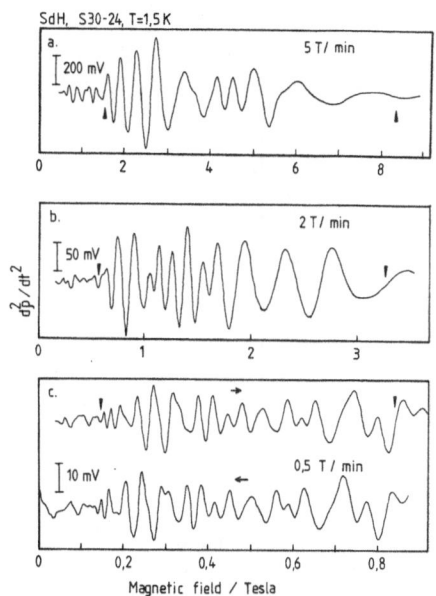

Fig. 5. Shubnikov-de Haas oscillations in a $Hg_{.8}Cd_{.2}Te$
M.I.S. surface accumulation
layer,[7].

Fig. 6. Second derivative traces of
Shubnikov-de Haas oscillations
in a $Hg_{.8}Cd_{.2}Te$ M.I.S.
structure, taken over
different field ranges at
different sweep speeds.

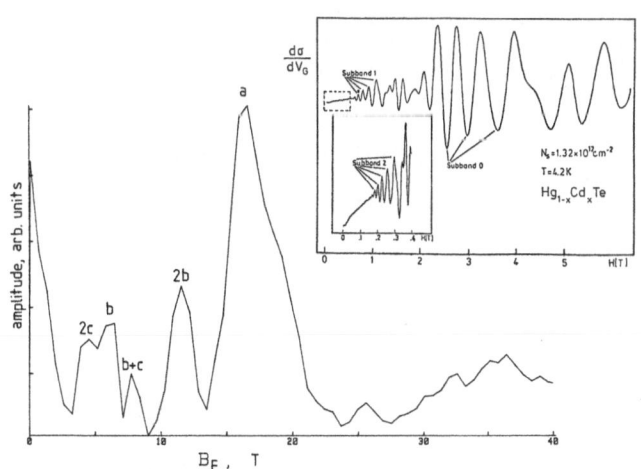

Fig. 7.   Shubnikov-de Haas oscillations in the transconductance (inset) of a HgCdTe M.I.S.
structure (after 8), and the results of a
Fourier analysis of the same trace below.

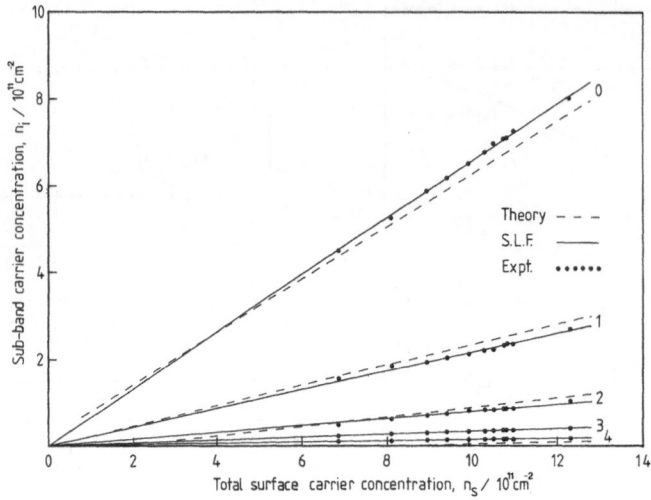

Fig. 8. Experimental results for the population of the first
5 subbands in $Hg_{.8}Cd_{.2}Te$ accumulation laers, compared
with theory,[9][10].

electron concentration the oscillation positions (B_N) are defined by the
condition

$$n_e = 2e(B_N N)/h \tag{12}$$

which gives a periodicity B_F

$$B_F = 1/\Delta(1/B) = NB_N = hn_e^i/2e \tag{13}$$

For multiple subband occupancy at low fields each subband contributes a
separate series of oscillations defined by equation (13), with the total
carrier concentration $n_e = \Sigma n_e^i$. At high fields when the Landau levels are
well separated then the oscillations from different subbands mix, and we
find combination bands given by $B_F = B_F^i + B_F^{i'}$, and also as the spin states
become resolved then the periodicity doubles and second harmonics appear.
This can be seen in fig. 7, where peak a corresponds to n^o, peak b is n^1,
peak b+c is the combination $n^1 + n^2$, and peaks 2b and 2c are the second
harmonics $2n^2$, and $2n^1$. As a result of this analysis, it is then possible
to deduce the distribution of carriers between different electric subbands,
and compare this with the predictions of theory,[9,10] as shown in fig. 8.

PARALLEL FIELD DEPOPULATION

Another very useful way of analysing and characterising systems with
a multiple subband occupancy is to apply a magnetic field in the plane of
the 2-D system. In this case the effect of the parallel field component
is to add additional terms to the Hamiltonian describing the system, which
act to change the energies of the electric subbands. The parallel field
component acts most strongly on electric subbands with a large spatial
extent, so that for an accumulation layer the higher subbands are pushed up
in energy relative to the positions of the lower subbands. The result of
this is that they become progressively depopulated. This depopulation can
be seen directly in the resistivity of the system, through its influence
on the scattering mechanisms. Fig. 9 shows schematically the energy levels

and dispersion relations for a system with two populated subbands. In this case scattering of the carriers can take place both within the individual electric subbands, and between subbands. Once the bottom of the higher subband rises above the Fermi energy, and is depopulated, then the elastic and quasi-elastic processes which cause inter-subband scattering can no longer take place, and the mobility of the system increases, leading to a decrease in resistivity. This behaviour is shown in fig. 10 for one of the $Hg_{0.8}Cd_{0.2}Te$ M.I.S. accumulation layers,[7] where the steady depopulation of four higher subbands can be seen, each resulting in a decrease in the resistivity of the system. This phenomenon is remarkably sensitive to the population of higher subbands with very small carrier concentrations. In the example shown, the two uppermost levels are estimated to have populations of 1 and 3×10^{10} cm^{-2}, which were undetectable in the Shubnikov-de Haas measurements with perpendicular magnetic fields.

The exact field positions of the depopulation structures are identified as the mid point of the resistivity decrease, which is broadened due to an energy broadening of the subband edge. At higher fields two depopulation structures can be seen for each subband, due to the spin splitting of the levels, which causes a different energy shift for the levels of opposite sign. In addition there is also a positive magnetoresistance, even after all of the carriers are accommodated into a single subband. This is thought to be due to a change in the subband wavefunction caused by the parallel field component, which in turn alters the scattering processes.

Using a perturbation approach it can be shown (Ando, Fowler and Stern,[11])that the parallel field adds a phase factor to the wavefunction of

$$\exp \left[ieB_y<z>x/\hbar \right] \tag{14}$$

and adds two additional terms to the Hamiltonian, H_2 and H_3, given by

$$H_2 = eB_y/m^* \ (z - <z>)\rho_x$$

$$\tag{15}$$

$$H_3 = e^2B_y^2/2m^* \ (z^2 - <z>^2 - 2z<z>)$$

where $<z>$ is the mean spatial extent of the quantised wavefunction. In this approximation the effect of H_2 is small, and the Hamiltonian remains separable. The energy of any particular level shifts under the influence of H_3 by an amount

$$E_i = E_i^o + e^2B_y^2/2m^* \left[<\Delta z>_i^2 \right] \tag{16}$$

where $<\Delta z>^2 = <z^2> - <z>^2$. If we consider the case of the depopulation of a system with two occupied subbands, then the upper level will be depopulated when the relative energy shift of the levels is twice the original Fermi level in the upper subband, giving

$$E_F^i \ (B = 0) = e^2B_y^2/m^* \left[<\Delta z>_i^2 - <\Delta z>_o^2 \right] \tag{17}$$

We have seen above that in accumulation layers it is generally the case that $n_i \ \alpha \ n_e$ (e.g. fig. 8), and in addition surface space charge layers in a self consistent potential obey the approximate rule that the z-wavefunction scales as $n_e^{-1/3}$, through the self-consistency of the potential,[11]. Taking these factors into account, we would then expect that the depopulation field should give

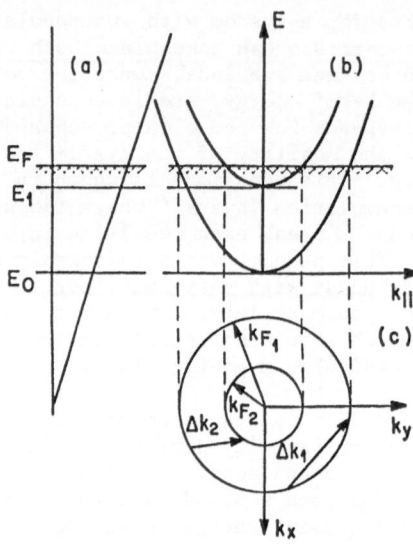

Fig. 9. A schematic picture of the scattering
processes in a 2 subband electron gas.
(b) shows the dispersion relation and
'Fermi discs' for the two subbands,
with intra-subband scattering (Δk_1)
inter-subband scattering (Δk_2).

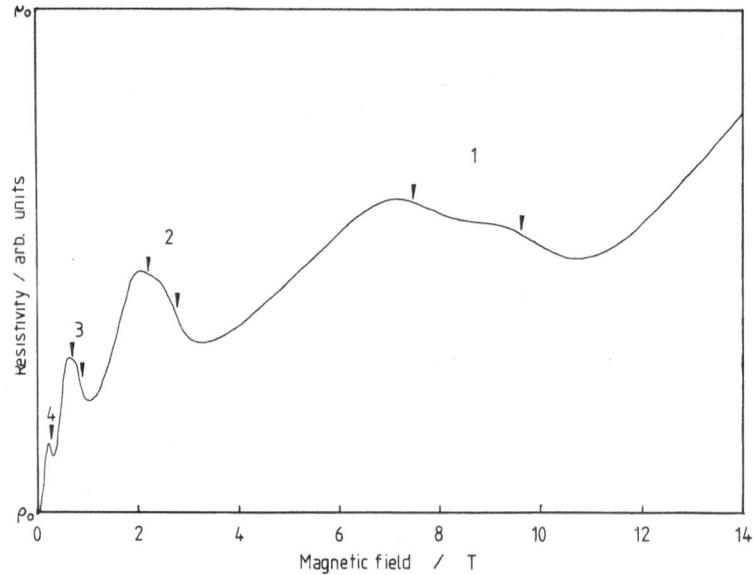

Fig. 10. The parallel field magnetoresistance of a $Hg_{.8}Cd_{.2}Te$
accumulation layer ($n \sim 9 \times 10^{11} \text{ cm}^{-2}$). The numbers
indicate the successive depopulation of the electric
subbands, which are also spin split.

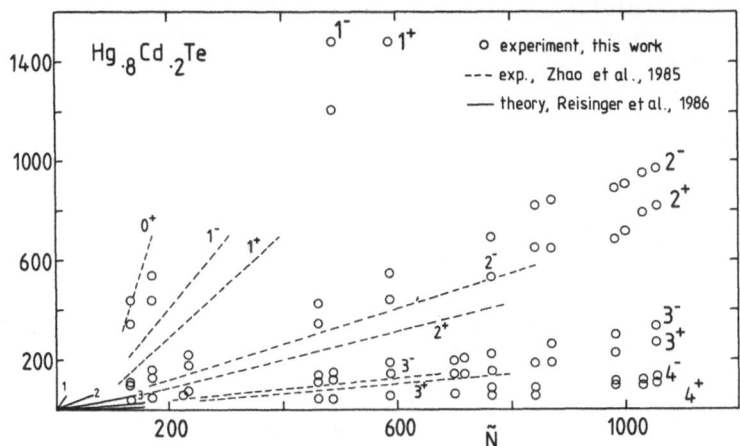

Fig. 11. Plots of the depopulation field versus carrier concentration for a variety of samples of Hg.8Cd.2Te accumulation layers. The fields and carrier concentrations are plotted in reduced units (see text). Linear extrapolations of the numerical models[12] agree well with the experimental results.

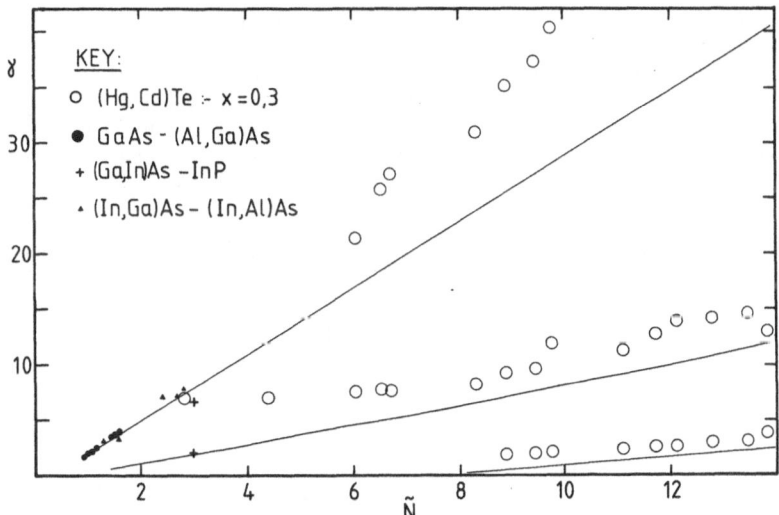

Fig. 12. Plots of the depopulation field versus carrier concentration in reduced units for a variety of different systems; Hg.7Cd.3Te (13), GaAs-GaAlAs (14), GaInAs-InP (6) and GaInAs-AlInAs (5). This shows good agreement with the theory of Reisinger and Koch[12].

$$B_y \propto n_e^{5/6} \tag{18}$$

Some examples of depopulation fields are shown in fig. 11 for the $Hg_{0.8}Cd_{0.2}Te$ surface space charge layers, plotted as a function of carrier concentration. These show an almost exactly linear dependence upon n_e, as might be expected from the reasoning given above. In this figure the field and carrier concentration values are in fact plotted in terms of the reduced units

$$B_y \text{ in units of } \hbar\omega_c^{11}/R^* \ , \quad (R^* = e^2/8\pi\varepsilon_s\varepsilon_o a^*) \tag{19}$$

and n_e in units of n_s/a^{*2} , $(a^* = 4\pi\varepsilon_s\varepsilon_o \hbar^2/m^*e^2)$ (20)

as first introduced by Reisinger and Koch,[12]. These authors used numerical calculations to show that the parallel field depopulation could be written as a universal function of n_e when written in these units, and the results of their calculations are shown as solid lines in fig. 11, together with the results of some earlier measurements by Zhao et al.,[8].

Although only performed for relatively lower electron concentrations, a linear extrapolation of the theoretical results appears to give remarkably good agreement with the experimental data, although some discrepancies appear between the two sets of data,[8,13]. This is partially due to an incorrect analysis of the Shubnikov-de Haas data by Zhao et al.,[8]. The agreement with theory is even more remarkable, when it is realised that the calculation takes no account of non-parabolicity, which can alter the effective mass by more than a factor of 5, for the lowest subband at high electron concentrations,[7].

Further evidence for this universal behaviour is shown in fig. 12, where the calculations are compared to results for GaAs,[14], GaInAs-InP,[6], GaInAs-AlInAs,[5], and $Hg_{0.7}Cd_{0.3}Te$,[13]. The systems chosen were all accumulation layers, or as in the case of the GaAs-GaAlAs have very low depletion charge densities. The agreement with theory is seen to be extremely good, giving a linear dependence over approximately three orders of magnitude.

REFERENCES

1. A.B. Fowler, F.F. Fang, W.E. Howard and P.J. Stiles, Phys. Rev. Lett. 16 901 (1966).
2. K. von Klitzing, G. Dorda and M. Pepper, Phys. Rev. Lett. 45 494 (1980).
3. R.B. Dingle, Proc. R. Soc. London, Ser. A211 500 (1952).
4. R.J. Nicholas, R.J. Haug, K. von Klitzing and G. Weimann, Phys. Rev. B, to be published (1987).
5. J.C. Portal, R.J. Nicholas, M.A. Brummell, A.Y. Cho, K.Y. Cheng and T.P. Pearsall, Solid State Commun. 43 907 (1982).
6. M. Razeghi, J.P. Duchemin, J.C. Portal, L. Dmowski, G. Remenyi, R.J. Nicholas and A. Briggs, Appl. Phys. Lett. 48 712 (1986).
7. J. Singleton, R.J. Nicholas, F. Nasir and C.K. Sarkar, J. Phys. C: Solid State Phys. 19 35 (1986).
8. W. Zhao, F. Koch, J. Ziegler and H. Maier, Phys. Rev. B31 2416 (1985).
9. T. Ando, J. Phys. Soc. Japan 54 2676 (1985).
10. Y. Takada, K. Arai, N. Uchimura and Y. Vemura, J. Phys. Soc. Japan 49 1851 (1980).
11. T. Ando, A. Fowler and F. Stern, Rev. Mod. Physics 54 437 (1982).
12. H. Reisinger and F. Koch, Surf. Sci. 170 397 (1986).
13. F. Nasir, J. Singleton and R.J. Nicholas, to be published (1987).
14. J.J. Harris, D.E. Lacklison, C.T. Foxon, F.M. Selten, A.M. Selten, A.M. Suckling, R.J. Nicholas and K.W.J. Bornham, to be published (1987).

PHYSICS AND APPLICATIONS OF THE QUANTUM HALL EFFECT

Klaus v. Klitzing

Max-Planck-Institut für Festkörperforschung, Heisenbergstr. 1
D-7000 Stuttgart 80, FRG

INTRODUCTION

The most fascinating property of the Quantum Hall Effect (QHE) is the
phenomenon that from a relatively simple experiment on a semiconductor a
new type of electrical resistor R_0 can be deduced which is independent of
microscopic properties of the semiconductor and reproducable at a level of
better than 10^{-6}. In the recent publication of "The 1986 Adjustment of the
Fundamental Physical Constants"[1] one finds a new general constant, the
quantized Hall resistance with a recommended value
$R_0 = (25812.8056 \pm 0.00012)$ Ohm. This value is identical with the ratio h/e^2,
the ratio between the Planck constant h and the square of the electron
charge e. The surprising result is, that this universal constant R_0 can be
measured directly on a macroscopic system. In principle all Hall effect
measurements in strong magnetic fields and low temperatures on a two-
dimensional electron gas show the QHE. However, for the experimental realization of this quantum phenomenon one has to specify the condition "low
temperature", "high magnetic field" and "two-dimensional system". The
inversion layer at the $Si-SiO_2$ interface of a silicon MOS fieldeffect
transistor is the classical example for a two-dimensional system[2] but in a
more general way all structures with a vanishing conductivity in one direction may be called two-dimensional electron gas. Even a superlattice
with a macroscopic thickness of some micrometers can show two-dimensional
properties if the periodicity of the superlattice leads to well separated
minibands[3]. In this respect most of the structures discussed at this
school are related to the physics of two-dimensional systems and can be
used for a discussion of the quantum Hall effect. The reduced dimensionality is necessary in order to obtain gaps in the electronic spectrum.

Theoretically all finite systems which consist in principle of a discrete energy spectrum should show the QHE. Therefore it is not surprising that some theoretican claim that the QHE is in principle present in any finite electronic system[4], but in reality only few seminconductor systems are used for an experimental study of the QHE. At present the simplest one seems to be the GaAs-heterostructure since the mobility of the electrons in the plane of the two-dimensional layer is so high that already at relatively low magnetic fields perpendicular to the layer a closed cyclotron orbit with discrete Landau energies E_n is present. This Landau quantization together with the size quantization of the two-dimensioal system leads to the desired energy gaps ΔE which form the basis for the simplest explanation of the QHE. The quantum Hall effect becomes more and more pronounced if the ratio $\Delta E/kT$ increases which means low temperatures and high magnetic fields since the energy gap ΔE increases approximately linearly with the magnetic field. This indicates already that all measurements at finite temperature are measurements under nonideal conditions.

After the Nobel Prize in Physics in 1985 for the discovery of the QHE different review articles and books with more than 500 references were published on this subject[5-8]. In the following not all aspects of the QHE can be covered and only a tutorial introduction together with a summary of recent work in this field will be given including the applications in metrology. The realization of a two-dimensional electron gas in field-effect transistors, heterostructures and quantum wells is the subject of a large number of contributions in this book and will not be discussed in this article.

HALL EFFECT

The textbook interpretation of the Hall effect is usually based on a discussion of the electron motion within a long sample (where the current direction I_x is fixed) in a transverse magnetic field B_z. The Lorentz force on the electrons moving with a velocity v_x is perpendicular to the current and magnetic field direction and has to be compensated by the force on electrons in an electric field E_H. This Hall field E_H is built up perpendicular to the magnetic field and current direction since the electrons have to flow finally in a fixed direction (x-direction). The resulting Hall voltage U_H in y-direction is inversely proportional to the

carrier density n_{3d} and the thickness d of the sample and increases
linearly with the current I and the strength of the magnetic field B

$$U_H = \frac{B_z \cdot I_x}{n_{3d} \cdot d \cdot e} \qquad (1)$$

The product of the three-dimensional carrier density n_{3d} with the
thickness d of the sample is equivalent to a two-dimensional electron
density $n_{2d} = n_{3d} \cdot d$ which corresponds to a projection of the carriers into a
plane perpendicular to the magnetic field direction. With the definition
of a Hall resistance $R_H = U_H / I_x$ one obtains

$$R_H = \frac{U_H}{I_x} = \frac{B}{n_{2d} \cdot e} \qquad (2)$$

It should be noted that the width of the sample does not appear in
Eq. (2). Based on the tensor relation between the current density j and
the electric field E

$$E = \rho \cdot j \qquad (3)$$

with the resistivity tensor ρ one can identify the Hall resistance R_H
directly as the resistivity component ρ_{xy} and the measured voltage drop in
the direction of the current is proportional to the diagonal component ρ_{xx}
of the resistivity tensor.

$$E_y = \rho_{yx} \cdot j_x \qquad (4a)$$

$$E_x = \rho_{xx} \cdot j_x \qquad (4b)$$

The Eqs. (4a) and (4b) are only correct if the current density j_y is zero.
This is not the case close to the current contacts of a Hall device since
the current direction in the presence of a magnetic field is not perpendicular to equipotential lines (= boundary of the metal contacts) so that
the current component j_y is not zero close to the current contacts.
Therefore potential probes are necessary for an experimental determination
of the resistivity components as shown in Fig. 1. For measurements without
magnetic field it is usually not very important whether the potential drop
in the direction of the current is measured by a two terminal method
(potential measurement between the current leads) or by a four terminal
method which is only used at small resistance values in order to avoid

problems with the contact resistance. However in a magnetic field it is absolutely necessary to use potential probes, especially in measurements of the quantum Hall effect. In this case the voltage U_x in Fig. 1 becomes zero whereas the voltage drops between the current contacts is identical with the Hall voltage U_H. The problem of the current and potential distribution in a magnetic field of a homogeneous system with Hall geometry has been solved by using different techniques[9-12] but these calculations do not include a solution of the Poisson equation.

Whereas Hall devices are used for the determination of the resistivity components ρ_{xx} and ρ_{xy} another geometry, the so-called Corbino geometry is used for a direct measurement of the conductivity tensor component σ_{xx}. In this case the sample has a circular geometry with one electrical contact in the center of the disk and the other contact at the outer boundary

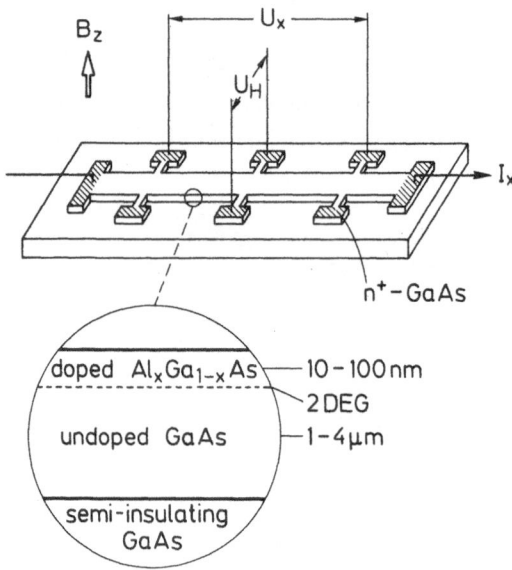

Fig. 1. Typical geometry of a sample used for Hall effect measurements. The formation of a two-dimensional electron gas (2DEG) in a GaAs heterostructure is shown in the enlargement of the cross section. The Hall voltage U_H and the voltage drop U_x are measured under constant current condition I_x=const as a function of the magnetic field B_z perpendicular to the 2DEG

232

of the sample. Since for this geometry a fixed direction for the electric field is given, the equation

$$j = \sigma \cdot E \qquad (5)$$

is more useful than Eq. (3) because the measured current at a fixed applied voltage is directly proportional to the component σ_{xx} of the conductivity tensor. However, the current flow perpendicular to the applied electric field, which is connected with the Hall conductivity σ_{xy}, cannot be measured. Therefore all measurements of the QHE are done on devices with Hall geometry as shown in Fig. 1.

The classical expressions for the resistivity and conductivity components of a homogeneous degenerate system as a function of the magnetic field are

$$\rho_{xx}(B) = \rho_0 = \text{const} \qquad (6a)$$

$$\rho_{xy}(B) = \frac{B}{n_{2d} \cdot e} \qquad (6b)$$

$$\sigma_{xx}(B) = \frac{\sigma_0}{1+\mu^2 B^2} \qquad (6c)$$

with $\sigma_0 = \rho_0^{-1} = n_{2d} \cdot e \cdot \mu$

In general, the relations between the resistivity and conductivity tensor components are the following:

$$\begin{pmatrix} \sigma_{xx} & \sigma_{xy} \\ \sigma_{xy} & \sigma_{yy} \end{pmatrix} = \frac{1}{\rho_{xx}^2 + \rho_{xy}^2} \begin{pmatrix} \rho_{xx} & -\rho_{xy} \\ \rho_{xy} & \rho_{xx} \end{pmatrix} \qquad (7a)$$

or

$$\begin{pmatrix} \rho_{xx} & \rho_{xy} \\ \rho_{xy} & \rho_{yy} \end{pmatrix} = \frac{1}{\sigma_{xx}^2 + \sigma_{xy}^2} \begin{pmatrix} \sigma_{xx} & -\sigma_{xy} \\ \sigma_{xy} & \sigma_{xx} \end{pmatrix} \qquad (7b)$$

Qualitative curves for the measurable quantities on a Hall device (ρ_{xx}, ρ_{xy}) and on a Corbino device (σ_{xx}) are shown in Fig. 2 for both the classical situation (dotted lines) and the situation where quantum pheno-

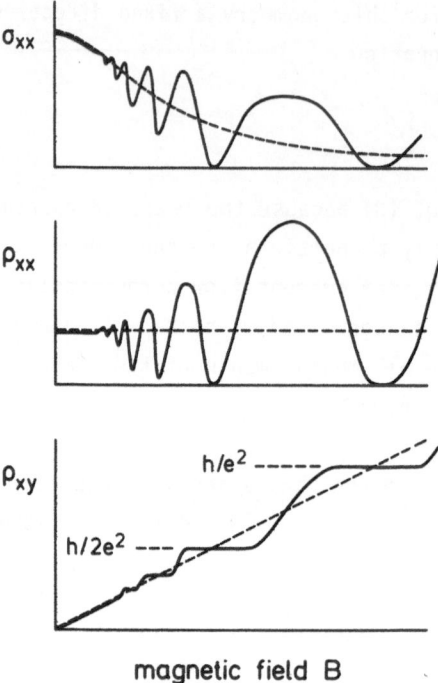

σ_{xx}

ρ_{xx}

ρ_{xy}

h/e^2 ----

$h/2e^2$ ---

magnetic field B

Fig. 2. Qualitative behaviour for σ_{xx}, ρ_{xx} and ρ_{xy} of a two-dimensional
electron gas with a fixed carrier density as a function of the
magnetic field. The dotted lines represent the classical curves. A
spin degeneracy is not included

mena are visible (full lines). The deviations from the classical curves
are observed if the energy spectrum is changed due to the Landau quantization. The following section discusses this quantization within the simple
picture of a free electron in a magnetic field. The electron-electron
interaction which seems to be responsible for the fractional Quantum Hall
Effect (see contribution of R.J. Nicholas in this book), is not included
in the following discussion.

LANDAU QUANTIZATION

The quantum Hall effect is characterized by an energy gap at the
Fermi energy at certain magnetic field values so that elastic scattering
processes are not possible. This leads to a vanishing conductivity σ_{xx} as
shown in Fig. 2 since scattering processes are necessary for a diffusion
of electrons in the direction of the electric field. Without scattering

the electrons move like free electrons in crossed electric and magnetic fields with a velocity $v_x = E_y/B_z$. Under the condition of the quantum Hall effect Eq. (7) reduces to

$$\sigma_{xx} = \rho_{xx} = 0 \qquad\qquad\qquad (8a)$$

$$\sigma_{xy} = -\frac{1}{\rho_{xy}} = \frac{-n_{2d} \cdot e}{B} \qquad\qquad\qquad (8b)$$

The existence of energy gaps in the spectrum of a two dimensional electron gas in a strong magnetic field can be understood from a discussion of the properties of an electron in a magnetic field. The absolute value of the energy gap is a very complicated function of the bandstructure but is not important for the discussion of the QHE. Only the existence of a gap is important and the interpretation of the QHE on the basis of gauge invariance arguments[13] demonstrates that a detailed knowledge of the energy spectrum is not necessary. However, for a microscopic picture of the QHE it is useful to start with an analysis of the properties of an electron in a magnetic field.

In a classical description one uses a picture where the electron is moving on a circle with radius ℓ around a center (X,Y). Without solving the Schrödinger equation of the problem one can show[14] that the commutator for the center coordinates of the cyclotron orbit $[X,Y]=i\ell^2=i\hbar/eB$ is finite, which is equivalent to the result that each electronic state occupies in real space the area

$$F_0 = \frac{h}{eB} \qquad\qquad\qquad (9)$$

This result is independent of the form of the wavefunction. The wavefunction itself depends on the gauge, on the orientation of the vector potential A. Depending on the symmetry of the problem one uses different gauges. The symmetric gauge is useful for a discussion based on a Corbino disk geometry whereas most of the calculations related to the Hall geometry use the Landau gauge with a vectorpotential in the direction of the electrical current. In this case the wavefunction ψ can be written as a plane wave in the direction of the current multiplied with the solution ϕ_n

of the harmonic oscillator equation

$$\psi = e^{ikx} \cdot \phi_n(y-Y) \tag{10}$$

In this representation, the center coordinate Y of the cyclotron motion is a good quantumnumber and directly related to the k-vector in x-direction.

$$Y = \hbar k/eB = k\ell^2 \tag{11}$$

For a finite length L_x of the sample in x-direction (with periodic boundary conditions) only discrete k-values with $\Delta k = 2\pi/L_x$ are permitted. This means (see Eq. (11)) that the center coordinates are separated by an amount $\Delta Y = (2\pi/L_x) \cdot \ell^2$ so that the area $F_0 = \Delta Y \cdot L_x$ occupied by one state with the quantumnumbers (k,n) is $F_0 = 2\pi\ell^2 = h/e \cdot B$. The energy E(k,n) is determined by the Landau quantumnumber n=0,1,2... if the electrostatic potential in y-direction is constant

$$E_n = E(k,n) = (n+\tfrac{1}{2})\hbar\omega_c \tag{12}$$

The cyclotron energy $\hbar\omega_c = eB/m_c$ depends on the mangetic field B and the cyclotron mass m_c and determines within the ideal one-electron picture the energy gap ΔE in the spectrum of a two-dimensional electron gas in a magnetic field.

The number N of electrons which can occupy each level E_n corresponds to the number of areas F_0 available within the area F of the device

$$N = \frac{F}{F_0} = F \cdot \frac{e \cdot B}{h} \tag{13a}$$

This result is identical with the number of flux quanta within the area F of the sample and can be interpreted as a condensation of all the electronic states without magnetic field (density of states $D = F \cdot (2\pi m_c/h^2)$ if the spin is not included) within the energy range $\Delta E = \hbar\omega_c$:

$$N = D\hbar\omega_c = \frac{F \cdot 2\pi \cdot m_c}{h^2} \cdot \frac{\hbar eB}{m_c} = \frac{F \cdot e \cdot B}{h} \tag{13b}$$

An integer number i of fully occupied energy level E_n corresponds to a carrier density of n=i·(eB/h). If this relation between carrier density n and magnetic field B is fulfilled, the Hall resistance R_H becomes (see Eq. (2))

$$R_H = \frac{h}{ie^2} \qquad (14)$$

Under the condition that the two-dimensional carrier density n_{2d} is fixed, as usually assumed in measurements on selectively doped heterostructures, a quantized Hall resistance h/ie^2 is expected at well defined magnetic field values

$$B_i = \frac{n_{2d} \cdot h}{i \cdot e} \qquad (15)$$

However, this result is not very exciting in connection with an application of the quantum Hall effect. The expected result is shown in the upper part of Fig. 3 where the Hall resistance R_H as a function of the magnetic field is plotted. R_H increases linearly with B and at very special values B_i of the magnetic field (see Eq. 15) the special value for the Hall resistance $R_H = h/ie^2$ is expected. The corresponding occupation of the Landau levels with electrons (a constant number of 180 electrons is assumed in this example) is sketched in the lower part of Fig. 3 where the magnetic field values are chosen in such a way that just one or four Landau levels are fully occupied and therefore the quantized Hall resistance values h/e^2 and $h/4e^2$ are expected. The size of the sample corresponds to the area $F = L_x \cdot L_y$ and the maximum number of electrons occupying one energy level is equivalent to the number of squares $F_0 = h/eB$ within the area F of the sample. It should be noted that for typical magnetic field values of about 4 Tesla the area F_0 is only $10^{-11} cm^2$, so that the size $L_x \cdot L_y$ of the samples shown in Fig. 3 is unrealistically small. The energy difference between the Landau levels increases linearly with the magnetic field.

This discussion of an ideal two-dimensional system in a magnetic field leads to the conclusion that for an accurate measurement of the quantized Hall resistance an accurate determination of the magnetic field is necessary. Fortunately, this is not the case. Experimentally well defined plateaus in the Hall resistance are observed so that a calibration of the magnetic field is not necessary. Different mechanisms may explain a constant value for ρ_{xy} within a certain magnetic field range and even today different groups prefer different interpretations. Some theories try to explain the Hall plateaus on the basis of the assumption that a reservoir

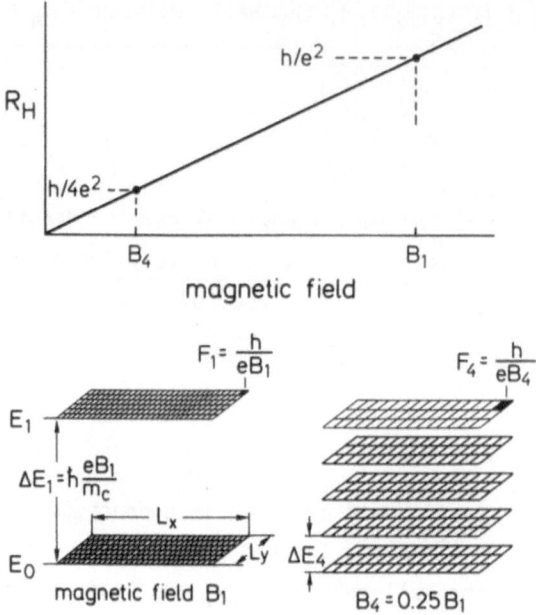

Fig. 3. For an ideal two-dimensional system in a magnetic field the Hall
resistance R_H increases linearly with the magnetic field B. The
quantized values $R_H=h/ie^2$ are obtained at fixed magnetic fields
$B_i=n_{2d} \cdot h/i \cdot e$. In the lower part of the figure the occupation of
the Landau levels E_n with 180 electrons at the magnetic fields B_1
and B_4 (where exactly one or four Landau levels are occupied) is
sketched. For simpicity, the spin degeneracy is not included. The
squares F_i surrounded by strong (weak) lines characterize occupied
(empty) electronic states

of electrons outside the two-dimensional system allows a change in the
electron concentration in such a way that a Landau level remains fully
occupied if the magnetic field is changed[15]. The reservoirs may be connected with interface states, metal contacts or the depletion layers in
the semiconductor. However, the majority of publications discuss the origin of the Hall plateaus in connection with a discussion of localized
states in the tails of the Landau levels[16-19]. Any disorder in the sample
like impurities, interface states or even the boundary of the sample may
lead to a change in the energy spectrum in such a way that Hall plateaus
appear. Some aspects of the Quantum Hall Effect in the presence of potential fluctuations will be discussed in the following section.

238

QUANTUM HALL EFFECT UNDER REAL CONDITIONS

A calculation of the QHE under real conditions is so complicated that a microscopic interpretation is not available. The finite temperature is usually not included in the calculations and the influence of the complicated boundaries of the sample on the QHE is an unsolved problem. A large number of theoretical papers discuss the influence of impurities on the quantum transport properties and the self-consistent Born approximation has been used for a determination of the density of states $D(E)$, the conductivity σ_{xx} and the deviation $\Delta\sigma_{xy}$ from the classical expression for the Hall conductivity [20],[21]. Within this approximation and short range scatterers, the following results are obtained. The density of states has a semi-elliptic shape with a broadening Γ proportional to $\sqrt{B}/\mu$, the conductivity σ_{xx} is mainly proportional to the square of the density of states at the Fermi energy and the expression for $\Delta\sigma_{xy}$ shows similarities to the classical calculations $\Delta\sigma_{xy}=\sigma_{xx}/\mu B$. However the main feature of the QHE, the existence of plateaus $\rho_{xy}(B)=\text{const}$ and $\rho_{xx}(B)=0$ could not be explained.

The experiments by Kawaji et al[22] indicated already that the standard picture within the mean field theory breaks down and that the localization of states due to disorder should be included. The numerical calculations[23] by Aoki shown in Fig. 4 demonstrated that the localization length α^{-1} of most of the states within one Landau level is finite but the conclusions regarding the influence of localized states on the Hall effect were incorrect[24]. After the experimental verification of the quantized Hall resistance in 1980 with plateau values independent of the amount of localized states[25], a large number of theoretical publications came to the conclusion that the occupation of localized states in the tails of the Landau level leads to a vanishing conductivity σ_{xx} and the Hall conductivity becomes stabilized at a value $i \cdot (e^2/h)$ corresponding to the ideal value of a fully occupied Landau level[16-19]. This result can be easily understood in the percolation picture[26-28] which is a good approximation in the limit of infinite high magnetic fields and long range potential fluctuations where the electron behaves like a classical particle with a motion on equipotential lines. A closed equipotential line within the area of the sample means that the corresponding electronic state is localized. This type of localization is quite different from the localization without magnetic field.

Fig. 5 represents a simple picture of the Landau energies like in Fig. 3 but with a repulsive and an attractive potential fluctuation (for example impurities) within the area of the sample. Depending on the strength of the fluctuations more or less pronounced "hills" and "lakes" within each energy plane E_n appear and the electrons surrounding the hill or the lake on an equipotential line are localized states. Starting from

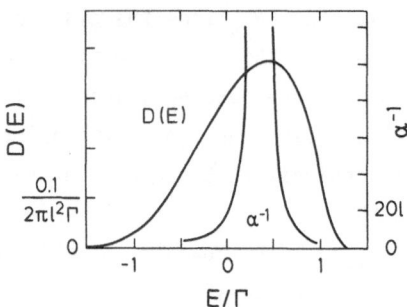

Fig. 4. Density of states D(E) and localization length α^{-1} calculated for the lowest Landau level E_0[23]. Attractive, short range scatterer with a concentration of 5 centers within a cyclotron orbit are assumed. Only states close to the maximum of the density of states are extended ($\alpha^{-1} \to \infty$). The asymmetry in D(E) with more localized states on the low-energy side originates from the assumption of attractive scattering centers whereas repulsive scattering centers give an asymmetry with more localized states on the high-energy side

the situation where the lowest Landau level is fully occupied and the higher levels are empty (which corresponds to the magnetic field value B_1 in Fig. 3 and a quantized Hall resistance $\rho_{xy}=R_0=h/e^2$), one can understand that the Hall resistance remains constant if the magnetic field differs slightly from the value B_1. With increasing magnetic field $B=B_1+\Delta B$ the degeneracy of the Landau levels increases so that under the condition of a

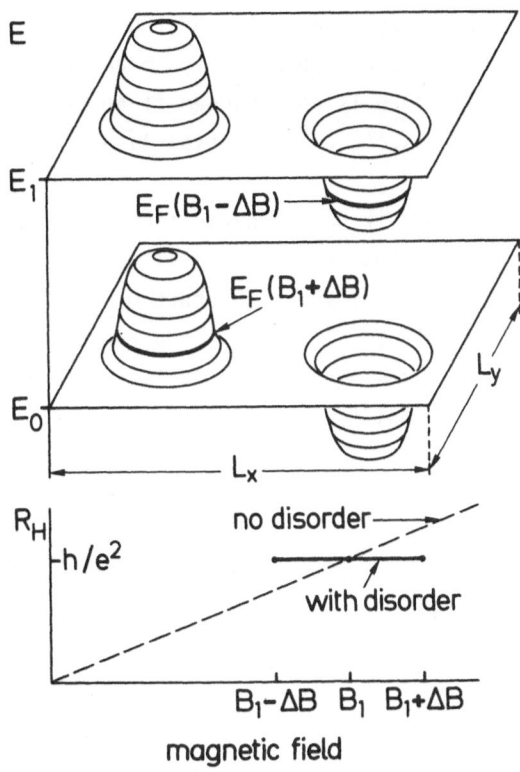

Fig. 5. Energy of the Landau levels E_0 and E_1 within the area $L_x \cdot L_y$ of the sample if a smooth attractive and repulsive potential is present. The variation of the Fermi energy $E_F(B)$ relative to the Landau levels is shown for different magnetic field values if the electron density is kept constant. At the magnetic field $B=B_1$ the Fermi energy is assumed to be in the gap between E_0 and E_1 so that a fully occupied lowest Landau level E_0 with a quantized Hall resistance R_0 is present. The value of the Hall resistance remains constant if the magnetic field is changed as long as only the occupation of localized states is changed

constant number of electrons some states close to the top of the hill in the lowest Landau plane are unoccupied. This situation corresponds to a sample with a hole (no electrons within the hole) where all states outside the hole are occupied like for a fully occupied Landau level. The Hall resistance can be calculated on the basis of two devices in parallel with fully occupied Landau levels as shown in Fig. 6 and the result is (see Eq. (2)) that the Hall resistance is not changed if holes are added as long as the boundaries of the holes are equipotential lines.

Similar arguments hold if the magnetic field is reduced to a value $B=B_1-\Delta B$. In this case, the lowest Landau level remains fully occupied but in addition localized states close to the bottom of the lakes of the next higher Landau level E_1 will be filled up since the degeneracy of the lowest Landau level decreases with decreasing magnetic field. The position of the Fermi energy $E_F(B_1-\Delta B)$ is shown in Fig. 5. However, the localized states in the Landau level E_1 do not contribute to the conductivity σ_{xx} and to the Hall effect so that from the experimental point of view the

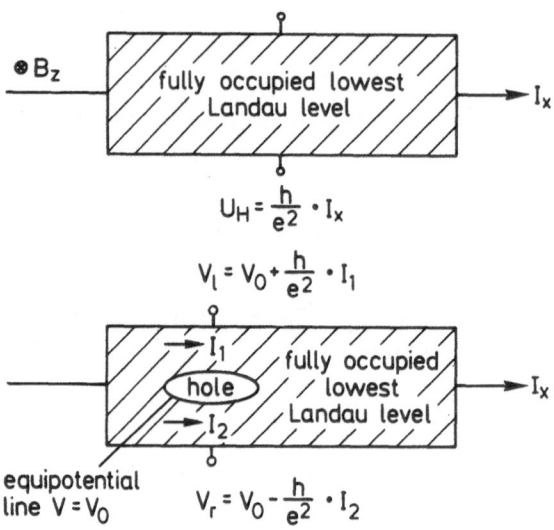

$$U_H = \frac{h}{e^2} \cdot I_x$$

$$V_l = V_0 + \frac{h}{e^2} \cdot I_1$$

$$V_r = V_0 - \frac{h}{e^2} \cdot I_2$$

Hallvoltage $U_H = V_l - V_r = \frac{h}{e^2}(I_1 + I_2) = \frac{h}{e^2} \cdot I_x$

Fig. 6. The Hall voltage of a 2DEG with a fully occupied lowest Landau level is h/e^2 times the current I_x flowing between the Hall probes (upper part). A hole in the sample surrounded by an equipotential line as shown in the lower part of this figure does not influence the Hall voltage as long as the lowest Landau level remains fully occupied in the rest of the sample

condition of a fully occupied lowest Landau level with $\sigma_{xx}=0$ and $\sigma_{xy}=e^2/h$ remains unchanged. This result that a variation of localized states does not influence the Hall effect has been confirmed by different quantum mechanical calculations, too[16-19], so that this process seems to be the main origin of the Hall plateaus.

A discussion of the QHE on the basis of the percolation picture shown in Fig. 5 allows also some predictions about differences in the experimental results if attractive or repulsive potential fluctuations dominate. For attractive centers, localized states on the low-energy side of the main Landau level E_n exists which leads to Hall plateaus above the classical curve $\rho_{xy}(B)$ whereas repulsive scatterers lead to Hall plateaus below the classical curve. Very recently such differences have been observed experimentally by adding ionized donors or acceptors close to the two-dimensional electron gas[29]. Fig. 7 is an example of such a measurement on a GaAs-AlGaAs heterostructure with about $2\times10^{10}\text{cm}^{-2}$ silicon donors close to the interface. As expected, the Hall plateaus are shifted to the low-magnetic field side whereas for a sample with Be-acceptors the Hall plateaus are shifted to such high magnetic fields that the plateau $R_0=h/e^2$ does not cross the classical curve $\rho_{xy}=B/n_{2d}\cdot e$. For a quantitative explanation of this result one has to include the spin splitting of each Landau level E_n but his is another complicated subject not included in this article since on the one hand the g-factor of electrons in a GaAs heterostructure determined from electron-spin-resonance[30] is about g=0.33 and therefore smaller than the bulk value whereas on the other hand from an analysis of magnetotransport data g-factors up to g=6 are observed[31]. These results can be explained by nonparabolicity effects and an exchange enhancement which has a maximum if one spin level is fully occupied corresponding to the situation of the QHE at odd integers i=1,3,... Under ideal conditions the quantized Hall resistance is independent of the value of the spin splitting but in measurements at finite temperatures a finite value for the resistivity ρ_{xx} becomes visible which depends on the energy spectrum[32]. Experimentally one has found that the quantized Hall resistance deviates from the value h/ie^2 if ρ_{xx} is finite and the following relation is found[33]:

$$\Delta\rho_{xy} < 0.5\ \rho_{xx}^{min} \qquad (16)$$

Since the resistivity ρ_{xx}^{min} at integer filling factors can be reduced to unmeasurably small values by reducing the temperature or using Hall plateaus at higher magnetic fields, the correction mentioned in Eq. (16) are at present not important in high precision measurements of the quantized Hall resistance.

However, a microscopic transport theory which allows predictions of corrections to the QHE under real experimental conditions is not available. Even such a question like the current distribution within a Hall

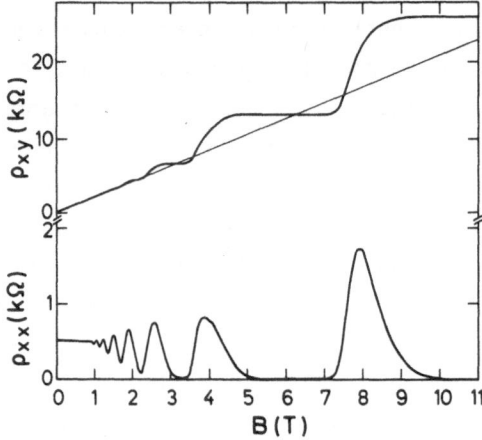

Fig. 7. Resistivity ρ_{xx} and Hall resistance $R_H = \rho_{xy}$ as a function of the magnetic field for a GaAs heterostructure with $2 \times 10^{10} cm^{-2}$ positively charged Si-donors close to the 2DEG. The Hall plateaus are asymmetric relative to the crossing point with the classical curve

device at zero temperature has not been solved exactly[34-36] so that a theoretical discussion of the real experimental situation including fluctuations in the carrier density within the area of the sample[37] cannot be expected in the near future. First attempts were made to include inhomogeneities in the carrier density in calculations of the density of states[38] (see contribution of E. Gornik in this book) but a transport theory is still much more complicated.

APPLICATIONS OF THE QUANTUM HALL EFFECT IN METROLOGY

The applications of the QHE are very similar to the applications of the Josephson effect, which can be used for the determination of the fundamental constant h/e or for the realization of a voltage standard. In analogy, the QHE can be used for a determination of h/e^2 or as a resistance standard[39].

In principle, corrections to the quantized Hall resistance measured under real experimental conditions are always present but all data indicate that in the limit $\rho_{xx}=0$ (for example, extrapolation to T=0 K) a quantized value $R_H=h/ie^2$ is expected. Already at a finite temperature of T=2 K and magnetic fields of about B=10 Tesla, the corrections found in measurements on GaAs heterostructures are smaller than the experimental uncertainty of 10^{-8}. The good agreement in the value for the quantized Hall resistance[40],[41] measured in different countries on different samples and different materials (see table 1) indicates that the quantized Hall resistance is really a stable and device independent phenomenon. Therefore, one of the applications of the QHE is the determination of the drift coefficient of the standard resistors kept at the national laboratories, since the quantized Hall resistance is more stable and more reproducible than any wire resistor. The National Physics Laboratory (NPL) in Teddington (GB) obtained the following result for their NPL-Ohm:

$$\frac{dR(1\ \Omega)}{dt} = (-0.0478 \pm 0.0074)\ \frac{\mu\ \Omega}{year}$$

As a consequence, the Comité Consultatif d'Electricité (CCE) which is responsible for the propagation and improvement of the electrical units throughout the world, adopted a declaration concerning the QHE for maintaining a representation of the Ohm (Declaration E2 (1986), 22.09.1986).

Table 1. Experimental data for the quantized Hall resistance at the plateau i=1[40],[41] (1.9.1986)

BIPM (Paris):	25812.809 ± 0.003 Ω	NPL (GB):	.811 ± 0.002 Ω
EAM (CH):	.809 ± 0.004 Ω		.803 ± 0.002 Ω
ETL (J):	.804 ± 0.008 Ω	NCR (Can.):	.814 ± 0.006 Ω
	.804 ± 0.014 Ω	PTB (FRG):	.802 ± 0.003 Ω
LCIE (F):	.810 ± 0.001 Ω	NVIIM (UdSSR):	.805 Ω
NBS (USA):	.810 ± 0.002 Ω	VSL (NL):	.802 ± 0.005 Ω
NML (Austr.):	.810 ± 0.002 Ω	mean value:	.807 Ω

One result of this declaration is the decision, that the CCE will meet in September 1988 "with a view to recommending the value of R_H to come into effect on 1st January 1990. The value of R_H will be based upon all relevant data collected by a Working Group that become available up until 15th June 1988".

It is not clear whether really a value for R_H will be fixed or whether a statement will be accepted that the quantized Hall resistance is identical with the fundamental constant h/e^2, so that resistance calibrations in SI-units on the basis of the recommended value for h/e^2 will be possible. In this respect, the situation for the Josephson effect differs from the situation for the QHE. The fundamental constant h/e, which determines the uncertainty of the Josephson-Volt, is only known with a relative uncertainty of 3×10^{-7} whereas the instability of the Josephson-Volt is less than 10^{-9}. Therefore a fixed value for $2e/h$ has been adopted with a value of $483594,0$ GHz/V_{76-BI} in order to reproduce everywhere in the world the same voltage (however not expressed in SI-units) within an uncertainty of 10^{-9}. In contrast, the fundamental constant h/e^2 of the quantized Hall resistance is already relatively well known (uncertainty 4.5×10^{-8}) so that the fixed value for h/e^2 will not improve drastically the stability of a resistance reference system. The high accuracy for h/e^2 in SI units originates from the fact that h/e^2 is (beside a fixed number) identical with the inverse finestructure constant which can be calculated from measurements of the electron magnetic moment anomaly combined with the extensive quantum electrodynamic (QED) calculations of the theory. However, it is not clear at which level of accuracy a breakdown of the QED-theory is expected. Therefore changes in the recommended value for the finestructure constant due to systematic errors may be necessary. If these changes are expected to be larger than the stability of the quantized Hall resistance, than the CCE may come to the conclusion that for applications in metrology everyone should use a fixed value for the quantized Hall resistance, for example $R_H = 25812.8000$ $\Omega_K = 1$ Klitzing[42]. The index K indicates that this resistance has in principle nothing to do with the SI-units but the value is chosen in such a way that the difference 1 $\Omega_K - 1$ Ω is as small as possible.

ACKNOWLEDGEMENTS
    This work was supported by the Bundesministerium für Forschung und Technologie.

REFERENCES

1   E.R. Cohen and B.N. Taylor, Codata Bulletin 63, 1 (1986)

2   for a review see:
     T. Ando, A.B. Fowler, and F. Stern, Rev. Mod. Phys. 54, 437 (1982)

3   H.L. Störmer, J.P. Eisenstein, A.C. Gossard, K.W. Baldwin, and
     J.H. English, Proc. 18th Int. Conf. on the Physics of Semiconductors,
     Stockholm 1986, ed. Olof Engström, World Scientific Publishing Co.
     Pte Ltd., Singapore

4   J. Avron and R. Seiler, Phys. Rev. Lett. 54, 259 (1985)

5   K. v. Klitzing, Rev. Mod. Phys. 58, 519 (1986)

6   The Quantum Hall Effect (Graduate Texts in Contemporary Physics,
     Springer Verlag New York, 1987), ed. by R.E. Prange and S.M. Girvin

7   H. Aoki, Reports on Progress in Physics (1987)

8   E.I. Rashba and V.B. Timofeev, Sov. Phys. Semicond. 20 , 617 (1986)

9   R.F. Wick, J. Appl. Phys. 25, 741 (1954)

10   K. v. Klitzing, Festkörperprobleme 21, 1 (1981)

11   R.W. Rendell and S.M. Girvin, Phys. Rev. B23, 6610 (1981)

12   B. Neudecker and K.H. Hoffmann, Solid State Commun. 62, 135 (1987)

13   R.B. Laughlin, Phys. Rev. B23, 5632 (1981)

14   R. Kubo, S.J. Miyake, and N. Hashitsume, Solid State Physics 17, 269
     (1965), (Academic Press, New York), ed. by F. Seitz and D. Turnball

15   G.A. Baraff and D.C. Tsui, Phys. Rev. B24, 2274 (1981)

16   R.E. Prange, Phys. Rev. B23, 4802 (1981)

17   H. Aoki and T. Ando, Solid State Commun. 38, 1079 (1981)

18   W. Brenig, Z. Phys. B50 305 (1983)

19   J.T. Chalker, J. Phys. C16, 4297 (1983)

20   T. Ando and Y. Uemura, J. Phys. Soc. Japan 36, 959 (1974)

21   T. Ando, J. Phys. Soc. Japan 36, 1521; 37, 622; 37, 1233 (1974)

22   S. Kawaji and J. Wakabayashi, Surf. Sci. 58, 238 (1976)

23   H. Aoki, J. Phys. C10, 2583 (1977)

24   H. Aoki and H. Kamimura, Solid State Commun. 21, 45 (1977)

25   K. v. Klitzing, G. Dorda, and M. Pepper, Phys. Rev. Lett. 45, 494
     (1980)

26   S.V. Iordansky, Solid State Commun. 43, 1 (1982)

27   R.F. Kazarinov and S. Luryi, Phys. Rev. B25, 7626 (1982)

28   S.A. Trugman, Phys. Rev. B27, 7539 (1983)

29   R.J. Haug, R.R. Gerhardts, K. v. Klitzing, and K. Ploog, to be
     published

30   M. Dobers, private communication. See also: G. Lommer, F. Malcher,
     and U. Rössler, Phys. Rev. B32, 6965 (1985)

31 R.J. Haug, K. v. Klitzing, R.J. Nicholas, J.C. Maan, and G. Weimann, to be published in Surf. Sci. (1988), Proc. Int. Conf. Electronic Properties of Two-Dimensional Systems

32 K. v. Klitzing, G. Ebert, N. Kleinmichel, H. Obloh, G. Dorda, and G. Weimann, Proc. 17th Int. Conf. Physics of Semiconductors, p. 271 (Springer-Verlag Berlin, Heidelberg, New York, Tokyo) ed. by D.J. Chadi and W.A. Harrison

33 M.E. Cage, B.F. Field, R.F. Dziuba, S.M. Girvin, A.C. Gossard, and D.C. Tsui, Phys. Rev. B30, 2286 (1984)

34 A.H. MacDonald, T.M. Rice, and W.F. Brinkman, Phys. Rev. B28, 3648 (1983)

35 J. Riess, J. Phys. C17, L849 (1984)

36 O. Heinonen and P.L. Taylor, Phys. Rev. B32, 633 (1985)

37 G. Ebert, K. v. Klitzing, and G. Weimann, J. Phys. C18, L257 (1985)

38 V. Gudmundsson and R.R. Gerhardts, Phys. Rev. B35, May 1987

39 K. v. Klitzing and G. Ebert, Metrologia 21, 11 (1985)

40 For a summary see:
    Conference on Precision Electromagnetic Measurements, IEEE Trans. Meas. IM-34, 301-327 (1985)

41 E. Braun, private communication

42 High precision wire resistors in units of 1, ½, and 1/4 Klitzing are available from: Burster Präzisionsmeßtechnik, D-7562 Gernsbach

HIGH FIELD MAGNETOTRANSPORT-LECTURE III: THE FRACTIONAL

QUANTUM HALL EFFECT

R.J. Nicholas, R.G. Clark, A. Usher, J.R. Mallett and
A.M. Suckling

Clarendon Laboratory, Parks Rd., Oxford OX1 3PU, UK

J.J. Harris and C.T. Foxon

Philips Research Laboratories, Redhill, Surrey, UK

INTRODUCTION

The existence of the fractional quantum Hall effect (FQHE) is taken
to be evidence for the formation of a new highly correlated ground state
of a two dimensional electron gas. This occurs at very low temperatures,
in high magnetic fields, and in systems where there is only a very small
amount of disorder present. The main experimental observations are that
minima are observed in the electrical resistivity component ρ_{xx}, at fractional Landau level occupancies $\nu = nh/eB = p/q$, where p is an integer and q
is an odd integer [1-9]; while corresponding Hall plateaus are seen at
quantized Hall resistivity values of $h/\nu e^2$. To date fractional states have
been reported at $\nu = 1/3$, 1/5, 2/5, 2/7, 3/7 and 4/9, and the equivalent
'hole' analogous of these states have been observed at occupancies $\nu =
1-(p/q)$. These states occur when all of the electrons lie in the lowest
spin split Landau level, but it has recently been shown that they can
exist in a similar manner in the upper spin state at occupancies of the
form $\nu = 1+(p/q)$. Once $\nu > 2$ the electrons occupy the second Landau level. At
this point the experimental position becomes less clear, with some reports
of the observation of 7/3 and 8/3 states [3,4], and some suggestions that
even denominator fractions may occur [8,9]. The significance of these
results is that the existence of minima in the resistivity and quantized
Hall plateaus may be shown, by using the gauge invariance arguements of
Laughlin [10], to result from the formation of a mobility gap in the density of states. In other words the degeneracy of the individual Landau
levels for isolated electrons has been lifted by the residual Coulomb
interactions, leading to the formation of an energy gap between the ground
and excited states of the system.

The theoretical treatments of the phenomenon break into three main
groups. Firstly there is the original quantum fluid picture developed by
Laughlin [11-13], which is based on a postulated trial ground state wavefunction. This posseses quasi-particle excitations which have fractional
charge of unit e/q. The higher fractional states are thought to result
from a hierarchy [12,14,15] in which the ground state of each succeeding
fraction (say 2/5), is the result of a condensation of the quasi-particles
associated with the preceding level (1/3). These quasi-particles exhibit
fractional charge, corresponding to the denominator of the fraction con-

cerned. This picture implies that no 'daughter' state may exist unless its parent state exists - a prediction which is brought into question below. Further extensions of this theory include the calculations by MacDonald et al. [16] for the N=1 Landau level, in which it is found that the fractional states repeat themselves, but that the quasi-particle energy gap associated with the 1/5 states is comparable with that of the 1/3 state. Several calculations have been made of the energy gap for the 1/3 state, i.e. the energy needed to create a separated quasi-electron - quasi-hole pair. These give numbers of order $Ce^2/4\pi\epsilon l_B$, with $C\sim0.1$ [17-19], where l_B is the cyclotron radius ($l_B= \sqrt{h/eB}$). If the quasi-particles may be approximated to point charges, then the energy gap will scale as $q^{-2.5}$ [14]. Calculations of the dispersion relation for the quasi-particle pair excitation [20,21] suggest that there may be a rather smaller 'indirect gap' at finite wavevectors, close to the reciprocal lattice constant for a Wigner crystal.

The second approach is that originated by Tao and Thouless [22,23], who used the Landau gauge as a starting point, in which the single particle states may be thought of as a set of parallel tracks. They suggested that a collective state could be formed by an ordered filling of the tracks. Quasi-particle excitations then consist of a 'defect' with one extra track filled or unfilled, while a quasi-particle pair excitation would consist of the translation of a single occupied track to its neighbour. This model also suggests that even denominator fractions may occur, but calculations of the energy gaps for both the odd and even fractions give values considerably greater than are found from the Laughlin approach, or by experiment.

More recently there have been several calculations made based on a Wigner crystal ground state [24-25], which predict a lower energy ground state than that found by Laughlin. Kivelson et al. [24] have found large contributions to the energy from cooperative ring exchange, in which electrons move coherently along a closed path in the crystal lattice. These contributions are enhanced at rational filling factors, again leading to an energy gap and quasi-particle excitations. The energy gap values are comparable with those predicted for the Laughlin ground state.

THEORY

The theoretical treatments of the fractionally quantized states involve the use of rather complex formalism, however it is worthwhile to attempt to give an intuitive picture of the problem. A further complicating factor is the ability to describe the system in a number of different gauges. The most widely accepted and successful theory is that developed by Laughlin [11-13], who works in the symmetric gauge. The starting point for his work are the single particle wavefunctions of the form

$$|n> = \frac{1}{\sqrt{2^{n+1}\pi n!}} \, z^n \exp(-1/4|z|^2) \qquad (1)$$

where $z_j=(x_j-iy_j)/l_B$, and n is the angular momentum quantum number which gives the degeneracy of the state. The next step is to introduce a variational many body wavefunction of the form

$$\psi_m(z_1,\ldots,z_N) = \prod_{j<k}^{N} (z_j-z_k)^m \exp(-1/4 \sum_1^N |z_l|^2) \qquad (2)$$

where m is an odd integer. Laughlin [11,12] was led to the use of this form by requiring it to be an eigenstate of angular momentum and by the need to make the polynomial part antisymmetric, which leads to the requirement that m should be odd. He was then able to show that the quasi-particle excitations of this state could be created by piercing the ground state with a single flux quantum. The resulting particle has an effective charge of -1/m, due to partial screening of the induced charge. The excitation of a quasiparticle across the energy gap separating the ground and excited states then requires the creation of a quasi-electron- quasi-hole pair. He has calculated that this will require energies of 0.056 and 0.014 $e^2/4\pi\epsilon l_B$ for the 1/3 and 1/5 states respectively.

In order to deduce the existence of quantized Hall plateaus from this calculation it is then necessary to introduce the idea of some disorder into the system, which then creates a mobility gap separating the ground and excited states. It is then possible to use the arguements of gauge invariance, as first introduced by Laughlin [10] to describe the integer quantum Hall effect, to predict the existence of plateaus. This involves the introduction of the rather unphysical geometry of an annulus or ring, which contains a solenoid in order to introduce flux changes. This causes the Hall current to flow around the ring, while at the same time transfering electrons from one side to the other through the Hall potential V. The fact that the conductivity and resistivity are zero at particular filling factors requires the wavefunction to be periodic around the ring, leading to flux quantization. The quantized Hall resistivity is then the ratio of the flux quantum to the electron charge. In the case of the fractional effect the reasoning is not quite so clear since it is not possible to introduce single quasi-particles externally. Instead one is led to the conclusion that motion of a single electron results in the introduction of m flux quanta. This seems a natural consequence of the ground state wave-function [2], which is invariant upon a $2\pi/m$ rotation of the system. The periodic boundary condition will then introduce a $2m\pi$ phase change upon a complete rotation of the system, coresponding to motion of the single electron across the ring. The conclusion is then that the size of the flux quantum has changed, and the ground state is triply degenerate.

An important feature of the Laughlin approach to the ground state is that it is a quantum liquid, which is formed through the Coulomb inter-actions between the electrons, and therefore correlates their motion. Since the sytem remains a liquid there is only short range order, in contrast to the Wigner solid, and he has calculated [2] that the correlation lengths are of order 5 l_B.

An alternative approach has been given by Tao and Thouless [22], who have worked in the linear gauge where

$$\psi_{k,n} = e^{ikx}\phi_n(y-y_0) \tag{3}$$

with k as the degenerate quantum number. The single particle wavefunctions may be thought of parallel tracks of width l_B and centred at

$$y_0 = \hbar k/eB - E_0 \tag{4}$$

For a system of length L the track separation is

$$\delta y_0 = \hbar \delta k/eB = 2\pi l_B^2/L \tag{5}$$

which is obviously much less than the width of any individual track. The arguement of Tao and Thouless is that in the fractional state the system can lower its energy by an ordered filling of tracks such that every mth single particle state is filled. To show the correspondence with the Laughlin approach we take the linear geometry and bend it around to form a ring or annulus of diameter L. In this case the ground state is a set of

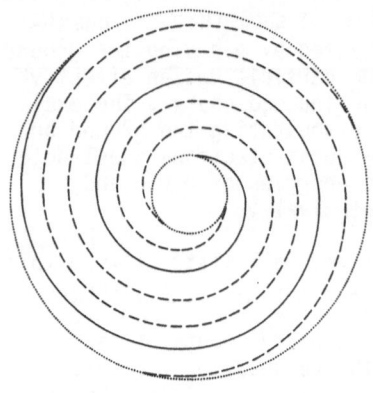

Fig.1: A schematic view of the application of a gauge transformation for the m=3 ground state, based on the Tao and Thouless approach. The occupied state is the solid line, and the spiral corresponds to the progressive application of the gauge potential around the ring, such that one occupied state is mapped into the next.

set of concentric rings. If we now introduce a flux into the ring then we produce an azimuthal change in magnetic vector potential $\Delta A = \Delta\psi/L$. This introduces a progressive phase change around the ring of $\exp(ie\Delta Ax/\hbar)$, which gives a shift in the track centre of

$$\Delta y_0 = \Delta A/B = \hbar \Delta k/eB \qquad (6)$$

The occupied tracks have now been shifted outward by Δy_0, and in order for the state to transform into the next occupied track we must have a shift of

$$\Delta y_0 = m \, \delta y_0 \qquad (7)$$

leading to a flux quantum of mh/e. The process of transformation is shown schematically in Fig. 1 as a spiral motion, for the m=3 state. This picture also has the $2\pi/m$ symmetry, corresponding to the m different starting points for the spiral.

THE EXPERIMENTAL PICTURE

Some typical experimental results are described here which were taken using Hall bridge specimens with channel widths of 50-150 μm. These were modulation doped GaAs-Ga$_{0.68}$Al$_{0.32}$As heterojunctions grown by MBE at Philips Research Laboratories, Redhill [26], using spacer layers of 400 and 800 Å, and with resulting electron concentrations in the range 0.6-4×10^{11}cm^{-2}. The sample mobilities ranged from 0.1 to 2.1×10^6cm^2/Vs, depending upon the sample and electron concentration. For any one sample it was possible to change the electron concentration by factors of 2-3 by excitation of persistent photoconductivity using a red LED. The samples were cooled to temperatures as low as 20 mK in a dilution refrigerator.

Fig. 2 shows a typical recording of the resistivity ρ_{xx} and Hall component ρ_{xy}, for the highest mobility sample G63 at an electron concentration of 1.9×10^{11}cm^{-2}, following the photo excitation of the majority of the carriers. The current density is 6.6×10^{-4} A/m. This shows what is probably the most comprehensive set of fractional states observed to date. The region above 8 T corresponds to the incomplete filling of the lower spin state of the N=0 Landau level, and clear features can be observed at 2/3, 3/5, 4/7, with a weak feature at 5/9. There is also a weaker minimum at 4/5 and associated Hall plateau, which has been observed more clearly at lower current densities [27]. This is the furthest into the Landau level tail that a fractional state has been detected. The fractional features can also be seen to repeat themselves in the region 1<ν<2, with a

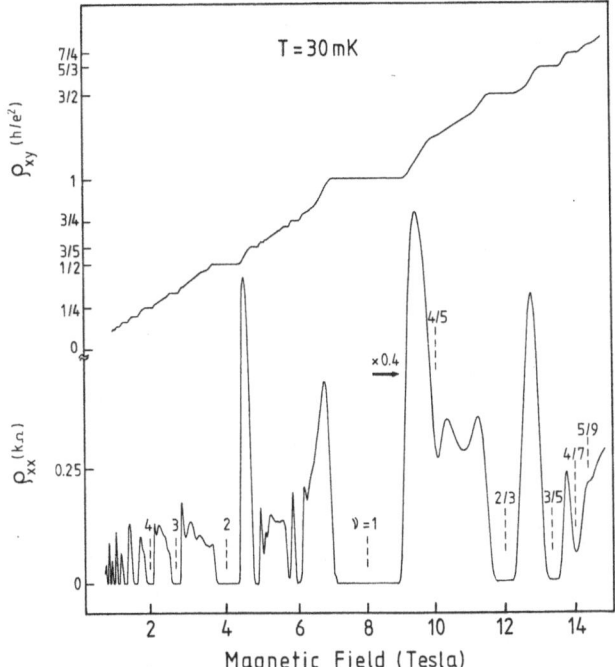

Fig. 2: The resistivity and Hall voltage (upper trace) for sample G63 at 30 mK. The fractional occupancies for $\nu < 1$ are indicated on the trace. The Hall resistivity is given in units of (h/e^2). The current density is 6×10^{-4}A/m

similarly regular set of q= 3,5,7 fractions, which is obviously symmetric about the half occupied level. It is possible to observe fractionally quantized states at magnetic fields as low as 3 T in such highly pure systems, as shown in Fig. 3, where the same sample is shown at a lower and a slightly higher carrier concentration. The systematic behaviour in such high purity layers is entirely consistent with the picture of a series of states becoming progressively more bound as the denominator decreases or as they move towards the centre of each level, where the influence of disorder is least. Considerable changes from this picture can be seen however when the electron concentration is varied and more disordered samples are studied.

Figure 4 shows the electron concentration (n) dependence of the FQHE in a rather more disordered sample G29, with n varying from 1.9 to 3.4×10^{11}cm^{-2}. The mobility variation in this sample is extremely rapid ($\sim n^{4\cdot5}$; Foxon et al [26]), which is thought to be due to the presence of long range potential fluctuations. At the lowest concentration the 5/3 and 4/3 states are hardly visible as weak minima in the resistivity, and the traces are dominated by integer quantization. On illumination however, the 5/3 feature rapidly evolves to be a well defined Hall plateau and resistivity minimum. In contrast the 4/3 state only ever gives rise to a weak resistivity minimum, but still shows a Hall plateau. On closer inspection, the Hall resistivity of this plateau is found to be a few per cent too high. The very surprising feature is the appearance of a stronger minimum at $\nu = 7/5$, which does have an accurately quantized Hall plateau, obviously dominating the 4/3 features. An even more pronounced example of this trend is shown in Fig. 5, from the earlier report of CLARK et al. [28], who were using more disordered samples grown in a completely different growth kit. In this case (n=4.3×10^{11}cm^{-2}, μ=600,000 cm^2/Vs), the 4/3 state has totally disappeared and been replaced by the 7/5 state in both the resistivity and

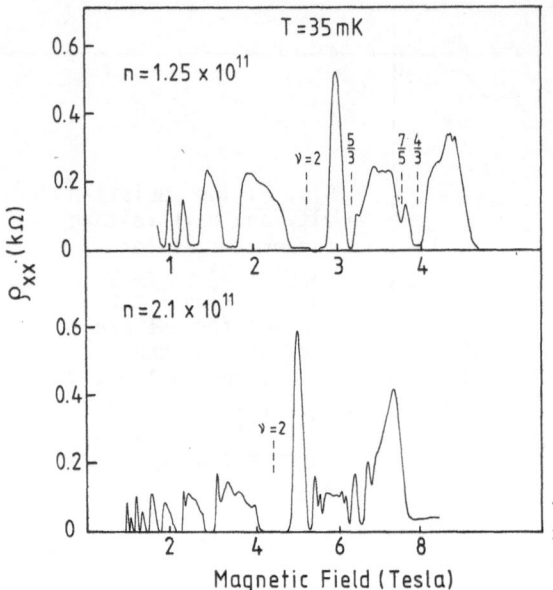

Fig. 3: Low field resistivity in sample G63 at 35mK for two different electron concentrations

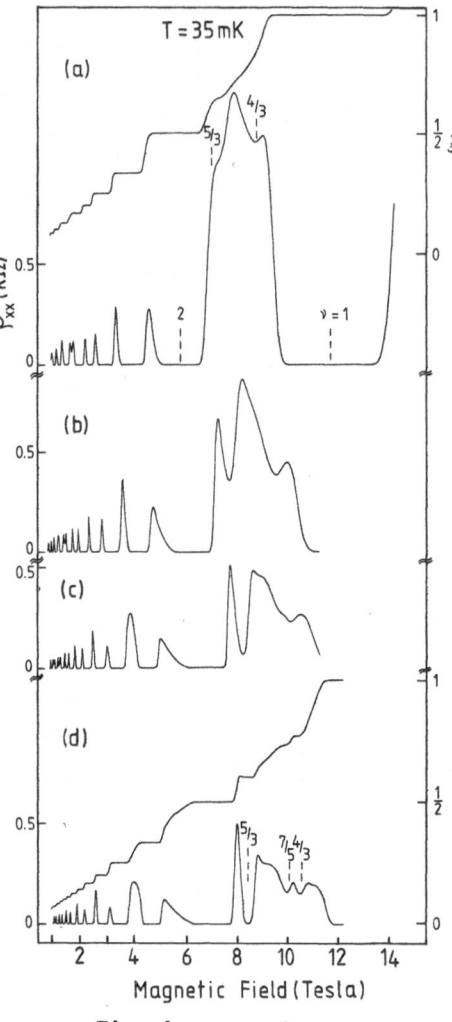

Fig. 4: ρ_{xx} and ρ_{xy} traces for sample G29 at 35mK. Four different carrier concentrations are shown, as produced by photoexcitation

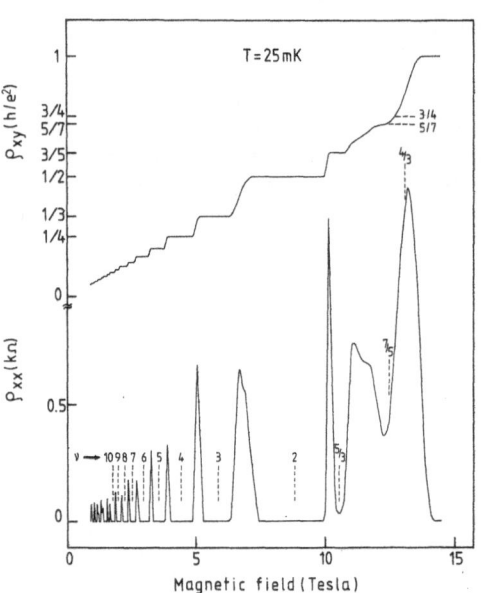

Fig. 5: ρ_{xx} and ρ_{xy}, in a more disordered GaAs-GaAlAs heterojunction at 25 mK. A wide plateau in the Hall resistivity is formed at a value $3/5(h/e^2)$. In constrast, only a weak feature occurs at $5/7(h/e^2)$. (Clark et al. [28])

Hall plateau following photoexcitation, but well behaved 5/3 structures remain.

It would thus appear that there is a systematic change in the character of the phenomenon as the influence of disorder becomes more pronounced. The common feature in all of these studies is that the inversion in strength of the 7/5 and 4/3 states is brought about by the photoexcitation of additional carriers into the 2DEG. At the same time the 5/3 state becomes more strongly favoured in even the highest mobility samples (for G63 the activation energy for 5/3 is substantially larger than for 4/3 [8]). It would appear to be the photoexcitation process which is critical, rather than the increase in electron concentration, as earlier results on samples of comparable concentration do not show such behaviour. In a recent measurement in which the electron concentration was increased by the use of a back electrode, Boebinger et al. [6] observed the opposite behaviour; at high concentrations a very weak 7/5 state was completely suppressed by a broad 4/3 resistivity minimum at low temperatures, although the high electron concentrations were again found to favour the 5/3 state.

The photoexcitation mechanism is associated with the presence of deep traps in the GaAlAs doped layer, and therefore once these have been excited they will act as positively charged remote scattering centres which alter the disorder in the system. The predominant sign of these scattering centres would seem to be the most likely cause of the systematic difference in behaviour between the results for the region $\nu<1.5$, where the Landau level is electron-like and therefore under the influence of an attractive potential, and for $\nu>1.5$, where the hole-like states will be repelled from the scattering centres. This may also be the reason why the q=3 states, with larger charge units, are suppressed relatively to q=5 while they are electron-like. Strong asymmetry has been noted for the conductivity in the integer quantum Hall regime [29], which has also been attributed to predominantly attractive scattering centres.

The final, and probably the most significant, conclusion from this behaviour is that considerable difficulties exist to explain this in terms of a hierarchical model of the FQHE, in which the existence of fractional states at lower order is a necessary prerequisite for the formation of successive orders. A 7/5 state should be one generation on from 4/3. Apparently it is not.

Another important prediction of current theoretical models of the FQHE is that only odd denominator states should exist. In our earlier works [8,9] we found the appearance of resistivity minima in the region 2<ν<4, with apparently even denominator fractions, although no Hall steps were seen. This data is shown in Fig. 6, together with data taken from the same sample tilted at 60° to the magnetic field direction. The minimum at 2 1/2 is clearly visible in both traces, but the effect of tilting the sample is to reduce the strength of the features at 2 1/4 and 2 3/4, and to produce clear minima at 2 1/3 and 2 2/3. The effects of tilting the field (and hence introducing a parallel field component) may well be rather complex, since at least three different factors will influence the states:
1) the total field corresponding to a given filling factor will increase, leading to an increased spin splitting and hence less overlap of adjacent spin states,
2) the parallel field component will alter the energies and wavefunctions in the z-direction. This will alter the magnitude of the exchange interaction, since the wavefunction will become more confined leading to larger energy gaps [30],

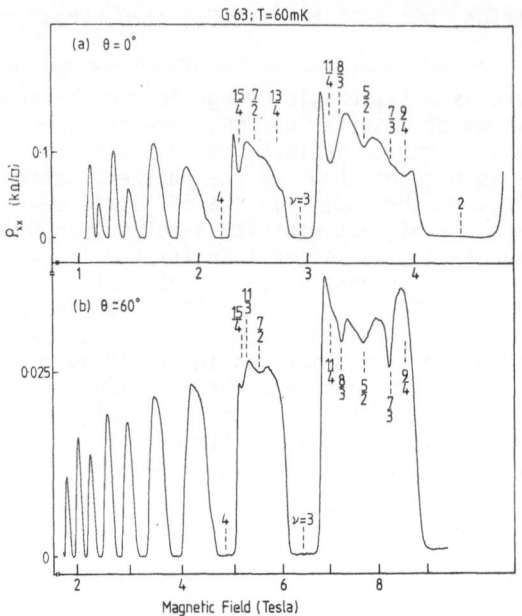

Fig. 6: Resistivity traces for sample G63 with i=100nA and the magnetic field at an angle (a) Θ=0° and (b) Θ=60° to the plane normal; μ= $2.15\times10^6 cm^2/Vs$, n= $2.1\times10^{11}/cm^2$. Landau level occupancy ν=nh/eB

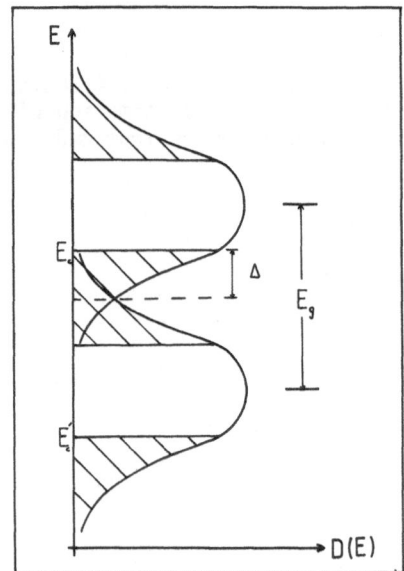

Fig.7: A schematic picture of the density of states for the many-body ground state. The mobility edges E_c are shown as is the activation energy Δ

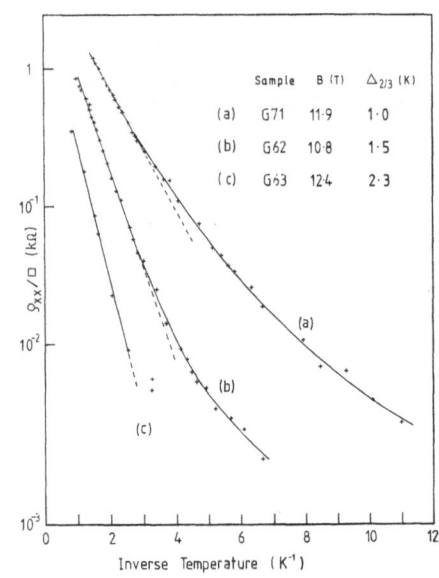

Fig. 8: Activation plots of the resistivity for the 2/3 minimum in three different samples

3)   the influence of the scattering centres may be changed, leading to a change in the effective disorder present.
In a rcent study of the activation energies as a function of tilt angle, Haug et al. [31], have found that the 1/3 state was weakened by tilting, while 2/3 became stronger.

The results shown in Fig. 6, however, suggest that there is no evidence as yet for the formation of an energy gap at even fractional occupancies. The appearance of features in the resistivity may nevertheless be caused by electron-electron interactions, since the calculations of Halperin [15] suggest that the system will have a maximum in energy at even filling factors, which may lead to an unstable charge distribution.

ENERGY AND MOBILITY GAPS

The most clear predictions from the theoretical descriptions of the FQHE described above concern the magnitude of the energy gap associated with the formation of the collective ground state. This is usually measured by studying the activated behaviour of the resistivity minima associated with the formation of the fractional states. Once an energy gap has been formed the density of states may be represented schematically as shown in Fig. 7, with a mobility gap separating the conducting states of each type of particle, which will begin at the 'mobility edges' E_c and E_c'. When the effects of disorder become small then this mobility gap will be close to the energy required to form a quasi-electron quasi-hole pair. At temperatures below the gap energy one would expect an exponentially activated resistivity, with an energy (Δ) equal to half the mobility gap. Such behaviour is shown in Fig. 8, for the samples G63, G62 and G71, which are in order of increasing 'dirt', as judged by mobility and quality of the FQHE states observed. The data are taken for the resistivity minimum associated with $\nu= 2/3$, and with electron concentrations adjusted by photoexcitation so that the minima all occur at approximately the same magnetic field. It can be seen that the effects of disorder are quite clearly to cause a significant decrease in the magnitude of the activation energy. This may be due to either one or a combination of two different effects. The energy gap of the many body state may have been reduced, and in addition a broadening of the density of states may bring the two mobility edges closer together, thus reducing the measured mobility gap. The maximum value of Δ is 2.3 K, for G63, which is larger than other values reported for similar fields [6,7], but still approximately a factor of three smaller than most theoretical predictions [17-19]. This is probably due to some residual disorder left in the system, to the possibly rather lower 'indirect gap' for quasi-particle excitation [20,21] and to the influence of the finite extent of the wavefunction out of the plane. ZHANG and DAS SARMA [30], and YOSHIOKA [32], have shown recently that there is an almost two-fold reduction in the gap energy when the finite z-extent of the electrons is considered.

Another question mark over the use of activation plots to determine the gap energies is the exact functional form of the temperature dependent conductivity. The width of the extended state region can be estimated from the minimum field required to produce well defined resistivity oscillations and Hall steps (of order 0.1-0.2 T), which will give widths of 1-3 K. Within this region the conductivity is apparently almost constant, and temperature independent. Since the width is comparable to, or larger than, the temperatures used, then using Fermi statistics we would deduce a thermally excited quasi-particle population of

$$n \sim D(E_c) \, kT \, \exp(E_F - E_c)/kT \qquad (8)$$

for one level. If this is dominated by the exponential term, we have for the conductivity:

$$\sigma \approx \sigma_c \, \exp(E_F - E_c)/kT \qquad (9)$$

for activated conduction to the mobility edge. If the mobility in the extended states is temperature independent, then the prefactor T should be included and may introduce some significant systematic errors into the determination of the energy gaps.

At still lower temperatures (of order 300 mK) there is a deviation from the Arrenhius plots, apparently due to the onset of hopping conduction. Previous workers have used both a second exponentially activated conductivity [7], and a high field hopping formula (6) derived by Ono [32], to give good descriptions of the lower temperature region. An approximate fit to our data gives $\rho \sim \exp(T^\alpha)$, with α close to 1/4, as expected for bulk materials, rather than 1/2 as predicted for 2D-systems in high magnetic fields [33]. This may be due to the presence of some leakage through the doped layers in the structure.

REFERENCES

1. D.C. Tsui, H.L. Stoermer and A.C. Gossard, Phys. Rev. Lett. 48, 1559 (1982)
2. A.M. Chang, P. Berglund, D.C. Tsui, H.L. Stoermer and J.C.M. Hwang, Phys. Rev. Lett. 53 997 (1984)
3. E.E. Mendez, L.L. Chang, M. Heiblum, L. Esaki, M. Naughton, K. Martin and J. Brooks, Phys. Rev. B30, 7310 (1984)
4. G. Ebert, K. v. Klitzing, J.C. Maan, G. Remenyi, C. Probst, G. Weimann and W. Schlapp, J. Phys. C17, L775 (1984)
5. G.S. Boebinger, A.M. Chang, H.L. Stoermer and D.C. Tsui, Phys. Rev. Lett. 55, 1606 (1985)
6. G.S. Boebinger, A.M. Chang, H.L. Stoermer and D.C. Tsui, Phys. Rev. B32, 4268 (1985)
7. J. Wakabayashi, S. Kawajii, J. Yoshino and H. Sakaki, Surf. Sci. 170, 136 (1986)
8. R.G. Clark, R.J. Nicholas, A. Usher, C.T. Foxon and J.J. Harris, Surf. Sci. 170, 141 (1986)
9. R.J. Nicholas, R.G. Clark, A. Usher, C.T. Foxon and J.J. Harris, Solid State Commun. 60, 183 (1986)
10. R.B. Laughlin, Phys. Rev. B23, 5632 (1981
11. R.B. Laughlin, Phys. Rev. Lett. 50, 1395 (1983)
12. R.B. Laughlin, Surf. Sci. 142, 163 (1984)
13. R.B. Laughlin, in Solid State Sciences (Springer Verlag) 53, p. 279 (1984)
14. F.D.M. Haldane, Phys. Rev. Lett. 51, 605 (1983)
15. B.I. Halperin, Phys. Rev. Lett. 52, 1583, 2390 (1984)
16. A.H. MacDonald and S.M. Girvin, Phys. Rev. B33, 4414 (1986)
17. S.M. Girvin, Phys. Rev. B30, 558 (1984)
18. A.H. MacDonald, G.C. Aers and M.W.C. Dharma-wardana, Phys. Rev. B31, 5529 (1985)
19. R. Morf and B.I. Halperin, Phys. Rev. B33, 2221 (1986)
20. S.M. Girvin, A.H. MacDonald and P.M. Platzman, Phys. Rev. B33, 2481 (1986)
21. F.D.M. Haldane and E.H. Rezayi, Phys. Rev. Lett. 54, 237 (1985)

22. R. Tao and D.J. Thouless, Phys. Rev. B28, 1142 (1983)
23. R. Tao, Phys. Rev. B29, 635 (1984)
24. S. Kivelson, C. Kallin, D.P. Arovas and J.R. Schrieffer, Phys. Rev. Lett. 56, 873 (1986)
25. R. Keiper and O. Zeip, Phys. Stat. Solidi 133b, 769 (1986)
26. C.T. Foxon, J.J. Harris, R.G. Wheeler and D.E. Lacklison, J.Vac.Sci, and Technol. B4, 511 (1986)
27. R.J. Nicholas, R.G. Clark, A. Usher, J.R. Mallett, A.M. Suckling, J.J. Harris and C.T. Foxon, in "Two-Dimensional system: Physics and New Devices", Ed. G. Bauer, F. Kuchar and H. Heinrich, Solid State Sciences 67, p. 194 (Springer Verlag) 1986
28. R.G. Clark, R.J. Nicholas, M.A. Brummell, A. Usher, S. Collocott, J.C. Portal and F. Alexandre, Solid State Commun. 56, 173 (1985)
29. R.J. Haug, K. v. Klitzing and K. Ploog, Phys. Rev. B, in press (1987)
30. F.C. Zhang and S. Das Sarma, Phys. Rev. B33, 2903 (1986)
31. R.J. Haug, K. v. Klitzing, R.J. Nicholas and G. Weimann, to be published (1987)
32. D.J. Yoshioka, J. Phys. Soc. Japan 55, 237 (1982)
33. Y. Ono, J. Phys. Soc. Japan 51, 237 (1982)

OPTICAL PROPERTIES OF QUANTUM WELLS

C. Weisbuch

Laboratoire Central de Recherche, Thomson CSF

B.P. 10 - ORSAY, 91401 FRANCE

I - INTRODUCTION

The purpose of the present set of lectures is to introduce students to the field of the optical properties of quantum wells (and superlattices ?). As this field has grown out of proportions so as to be covered in three lectures, we have choosen to focus on three areas of the subject which seem to us more appropriate to the aims of the school, namely :

(i)   The specific aspects of 2D systems concerning optical properties.

(ii)   A description of the techniques of optical spectroscopy so widely used in 3D systems, and their relative qualities when applied to the field of quantum wells.

(iii)   The specific design rules and properties of quantum-well lasers (QWLs). As will be seen below, they are quite dominated by "subtle" 2D effects, and QWLs therefore represent a good laboratory to study what are the pros and cons of **2D devices.** The discussion of the difficulties encountered with 2D QWLs will then be briefly extended to 1D and 0D devices.

This set of lectures therefore does **not** discuss many topics concerning the optical properties of quantum wells : calculation of energy levels[1], optical properties of type II quantum wells and superlattices[2], strained-layer superlattices[3], nipis[4], $Ge_x Si_{1-x}$ superlattices[5], II-VI quantum wells and superlattices (both wide[6] or narrow gap[7]), light-scattering phenomena[8], electric[9] and magnetic field perturbations[10] of optical spectra, hot-electron phenomena as studied by optical techniques[11] etc... Excellent coverage of these fields can be found in other lectures in this school or in recent reviews. Time-resolved spectroscopy will only be briefly discussed in direct relation with other topics presented in these lectures.

As is clear from the preceeding discussion of what is not included here, these lectures are only concerned with the **optical properties of type I quantum wells**, even so restricting us to **interband transitions**.

II - THE OPTICAL TRANSITION PROBABILITY

All interband optical phenomena near a band extremum can be expressed in terms of the dielectric function :

$$\epsilon(\omega) = \epsilon_1(\omega) + i\epsilon_2(\omega) = \epsilon_0 + \frac{4\pi\beta\omega_0^2}{\omega^2 - \omega_0^2 + i\omega\Gamma} \tag{1}$$

where ϵ_1 and ϵ_2 are the real and imaginary part of the dielectric function respectively, ϵ_0 is the background dielectric functions (involving all crystal quantum states <u>but</u> the quasi-resonant state), β is the polarizability of that resonant state with energy ω_0 and damping constant Γ. Such a description of optical properties through the dielectric function has been very widely used to describe modulation spectroscopy of 3D-systems[12]. A large amount of information is contained in β such as the description of quantum states in uncorrelated one-electron states or in correlated exciton states, the dimensionality of the Density of States (DOS) and type of the transition (number of negative effective masses in the joint DOS) etc... To our knowledge, no **detailed quantitative** analysis of QW phenomena has been performed in the dielectric function framework, but we will use it below for the description of some experiments. The reader is deferred to the very good reviews of the field[12] for the description of the properties of $\epsilon(\omega)$, its quantum-mechanical calculation and its relation to various optical properties. It suffices to recall here that ϵ_1 and ϵ_2 are related through the Kramers-Kronig relations, and that $\epsilon_2(\omega)$ is very directly proportional to the absorption coefficient $\alpha(\omega)$ as calculated from time-dependent perturbation theory (Fermi's golden rule). We can therefore write :

$$\alpha(\omega) \sim \epsilon_2(\omega) \sim |<f|\vec{\epsilon}.\vec{p}|i>|^2 \varrho(\omega) \tag{2}$$

where i and f are initial and final states of the optical transition, $\vec{\epsilon}$ is the light polarization vector, $\vec{p}$ is the momentum operator and $\varrho(\omega)$ is the joint density of states at energy ω. It has been assumed that the matrix element is independant of ω.

Uncorrelated electron-hole pair transitions

In the enveloppe wavefunction approximation, the electron wavefunctions take the simple normalized form :

$$\psi^{i,f}(\vec{r}) = \chi^{e,h}(z) e^{i\vec{k}_\perp \cdot \vec{r}_\perp} u_{c,v,\vec{k}}(\vec{r}) \qquad (3)$$

where $\chi^{e,h}(z)$ represents the confined electron (or hole) enveloppe wave-function, $\vec{k}_\perp$ and $\vec{r}_\perp$ are the transverse momentum and position, $u_{c,v}$ are the usual periodic part of the Bloch wavefunctions.

The matrix element appearing in eq. (2) can be then factorized along the usual procedure[13] into the integral over the unit cell of the fast varying part of the wavefunction (the u_c's and u_v's) and a sum at lattice points R_i of the slowly varying functions :

$$<f|\vec{\epsilon}\cdot\vec{p}|i> = \sum_{R_i} \chi^e(\vec{R}_i)\chi^h(\vec{R}_i) e^{i(\vec{k}_{e\perp}-\vec{k}_{h\perp})\cdot\vec{R}_i} \int_\Omega u_{c,\vec{k}_{e\perp}} \vec{\epsilon}\cdot\vec{p}\ u_{v,\vec{k}_{h\perp}}\ d^3r \quad (4)$$

The latter integral is independant of R_i and is the usual bulk matrix element P which contains the selection rules due to band symetry and light polarization[13]. The former sum yields, after transformation back to an integral and taking into account normalization factors :

$$<f|\vec{\epsilon}\cdot\vec{p}|i> \sim L_z^{-1/2} <\chi^e|\chi^h> P\ \delta_{\vec{k}_{e\perp},\ \vec{k}_{h\perp}} \qquad (5)$$

As usual transitions are vertical ($\vec{k}_e = \vec{k}_h$). Recalling that the 2D joint (DOS) is a constant ($\mu*/\pi\hbar^2$, where $\mu*$ is the reduced effective mass $\mu*^{-1} = m_e^{-1} + m_h^{-1}$), one finds the important result :

$$\alpha(\omega).L_z \sim \epsilon_2(\omega).L_z \sim cst\ |<\chi^e|\chi^h>|^2\ p^2 \sim cst\ P^2 \qquad (6)$$

per allowed transition. The absorption probability is independant of QW thickness ! As P^2 is almost the same for all semiconductors ($P^2/2m_o \approx 23eV$) the transition rate per layer is almost a universal constant (assuming equal μ's) for type I QWs, with a value of 6.10^{-3} for GaAs[14].

Exciton effects

Taking into account the effect of electron-hole correlation, the initial and final states of the crystal are two-particle states, with either 0 or 1 occupancy factor. The absorption coefficient is then :

$$\alpha(\omega) \sim \varepsilon_2(\omega) \sim \beta \sim |<o|\vec{\varepsilon}.\vec{p}|\Psi_{exc}>|^2$$

$$\sim |<u_{c,k_e}|\vec{\varepsilon}.\vec{p}|u_{v,\vec{k}_h}>|^2|\phi^{(n)}(o)|^2\delta_{o,\vec{K}} \qquad (7)$$

where the first matrix element is the usual Bloch optical matrix element, $\phi^{(n)}(o)$ is the value of the exciton enveloppe wavefunction at zero relative motion, i.e. represents the overlap of the electron and hole wavefunctions. $\delta_{o,\vec{K}}$ ensures the conservation of total momentum in the transition. More rigourously, $\vec{K}$ is equal to $\vec{K}_{ph}$, incoming photon momentum.

It has been discussed at length in 3D that the mere existence of exciton effects increases strongly the strength of the interaction between light and solids[15,16]. This is well-exampllfied in 3D physics, but also in 2D systems (see the lectures by Chemla[9]). For our purpose here, we need only the following properties of 2D excitons :

(i)   Exciton appear as peaks in the absorption spectra due to the K-conservation rule. Only the exciton state which matches the light momentum can be coupled to incoming photons[16]. This is however only true as long as the exciton momentum is a good quantum number. Due to interface disorder, this conservation rule might be somewhat relaxed, but even with such an homogeneous disorder-induced broadening one retains **exciton peaks**.

(ii)   The exciton binding energy is increased in 2D (see Bastard's lectures[1]). This increase of the binding energy with 2D character leads to a very unique situation in semiconductors[17] : usually, large exciton binding energies require large reduced masses[15]. These exist only for large gap materials (effective mass theory) which in turn are linked to a large ionicity of the material (Phillips theory of covalent bonding)[18], which then implies strong LO phonons inter-

264

actions. Therefore, in 3D, large exciton binding energies are only observed in strongly polar materials (such as CdS, ZnS...) with such LO phonon couplings so as to destroy excitons at room temperature. On the contrary, excitons in 2D quantum wells have sizeable energies with marginally modified LO-phonon scattering rates as compared to 3D[19] and lead therefore to those large observable features at room temperature (RT)due to the strong light-exciton coupling[9].

(iii)   A decrease in exciton Bohr radius a_B is associated with the increase in binding energy. From eq. (7), this leads to an increased exciton-photon coupling as compared to 3D which will appear in many experimental situations such as reflectivity, luminescence, non-linear mixing etc...

(iv)   As in 3D the hydrogenic continuum states of the uncorrelated electron-hole pairs lead to an enhancement of the electron-hole wavefunction overlap and therefore of the absorption coefficient (so-called Sommerfeld factor)[15]. This enhancement doubles the absorption at the band edge and decreases over a few Rydbergs[20].

(v)   The excitonic levels and radii depend on the reduced mass and on the confined electron and hole wavefunctions. One then expects them to vary with the type of hole and with the confinement quantum number. A basic complicacy arises from valence band mixing : the exciton enveloppe wavefunction is a linear combination of electrons and hole wavefunctions with k's in an extension a_B^{-1} around 0[15]. For such values of k (up to a few $10^6 cm^{-1}$), strong mixing has been predicted.

Selection rules

Zero-order selection rules on the optical matrix element P and radiation patterns can easily be deduced from atomic-physics analogies as it is known from the correspondance principle that **quantum-mechanical electric dipoles radiate like their classical counterparts**[21]. One can then deduce the radiation pattern for a transition between two quantum states, provided one calculates the electric-dipole matrix-element between these two states. Figure 1a shows the dipole moment between an electron **s-state** and a hole **p-state**, and 1b represents the resulting radiation pattern. If one neglects band-mixing in quantum wells at zero wavevector, the dipole moments are shown in fig. 1c. The emission diagram is simple to calculate and is shown in fig. 1d. This description is however somewhat oversimplified as one has to sum possible transitions over all directions [22], but this summation yields the results given in the atomic picture of fig. 1c.

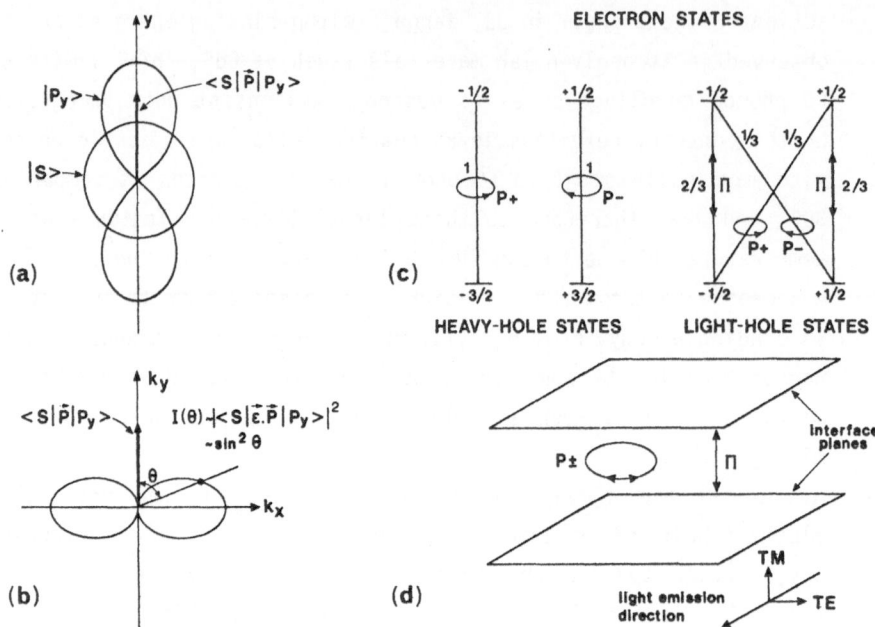

Fig. 1 - Optical selection rules for absorption and luminescence between atomic-like states (Bloch States) of
valence and conductions band.

   a Dipole matrix element between an s and a p state.

   b Emission diagram of that dipole according to the
     correspondance principle.

   c Possible dipole moments between conduction and
     valence band states ; p+(-) indicate rotating dipole
     moments, which emit circularly polarized light ;
     π indicates a linearly polarized dipole. Relative
     dipole strengths are indicated.

   d Geometry of the dipoles in the quantum well situa-
     tion. One can see that heavy-holes only emit TE
     polarized light in the x-direction.

In semiconductors, for transitions occuring with **non-zero transverse
wavector**, the anisotropic heavy-hole to electron matrix element introduces
a gradual change of selection rules with kinetic energy when one sums all
possible transitions for all k-directions[23]. One should in principle also
introduce heavy and light-hole band mixing.

At zero order, **excitons** tend to retain these selection rules as long as
band mixing effects are neglected. This is usually done as the main features
observed in polarized luminescence experiments (see below) reproduce

qualitatively the expected polarization behaviour.

An important set of selection rules originates in the overlap integral $\langle \chi^e | \chi^h \rangle$ of equation (6). In the **infinite well** approximation, χ^e and χ^h are usually orthogonal, i.e. their integral is zero unless they have equal quantum numbers, hence the $\Delta n = 0$ selection rule of Dingle. For **finite quantum wells**, the χ'_s are no more exactly orthogonal as the penetration of the wavefunction in the barrier material depends on the quantum number. However, the single-particle wavefunctions retain they parity (even or odd), due to the symetric confinement potential and one therefore only expects weak $\Delta n \neq 0$ transitions with both n's being either even or odd. When the single particle picture breaks down, such as at high densities, other symmetry-breaking transitions can be observed.

III - TECHNIQUES OF OPTICAL SPECTROSCOPY

The main optical techniques are photoluminescence, absorption, excitation spectroscopy, reflectivity and modulation spectroscopy. We compare below the various qualities of these techniques.

Photoluminescence

This is the most widely used technique[24] due to the ease of its implementation, but interpretation of results is not straight-forward : luminescence occurs as the result of the creation of elementary excitations in the crystal, their thermalization in free and bound states, and their radiative recombination. The observed emission is thus the result of a complex cascade of events and represents a more or less complete thermalization of excitations into some energy-distributed radiative states. As such it does not directly label a characteristic energy as was already pointed out on very fundamental reasons by Hopfield in the case of 3D excitons[25]. The **coincidence** of absorption and luminescence peaks is not in itself a proof of the **equivalence** of the species involved as a state observed in absorption can well be relaxed before emitting light (in this respect see the controversy about the question of excitonic luminescence at RT in QW's[26]). Great care must therefore be devoted when one wants to use photoluminescence to label energy levels and luminescent species.

Absorption spectra

Absorption spectra directly probe oscillator strengths and DOS. Usually

(unless the DOS exhibits a divergence), absorption peaks are the signature
of excitonic phenomena in 2 and 3D, as uncorrelated electron-hole absorption yields a featureless spectrum representing the DOS. The exact lineshape of the exciton absorption peak is not simple to analyze, like in 3D, as
an absorption event requires the transformation of a photon into an exciton
state followed by a scattering into another exciton state not directly
equivalent to the incoming photon state (see the various discussions of
polariton phenomena in 3D[16,27]). In quantum wells, the homogeneous exciton
linewidth is usually masked by the spatially inhomogeneous distribution of
exciton levels due to interface roughness which gives rise to an exciton
absorption lineshape determined by spatial variations of the exciton energy.

Absorption measurements require either a transparent substrate (case of
In-based infrared materials grown on InP) or the removal of the substrate
(GaAs/GaAlAs case). In this latter case, substrate-free layers often exhi-
bit inhomogeneous strains[28] (figure 2).

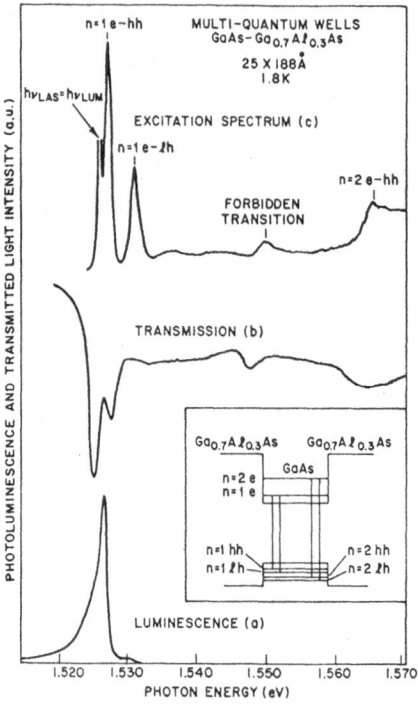

Fig. 2 - Luminescence (a), Transmission (b) and PLE spectrum
(c) of 188Å GaAs multiquantum wells. Note the strain-
induced shift and broadening of the transmission curve. The line marked as forbidden was later attributed
by R.C. Miller as due to the 1e-3hh transition.
(From Weisbuch et al.[28]).

Another drawback of absorption experiments is the required number of layers in order to reach a significant absorption probability : usually, the absorbed fraction of incident light ($\approx 6.10^{-3}$ per layer) is too small to yield fine structure in the spectra and one has to grow multilayers, with the difficulty associated with the necessary layer-to-layer reproducibility.

Excitation spectra

The photoluminescence excitation (PLE) method is based on the variation of the photoluminescence intensity at a given energy on the exciting light energy. As mentioned above, the luminescence at a given energy is the result of the long history of relaxation towards the light emitting state and some care must be exercised in order to yield significant information.

In an optically thin sample (total thickness $<\alpha^{-1}$), the luminescence intensity I_{lum} can be written :

$$I_{lum} = \eta \; \alpha \; (1-R)I_{inc} \tag{8}$$

where I_{inc} is the incoming light intensity, R is the reflectivity for the incoming light, $\alpha \; (1-R)I_{inc}$ therefore represents the absorbed intensity, and η is the efficiency of relaxation towards the light-emitting stage. In 3D, a wide variety of PLE spectra has be generated due to the spectral variation of the various factors in (8). This is particularly true in 3D II-VI compounds. It was shown in CdTe[29] that various lines displayed different oscillatory PLE spectra due to η variation with LO phonon phenomena associated with hot electron, hot exciton or resonant Raman effects. In CdS, the reflectivity factor R could be observed as an important feature[30]. In 3D III-V compounds, the main feature observed was a smooth oscillation due to more or less efficient thermalization of carriers by LO phonons[31].

The striking feature of PLE spectra in III-V quantum-wells is that they only display features associated with the absorption coefficient, i.e. they show no variation of η with exciting light energy. As we know that LO phonon scattering phenomena are similar to those in 3D, it means that relaxation toward the emitting state has no competing channel which would usually, like in 3D, have a different energy dependance on exciting light energy. This tends to prove that non-radiative channels are relatively unimportant in quantum wells and that quantum efficiency could reach unity. In II-VI QWs, LO-phonon oscillations have been observed, however quite small as compared to the 3D case, presumably because of the less important competing recombination channels.

An important advantage of the PLE technique is its very high sensitivity : as will be discussed later, QWs have excellent quantum efficiencies. Therefore, even though only a small fraction of incident light is absorbed in a single quantum well, the emitted light intensity is readily observable. This allows to study PLE spectra of single QWs even when excited with a monochromator-filtered quartz-iodine lamp ($I_{inc} \lesssim 1\,\mu W/cm^2$), which is of paramount importance to perform PLE measurements in those spectral ranges where no dye laser is available[32].

The PLE technique also provides selectivity : the efficiency depends on the path followed by the crystal excitations from the photocreated state to the light-emitting state. One can therefore check the identity of that level knowing the initial state and the possible connections to the various luminescent channels (see e.g. the 3D uses of this concept[29]). Spin-polarization memory experiments also can help to ascertain luminescent levels as will be discussed below for the assignment of the free exciton luminescence of QWs.

Photoreflectance experiments

It has been well established in 3D that reflectivity modulation techniques provide a powerful tool to determine energy levels and joint DOS throughout the Brillouin zone[12]. The strength of the exciton-photon coupling in QWs and the existence of a small-enough damping of excitons at room temperature make such experiments very easy in QWs where one can obtain very large signals, comparable with 3D features, on single QWs.

The normalized reflectivity changes are directly proportional to dielectric constant changes :

$$\Delta R/R = \alpha\,\Delta\epsilon_1 + \beta\,\Delta\epsilon_2 \qquad\qquad (9)$$

where α and β are the Seraphin coefficients. The photoreflectance (PR) technique is the modulated-reflectivity method which seems to become widely used due to its simplicity of implementation (no sample preparation, no electrolyte required etc...) and the apparent universality of the phenomenon : one measures the changes in the reflectivity spectrum under simultaneous modulation of a strong pump beam with energy above the bandgap. By a careful comparison of photoreflectance spectra and of the derivatives of the dielectric function (figure 3) Shanabrook et al.[33] demonstrated that the photoreflectance signal is due to the modulation of the bandedge energy by the pump beam. The origin of this modulation might be in the changes in surface electric field unduced by the pump beam. It is well known that : (i) Photoexcitation tends to diminish space-charge induced electric fields as photocarriers are very efficient in photoneutralizing ionized impurities. (ii) Electric fields can shift energy levels in quantum wells (Stark effect) without destroying exciton effects (see Chemla's lectures[9]).

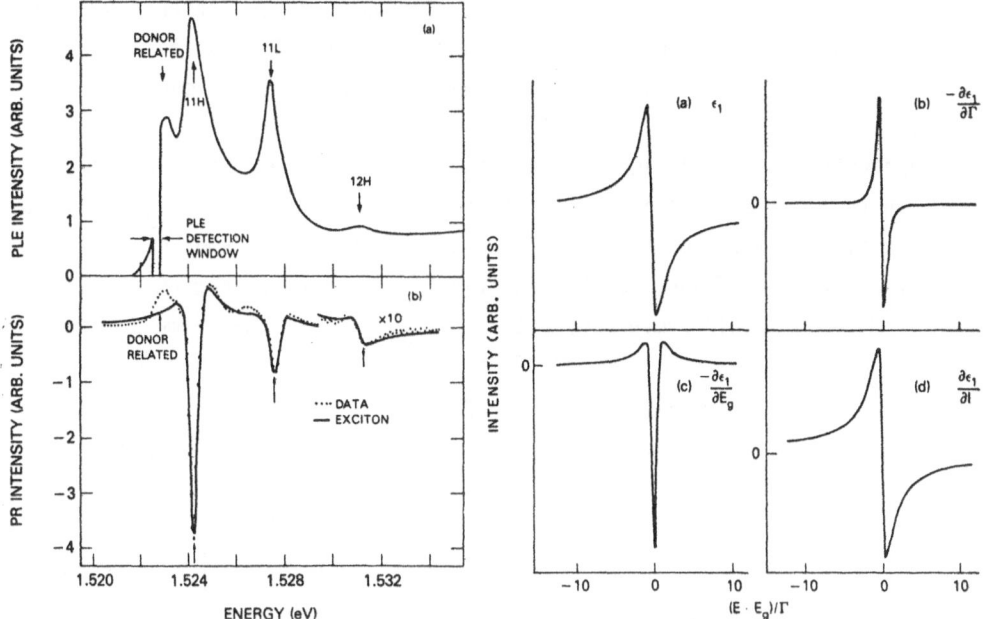

Fig. 3 - Experimental 6K GaAs QW PLE (left a) and photoreflectance (PR) (left b) spectra and comparison with
the derivatives (right) of the real part of the
dielectric function. Coïncidence of PLE and PR peaks
establish that PR is due to exiton effects. Comparison
of lineshapes show that 1-1H and 1-1L peaks are due to
gap energy modulation while the 1-2H forbidden tran-
tion has a mixed character (from Shanabrook et al.[33]).

Transient optical measurements

We discussed up to now only the various merits of the optical techniques based on static measurements.As optics now appears more and more as
the premium way to perform ultrafast measurements down to the subpicosecond
regime, it is worthwile emphasizing the transient performance of the
various optical techniques.

Absorption and reflectivity measurements are instantaneous measurements, in that one probes unrelaxed crystal excitations. The response time of the physical systems under an external perturbation is just the usual dielectric response time, usually in the femtosecond range. As a very convenient way to excite a crystal is also provided by optical pumping, all optical excite-and-probe experiments have recently seen a huge development : pump-probe absorption experiments yield very precise information on carrier dynamics, carrier-carrier and carrier-exciton interactions, bandgap renormalization[11]... Actually, thanks to the damping of exciton states at high carrier concentrations, time-delayed absorption spectra yield exciton--less absorption features evidencing the 2D-square stepped absorption spectra[34] (figure 4). The very recent development of the electric field generation method by optical rectification opens the way to transient electrical response measurements in the subpicosecond regime[35]. Ultrafast reflectivity measurements were widely used in 3D studies such as the reflectivity studies of the a-Si cristallization[36]. That case is a good illustration of the difficulty associated with the analysis of the sometimes ambiguous reflectivity spectra. However, due to sensitivity and speed of response, reflectivity methods should prove very useful in QWs. When lower time resolution is acceptable, combined electrical excitation (either direct or through photoconductive switches) and optical probe can be used.

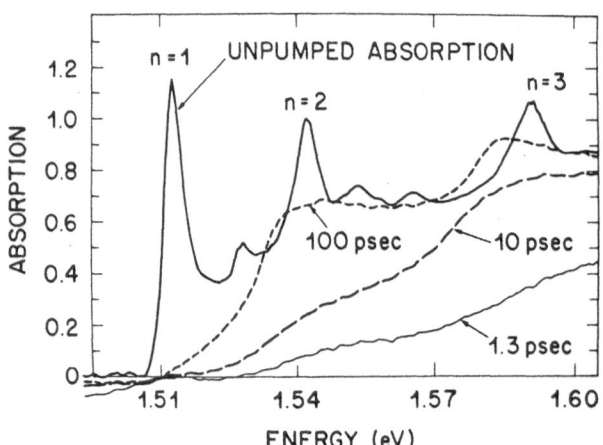

Fig. 4 -- Pump-and-probe transient absorption of a 250A GaAs MQW sample. Carrier density is $10^{12}cm^{-2}$. One observes strong bandfilling and reduced absorption at short times. At longer times and lower bandfilling, the almost square absorption edge due to the 2D DOS is observable as excitons are screened. Notice the shift of n = 3 absorption edge due to bandgap renormalization (from Shank et al.[34]).

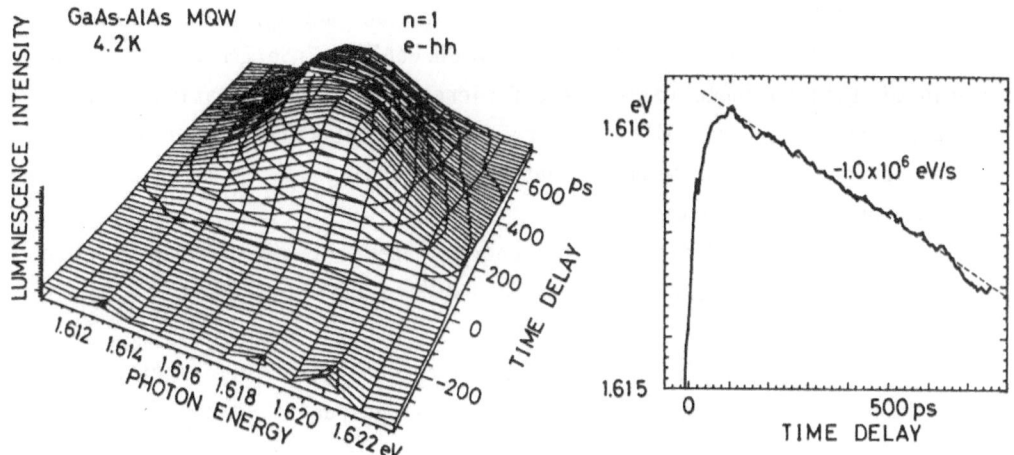

Fig. 5 - Time-evolution of GaAs 76Å MQW exciton luminescence
at 2K (left) and time evolution of the average excitons energy (right). Notice the change in spectral
shape of the exciton band. The measured energy relation rate is much slower than the calculated homogeneous rate, due to the required spatial motion in
the disordered exciton band (from Matsumoto et al.[37]).

Transient luminescence spectra are usually not simple to analyse as
the rise and fall-time of a given emission peak depends on the history of
the excitations and on the actual lifetime of the emitting state. The case
of the GaAs/GaAlAs exciton emission is very clear in this respect : due to
the site-to-site migration of excitons, the time evolution of the exciton
band is not homogeneous (figure 5) and the energy relaxation of excitons
depends in an essential way on the spatial migration[37]. In some cases, one
can analyse luminescence build-up and decay times with simple cascade
models. Exciting at characteristic energies of the physical system, one can
then determine separately the various decay and transfer times (this is the
equivalent of time-resolved PLE spectra). This has been performed by Göbel
et al.[38] on GaAs/GaAlAs QWs and allows to determine capture times, intersubband relaxation and exciton lifetime.

IV - ILLUSTRATIVE APPLICATIONS OF OPTICAL TECHNIQUES

The overwhelming property of quantum wells : the very high quantum efficiency

Whereas the bulk of attention has been focussed on the size quantization of energy in QWs light emission, the most evident and ubiquitous
effect is the much larger quantum efficiency than in 3D reported in all

273

materials systems studied up to now : GaAs/GaAlAs, GaInAs/InAlAs, GaSb/GaAlSb, GaInAs/InP, CdTe/CdMnTe, GaAsSb/GaAlSb, ZnSe/ZnMnSe etc... The origin of this enhanced radiative efficiency is not **quantitatively** worked out and might be the result of one or several of the following effects, on radiative or non-radiative recombination, depending on the materials systems and on growth conditions :

- Due to the larger exciton binding energy, correlation effects in light emission can be stronger (increased e-h overlap) and exist at higher temperatures.

- As evidenced by Petroff et al.[39] interfaces can act as efficient getters for impurities which would otherwise act as non-radiative centers (figure 6). The smoothing action of interfaces also diminishes structural defects and therefore non-radiative recombination centers. The improvement in quantum-efficiency of material grown after a superlattice buffer layer has been widely reported.

- Dislocations in QWs have been shown to be inactive as non-radiative centers, although selective etching reveals that dislocations are present[40]. No complete explanation of this phenomenon has been given which could either be due to inefficient capture by dislocations in 2D or by diminished gettering of impurities by dislocations due to the more efficient gettering by interfaces.

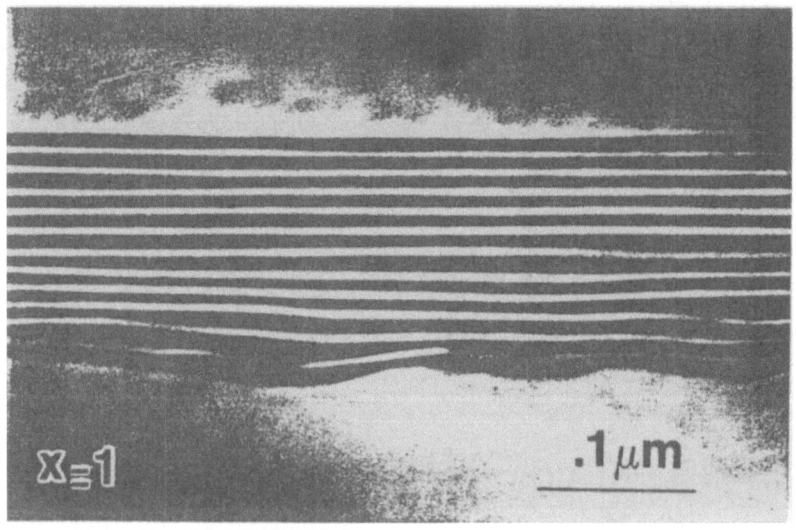

Fig. 6 - Dark-field TEM picture of a substrate - GaAs Interface. Notice the smoothing action of the MQW. Scanning cathodoluminescence spectra evidence a large density of non-radiative centers in rough areas (after Petroff et al.[39]).

Inactivity of impurities : as will be discussed below, extrinsic recombination processes involving impurities (electron to neutral acceptor recombination, bound excitons, etc...) only appear in QWs at significantly higher impurity concentrations than in 3D. This might be due to the more efficient intrinsic radiative recombination mechanisms or to inefficient 2D impurity capture mechanism , although this last point has not been theoretically evaluated.

Free exciton luminescence

In high-purity QW's, the photoluminescence spectra consist in only few lines, in sharp constrast with bulk material of similar purity[28]. These lines were shown in GaAs QWs to be due at low temperature to free exciton recombination by the following analysis[28] :

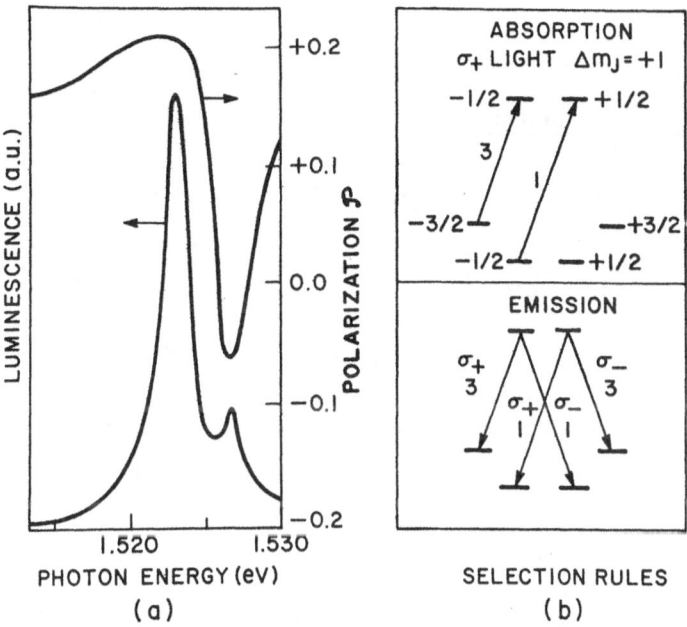

Fig. 7 - Luminescence, spectrum and circular polarization
spectrum under circularly polarized excitations (a)
and transition probabilities.(b) σ_+ excitation creates
more -1/2 electron spins which recombine with heavy
or light-holes by emitting circularly polarized light
with opposite polarizations. This observation of opposite polarizations on the two peaks in (a) together
with other experiments establish that the luminescence lines are due to free excitons (after Weisbuch et
al.[28]).

- The quasi-coïncidence of the luminescence lines with PLE peaks indicate that they are due to exciton-related features and not to band-to-band recombination (figure 3).
- The temperature and exciting light intensity dependance of the high-energy slope of these lines show that they are due to free-moving particles with associated kinetic energy.
- Spin memory experiments involving the analysis of the circular polarization of emitted light under circularly polarized excitation demonstrates that the two lines in figure **7** are due to two-particle excitation (i.e. free excitons) involving heavy and light holes.

It appears that many of the QW systems should luminesce through free exciton emission, but this has not be usually proven by the same detailed analysis as in the low-temperature GaAs/GaAlAs case.

Energy level determinations

The exciton energy level is given by :

$$E_X = E_G + E^e_{conf} + E^h_{conf} - R_X \qquad (10)$$

where E^e_{conf} and E^h_{conf} are the confining energies for electrons and holes and R_X is the exciton binding energy. Although the exciton luminescence energy only gives an approximate value for the exciton energy, it has been sometimes used to evaluate the various terms in equation (10). More often, absorption, PLE or PR peaks have been used to measure energy levels in QWs[41]. Exciton binding energies have been determined from the distance between $n = 1$ and $n = 2$ absorption peaks[41] or from PLE measurements in magnetic fields[10]. The determination of $E^e_{conf} + E^h_{conf}$ through (10) has been widely used to assess bandgap discontinuities[41]. Although this might appear as a simple task, it is a rather involved problem due to the number of fitting parameters which can be used : $Q = \Delta E_C / \Delta E_G$, electron and hole effective masses, R_X. Interband energies do not suffice to unambiguously determine Q, as was shown by Miller et al.[42] who would fit all GaAs/GaAlAs QW data with different Q's by adjusting hole masses[42]. Forbidden transitions, by allowing an intraband energy to be evaluated, allow a much sharper determination of Q[42]. The precision of the optical methods is however still a matter of controversy[43].

It was argued early that inefficient carrier capture in narrow quantum wells might be a limiting process in the performance of QW lasers. The various measurements, either C.W. or transient, show that this is not the case (figure 8). The capture time is of the order of 0.1 ps, which cannot influence laser operation. A very simple proof is provided by the usually small or unobservable C.W. luminescence from barrier materials, which shows that barrier material states have a very low probability of occupancy and are depleted by carrier capture into the wells. Theoretical calculations have predicted smooth oscillations in the efficiency of carrier capture into wells as a function of well thickness[45]. This was explained by the **increased** efficiency of LO-phonon assisted capture whenever a quantum well state reaches the top of the well. However, the only experiment reported so far evidences a **decrease** in capture efficiency explained by increased thermal excitation out of the well whenever such a situation develops[46].

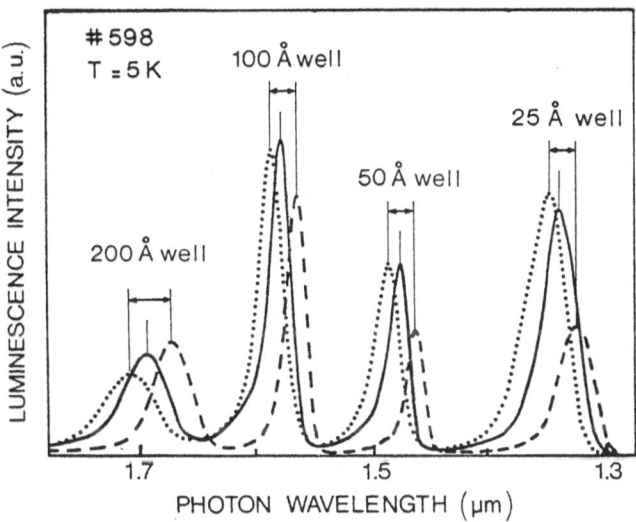

Fig. 8 - Luminescence of a GaInAs/InP sample with single QWs
         with indicated thicknesses. Such samples were first
         introduced by Frijlink and Maluenda[44] and are very
         useful to compare QWs grown at once under constant
         conditions. One observes here that carrier capture
         into wells depends only weakly on well thickness. The
         constant measured energy shifts while one observes
         spectra at different points on the wafer indicate that
         they originate in macroscopic variations of QW alloy
         composition, as changes in QW thickness would lead to
         unequal shifts (after Razeghi et al.[32]).

Interface disorder

The systematic study of **MBE** GaAs QWs with varying thicknesses shows that linewidth increases with decreasing layer thickness[47]. Due to the spatial selectivity of PLE spectra, it is possible to assign in MQW samples such a linewidth to **intrawell** thickness fluctuations rather than to **layer-to-layer** thickness differences. One certainly expects the local layer thickness to be defined at most to $\pm$ a/2 precision, where a is a monolayer. If the well-defined portions of a layer had a lateral extension large as compared to the exciton diameter, one would observe discrete exciton energies along (10), corresponding to the various values of $E_{conf}^e + E_{conf}^h$ when scanning the layer. This has been indeed sometimes observed for continuously grown samples and gives rise to sets of sharp luminescence peaks[48]. Usually, the "island" structure of the interface occurs on a scale of the order of the exciton diameter ($\sim 300\overset{\circ}{A}$). In that case the energy fluctuation will depend on the ratio of island to exciton diameter[49]. One then observes a quasi-continuum of exciton levels corresponding to the various islands, with a linewidth determined by the average energy fluctuation. The fit of measured values[47] yields a layer fluctuation equal to that predicted from thickness fluctuations of a single monotonic layer, which evidences that the island have an extension of $\sim 300\overset{\circ}{A}$, which is actually confirmed by X-ray[50] and TEM[51] measurements. In **MOCVD** GaAs samples, one observes narrower spectra, which even get sharper with increasing deposition temperature[52]. As no discrete line is observed, this is explained by assuming that the interface island size is significantly smaller than the exciton diameter, leading to thickness averaging over the exciton wavefunction.

The avaibility of layer growth control provided by RHEED oscillations in MBE chambers has led to the concept of interface remodelling through growth interruption at interfaces[53] : this leaves time for atoms to migrate to energy-favorable nucleation sites on the interfaces. In this manner, extremely sharp luminescence peaks have been observed, corresponding to atomically-flat island sizes in the micron range[54]. The improvement on interface quality has been shown to depend strongly on growth condition. Several authors have stressed that the improvement was greater on the usually "bad" inverted interface.

The most widely studied systems besides the GaAs/AlGaAs materials system have been the GaInAs/InP and GaInAs/InAlAs. In addition to the usual broadening mechanisms encountered in binary semiconductors, one has in addition to evaluate the influence of alloy disorder[55]. This has not been completely worked out yet in the bulk Ternary semiconductors, although the work by Cohen[56], in particular in the indirect-gap GaAlAs case, has been quite successful in pointing out some of the aspects of alloy luminescence. One of the major issues is the amount to which one can consider an alloy as homogeneously random, whereas there is strong evidence both to microscopic[57] and macroscopic variations of the alloy composition (figure 8).

Dynamics of excitations

The existence of spatial disorder in QWs makes them very interesting as a prototype 2D disordered system where the strong exciton-photon coupling makes optical techniques very sensitive. Very detailed and elegant pionneering experiments were performed by Hegarty et al.[58]. in order to sort out the dynamical properties of such a system. We refer the reader to a review article by Hegarty[58] on the comparison and results of the various techniques (transient gratings, hole-burning measurements, degenerate four-wave mixing, photon echoes etc...) and only discuss here the most puzzling measurement, the resonant Rayleigh scattering intensity[59]. Shown in figure 9 is the result of such measurements together with the absorption (as revealed by PLE) and luminescence spectra. In the case of an inhomogeneous broadening, formula (1) represents the local dielectric constant with eigenenergy ω_o and **homogeneous** broadening Γ_h, characteristic of that location and energy. The displacement of the Rayleigh scattering curve versus the DOS as measured by the absorption curve evidences that Γ_h incrases from the low energy side of the DOS to the high energy side, showing the existence of a mobility edge in this 2D disordered system : above that edge, located in the middle of the exciton DOS, excitons move freely throughout the layer and are therefore strongly damped by various interactions. Below that edge, excitons are fixed and weakly damped. In terms of scattering efficiency, if one considers excitons as classical oscillators[16], the picture is very simple : above the mobility edge, strong damping leads to optical-field driven oscillators with a small amplitude, thus to a small scattered intensity per oscillator. Below the mobility edge, each oscillator reaches a far greater oscillation amplitude. This explains why the maximum of scattered light occurs below the mobility edge, even though this corresponds to a smaller DOS. The luminescence is further displaced towards low energies as it corresponds to energy-relaxed excitons in the disordered band. All other transient measurements support the description in terms of mobility edge, in particular transient grating experiment which directly probe spatial exciton transport[58].

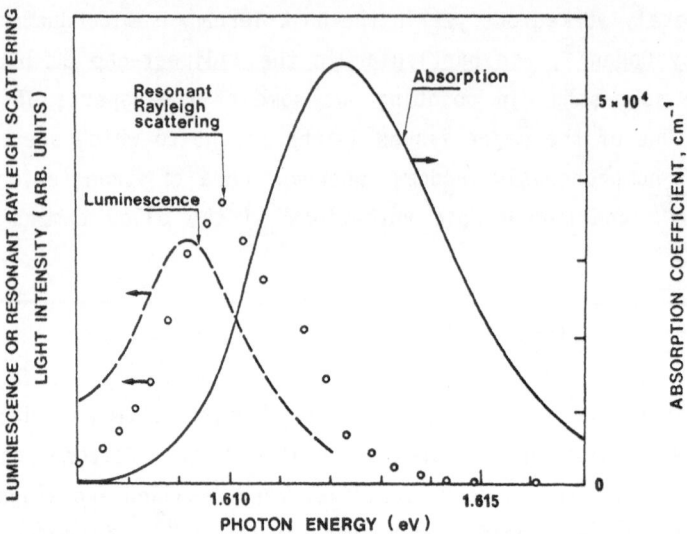

Fig. 9 - Absorption (full line), luminescence (dashed line) and intensity of elastically scattered light (Rayleigh scattering) of a 51Å GaAs MQW sample at 5.7K (from Hegarty et al.[59]).

When dealing with atomically flat QWs grown by the interrupted-process, excitons do not experience inhomogeneity of the layer thickness. In that case measured linewidths only evidence an homogeneous broadening in degenerate four-wave mixing experiments[60]. Luminescence peaks also exactly coïncide with excitation spectra peaks.

Impurity-related luminescence

Although pure samples exhibit intrinsic recombination lines, purposely or inadvertently doped samples display a variety of impurity related lines[61]. The situation is complicated by the variation in impurity binding energy due to the location of the impurity ion relative to the well interface. One of the most striking impurity-related luminescence occurs for growth-interrupted samples in which a large carbon acceptor concentration builds up at the interrupted interfaces. With purposely-doped samples in which the dopant location was well-defined, it has been possible to map the impurity energy levels as a function of position in the well[61].

Modulation-doped samples

The luminescence of modulation-doped sample has only recently received attention, although they give very useful information on such structures[62] : figure 10 shows the schematics of light recombination in such sam-

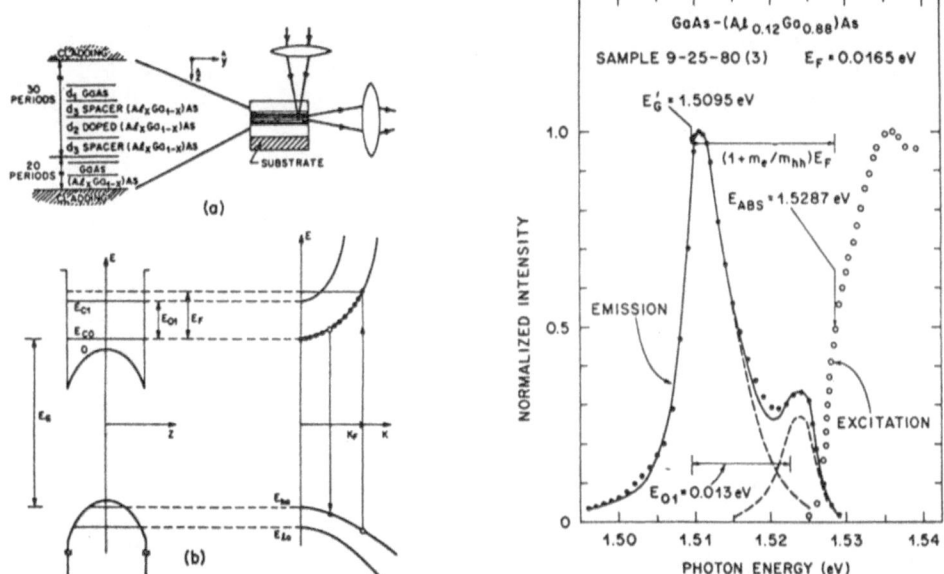

Fig. 10 - Luminescence and PLE spectrum of a 221Å modulation
doped GaAs MQW sample at 2K
Left : (a) Schematics of experiment ; (b) Schematics
of transitions in E-k space.
Right : Observed spectrum in the z-direction perpendicular to layers (from Pinczuk et al.[62]).

ples and the recorded spectrum for a sample with two occupied electron
subbands. By analysing the spectrum, Pinczuk et al.[62] deduced the Fermi
energy of the electron gas, the position of the two subbands and the bandgap renormalization due to the high density of carriers. The peak in the
PLE spectrum shows the existence of excitons at high densities.

Figures 10 a and b explain how one can perform the k-space spectroscopy
of hole levels : as electron states are filled up to the Fermi energy,
recombination intensity probes the occupancy of hole levels and the transition probability. As holes tend to be rather heavy, one can in first approximation assume an even distribution of holes in k-space. Intensity changes
are then only due to matrix element and DOS changes. In the geometry of
emission of light propagating in the plane of the layer, the zero-order
selection rules shown in figure 1 forbid any TM emission from the heavy
hole band. Any TM emission is therefore due to hole band mixing. By
complete calculation of band mixing and transition probabilities,
Sooryakumar et al.[63] were able to calculate the forbidden luminescence
spectrum. The excellent prediction of the band-mixing changes induced by a
uniaxial pressure definitely proves the accuracy of the calculation (fig.11).

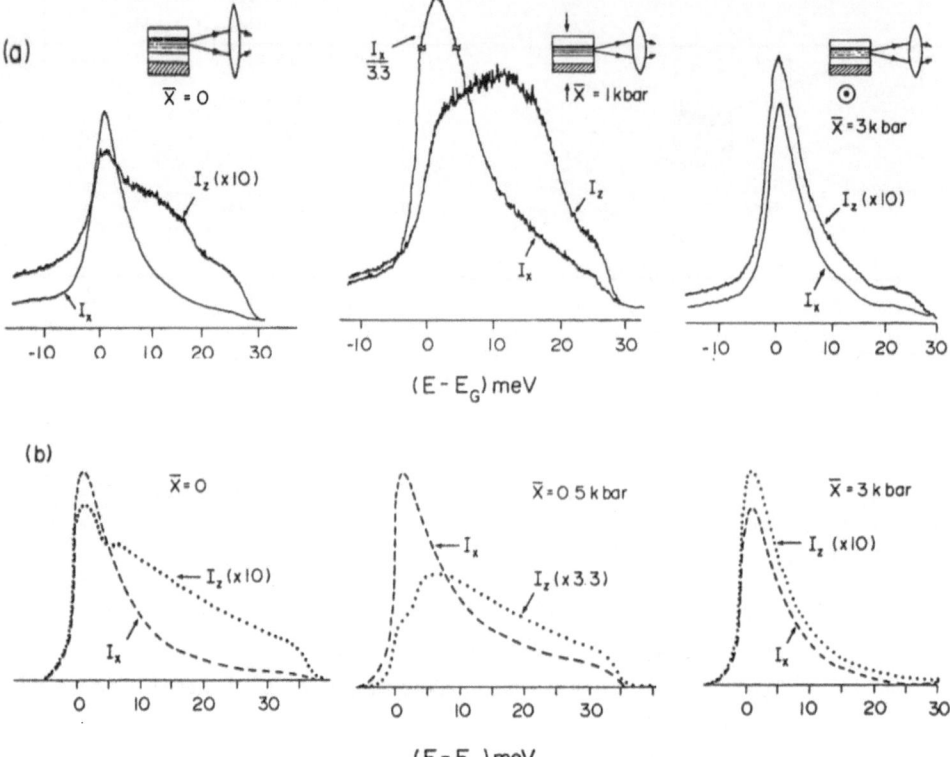

Fig. 11 - Luminescence along the y-direction in layer plane
of a 166Å modulation-doped GaAs MQW sample at 2K.
I_x and I_z are the two linear polarizations of the
electric field of the optical wave (TE and "forbidden" TM modes respectively). The three (a) curves are obtained under zero and non-zero uniaxial
stress in the two indicated directions. The (b)
curves show the **calculated** luminescence curves
(from Sooryakumar et al.[63]).

V - PRINCIPLES OF LASER ACTION IN QUANTUM WELLS

Background on semiconductor lasers[13]

i)   Lasers are based on self-oscillations of light emitting systems
     through stimulated emission. In order to have gain overcome
     absorption, one needs to have **inversion** of the quantum states between
     which the laser transition is to occur. This means that one has to
     fill the conduction band states of a semiconductor laser over kT, or
     actually a few kT.

ii) One uses emission in an optical cavity in order to **concentrate the**
     emitted photons in only a few modes through reflections back and forth

on the mirrors. As the quantum stimulated emission probability is proportional to the number of photons per mode, one increases in that way the efficiency of emitted photons.

iii) A main parameter is the **optical confinement factor** Γ which represents the fraction of the optical wave in the active layer material, i.e. the efficiency of an emitted photon to interact with another e-h pair in order to further induce stimulated emission. This confinement factor depends on the active layer thickness and on the difference in index of refraction between the active layer and confining materials.

As indicated in figure 12, the d^2 variation of Γ for simple hetero-structures (because of **diminished overlap** of an optical wave which becomes **wider** with decreased active layer material) leads to extremely small values of Γ in quantum wells, typically 4.10^{-3} for $d = 100\text{Å}$. Using the separate confinement heterostructure (SCH) scheme, with a fixed cavity to confine the optical wave separately from the electron wave (quantum well), one obtains $\Gamma \sim 3.10^{-2}$ in a 100Å GaAs/GaAlAs QW.

The medium gain g_{th} at threshold is obtained by stating that the opti-cal wave intensity after a roundtrip in the cavity must stay equal, under the opposite actions of losses and gain. This is conveniently written as :

$$I_0 \, R^2 e^{2(\Gamma g_{th} - \alpha)L} = I_0 \tag{11}$$

where R is the facet reflectivity, Γg_{th} represents the modal gain per unit length of the optical wave, α sums all the various loss mechanisms such as free carrier absorption, light scattering by waveguide imperfections, barrier material absorption... and L is the laser length. Equation (11) can be rewritten in the form :

$$\Gamma g_{th} = \alpha + 1/L \, \text{Log} \, 1/R \tag{12}$$

In GaAs/GaAlAs lasers, the first term is usually 10 cm^{-1} and the second 40 cm^{-1} for a 300μ long laser and uncoated facet reflectivity of 0.3.

The threshold current is known once the gain versus injected current is determined. This can be readily done by using the carrier density as an input parameter. One can both calculate for that density the maximum gain and the required injection current from the known radiative and non-radia-tive recombination channels. As mentioned above, gain occurs only once a significant inversion has been achieved.

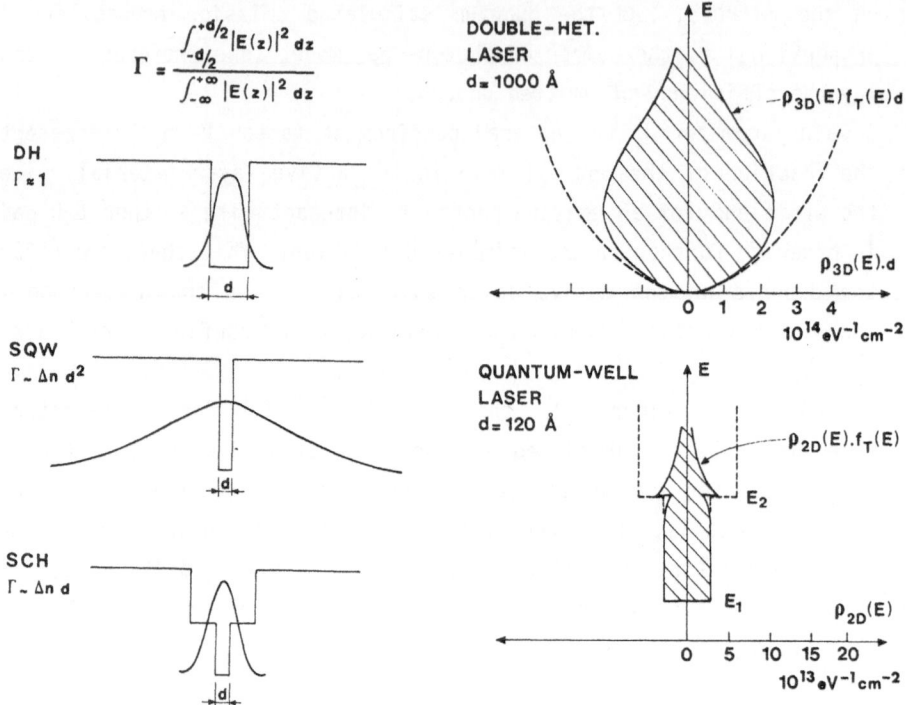

$$\Gamma = \frac{\int_{-d/2}^{+d/2}|E(z)|^2\,dz}{\int_{-\infty}^{+\infty}|E(z)|^2\,dz}$$

DH
$\Gamma \approx 1$

SQW
$\Gamma \sim \Delta n\, d^2$

SCH
$\Gamma \sim \Delta n\, d$

DOUBLE-HET. LASER
$d = 1000$ Å

$\rho_{3D}(E)f_T(E)d$

$\rho_{3D}(E).d$

0 1 2 3 4
$10^{14}.eV^{-1}cm^{-2}$

QUANTUM-WELL LASER
$d = 120$ Å

$\rho_{2D}(E).f_T(E)$

E_2

E_1 $\rho_{2D}(E)$

0 5 10 15 20
$10^{13}.eV^{-1}cm^{-2}$

Fig. 12 - Comparison of confinement factors and density-of-
states (DOS) in DH and QW structures. **Left** : Schematics of conduction band extrema and corresponding
optical wave for a double-heterostructure (DH) (top),
single quantum well (SQW) (middle) and Separate Confinement Heterostructure (SCH)(bottom).Note the width
of the optical wave in the SQW as confining waguide is
too thin, whereas in the SCH case the optical waveform
is independant of the quantum well thickness as it is
confined by the intermediate composition layer.
Right : Schematics of the DOS (dashed curves) and occupied states (cross-hatched) for DH (top) and SQW
(bottom). Note the change of horizontal scale, due to the
numerous allowed k_z states in the DH.

Single quantum well operation

There are two main effects in opposite directions which occur when
comparing QW and DH lasers :
(i) The DOS to be inverted decreases as one has only one or a few per-
pendicular-momentum quantum values (as many as populated states) vs a
few tens in a typical 1500 Å thick DH laser (figure 12).
(ii) The confinement factor is decreased (typical DH value of 0.4).

284

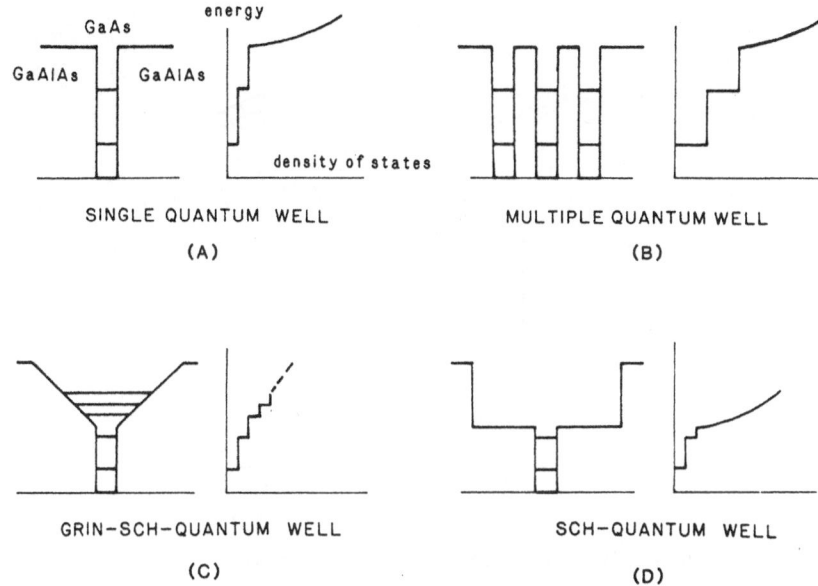

energy

GaAs

GaAlAs   GaAlAs

density of states

SINGLE QUANTUM WELL

(A)

MULTIPLE QUANTUM WELL

(B)

GRIN-SCH-QUANTUM WELL

(C)

SCH-QUANTUM WELL

(D)

Fig. 13 - Various quantum well laser structure schematically
depicted by their conduction band edge space variation (left-side of each figure) and their 2D density-
of-states (DOS) (right side).

a) : Single Quantum Well. Each quantized well state
introduces a 2D DOS equal to $m*/\pi\hbar^2$, while the
onset of 3D states at the top of the well introduces a much larger DOS.

b) : Multiple Quantum Well (MQW) : Each quantized
state introduces a 2D DOS equal to $N\ m*/\pi\hbar^2$, N
being the number of wells.

c) : Graded-Index Separate Confinement Heterostructure GRIN-SCH laser. Note the ladder of quantum
states in the graded region.

d) : Separate-Confinement Heterostructure (SCH) laser : The intermediate-composition layers introduce a large DOS, not as far apart from the
ground state as in the DH, SQW or MQW cases.

As discussed above, the single QW DH structure leads to $\Gamma \sim 4.10^{-3}$, an
unacceptable value and one has to resort to the various structures shown on
figure 13[64]. Focussing on the single QW structures 13c and 13d, one can
show that the decrease in confinement factor almost exactly cancels the
decrease in quantum states to be inverted at threshold.

The good operation of single QW lasers therefore relies on some more
subtle effects[65] :

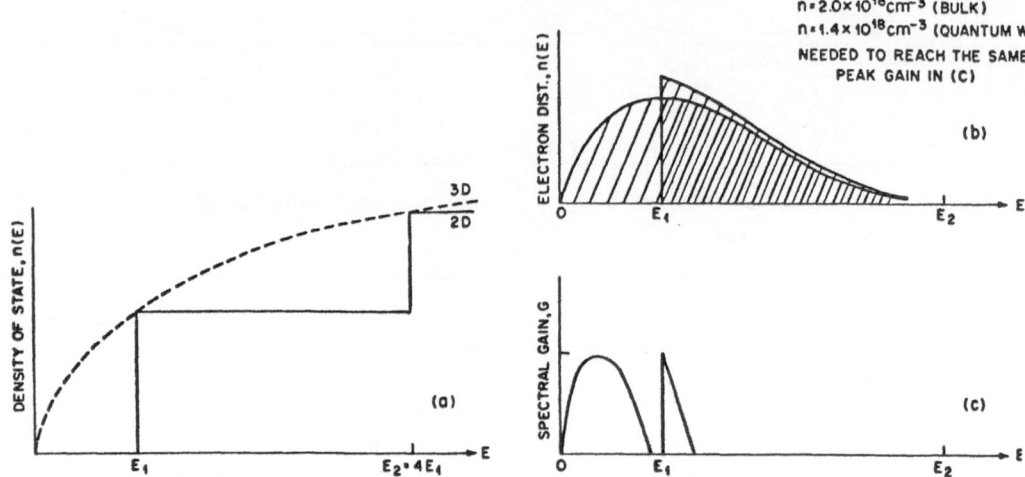

Fig. 14 - Schematics of the required carrier population to
achieve a given gain in 2D and 3D systems.
a) : DOS in 2D and 3D systems **with equal thicknesses**
b) : Required numbers of electrons and populated
states in order to achieve equal maximum gains
in the gain curves shown in (c) (from Tsang[64]).

(i) The square DOS is of course advantageous to build a more significant
gain for a given injection, the maximum gain will be obtained at the
bottom of the band and not like in 3D at the maximum of the product of
the 3D DOS times the Fermi-Dirac occupancy factor (figure 14). This
makes QW lasers extremely efficient at low temperatures where one
retains finite gain even at vanishingly small injection current. There
is however a price to pay for this high injection efficiency : the
gain saturates once complete band filling is achieved. The only way to
recover more gain and eventually reach threshold is either to reach
and populate another confined level or to use multiple quantum wells :
one multiplies the saturated value of the single QW laser gain by the
number of states which can be populated. However one has first to
invert the new states at the bottom of the bands (figure 15).

(ii) As mentionned above, allowed transitions occur for a selection over
k-states which forbid TM emission for heavy-holes. It also leads to a
matrix element which is 50% larger than the k-space averaged 3D matrix
element[23,66-68].

(iii) Some electron-hole interaction which could enhance the radiative
recombination probability might be present at the carrier densities
observed in QW lasers.

(iv) The high radiative efficiency usually observed means that a lower

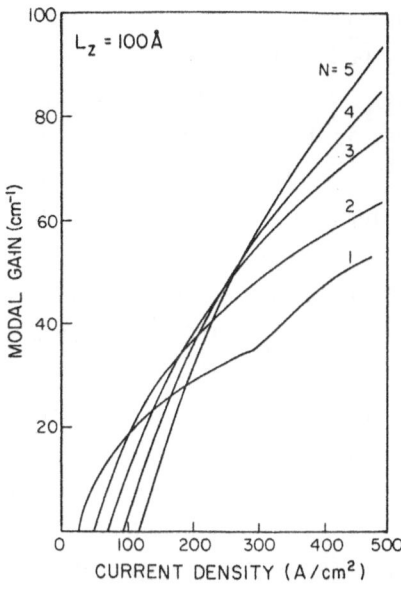

Fig. 15 - The modal gain (Γg) variation as a function of
injected current density for various numbers of
quantum wells. Note the saturation of the n = 1
quantum state for the single QW laser above
~ 100 A/cm^2 and the onset of n = 2 quantum state
transitions leading to higher gain at 300 A/cm^2.
Net gain starts at higher currents for increasing
number of wells because of the increasing number
of states to be inverted. Well thickness is 100 Å
(from Arakawa and Yariv[73]).

injection current will be required to achieve the carrier density at
threshold.

Through detailed calculations all the observed features of QW lasers
have been explained[65,69]. In particular the advantage of the GRIN-SCH
(figure 13c) over the SCH (figure 13d) is well-explained by the important
population of the confining layer quantum states under normal QW laser
operation. Also, the difference in temperature dependances of the threshold
current, described by an exponential variation with a characteristic factor
T_0 has been justified : The T_0 decreases in GaAs/GaAlAs lasers from
excellent values for MQW lasers (figure 13b) to poor values in SCH lasers,
with GRIN-SCH lasers having intermediate values. This is mainly due to
increased carrier leakage out of the "useful" QW states into confining
layer states.

One of the puzzling results has been the very poor operation of GaInAsP-based single QW lasers in the 1.3 and 1.5µm ranges. Experiments showed that SCH single QW laser could not lase on a quantum well transition. What happens is that carrier temperature increases so rapidly with injection that the maximum gain actually decreases with increased carrier density and a large cavity population builds up, until laser emission occurs for the cavity material[65,70]. This can be readily understood by detailed calculation : as QW lasers operate with poor Γ's, they require a high volume gain g_{th}. This requires a higher carrier density in QWs than in DHs, which leads to higher Auger recombination[65,71], even though the Auger coefficient has been measured to be the same in 2 and 3D[72]. This high-density operation of QW lasers puts severe conditions on the optimization of laser operation whenever a detrimental non-linear effect is present. One must therefore examine in details the pros and cons of 2D operation for each materials system. Multiple QWs lead to lower densities, but at the same time lead to some loss of the 2D advantage as more and more states have to be inverted before obtaining any gain (figure 15).

Additional properties of quantum well lasers

Besides their good current threshold and T_o, QW lasers possess a number of additional important features which make them highly useful. As can be seen in figure 14, at low injection, gain increases faster with carrier density in 2D QWs than in DHs. Therefore, QW lasers have a high-differential gain dg/dI where I is the injected current. Due to it, Arakawa et al.[73] have shown that an increase in speed response of a factor ≈ 2 could be expected from QW lasers over DH lasers. This was confirmed by Uomi et al.[74]. From the same high-differential gain in QW lasers, Arakawa et al.[73] and Burt[75] have predicted narrower linewidths than in 3D DH lasers. This improvement has been directly measured by Noda et al.[76] in DFB lasers.

The QW metallurgy also allows **new manufacturing techniques**. Impurity-induced interdiffusion of quantum wells has been evidenced in many materials systems with various n and p-type implanted species[77]. It allows to transform an implanted region with thin wells and barriers into an alloy with the average composition. Such an alloy has the usual properties of light and carrier confinement. Therefore, one can use such an effect to construct a variety of stripe-geometry lasers, window lasers (i.e. lasers where the active regions does not extend to the external facets) etc... Excellent results have been obtained by such techniques which permit to make complex devices with simple planar-type operations while one would otherwise require multiple-step etch and deposition proccesses[78]. In particular, a three-fold increase in catastrophic-damage threshold was demonstrated in window lasers[79]. The spatial precision of the inter-diffusion process is in a spectacular way attested by the recent fabrication of 1D quantum wires and 0D quantum boxes[80] with lateral definition below 1000 Å.

Although relatively few degradation studies have been reported on QW lasers, they all point to very good performances[64]. This might be due to several factors, including inactivity of dislocations, stabilization of defects by interfaces, smallness of the carrier-induced facet degradation as the active layer is much thinner than the optical cavity. The catastrophic failure due to crater formation at the cleaved facet at high powers does not then lead to the deterioration of the quality factor of the optical cavity.

Applications of QWs to integrated optics systems with high performance is also very promising[81]. As was described in details by Okamoto et al.[81], optical absorption of passive MQW waveguides is very low at the laser wavelength of the same MQW structure due to the low 2D absorption tail at the energy-shifted laser emission. QWs also lend themselves easily to wavelength modifications through thermal or impurity-induced interdiffusion[77,78,81]. The recent demonstration[69,82] of the shift from n=1 to n=2 QW laser emission when changing cavity losses should open the way to wavelength-agile lasers once active modulation of cavity losses are achieved such as in a C^3 laser.

VI - EVOLUTION TOWARDS 1 and 0D

Several papers have theoretically shown that 1 and 0D physical systems should in principle have even better performance than the 2D QW laser system thanks to their advantageous DOS[83,84]. Although this is in principle true, we would like to emphasize two important difficulties associated with low-dimensional systems.

The importance of fluctuations

Interface disorder is limiting recombination line sharpness in quantum wells, although some recent interrupted-growth experiments evidence excellent improvement (however at the expense of high-impurity incorporation). In a series of very demonstrating experiments, Okamoto and his group[81] have shown that in order to reach short wavelengths with quantum wells, the use of ultra-narrow quantum wells led to unacceptable linewidth increase, leading in turn to inhomogeneously-broadened gain curves. It is only by an optimization of quantum well thickness and alloying that a short red wavelength was obtained with a reasonable threshold. From this example, one sees the importance of size fluctuations in 2D. It is at the moment difficult to imagine 1 and 0D systems which would not exhibit significantly broadened homogeneous and inhomogeneous linewidths, as a large number of 1 and 0D systems have to be operated together.

The required high density of states, i.e. high number of 1 and 0D systems

As is evident from the above discussion of the 2D QW laser, one cannot diminish too much the active material volume for two reasons :

(i) To achieve significant gain, required carrier densities per 1 or 0D well would be so high that total band filling would occur before enough gain would be obtained.

(ii) The confinement factor would be vanishingly small.

Therefore, one definitely needs constant confinement factor and constant total number of states when going from 3 to 2 to 1 to 0D, in order to obtain the advantages predicted theoretically (figure 16). One has however to remember that this then represents millions of 0D quantum boxes to be fabricated for each laser and with little or no size fluctuation ! The systems which have been fabricated so far are those of Petroff et al.[80], Reed et al.[85], Kash et al.[86] which were studied by photo-or electro-luminescence. Those of Miyamoto et al.[87] were current injected. All those systems were fabricated by lithographic techniques and one might wonder wether this could ever lead to an industrial fabrication method. Another way to obtain 0D systems is to fabricate them into spheres by condensation. This method recently reviewed by Brus[88] could be more manageable.

The full 0D quantization occuring for 2D systems when placed in a strong, perpendicular, magnetic field is unique[89] ; The magnetic field creates at once millions of quantum boxes with sizes equal to the cyclotron orbit radius : one retains all the 2D quantum states, as it is well-known that the degeneracy of a Landau level is exactly equal to the number of states which were previously situated in the energy range between two

290

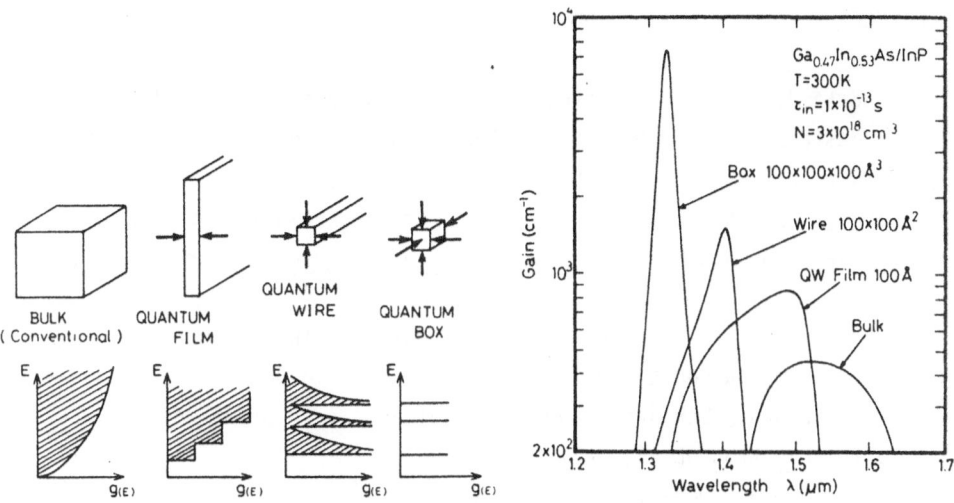

Fig. 16 - Schematics of 3D, 2D, 1D, 0D quantum systems and
corresponding DOS's (left) and gain curves (right).
The same number of carriers is injected in the
systems and an inelastic broadening time of 10^{-13}s is
assumed in all four cases. One should bear in mind
that the modal gain entering eq.(12) is the product
of the gain shown here times the confinement factor,
and that lower D systems are better only insofar
they retain a sufficiently high confinement factor
(from Asada et al.[84]).

adjacent Landau levels. Therefore, the number of states and the confinement
factor are the same in the 0D system created in a strong magnetic field as
in the original 2D system. The experiments performed up to now have however
never evidenced huge effects, just an anisotropy of the spontaneous
emission spectrum, which shows the creation of 0D quantum states[89]. The
quantum-wire system obtained when placing a 3D DH laser in a strong
magnetic field has shown the expected enhanced bandwidth[90].

ACKNOWLEDGMENTS

Early portions of this work were performed at Bell Laboratories with
the collaboration and support of R. Dingle, H. Störmer, V. Narayanamurti,
P. Petroff, J. Hegarty, M. Sturge, A. Pinczuk, C. Shank, A. Gossard, W.
Tsang, A. Cho, W. Wiegmann, R. Kopf, R. Miller. In Thomson-CSF, J. Nagle
has made major contributions to this work. Other contributors include B.
Vinter, T. Weil, F. Vallet, N. Vodjdani, M. Razeghi, S. Hersee and J.P.
Hirtz. This work has been partially supported under ESPRIT N° 514.

REFERENCES

1 G. Bastard, Electronic States in Heterostructures, this volume.
   G. Bastard and M. Voos, "Wave Mechanics Applied to Heterostructures",
   Editions de Physique, Les Ulis (1987).
2 See e.g. L. Esaki, a Bird's Eye View on the Evolution of Semiconductor
   Superlattices and Quantum Wells, IEEE J. Quantum Electron QE-22, 1611
   (1987).
   L. Esaki, The Evolution of Quantum Structures, this volume.
3 E. Kasper, Strained-Layer Superlattices, this volume.
   G.C. Osbourn, Strained-Layer Superlattices : A Brief Review, IEEE J.
   Quantum Electron QE-22, 1677 (1986).
4 See e.g. G.H. Döhler, Doping Superlattices ("n-i-p-i Crystals"), IEEE
   J. Quantum Electron QE-22, 1682 (1986).
5 E. Kasper, Strained-Layer Superlattices, this volume.
   R. People, Physics and Applications of GeSi/Si Strained-Layer Heterostructures, IEEE J. Quantum Electron QE-22, 1696 (1986).

6 A.V. Nurmikko, R.L. Gunshor and L.A. Kolodziezski, IEEE J. Quantum
   Electron QE-22, 1785 (1986).
7 J.P. Faurie, Molecular Beam Epitaxial Growth and Properties of Hg-Based
   Microstructures, this volume.
   J.P. Faurie, Growth and Properties of HgTe-CdTe and other Hg-based
   Superlattices, IEEE J. Quantum Electron QE-22, 1656 (1986).

8 G. Abstreiter, Light Scattering in Heterostructures, this volume.
   G. Abstreiter, R. Merlin and A. Pinczuk, Inelastic Light Scattering by
   Electronic Excitations in Semiconductor Heterostructures, IEEE J. Quantum Electron QE-22, 1771 (1986).
   M.V. Klein, Phonons in Semiconductor Superlattices, IEEE J. Quantum
   Electron QE-22, 1760 (1986).
9 D.S. Chemla, Non-Linear Optics and Optoelectronics in Semiconductor
   Quantum Wells, this volume.
10 J.C. Maan, Magneto-Optical Properties of Heterojunctions, Quantum Wells
   and Superlattices, this volume.
11 E. Gornik, Energy Relaxation Phenomena in GaAs/GaAlAs Structures, this
   volume.
   J. Shah, Hot Carriers in Quasi-2D Polar Semiconductors, IEEE J. Quantum
   Electron QE-22, 1728 (1986).
12 See e.g. DE.E. Aspnes, Modulation Spectroscopy/Electric Field Effects
   on the Dielectric Function of Semiconductors in : "Handbook on Semi-

conductors" vol. 2, M. Balkanski vol. ed., T.S. Moss Series ed. North-Holland, Amsterdam (1980).

M. Cardona, "Modulation Spectroscopy", <u>Solid State Phys. Suppl. 11</u>, Academic, New-York (1969).

13 See e.g. H.C. Casey Jr. and M.B. Panish, "Heterostructure Lasers,Part A : Fundamental Principles", Academic, New-York, 1978.

14 P. Voisin, Optical and Magnetooptical Absorption in Quantum Wells and Superlattices, <u>in</u> "Heterojunctions and Semiconductor Superlattices", G. Allan, G. Bastard, N. Boccara, M. Lannoo and M. Voos, Springer, Berlin (1986).

15 See e.g. Dimmock, Introduction to the theory of Exciton States in Semiconductors, <u>in</u> "Semiconductors and Semimetals" Vol. 3, R.K. Willardson and A.C. Beer eds., Academic, New-York (1967).

16 J.J. Hopfield, Theory of the Contribution of Excitons to the Complex Dielectric Constant of Crystals, <u>Phys. Rev.</u> 112, 1555 (1958).

J.J. Hopfield, Excitons and Their Electromagnetic Interactions, <u>in</u>, Proc. of the International School of Physics, "Enrico Fermi", Course 42, Quantum Optics, Academic, New-York (1969).

17 D.S. Chemla, Quasi-Two Dimensional Excitons in GaAs/AlGaAs Semiconductor MQW Structures, <u>Helv. Physica Acta</u> 56,67 (1983).

18 J.C. Phillips, "Bonds and Bands in Semiconductors", Academic, New-York (1973).

19 F. Stern, Electrical Transport in Microstructures, this volume.

20 M. Shinada and T. Sugano, Interband Optical Transitions in Extremely Anisotropic Semiconductors : I Bound and Unbound Exciton Absorption, <u>J. Phys. Soc. Japan</u> 21, 1936 (1966).

21 See e.g. M. Born, "Atomic Physics" 6th ed., Blackie, London (1957).

L. Pauling and E.B. Wilson, "Introduction to Quantum Mechanics", Mc. Graw Hill, New York 1935.

W. Heitler, The Quantum Theory of Radiation, Dover, New-York (1984).

22 F. Stern, Elementary Theory of the Optical Properties of Solids, <u>in</u> "Solid State Physics vol. 15", F. Seitz and D. Turnbull eds., Academic, New-York (1963).

23. See e.g. M. Asada, A. Kameyama and Y. Suematsu, Gain and Inter-valence Band Absorption in QW Lasers, <u>IEEE J. Quantum Electron</u> QE-20, 745 (1984).

24 See e.g. H. Barry Bebb and E.W. Williams, Photoluminescence I : Theory, <u>in</u> "Semiconductors and Semimetals vol. 8", R.K. Willardson and A.C. Beer eds., p. 182, Academic, New-York (1972).

M. Voos, R.F. Leheny and J. Shah, Radiative Recombination, in "Handbook of Semiconductors vol. 2, M. Balkanski vol. ed., T.S. Moss series ed.,

North Holland, Amsterdam (1980).

25 J.J. Hopfield, Aspects of Polaritons, Proc. Int. Conf. on the Physics of Semiconductors, Kyoto (1966), J. Phys. Soc. Japan 21 suppl., 77 (1966).

26 P. Dawson, G. Duggan, H.I. Ralph and K. Woodridge, Free Excitons in Room Temperature Photoluminescence of GaAs/AlGaAs MQWs, Phys. Rev. B28, 7381 (1983).
E. Böttcher, K. Ketterer, D. Bimberg, G. Weimann and W. Schlapp, Excitonic and Electron-Hole Contributions to the Spontaneous Recombination Rate of Injected Charge Carriers in GaAs/AlGaAs MQW lasers at Room Temperature, Appl. Phys. Lett. 50, 1074 (1987).
Y. Arakawa, H. Sakaki, M. Nishioka, Y. Yoshino and T. Kamiya, Recombination Lifetime of Carriers in GaAs/GaAlAs QWs near Room Temperature, Appl. Phys. Lett. 46, 519 (1985).
J.E. Fouquet and R.D. Burnham, Recombination Dynamics in GaAs/AlGaAs Quantum Well Structure, IEEE J. Quantum Electron. QE-22, 1799 (1986).

27 C. Weisbuch and R.G. Ulbrich, Resonant Light Scattering by Excitonic Polaritons in Semiconductors, in "Light Scattering in Solids III", M. Cardona and G. Güntherodt eds., Springer, New-York (1982).

28 C. Weisbuch, R.C. Miller, R. Dingle, A.C. Gossard and W. Wiegmann, Intrinsic Radiative Recombination from Quantum States in GaAs/AlGaAs MQW structures, Solid State Commun. 37, 219 (1981).

29 A. Nakamura and C. Weisbuch, Resonant Raman Scattering Versus Hot Electron Effects in Excitation Spectra of CdTe, Solid State Electron. 21, 1331 (1978).

30 R. Planel, A. Bonnot and C. Benoit à la Guillaume, Excitation Spectra of CdS, Physica Status Solidi (b). 58, 251 (1973).

31 C. Weisbuch, Photocarrier Thermalization by Laser Excitation Spectroscopy, Solid State Electron. 21, 179 (1978).

32 M. Razeghi, J. Nagle and C. Weisbuch, Optical Studies of GaInAs/InP Quantum Wells, Int. Symp. GaAs and Related Compounds, Biarritz (1984), Inst. Phys. Conf. Ser. N° 74, p. 379, Adam Hilger, London (1985).

33 B.V. Shanabrook, O.J. Glembocki and W.T. Beard, Photoreflectance Modulation Mechanisms in GaAs/AlGaAs MQWs, Phys. Rev. B 35, 2540 (1987).

34 C.V. Shank, R.L. Fork, R. Yen, J. Shah, B.I. Greene, A.C. Gossard and C. Weisbuch, Picosecond Dynamics of Hot Carrier Relaxation in Highly Excited MQW Structures, Solid State Commun. 47, 981, (1983)

35 M.C. Nuss and D.H. Auston, Direct Subpicosecond Measurement of Carrier Mobility of Photoexcited Electrons in GaAs, Proc. Topical Meeting on Picosecond Electronics and Optoelectronics, Incline Village (1987), Springer, New-York (1987).

36 C.V. Shank, R. Yen and C. Hirlimann, Time-Resolved Reflectivity Measurements of Femtosecond-Optical-Pulse-Induced Phase Transitions in Silicon, Phys. Rev. Lett. 50, 454 (1983).

37 Y. Masumoto, S. Shionoya and H. Kawaguchi, Picosecond Time-Resolved Study of Excitons in GaAs/AlGaAs MQW structures, Phys. Rev. B29, 2324 (1984).
T. Takagahara, Localization and Energy Transfer of Quasi Two-Dimensional Excitons in GaAs/AlAs Quantum-Well Heterostructures, Phys. Rev. B31, 6552 (1985)

38 E.O. Göbel, J. Kuhl and R. Höger, Short Pulse Physics of Quantum Well Structures, J. Lumin. 30, 541 (1985).
E.O. Göbel, R. Höger and J. Kuhl, Carrier Dynamics in Quantum Well Structures, MRS-Europe, Strasbourg 1985, p. 53, Editions de Physique, Les Ulis (1985).

39 P.M. Petroff, R.C. Miller, A.C. Gossard and W. Wiegmann, Impurity Trapping, Interface Structure, and Luminescence of GaAs Quantum Wells grown by MBE, Appl. Phys. Lett. 44, 217 (1984).

40 P.M. Petroff, C. Weisbuch, R. Dingle, A.C. Gossard and W. Wiegmann, Luminescence Properties of GaAs/GaAlAs DH and MQW Superlattices grown by MBE, Appl. Phys. Lett. 38, 965 (1981).

41 See e.g. R.C. Miller and D.A. Kleinman, Excitons in GaAs Quantum Wells, J. Lumin. 30, 520 (1985).

42 R.C. Miller, D.A. Kleinman and A.C. Gossard, Energy-Gap Discontinuities and Effective Masses for GaAs/AlGaAs Quantum Wells, Phys. Rev. B29, 7085 (1984).

43 See e.g. K. Shum, P.P. Ho and R.R. Alfano, Determination of Valence-Band Discontinuity via Optical Transitions in Ultrathin Quantum Wells, Phys. Rev. B 33, 7259 (1986).

44 P.M. Frijlink and J. Maluenda, MOVPE Growth of GaAlAs/GaAs Quantum-Well Heterostructures, Jpn. J. Appl. Phys. 21, L574 (1982).

45 J.A. Brum and G. Bastard, Resonant Capture by Semiconductor Quantum Wells, Phys. Rev. B33, 1420 (1986).
J.A. Brum, T. Weil, J. Nagle and B. Vinter, Calculation of Carrier Capture Time of a Quantum Well in Graded-Index Separate Confinement Heterostructures, Phys. Rev. B34, 2381 (1986).

46 T. Mishima, J. Kasai, M. Morioka, Y. Sawada, Y. Katayama, Y. Shiraki and Y. Murayama, Determination of Bandgap Discontinuity in AlGaAs/GaAs system by Quantum Oscillations of Photoluminescence Intensity, Surf. Science 174, 307 (1986).

47 C. Weisbuch, R. Dingle, A.C. Gossard and W. Wiegmann, Optical Characterization of Interface Disorder in GaAs/GaAlAs MQW Structures, Solid

State Commun. 38, 709 (1981).

48 B. Deveaud, J.Y. Emery, A. Chomette, B. Lambert and M. Baudet, Single Monolayer Well Size Fluctuations in the Luminescence of GaAs/GaAlAs Superlattices, Appl. Phys. Lett. 45, 1078 (1984).
D.C. Reynolds, K.K. Bajaj, C.W. Litton, P.W. Yu, J. Singh, W.T. Masselink, R. Fischer and H. Morkoc, Determination of Interfacial Quality of GaAs/GaAlAs MQW Structures Using Photoluminescence Spectroscopy, Appl. Phys. Lett. 46, 51 (1985).
P.W. Yu, D.C. Reynolds, K.K. Bajaj, C.W. Litton, J. Klem, D. Huang and H. Morkoc, Observation of Monolayer Fluctuations in the Excited States of GaAs/GaAlAs MQW Structures using Photocurrent and Reflections Spectroscopies, Solid State Commun. 62, 41 (1987).

49 G. Bastard, C. Delalande, M.H. Meynadier, P.M. Frijlink and M. Voos, Low Temperature Exciton Trapping on Interface Defects in Semiconductor Quantum Wells, Phys. Rev. B29, 7042 (1984).

50 R.M. Fleming, D.B. McWhan, A.C. Gossard, W. Wiegmann and R.A. Logan, X-Ray Diffraction study of Interdiffusion and Growth in (GaAs)n/(AlAs)m Multilayers, J. Appl. Phys. 51, 357 (1980).

51 P.M. Petroff, Transmission Electron Microscopy of Interfaces in III-V Compounds Semiconductors, J. Vac. Sci. Technol. 14, 973 (1977).

52 N. Watanabe and Y. Mori, Ultrathin GaAs/GaAlAs Layers Grown by MOCVD and their Structural Characterization, Surf. Science 174, 10 (1986).

53 M. Tanaka, H. Sakaki, J. Yoshino and T. Furuta, Photoluminescence and Absorption Linewidth of Extremely Flat GaAs/AlAs QWs Prepared by MBE including Interrupted Deposition for Atomic Layer Smoothing, Surf. Science 174, 65 (1986).
T. Fukunaga, K.L.I. Kobayashi and H. Nakashima, Reduction of Well Width Fluctuation in GaAs/AlGaAs SQW by Growth Interruption during MBE, Surf. Science 174, 71 (1986).
W.T. Tsang and R.C. Miller, GaAs/AlGaAs Quantum Wells and Double Heterostructure Lasers grown by Chemical Beam Epitaxy, J. Cryst. Growth 77, 55 (1986).

54 R.C. Miller, C.W. Tu, S.K. Sputz and R.F. Kopf, Photoluminescence Studies of the Effect of Interruption during the growth of single GaAs/AlGaAs Quantum Wells, Appl. Phys. Lett. 49, 1245 (1968).

55 D.F. Welch, G.W. Wicks and L.F. Eastman, Luminescence Broadening Mechanisms in GaInAs/InAlAs Quantum Wells, Appl. Phys. Lett. 46, 991 (1985).

56 E. Cohen, Exciton Dynamics in Weakly Disordered Systems, in "Proc. Int. Conf. on the Physics of Semiconductors", San Francisco, 1984, D.J. Chadi and W.A. Harrison eds., p. 1221, Springer, New-York (1985).

57 T.S. Kuan, T.F. Kuech, W.I. Wang and E.L. Wilkie, Long-Range Order in

AlGaAs, Phys. Rev. Lett. 54, 201 (1985).

58 See the review by J. Hegarty and M. Sturge, Studies of Exciton Localization in QW Structures by Nonlinear Optical Techniques, J. Opt. Soc.
Am. B2, 1143 (1985).

59 J. Hegarty, M.D. Sturge, C. Weisbuch, A.C. Gossard and W. Wiegmann,
Resonant Rayleigh Scattering from an Inhomogeneously Broadened Transition : A New Probe of the Homogeneous Linewidth, Phys. Rev. Lett. 49,
930 (1982).

60 L. Schultheis, A. Honold, J. Kuhl, K. Köhler and C.W. Tu, Optical
Dephasing of Homogeneously Broadened 2D Exciton Transition in GaAs
Quantum Wells, Phys. Rev. B34, 9027 (1986).

61 See e.g. W.T. Masselink, Y.C. Chang, H. Morkoc, D.C. Reynolds, C.W.
Litton, K.K. Bajaj and P.W. Yu, Shallow Impurity Levels in AlGaAs/GaAs
Semiconductor Quantum Wells, Solid State Electron. 29, 205 (1986) and
references therein.

62 A. Pinczuk, J. Shak, R.C. Miller, A.C. Gossard and W. Wiegmann, Optical
Processes of 2D Electron Plasma in GaAs/GaAlAs Heterostructures, Solid
State Commun. 50,735 (1984).

63 R. Sooryakumar, A. Pinczuk, A.C. Gossard, D.S. Chemla and L.J. Sham,
Tuning of the Valence-Band Structure of GaAs Quantum Wells by Uniaxial
Stress, Phys. Rev. Lett. 58, 1150 (1987).

64 W.T. Tsang, Heterostructure Semiconductor Lasers Prepared by MBE, IEEE
J. Quantum Electronics QE-20, 1119 (1984).

65 C. Weisbuch and J. Nagle, The Physics of the Quantum Well Laser, in
"Optical Properties of Narrow-Gap Low-Dimensional Structures", C.M.
Sotomayor Torres, J.C. Portal, J.C. Maan and R.A. Stradling eds., NATO
Series ASI : Series B : Physics vol. 152, 251, Plenum, New-York (1987).

66 M. Yamanishi and I. Suemune, Comment on Polarization-Dependent Momentum
Matrix Elements in Quantum Well Lasers, Jpn. J. Appl. Phys. 23, L35
(1984).

67 M. Asada and Y. Suematsu, Density-Matrix Theory of Semiconductor Lasers
with Relaxation Broadening Model-Gain and Gain-Suppression in Semiconductor Lasers, IEEE J. Quantum Electron. QE-21, 434 (1985).

68 M. Yamada, S. Ogita, M. Yamagishi and K. Tabata, Anisotropy and Broadening of Optical Gain in a GaAs/AlGaAs MQW Laser, IEEE J. Quantum Electron. QE-21, 640 (1985).

69 J. Nagle, S. Hersee, M. Krakowski, T. Weil and C. Weisbuch, Threshold
current of SQW Lasers : The Role of the confining Layers, Appl. Phys.
Lett. 49, 1325 (1986).

70 J. Nagle, S. Hersee, M. Razeghi, M. Kratowski, B. de Cremoux and C.
Weisbuch, Properties of 2D Quantum Well Lasers, Surf. Science 174, 148

(1986).

71 J. Nagle, unpublished

72 B. Sermage, D.S. Chemla, D. Sivco and A.Y. Cho, Comparison of Auger
Recombination in GaInAs/AlInAs MQW structure and in Bulk GaAs IEEE J.
Quantum Electron. QE-22, 774 (1986).

73 Y. Arakawa, K. Vahala and A. Yariv, Quantum Noise and Dynamics in Quantum Well and Quantum Wire Lasers, Appl. Phys. Lett. 45, 950 (1984).
Y. Arakawa and A. Yariv, Theory of Gain, Modulation Response and Spectral Linewidth in AlGaAs Quantum Well Lasers, IEEE J. Quantum Electron.
QE-21, 1666 (1985).

74 K. Uomi, N. Chinone, T. Ohtoshi and T. Kajimura, High Relaxation
Oscillation Frequency (beyond 10 GHz) of GaAlAs MQW Lasers, Jpn. J.
Appl. Phys. 24, L 539 (1985).

75 M.G. Burt, Linewidth Enhancement Factor for Quantum Well Lasers, Electron. Lett. 20, 27 (1984).

76 S. Noda, K. Kojima, K. Kyuma, K. Hamanaka and T. Nakayama, Reduction of
Spectral Linewidth in AlGaAs/GaAs DFB Lasers by a MQW structure, Appl.
Phys. Lett. 50, 863 (1987).

77 R.D. Burnham, W. Streifer and T.L. Paoli, Growth and Characterization
of AlGaAs/GaAs QW Lasers, J. Cryst. Growth 68, 370 (1984).

78 See e.g. D.G. Deppe, G.S. Jackson, N. Holonyak Jr., R.D. Burnham and
R.L. Thornton, Coupled-Stripe AlGaAs/GaAs QW lasers defined by
Impurity-Induced (Si) Layer Disordering, Appl. Phys. Lett. 50, 632
(1987) and references therein.

79 Y. Suzuki, Y. Horikoshi, M. Kobayashi and H. Okamoto, Fabrication of
GaAlAs Window-Stripe MQW Heterostructure Lasers using Zn Diffusion-
Induced Alloying, Electronics Lett. 20, 383 (1984).

80 J. Cibert, P.M. Petroff, G.J. Dolan, S.J. Pearton, A.C. Gossard and
J.H. English, Optically Detected Carrier Confinement to One and Zero
Dimension in GaAs Quantum Well Wires and Boxes, Appl. Phys. Lett. 49,
1275 (1986).
P.M. Petroff, J. Cibert, A.C. Gossard, G.J. Dolan and C.W. Tu, Interfa-
ce Structure and Properties of Quantum Wells and Quantum Boxes, J. Vac.
Sci. and Technol. (1987).

81 See e.g. H. Okamoto, Semiconductor Quantum-Well Structures for Opto-
electronics - Recent Advances and Future Prospects - Jpn. J. Appl.
Phys. 26, 315 (1987) and references therein.

82 Y. Tokuda, N. Tsukada, K. Fujiwara, K. Hamanaka and T. Nakayama, Widely
Separate Wavelength Switching of SQW Laser Diode by Injection-Current
Control, Appl. Phys. Lett. 49, 1629 (1986).

83 Y. Arakawa and H. Sakaki, Multidimensional Quantum Well Laser and Tem-

perature Dependance of its Threshold Current, Appl. Phys. Lett. 40, 939 (1982)

Y. Arakawa and A. Yariv, Quantum Well Structures : Gain, Spectra, Dynamics, IEEE J. Quantum Electron, QE-22, 1887 (1986).

84 M. Asada, Y. Miyamoto and Y. Suematsu, Gain and the Threshold of Three-Dimensional Quantum-Box Lasers, IEEE J. Quantum Electron. QE-22, 1915 (1986).

85 M.A. Reed, R.J. Bate, K. Bradshaw, W.M. Duncan, W.R. Frensley, J.W. Lee and H.D. Shih, Spatial Quantization in GaAs/GaAlAs Multiple Quantum Dots, J. Vac. Sci. Technol. B4, 358 (1986).

86 K. Kash, A. Scherer, J.M. Worlock, H.G. Craighead and M.C. Tamargo, Optical Spectroscopy of Ultrasmall Structures Etched from Quantum Wells, Appl. Phys. Lett. 49, 1043 (1986).

87 Y. Miyamoto, M. Cao, Y. Shingai, K. Furuya, Y. Suematsu, K.G. Ravikumar and S. Arai, Light Emission from Quantum-Box Structure by Current Injection, Jpn. J. Appl. Phys. 26, L225 (1987).

88 L. Brus, Zero-Dimensional Excitons in Semiconductor Clusters, IEEE J. Quantum Electron. QE-22, 1909 (1986).

89 Y. Arakawa, H. Sakaki, M.N. Nishioka, H. Okamoto and N. Miura, Spontaneous Emission Characteristics of Quantum Well Lasers in Strong Magnetic Fields. An Approch to Quantum Box Light Source, Jpn. J. Appl. Phys. 22, L804 (1985).

90 Y. Arakawa, K. Vahala, A. Yariv and K. Lau, Enhanced Modulation Bandwidth of GaAlAs Double Heterostructure Lasers in High Magnetic Fields : Dynamic Response with Quantum Wire effects, Appl. Phys. Lett. 47, 1142 (1985).

RAMAN SPECTROSCOPY FOR THE STUDY OF SEMICONDUCTOR
HETEROSTRUCTURES AND SUPERLATTICES

Gerhard Abstreiter

Physik-Department
Technische Universität München
Garching, FRG

INTRODUCTION

Inelastic light scattering has been used widely to investigate numerous properties of bulk semiconductors, thin films, surfaces, interfaces, and superlattices. There exist many excellent review articles in the literature. The interested reader can find most of the recent developments in this field in special volumes on "Light scattering in solids" of the series "Topics in applied physics".[1] The present article can only scratch the surface of various applications of Raman spectroscopy for the investigation of semiconductor structures. It is neither intended to discuss all details nor to cover all interesting recent publications. We concentrate on GaAs and Si/Ge related structures. The paper is organized in the following way: Special aspects of allowed phonon scattering which give information on composition and strain of thin semiconducting films are discussed first. It is followed by a short overview of phonon properties of superlattices. The information of heterostructures as studied by forbidden LO-phonon scattering is described next. We close with a discussion of inelastic light scattering by free carriers, with special emphasis put on two-dimensional carrier systems.

ALLOWED PHONON SCATTERING

In first order Raman scattering only optical phonons close to the Brillouin zone center are observed. Pure, elementary semiconductors, like Si or Ge, which have diamond structure, exhibit only one sharp optical phonon mode at the energy of the triply degenerate zone center phonon. In polar, zincblende type semiconductors like GaAs the optical phonons are split into the doubly degenerate transverse (TO) mode and the longitudinal (LO) mode which appears at slightly higher

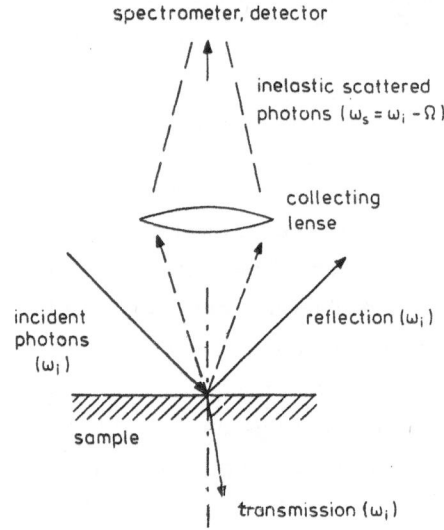

Fig.1: Raman back-scattering geometry (schematically)

energies. First order selection rules depend on scattering configuration. In back-scattering geometry from (100) surfaces, scattering by LO-phonons is allowed; TO phonons are forbidden. This is opposite in back-scattering from (110) surfaces. For laser lines in the visible, back-scattering is used, because most of the semiconductors are opaque for these energies. The geometry is shown schematically in Fig.1. The symmetry properties of phonon Raman scattering can be used to obtain information on the orientation of thin epitaxial films.

In mixed crystals like $Al_xGa_{1-x}As$ or Si_xGe_{1-x} the phonon spectra abserved in Raman spectroscopy are more complicated. The mixed III-V compound semiconductors can show a so-called one-mode or a two-mode behaviour. $Al_xGa_{1-x}As$ exhibits both GaAs and AlAs like optical phonons. Their positions depend on x and have been used to determine the composition of thin films of alloys.[2] Si_xGe_{1-x} has three characteristic modes, related to the Si-Si, Si-Ge and Ge-Ge vibrations. Examples for different compositions are shown in Fig.2. The positions of the Si-Si and Ge-Ge like modes as well as their intensity ratio depend critically on x. This can be used directly to determine the Si and Ge content.

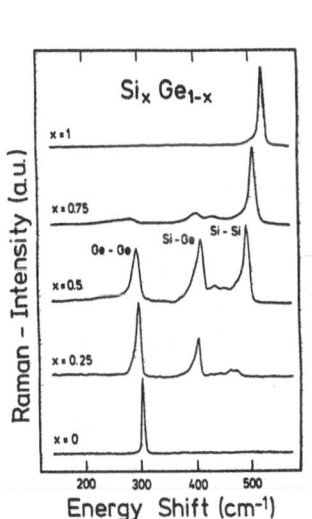

Fig.2: Raman spectra of
bulk Si_xGe_{1-x}

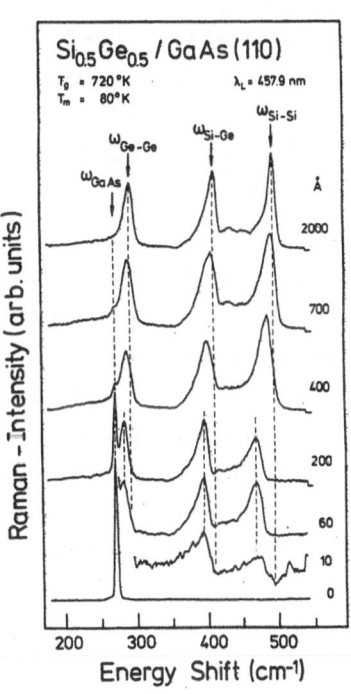

Fig.3: Raman spectra of
Si_xGe_{1-x} on GaAs

In ultrathin films, the exact position of the three modes may be shifted due to builtin strain. This is shown in Fig.3 for $Si_{0.5}Ge_{0.5}$ layers grown on (110) GaAs substrates.

The measurements have been performed "in situ" by interrupting the growth in a specially designed Si/Ge MBE machine.
The pure GaAs (110) surface exhibits only the TO phonon mode,
as expected from first order selection rules. At overlayer
thicknesses of only a few A, the characteristic $Si_{0.5}Ge_{0.5}$
modes appear already in the spectrum. They grow in intensity
with increasing thickness, concomitant with a decreasing
intensity of the GaAs TO-mode due to absorption of the light
in the overlayer. For thin films, however, the position of
the $Si_{0.5}Ge_{0.5}$ phonons are shifted downwards compared with
their expected energy. With growing thickness the peaks move
to higher energies, approaching the values observed for bulk
$Si_{0.5}Ge_{0.5}$ random alloys. The downward shift has been
explained by the built-in strain which exists in thin layers
grown epitaxially on substates with different lattice
constants.[3,4] $Si_{0.5}Ge_{0.5}$ has an average lattice parameter
which is about 2% smaller than that of GaAs. Up to a critical
thickness, the overlayer grows lattice matched to the
substrate, which leads to a large biaxial tension in the
$Si_{0.5}Ge_{0.5}$ layer. This commensurate growth causes the
downward shift of the Raman peaks. The magnitude is
determined quantitatively by the strain or the corresponding
lattice distortion. After a critical thickness, the strain
relaxes slowly, which leads to an upward shift of the
phonons. Finally they approach the unstrained situation. For
compressive strain, the shift is opposite. The phonon
energies have been used to investigate the strain
distribution in symmetrically and asymmetrically strained
Si/SiGe superlattices and multilayer structures. [3-5]

The few selected examples demonstrate the usefulness of
phonon Raman spectroscopy as a microscopic probe of various
properties of semiconductor films and superlattices. It is
sensitive to layer thicknesses of only a few monolayers. Many
more examples can be found in the literature which include
properties of various compound semiconductors, amorphous and
micro-crystalline semiconductors, metal layers including
silicides, and even thin insulating films. It goes beyond the
scope of this article to cover all the exciting recent
developments.

SUPERLATTICE PHONONS

Characteristic features of periodic multilayer structures are related to Brillouin-zone folding effects due to the new periodicity in superlattice direction. It affects both the electronic band structure and the lattice dynamics. The Brillouin-zone folding causes new gaps in the phonon spectrum which can be easily detected in the Raman spectra. Folded acoustic modes appear in the low energy region basically as doublets. Their dispersion is shown schematically in Fig.4. Up to a certain energy, the acoustic modes of different semiconductors overlap. This leads to propagating waves with an average sound velocity and energy gaps at the new Brillouin-zone boundary and at the zone center. Both the dispersion and the energy gaps can be well explained in the approximation of elastic continuum theory.[6] The most widely studied system up to now is $GaAs/Al_xGa_{1-x}As$.[7-9] However, results on various other semiconductors have been studied recently.[10-12] An example for a large period Si/Si_xGe_{1-x} strained layer superlattice is shown in Fig.5. In back- scattering geometry, the scattering wave vector q_s involved in the process can be of the order of the new Brillouin-zone boundary. It is possible to tune q_s experimentally from about $0,7 \times 10^6$ cm^{-1} to $1,5 \times 10^6$ cm^{-1} by use of different laser lines in the whole visible spectral range. The spectra shown in Fig.5 are the first example where q_s could be tuned even through the new Brillouin-zone boundary. A finite splitting of about 1 cm^{-1} has been extracted from these data.[12] It is in excellent agreement with predictions of the elastic continuum model. The frequencies of the folded acoustic phonons vary sensitively with the period length. This has been used for an accurate determination of layer thicknesses in different super-lattices.[11] The scattering wave vector q_s has to be fixed in those experiments. Additional information can be extracted from the intensity ratios of various doublets which are given by the photoelastic coefficients. The intensity fall-off of the higher order folded modes is sensitive to the interface sharpness. Such measurements can be used to study quantitatively interdiffusion at elevated temperatures.

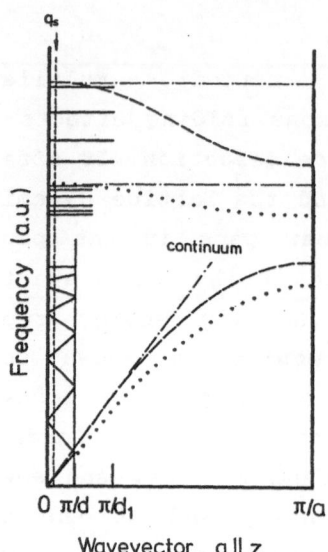

Fig.4: Phonon dispersion in
superlattices (schematic)

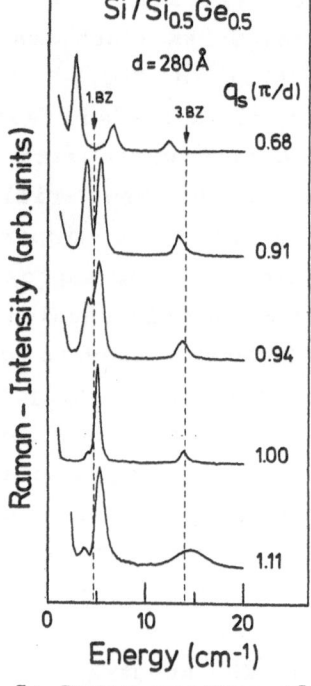

Fig.5: Raman spectra of
folded LA modes in
Si/Si$_{0.5}$Ge$_{0.5}$

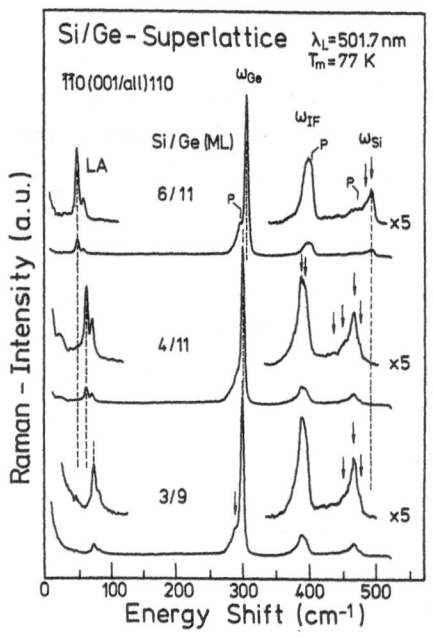

Fig.6: Raman spectra of ultrashort period Si/Ge superlattices

Raman spectra of ultrashort period Si/Ge strained layer superlattices are shown in Fig.6. The samples exhibit different period length $d=d_{Si}+d_{Ge}$. They are grown on (110) oriented Ge or $Si_{0.25}Ge_{0.75}$ buffer layers. Only one strong doublet is observed below 100 cm^{-1} which corresponds to folded LA modes. Additional weak structure appears at lower energies which is believed to be due to disorder activated scattering by LA modes. It corresponds to the density of states of the LA superlatttice phonon branch with a gap at the new zone boundary. It is observed most pronounced in the middle spectrum (dip marked by an arrow). The peaks at higher energies correspond to optical phonons. The dispersion curves of Si and Ge do not overlapp as shown schematically in Fig.4. Therefore confined modes in each layer are expected. They appear as sholders in the optical phonon peaks. In addition an interface mode shows up in between the Si and Ge confined modes.

Superlattice effects on optical phonons have been studied extensively in short period GaAs/AlAs structures.[13-16] Confined and dispersive modes have been observed. A finite dispersion is obtained in the energy region where the optical branches of the two materials overlapp in energy. Modes propagating along the layers are either interface modes or slab modes.[17,18] Confined modes can also exist in the high energy part of the acoustic branches where no overlapp exists anymore.

FORBIDDEN LO-PHONONS

It has been shown that under resonance condition the Raman selection rules can be violated due to higher order effects. As mentioned already, LO-phonon scattering is forbidden in backscattering geometry from (110) surfaces in polar III-V semiconductors. For laser energies close to resonance, however, forbidden, electric-field-induced LO-phonon Raman scattering (EFIRS) can be quite strong. The LO-phonon intensity was found to be proportional to the square of the macroscopic electric field. This method has been used to study the formation of Schottky barriers[19] and semiconductor heterostructures on GaAs (110).[20,21] Possibilities and limitations of the technique are discussed in Ref.22. The interface formation can be studied from the clean surface to overlayers of more than 200 A thickness. Forbidden LO-phonon scattering is shown in Fig.7 for Ge and $Si_{0,5}Ge_{0,5}$ overlayers on (110) GaAs surfaces which are prepared by cleaving a pre-oriented single crystal under ultra-high-vacuum conditions.

Ge is well lattice matched to GaAs. High quality epi-taxial growth is achieved at stubstrate temperatures between 600 K and 700 K. Raman measurements in the frequency range of the TO and LO phonon modes have been performed "in situ" at room temperature. The wavelength of the selected laser line (λ_L = 413,1 mm) is close to the E_1 energy gap of GaAs. This leads to forbidden LO-phonon Raman scattering. The LO intensity is strongly increasing already at submonolayers coverages. A maximum is reached at a few monolayers which is followed by a drastic drop with increasing overlayer thickness. This behaviour has been explained by a model which takes Fermi level pinning at chemisorption induced surface states into account.[21] It leads to an increasing surface barrier height already at submonolayer coverages. The surface electric field is responsable for the strong enhancement of forbidden LO-phonon scattering. After perfect monolayer

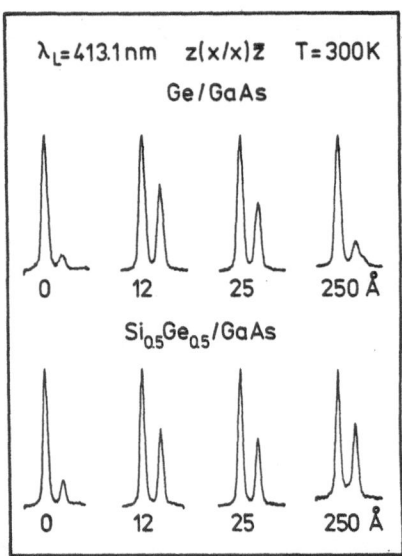

Fig.7: Forbidden LO and TO phonon Raman spectra of GaAs with different Ge and $Si_{0.5}Ge_{0.5}$ overlayers.

coverage no interface states are present anymore. The barrier height is then governed by Fermi level pinning at the surface of the overlayer and the conduction or valence band offset at the interface. A careful analysis of such EFIRS experiments on n- and p-type GaAs surfaces can be used to determine quantitatively barrier heights and band alignments.[23]

For strained overlayers the behaviour is very similar in the region of low coverage. Above the critical thickness, however, misfit dislocations are created at the interface, which cause a Fermi level pinning and consequently a different barrier height. This is directly reflected in a change of the forbidden LO-phonon intensity. An example is shown for $Si_{0.5}Ge_{0.5}$ on GaAs in Fig.7. EFIRS has been used to determine critical layer thicknesses for various Si_xGe_{1-x} overlayers with different lattice mismatch.[23,24]

The examples discussed show that forbidden resonant Raman scattering can be used successfully to investigate interface barrier heights, bandoffsets, and critical thicknesses with high spatial resolution from lowest coverages up to overlayers of more than 200 Å thickness.

ELECTRONIC LIGHT SCATTERING

In recent years, it has been shown that inelastic light scattering is a powerful experimental tool to study, besides phonons, also electronic excitations in semiconductors.[25-27] The application to two-dimensional systems was realized after detailed studies of the resonance behaviour of free carrier excitations in bulk GaAs[28], where it was demonstrated that scattering intensities are strong enough to allow the study of carrier densities of only a few times $10^{11}cm^{-2}$. Scattering mechanisms and selection rules were first discussed by Burstein et al.[29,30] Resonant light scattering experiments in two-dimensional electron systems in selectively doped GaAs/Al_xGa_{1-x}As heterostructures were reported shortly afterwards.[31,32] The developments in this field have been reviewed in several publications.[27,33-35] For details the reader is referred to these reviews which also contain all the original references.Inelastic light scattering in two

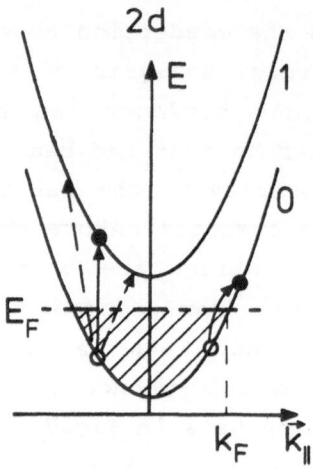

Fig.8: Inter- and intrasubband excitations in 2d-systems

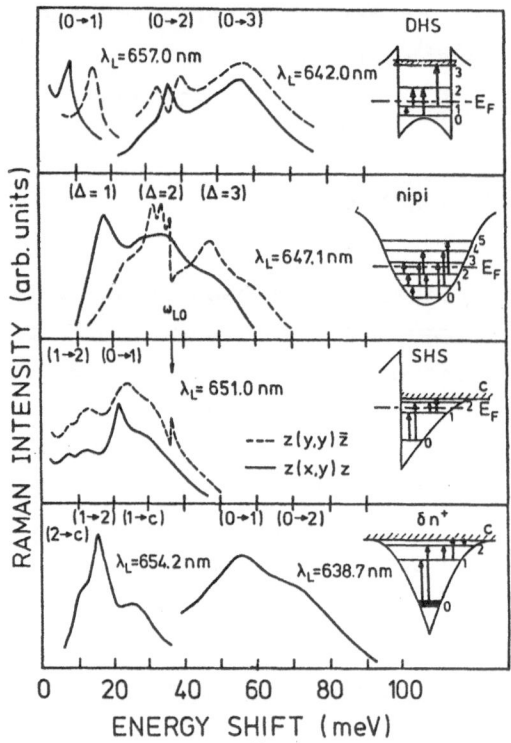

Fig.9: Raman spectra of single-particle (solid lines) and
collective (dashed lines) excitations in various
potential wells

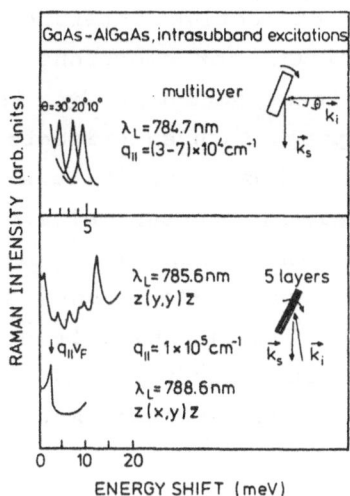

Fig.10: Raman spectra of intrasubband excitations in GaAs quantum wells (from Ref. 36 and 37)

dimensional systems differs from the three dimensional situation by the separation of motion perpendicular and parallel to the interface or quantizing layer. In the parallel direction a two-dimensional dispersion of the bands is maintained. Normal to the interface the carriers are quantized in elctric subbands. Possible elementary carrier excitations split into inter- and intra-subband excitations. This is shown schematically in Fig.8.

In back scattering perpendicular to interfaces or multilayers only excitations between subbands are observable. Intra-subband excitations require a finite scattering wave vector in parallel direction ($q_{\parallel}$). This can be achieved by special scattering geometries. In addition resonant light scattering allows the separation of single particle and collective excitations by analyzing the polarisation of the scattered light. The appeal of electronic light scattering goes far beyond its power as a spectroscopic tool that yields just energies of the electronic excitations. Collective excitations couple to LO-phonons due their macroscopic electric field. Electronic light scattering has been used to investigate energy levels and Coulomb interactions in various GaAs quantum well structures and superlattices. Examples for single particle and collective intersubband excitations in different potential wells for electrons are shown in Fig.9. The various wells are achieved by the combination of heteroepitaxy and doping. Such experiments have been extended to other systems including holes in GaAs and Si, as well as shallow impurity levels in quantum wells.[27,33]

Recently also intra-subband excitations have been studied by different groups.[36-38] Both single particle and quasi-two-dimensional plasmons have been observed. Examples are shown in Fig.10. The dispersion of layered plasmons can be analyzed by special scattering geometries.

CONCLUDING REMARKS

Various aspects of Raman spectroscopy relevant for the investigation of layered semiconductor structures have been sketched briefly. The versatility of light scattering is evident from the many different examples. Recent developments of Raman equipment with multichannel detection and microscope will further increase the wide applicability of this method in semiconductor physics. It will include time resolved spectroscopy of both phonon and electronic excitations, "in situ" studies during processing and the investigation of various aspects of microstructured devices like hot electron and hot phonon effects. Raman spectroscopy has become an important analytical tool. It certainly will develop further in this direction.

ACKNOWLEDGEMENTS

Most of the work which I have discussed at the school in Erice and which is summarized in this article is based on the work of many students and colleagues at the Technische Universität München. It is a pleasure to mention especially H.Brugger, K.Eberl, F.Schäffler and R.Zachai. Our group is supported financially by the Deutsche Forschungsgemeinschaft via SFB 128, by the Siemens AG via SFE, and by the Stiftung Volkswagenwerk. I also want to mention the fruitful and excellent cooperation with E.Kasper, AEG Ulm, K.Ploog, MPI Stuttgart and G.Weimann, FTZ Darmstadt and their coworkers.

REFERENCES

1. "Topics in Applied Physics" Vols.8, 50 and 54, "Light Scattering in Solids", eds M.Cardona and G.Güntherodt, Springer Verlag, Berlin Heidelberg

2. G.Abstreiter, E.Bauser, A.Fischer, and K.Ploog, Appl. Phys. 16, 345 (1978)

3. G.Abstreiter, H.Brugger, T.Wolf, H.Jorke, and H.J.Herzog, Phys.Rev.Lett.54, 2441 (1985)

4. G.Abstreiter and H.Brugger, Proc. of the 18th Int. Con. on the Physics of Semiconductors, Stockholm 1986, O.Engström (ed.), World Scientific, Singapore 1987, Vol.1, p.739

5. F.Cerdeira, A.Pinczuk, J.C.Bean, B.Batlogg, and B.A.Wilson, Appl.Phys.Lett.45, 1138 (1984)

6. S.M.Rytov, Akust. Zh.2, 71 (1956) (Sov.Phys.Acoust, 2, 68 (1956))

7. C.Colvard, R.Merlin, M.V.Klein, and A.C.Gossard, Phys.Rev.Lett.45, 298 (1980)

8. J.Sapriel, J.C.Michel, J.C.Tolédano, R.Vacher, J.Kervarec and A.Regrany, Phys.Rev. B28, 2007 (1983)

9. C.Colvard, T.A.Gant, M.V.Klein, R.Merlin, R.Fischer, H.Morkov, and A.C.Gossard, Phys.Rev. B31, 2080 (1985)

10. H.Brugger, G.Abstreiter, H.Jorke, H.J.Herzog, and E.Kasper, Phys.Rev. 33, 5928 (1986)

11. S.Venagopalan, L.A.Kolodziejski, R.L.Gunshor, and A.K.Ramdas, Appl.Phys.Lett. 45, 974 (1984)

12. H.Brugger, H.Reiner, G.Abstreiter, H.Jorke, H.J. Herzog, and E.Kasper, Superlattices and Microstructures 2, 451 (1986)

13. B.Jusserand, D.Paquet, and A.Regreny, Phys.Rev. B30, 6245 (1984)

14. A.K.Sood, J.Menendez, M.Cardona, and K.Ploog, Phys.Rev.Lett.54, 2111 (1985)

15. A.K.Sood, J.Menendez, M.Cardona, and K.Ploog, Phys.Rev.Lett. 54, 2115 (1985) and Phys.Rev. B32, 1412 (1985)

16. E.Molinari, A.Fasolino, and K.Kunc, Phys.Rev.Lett. 56, 1751 (c) (1986); B.Jusserand and D.Paquet, ibid., 1752 (c); A.K.Sood, M. Cardona, and K.Ploog, ibid., 1753 (c)

17. J.Zucker, A.Pinczuk, D.S.Chemla, A.Gossard, and W.Wiegmann, Phys.Rev.Lett. 53, 1280 (1984)

18. B.Jusserand and D.Paquet, in "Heterojunctions and Semiconductor Superlattices", Proceedings of the Winterschool, Les Houches, France 1985, G.Allan, G.Bastard, N.Boccara, M.Lannoo, and M.Voos (eds.), Springer Verlag, Berlin Heidelberg, p.108

19. F.Schäffler and G.Abstreiter, J.Vac.Sci.Technol.B3, 1184 (1985)

20. H.Brugger, F.Schäffler, and G.Abstreiter, Phys.Rev.Lett. 52, 141 (1984)

21. G.Abstreiter, in Festkörperprobleme (Advances in Solid State Physics), Vol.XXIV, p.291, and Vol.26, p.41, P.Grosse (ed), Vieweg Braunschweig, 1984 and 1986

22. F.Schäffler and G.Abstreiter. Phys.Rev. B34, 4017 (1986)

23. H.Brugger, doctoral thesis, TU München 1987

24. H.Brugger and G.Abstreiter in Proc. of the 3rd Int.Conf. on Modulated Semiconductor Stuctures, Montpellier France, 1987 (to be published)

25. A. Mooradian, in Festkörperprobleme (Advances in Solid State Physics), Vol IX, p.74, O.Madelung (ed.) Vieweg Braunschweig, 1969

26. M.V.Klein, in Ref.1, Vol 8, p.147 (1975)

27. G.Abstreiter, M.Cardona, and A.Pinczuk in Ref.1, Vol 54, p.5 (1984)

28. A.Pinczuk, G.Abstreiter, R.Trommer, and M.Cardona, Solid State Commun. 30, 429 (1979)

29. E.Burstein, A.Pinczuk, and S.Buchner, in Physics of Semiconductors, ed. B.L.Wilson (The Institute of Physics, Bristol, London, 1979) p.585

30. E.Burstein, A.Pinczuk, and D.L.Mills, Surf.Sci. 98, 451 (1980)

31. G.Abstreiter and K.Ploog, Phys.Rev.Lett. <u>42</u>, 1308 (1979)

32. A.Pinczuk, H.L.Störmer, R.Dingle, J.M.Worlock, W.Wiegmann, and A.C.Gossard, Solid State Commun. <u>32</u>, 1001 (1979)

33. G.Abstreiter, R. Merlin, and A.Pinczuk, JEEE <u>QE22</u>, 1771 (1986)

34. A.Pinczuk and J.M.Worlock, Surf.Sci.<u>113</u>, 69 (1982)

35. G.Abstreiter, in Molecular Beam Epitaxy and Heterostructures. L.L.Chang and K.Ploog (eds), Dordrecht, The Netherlands Nijoff 1985, p.425; see also Ref.18,p.99

36. D.Olego, A.Pinczuk, A.C.Gossard, and W.Wiegmann, Phys.Rev. <u>B25</u>, 7867 (1982)

37. G.Fasol, N.Mestres, A.Fischer, and K.Ploog, in Ref.4, p.679

38. A.Pinczuk, M.G.Lamont, and A.C.Gossard, Phys.Rev.Lett.<u>56</u>, 2092 (1986)

FAR-INFRARED SPECTROSCOPY IN TWO-DIMENSIONAL ELECTRONIC SYSTEMS

Detlef Heitmann

Max-Planck-Institut für Festkörperforschung,

Heisenbergstr. 1, D-7000 Stuttgart 80, West-Germany

ABSTRACT

These lecture notes include an introduction into far-infrared spectroscopic techniques, a treatment of the dielectric response of a two-
dimensional electronic system in layered structures and experimental
results on characteristic excitations, e.g., 2D plasmons, intersubband
resonances, cyclotron resonances and combined interactions.

INTRODUCTION

Two-dimensional electronic systems (2DES), which can be realized in
semiconductor heterostructures or in metal-insulator-semiconductor (MOS)
systems have attracted great interest in the last decade[1]. The interest
comes from many proposed and realized technical applications and — in an
interacting way — also from fundamental physical investigations. In these
systems many of the interesting physical properties can be varied over
wide regimes during the fabrication process or directly for a given sample
via electric or magnetic fields, stress, temperature etc. This allows very
detailed studies of fundamental physical properties and of interactions
that are unique for the 2D system. A great number of investigations have
been performed with spectroscopic methods. Since typical energies in the

2DES, e.g., subband spacing, 2D-plasmon and cyclotron resonance energies, are in the regime 1-100 meV, far-infrared (FIR) transmission spectroscopy [2-5] is widely used to study the 2DES. Similar information can also be extracted from emission spectroscopy (e.g.,Ref. 6) and from Raman spectroscopy[7], but all methods have their specific advantages and drawbacks.

In this lecture I will first give a short introduction into the experimental techniques. Then I would like to take the opportunity to give a more detailed introduction into an electrodynamic treatment of the dielectric response of 2DESs in layered structures. This is important to extract the microscopic physical properties from the experimental spectra. I will then give an overview of characteristic dynamic excitations in 2DESs, e.g., 2D plasmons, intersubband resonances (ISR), cyclotron resonances (CR) and combined resonances. It is not so much the goal of my lecture and these notes to give a complete survey, but rather an attempt to give a general understanding by the description of some selected examples. Most of the examples will be devoted to AlGaAs-GaAs heterostructures. I will, however, also report on experiments on the Si-MOS system. Due to the special surface bandstructure of Si, which arises from the indirect gap, unique excitations can be observed in Si, which are not directly observable for the GaAs system, but which are important for a fundamental understanding of spectroscopy in 2DESs.

EXPERIMENTAL TECHNIQUES

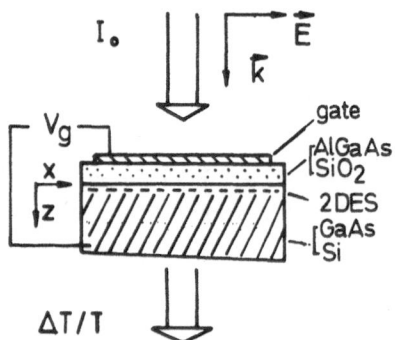

Fig. 1: Schematics for the transmission of FIR radiation through MOS- or heterostructure samples

The most common arrangement for FIR spectroscopy is shown schematically in Fig.1 where FIR radiation is incident normally onto the sample
and the transmitted light is detected. The samples are usually wedged
(2°-3°) to avoid interference effects which would arise from multi-reflec-
tions in plane parallel samples. If the signal is large enough, e.g.,
cyclotron resonance in high mobility GaAs samples, the transmission can be
recorded directly and the resonance position and linewidth can be analyzed. In many cases, however, differential techniques are necessary to
resolve also very weak responses with a sufficient signal-to-noise ratio.
Here it is most convenient if one can switch the 2D charge density ,N_S,
via a gate voltage which is applied to a semitransparent gate. Then the
differential transmission $\Delta T/T = (T(N_S) - T(N_S=0))/T(N_S=0)$ can be evaluated.
This is possible for Si-MOS structures where the gate voltage can be
switched between a certain value and the threshold voltage V_t with
$N_S(V_t) = 0$. For GaAs-heterostructures, where N_S is determined essentially by
the doping of the AlGaAs layer, this is not easily possible. Thus so far,
for most of the FIR spectroscopic investigations in this system, magnetic
fields have been applied to study excitations. These experiments are
described elsewhere in this volume[8,9]. Only very few experiments have been
reported on GaAs samples with front gates (e.g. Ref. 10). I will describe
here some results on the cyclotron resonance in GaAs heterostructures
where the charge density could be varied over a wide density regime via
front gates.

Many spectroscopic investigations are performed with FIR lasers. H_2O-
and HCN-lasers have characteristic strong lines at 118μ, 78μ and 337μ.
CO_2-laser pumped molecular gas lasers can be used with different active
gases to cover a wide range of spectral lines from 1600μ to 28μ. In a
laser experiment the magnetic field or, if possible, N_S is varied. Often
modulation techniques are used to improve the signal to noise ratio. One
drawback of laser experiment is that, in sweeping N_S or B, the properties
of the 2DES are changed, e.g. the subband spacing, the filling factor or
others. Thus it is highly desirable to investigate the sample at fixed N_S
and B in the frequency domain. Since broadband light sources (Hg-lamps,
Globar-lamps) have a weak spectral emission in the FIR, the technique of
Fourier transform spectroscopy is applied, where all spectral information
is measured simultaneously. The principle of the operation is that one
mirror of a Michelson type interferometer is moved with a constant velocity. Thus each spectral component S_ν of wavenumber ν is modulated with
cos $\nu \cdot x$, where x is the optical pathlength of the mirror. At the output of

the interferometer the signal $I(x) \propto \sum S_\nu \cdot \cos\nu \cdot x$ is measured. $I(x)$, the so-called interferogram, is digitized and the spectral components S_ν are computed from a numerical Fourier transform $S_\nu = FT\{I(x)\}$. In a rapid scan Fourier spectrometer a large number of interferograms, measured in different succeeding scans of the interferometer, are collected ("co-added") to improve the signal to noise ratio. In samples, where one can switch the gate voltage, one can apply the following differential technique. Alternatingly a small number of interferograms are taken at a gate voltage V_g and at threshold voltage, V_t. The interferograms are co-added in different files and are Fourier transformed separately at the end of the experiments. In this way, the effective delay between measuring the sample spectrum $I(V_g)$ and the reference spectrum $I(V_t)$ is small (1s to 10s) and very weak excitations (as low as 0.01% change in relative transmission) can be resolved without problems from long time drifts. These techniques are described in more details in Ref. 11. The typical spectral range of commercial FT-spectrometers is 10cm^{-1} to 4000cm^{-1}. The highest spectral resolution, which depends on the inverse optical pathlength of the moving mirror, is typically 0.03 cm^{-1}. Normally all experiments are performed at low temperatures $T < 10$ K.

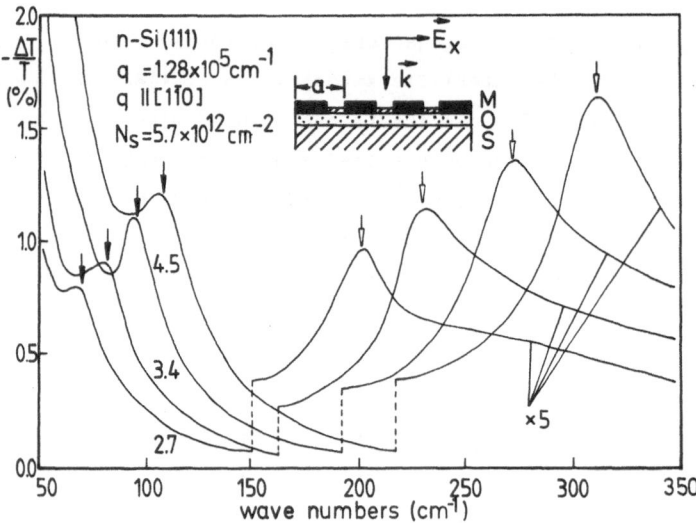

<u>Fig. 2:</u> Relative change of transmission $-\Delta T/T$ vs wavenumber in 2D electron accumulation layers of n-Si(111) at different charge densities N_S. Full and open arrows indicate, respectively, 2D plasmon- and intersubband resonance excitations (from Ref. 11)

320

As an example of differential spectroscopy on 2D systems I show in Fig. 2 spectra measured on electron-accumulation layers in a n-type Si-(111)-MOS system. The carrier density N_S has been varied via the gate voltage. The background of the spectra, which decreases with frequency, is caused by the non-resonant intraband absorption (=Drude absorption) of the 2DES. The resonances in the frequency regime 30-150cm^{-1} are 2D plasmon resonances which shift with increasing N_S to higher frequencies. The excitation of these resonances is possible due to the grating coupler effect of the gate which consists of periodic stripes of high and low conductivity materials. At high frequencies intersubband resonances are observed, resonant transitions from the lowest subband E_0 to the first excited subband E_1, which shift with increasing surface electric field and corresponding N_S to higher frequencies. These ISR can be excited with normally incident radiation due to the special subband structure of Si(111). The latter arises from the projection of volume energy ellipsoids which are tilted with respect to the (111) surface[12,1]. 2D plasmon resonances, intersubband resonances and various other kinds of excitations will be the topic of this paper.

THE DIELECTRIC RESPONSE OF A 2DES

In the following x and y denote the plane of the 2DES, z is perpendicular to this plane. The simplest model for the response of electrons to electromagnetic fields is the Drude model.

$$m^* \ddot{x} + m^* \dot{x}/\tau = e \cdot E_x \quad . \tag{1}$$

The electrons are accelerated by the electric field E_x in x-direction. The damping of the motion is characterized by a scattering time τ. If we assume a periodic time dependence for $E(t) \propto \exp(-i\omega t)$ (and thus, in the steady case, for $x(t) \propto \exp(-i\omega t)$) we find for the 2D current density

$$\vec{j} = N_S e \cdot \dot{\vec{x}} = \sigma(\omega) \cdot \vec{E} \tag{2}$$

with

$$\sigma = \sigma_0 (1-i\omega\tau)^{-1} \quad ; \quad \sigma_0 = N_S e^2 \tau/m^* \quad . \tag{3}$$

The real part of this Drude conductivity $\sigma_r = \sigma_0(1+\omega^2\tau^2)^{-1}$ is characterized by the DC conductivity σ_0. It decreases with frequency and drops to $\sigma_0/2$ at a frequency $\omega' = \tau^{-1}$.

Due to their confinement in a narrow potential well, the motion of the electrons in a 2DES is quantized in the z-direction. In combination with the free motion of the electrons parallel to the interface this leads to an energy spectrum that consists of different subband (index i) with dispersion $E_i(K) = E_i + (\hbar^2/2m_x) \cdot K_x^2 + (\hbar^2/2m_y) \cdot K_y^2$ for a parabolic system. Here K_x and K_y are quasi momenta in the 2D plane. The dynamic conductivity perpendicular to the interface, which characterizes the transitions be-tween different subbands, e.g., 0 and 1, is[13]:

$$\mathrm{Re}\,\sigma_{zz}(\omega) = \quad \cdot \frac{N_S e^2 \cdot \tau_{op}}{2 \cdot m_z} \cdot \frac{f_{10} \cdot (\hbar/\tau_{op})^2}{(\hbar\omega_{10} - \hbar\omega)^2 + (\hbar/\tau_{op})^2} \qquad (4)$$

$$f_{10} = \frac{2}{m_z} \cdot \frac{1}{\hbar\omega_{10}} \cdot \left| \left< f \left| \frac{\hbar}{i}\frac{d}{dz} \right| i \right> \right|^2 . \qquad (5)$$

It has a resonance at frequency $\hbar\omega_{10} = E_1 - E_0$ and the strength of the transi-tion is proportional to the matrix element f_{10}. The lineshape of the ISR, governed by τ_{op}, is discussed in detail in Ref. 13. We will see later that one has to be careful when using eq. (4). σ_{zz} is the response to the internal electric field. The screening of the external field leads to a shift of the resonance with respect to $\hbar\omega_{10}$.

A more general treatment of the dielectric response can be given within a quantum mechanical time dependent perturbation theory, the random phase approximation RPA[14]. Here one calculates the effective potential $V_{eff}(\omega,q) = V_{ext}(\omega,q)/\varepsilon(\omega,q)$ that an electron feels if an external spatial and time dependent potential is applied to the system. The dielectric function is found to be[14]

$$\varepsilon(\omega,\vec{q}) = 1 - V_c \cdot \sum_{i,K} \frac{f(\vec{K}) - f(\vec{K}+\vec{q})}{E(\vec{K}) - E(\vec{K}+\vec{q}) + \hbar\omega - i\hbar\alpha} . \qquad (6)$$

The sum has to be taken over the states in all subbands i and at all quasi momenta $\vec{K}$. $f(K)$ is the Fermi distribution. In the 3D case $V_c = e^2/\varepsilon_0 q^2$ is the 3D Fourier transform of the Coulomb potential. In a strictly 2D system

V_c is $e^2/2\varepsilon_o \cdot q$, the 2D Fourier transform of the potential. A full evalu-
ation of (6), which means that all excitations and interaction between
different excitations are taken into account, is not easily accomplished.
However, approximate solutions can be given in special restricted cases.
Let us assume, for example, only one parabolic subband in the limit that
$E(K+q)-E(K)<<\hbar\omega$. From an expansion for small q we find for the sum in (6)
$(N\cdot q^2)/(m^*\omega^2)$ (N= number of electrons per normalized volume (area), m*=
effective mass). This gives for the 3D case

$$\varepsilon(\omega,q) = 1-\omega_p^2/\omega^2 \tag{7}$$

with the 3D volume plasma frequency $\omega_p^2=N_v e^2/\varepsilon_o m^*$. For $\omega=\omega_p$ $\varepsilon(\omega,q)$ is zero
thus that from $V_{int}=V_{ext}/\varepsilon(\omega,q)$ we find excitations in the electronic sys-
tem without external fields. These eigensolutions are the well known
volume plasmons, longitudinal collective excitations of the 3DES at frequency ω_p which, in this approximation of small q, not depends on q. In
the 2D case we find in the same approximation

$$\varepsilon_{2D}(\omega,q) = 1- \frac{N_S e^2}{2\varepsilon_o m^* \cdot \omega^2} \cdot q \; . \tag{8}$$

Here the zeros of $\varepsilon_{2D}(\omega,q)$ define a 2D plasmon frequency

$$\omega_p^2 = \frac{N_S e^2}{2\varepsilon_o m^*} \cdot q \; . \tag{9}$$

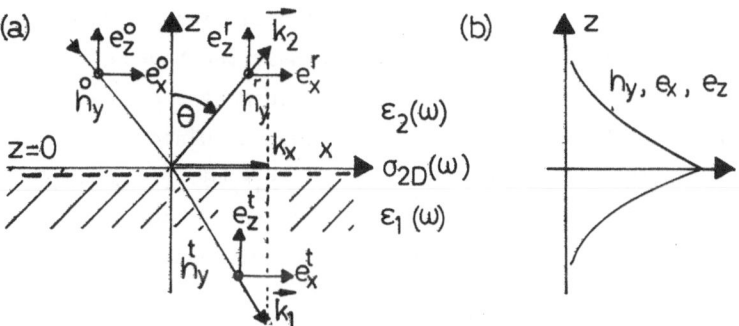

Fig. 3: (a) Calculation of reflected and transmitted fields for a 2DES at
a boundary between two dielectrics 2 and 1. (b) Sketch of the exponential
decay of the fields for nonradiative surface modes $|k_x| > \omega/c$

323

Thus there is the important difference that the 2D plasmon frequency depends on $\sqrt{q}$ and approaches zero for $q \to 0$.

In the following we will calculate the response of a layered system. We assume one boundary between two media 2 and 1 with dielectric functions $\varepsilon_2(\omega)$ and $\varepsilon_1(\omega)$. The 2DES at the boundary is characterized by a parallel conductivity $\sigma(\omega)$. We consider here p-polarised (TM) electromagnetic waves propagating in x-direction. Then it is convenient to use the magnetic field components of the exciting field which only have h_y components. The exciting field has the following dependence in space and time

$$\vec{h}^o = (0, h_y^o, 0) \cdot e^{i(k_x x - \omega t)} \cdot e^{-ik_{z2}z} \quad . \tag{10}$$

Since k_x is proportional to the inverse phase velocity in x-direction, all excited fields have the same x (and of course time) dependence which will not explicitly be shown in the following. However, there is a different z-dependence. The reflected field is $\vec{h}^r = (0, h_y^r, 0) \cdot \exp(+ik_{z2} \cdot z)$ and the transmitted field is $\vec{h}^t = (0, h_y^t, 0) \cdot \exp(-ik_{z1} \cdot z)$, where in the medium i the z-component of the wavevector for the electromagnetic fields is:

$$k_{zi} = \sqrt{\varepsilon_i \omega^2/c^2 - k_x^2} \quad . \tag{11}$$

Besides ω, k_x is the important quantity that determines the Fourier components of the electromagnetic fields. For $k_x^2 < \varepsilon_i \omega^2/c^2$ k_{zi} is real and we have an oscillating field dependence in the z-direction. Then it is $k_x = (\omega/c)\sin\Theta$, where Θ is the angle of the light propagation with respect to the surface normal. However, it is also possible to consider electromagnetic fields with $k_x^2 > \varepsilon \omega^2/k_x^2$. Then k_{zi} is imaginary, meaning that we have evanescent waves, i.e., fields that decay exponentially from the boundary. For certain ω and k_x we will then find eigensolutions of the electromagnetic fields. These fields are concentrated near the boundary and represent so-called non radiative surface modes. 2D plasmons are an example of such excitations. From Maxwells' equations we can calculate from h_y^o, h_y^r, h_y^t the corresponding e_x components $e_x^i = \pm k_{zi} \cdot h_y^i / (\varepsilon_i \varepsilon_o \omega)$. ($\pm$ corresponds to the sign in the z-dependence $\exp(\pm ik_{zi} \cdot z)$. To calculate the response, i.e., the unknown h_y^r, h_y^t, we use the boundary conditions at z=0

$$e_x^o + e_x^r - e_x^t = 0 \quad ; \quad h_y^o + h_y^r - h_y^t = -j_x = -\sigma \cdot e_x^t \quad . \tag{12}$$

Here the important point is that, due to the singular 2D surface current density at $z=0$, the tangential component h_y is discontinuous at the boundary. The discontinuity is determined by j_x which is driven by e_x. This strictly 2D approximation is valid as long as the spatial extend of the wavefunction d_e is small compared with the spatial extent of the fields $d_e << (\sqrt{\varepsilon}(\omega/c))^{-1}$ and $(k_x)^{-1}$.

The solution of eq. (12) gives for the unknown quantities h_y^r, h_y^t:

$$h_y^r = r_{p21} \cdot h_y^o = (D^r/D^o) \cdot h_y^o \quad ; \quad h_y^t = t_{p21} \cdot h_y^o = (D^t/D^o) \cdot h_y^o \quad , \tag{13}$$

$$D^o(\omega,k_x) = k_{z2}/\varepsilon_2 + k_{z1}/\varepsilon_1 + k_{z2} \cdot k_{z1} \cdot \sigma/(\varepsilon_2 \cdot \varepsilon_1 \cdot \varepsilon_0 \cdot \omega) \quad , \tag{14}$$

$$D^r(\omega,k_x) = k_{z2}/\varepsilon_2 - k_{z1}/\varepsilon_1 + k_{z2} \cdot k_{z1}\sigma/(\varepsilon_2 \cdot \varepsilon_1 \cdot \varepsilon_0 \cdot \omega) \quad , \tag{15}$$

$$D^t(\omega,k_x) = 2k_{z2}/\varepsilon_2 \quad . \tag{16}$$

r_{p21} and t_{p21} are, respectively, the Fresnel coefficients for the p-polarised reflected and transmitted amplitudes h_y. Note that both expressions include in the denominator $D^o(\omega,k_x)$, the determinant of the homogeneous system (eq. (12) with h_y^o, $e_x^o=0$).

This formalism can be expanded to calculate the response of multilayered systems with several 2D electron layers. For a two boundary system with three media 3,2,1 the transmission coefficient is for example:

$$t_{p321} = \frac{t_{p32} \cdot t_{p21} \cdot \exp(ik_{z2} \cdot d_2)}{1 - r_{p23} \cdot r_{p21} \cdot \exp(2ik_{z2} \cdot d_2)} \quad . \tag{17}$$

Here d_2 is the thickness of the medium 2, t_{pij} and r_{pij} are defined in eq. (13).

Transmission at normal incidence

Let us assume for a moment $\varepsilon_2=1$, $\varepsilon_1=\varepsilon=$real, and define $s= \sigma_o/\varepsilon_o c$, $((\varepsilon_o \cdot c)^{-1}= 377\ \Omega)$. Then we find at normal incidence ($k_x=0$) for the transmission coefficient of the amplitudes:

$$t_p = 2\sqrt{\epsilon}/(1+\sqrt{\epsilon}+s) \quad . \tag{18}$$

The transmitted intensity is $T=\left|t_p\right|^2$. In the experiment one is very often interested in the differential transmission $\Delta T/T=(T(s)-T(s=0))/T(s=0)$. We find:

$$\frac{\Delta T}{T} = \frac{-2(1+\sqrt{\epsilon}) \cdot s_r + s_r^2 + s_i^2}{(1+\sqrt{\epsilon}+s_r)^2 + s_i^2} \quad . \tag{19}$$

For small signals, if both the real and imaginary part s_r and s_i are small compared with $1+\sqrt{\epsilon}$, we find the well known result:

$$\Delta T/T = -2\sigma_r(\omega)/c \cdot \epsilon_o(1+\sqrt{\epsilon}) \propto Re(\sigma(\omega)) \tag{20}$$

Thus the relative change of transmission is directly proportional to the real part of the dynamic conductivity. This approximation can often be used to extract $\sigma_r(\omega)$ if $\Delta T/T$ is small. However one has to be careful in the case of large signals, e.g. cyclotron resonances at high mobility GaAs-heterostructures, and also in regimes where $\epsilon=\epsilon(\omega)$ depends strongly on the frequency, e.g., in the reststrahlen regime of polar materials. For such cases the full equation (19) should be used to extract $\sigma(\omega)$ from $\Delta T/T$ (see, e.g., Ref. 15).

Non radiative waves

In (13) r_{p21} and t_{p21} give the response with respect to an external exciting field h^o. A singular behaviour is expected if $D^o(\omega,k_x)$, the denominator, becomes zero. Then excitations without external field are possible. Let us first demonstrate this for a system without 2DES ($\sigma=0$). We assume $\epsilon_2=1$ and $\epsilon_1=\epsilon$ to be real. We find that it is possible to satisfy $D^o(\omega_{sp}, k_{sp})=0$ when

$$k_{sp} = \frac{\omega_{sp}}{c} \cdot \sqrt{\frac{\epsilon}{\epsilon+1}} \tag{21}$$

and $\epsilon<0$. These conditions can be found in metals at frequencies below the plasma frequency ω_p. These eigensolutions are surface plasmons (more accurately, surface plasmon polaritons). For a free electron gas with dielectric function $\epsilon(\omega)=1-\omega_p^2/\omega^2$ we find that, for $k_{sp} \to \infty$, ω_{sp} approaches the non-retarded surface plasmon frequency $\omega_{sp0}=\omega_p/\sqrt{2}$ with $\epsilon(\omega_{sp0})= -1$.

With decreasing k_{sp}, ω_{sp} decreases and approaches the light line ω/c.
However, since $|k_{sp}|>\omega/c$, both k_{z2} and k_{z1} are imaginary. Thus the surface
plasmons are accompanied by transverse electromagnetic field components
(h_y,e_z,e_x) which decay exponentially into the top dielectric and the
metal. The electromagnetic fields of the surface plasmon resonance are
strongly enhanced as can be seen from (13), where h_y^r and h_y^t, and similarly
the electric field components, increase to infinity if $D^o(\omega_{sp},k_{sp})\to 0$ (in a
real system this is of course limited by damping). The strong field enhancement gives rise to interesting applications in nonlinear optics and
for the giant Raman effect[16]. Similar types of surface waves, surface
phonon polaritons, exist in the reststrahlen regime of polar materials.

Let us consider now a 2D system with two adjacent dielectrics with
real ε_2 and ε_1. We consider the nonretarded regime $|k_x|>>\sqrt{\varepsilon}\cdot\omega/c$ and assume
$\sigma(\omega)=N_se^2\cdot i/(m^*\omega)$, an ideal Drude behaviour ($\tau\to\infty$). Then we find from (14)
for $D^o(\omega_p,k_p)=0$,

$$\omega_p^2 = \frac{N_se^2}{2\bar{\varepsilon}\cdot\varepsilon_o m^*} \cdot k_p \quad . \tag{22}$$

This is exactly the 2D plasmon dispersion that we have derived in RPA (9)
with $q=k_p$. In the treatment here we also calculate the influence of the
dielectric surrounding by the effective dielectric function $\bar{\varepsilon}=(\varepsilon_1+\varepsilon_2)/2$.

Whereas in a quantummechanical treatment 2D plasmons appear as quasi particles - quanta of the collective excitation of the 2DES - we have emphasized here the aspect, that 2D plasmons, analogous to the surface
plasmons (21), represent electromagnetic surface resonances. They are
accompanied by electromagnetic field components h_y,e_z,e_x which decay exponentially into the adjacent dielectrics. These field components allow us
to couple 2D plasmons with FIR radiation. This is in contrast to bulk
volume plasmons which have no h-components and do not couple with light.
Also the resonantly enhanced field strength of the 2D plasmon resonance is
observed in experiments, a strongly enhanced 2D plasmon - intersubband
resonance coupling[17] which will be discussed below.

Since the electromagnetic fields of the 2D plasmons extent into both
dielectrics, the plasmon frequency depends on the dielectric surrounding
and is govern by $\bar{\varepsilon}$. Very often the arrangement of a MOS-structure or a

gated GaAs heterostructure is used (see Fig. 1). Then the dielectricum ε_2 has the thickness d_2 and it is covered with an ideally screening gate $(\sigma_2(\omega)=i\infty, r_{p23}=1)$. We can then evaluate the zeros of the denominator of the multilayered system (17) and find

$$\bar{\varepsilon} = (\varepsilon_1 + \varepsilon_2 \cdot \coth(q \cdot d_2))/2 \quad . \tag{23}$$

In a similar way also the 2D plasmon dispersion for different arrangements including several layers of 2DESs can be calculated.

Grating coupler

2D plasmons are nonradiative modes in the sense that $q=k_p>\omega/c$ (taking into account retardation, also at small q and ω). Thus no direct coupling with radiation is possible. An effective way to excite these nonradiative modes are grating couplers[3,18,19]. A grating coupler may consist of periodic stripes (periodicity a) of high and low conductivity materials (Fig. 4a). (In principle, any periodic variation of the dielectric properties in the vicinity of the 2DES, e.g., a modulation of the 2DES itself, causes a grating coupler effect). Then the incident field, which has a homogeneous E_x component only, is shortened in the high conductivity regime. Thus in the near field of the grating, the electromagnetic fields are spatially modulated and consist of a series of Fourier components

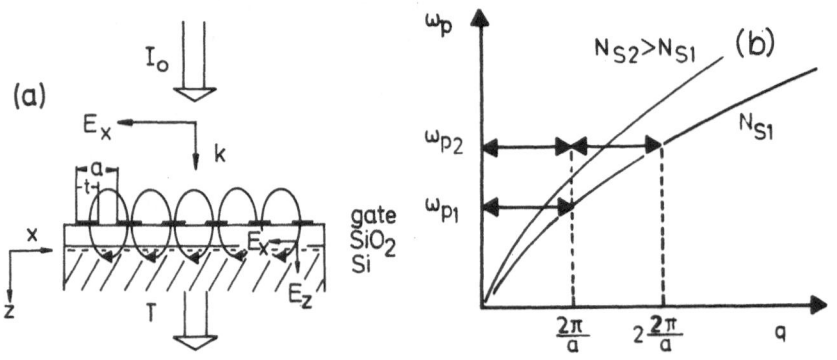

Fig. 4: (a) Grating coupler effect of a periodical structure. For normally incident FIR radiation spatially modulated parallel $(e_x(\omega,q))$ and perpendicular $(e_z(\omega,q))$ electric field components are induced. (b) Coupling of FIR radiation to 2D plasmons of wavevectors $q_1=2\pi/a$ and $q_2=2\cdot2\pi/a$

$$\vec{e} = \sum_{n=-\infty}^{+\infty} (e_x^n, 0, e_z^n) \cdot \exp(i(k_x^n \cdot x - \omega t)) \qquad (24)$$

with $k_x^n = n \cdot 2\pi/a$. Fourier components with wavevector k_x^n will couple with 2D plasmons if $q = k_x^n$ and ω satisfy the dispersion relation (Fig. 4b). It is also important that in the near field of the grating e_z components of the electric field are induced. These field components can be used to excite intersubband resonances[21,22]. A full general calculation of the dielectric response in the presence of a grating is a complicated numerical problem. Approximate treatments for the conditions here are given in Refs. 18-20. The effectiveness of the grating coupler depends on its design. The n-th Fourier components of the fields decays with $\exp(-|k_x^n| \cdot z)$ for $k_x \gg \omega/c$. Thus the distance d_2 of the 2DEG from the grating should be small compared with the periodicity $(d_2 < 0.1 a/n)$.

EXPERIMENTAL RESULTS

In the remaining part of these lecture notes I would like to discuss some typical examples of spectroscopic investigations of 2DESs. Within the available space I cannot explain all the details, instead I would like to give some general ideas. For a full understanding I refer the reader to the original papers.

2D plasmons

2D plasmons were first observed for electrons on the surface of liquid He.[23] In semiconductor systems 2D plasmons were investigated first on Si(100) [18,19]. Magnetoplasmons[24] have been observed too[25]. 2D plasmons were also studied in FIR emission experiments[26,27]. The first calculations of 2D plasmon dispersion, which also include the theoretical results derived here (eqs. (9),(22),(23)), were performed by Ritchie[28], Stern[29] and Chaplik[24]. Recent extended reviews on 2D plasmons are given in Refs. 20 and 30 (experiments) and Ref. 31 (theory).

2D plasmon resonances in MOS structures for Si(111) are present in the spectra of Fig. 2 in the frequency regime 50-150 cm^{-1}. As we expect from the 2D plasmon dispersion (22) the resonances shift with increasing N_S to higher frequencies. Recent extended studies on 2D plasmons for Si(100),(111) and (110) show that the experimental dispersion in general

agrees with the theoretical predictions[21,32]. For Si(110), e.g., the
plasmon dispersion is found to be anisotropic for different directions $\vec{q}$
with respect to, e.g., the [001] direction in the surface[32]. This follows
from (6) if we evaluate this expression for different $\vec{q}$ in an anisotropic
bandstructure $E(\vec{K})$. Also 2D plasmons in hole space charge layers of
Si(110) have been investigated[33]. Here the plasmon dispersion reflects the
nonparabolicity of the hole surface bandstructure and the anisotropy of
the Si(110) surface. A more detailed analysis of the plasmon dispersion
for electron inversion layers of Si shows that, particular at low N_S and,
for Si(100) also at high N_S, there are characteristic deviations, in general a decrease, of the experimental plasmon dispersion with respect to the
calculated dispersion. Most likely these deviations are caused by interaction of 2D plasmons with different excitations in the system, e.g. plasmons of different $\vec{q}$ and single particle excitations. A theoretical treatment of this coupling via extrinsic potential fluctuations for Si-MOS
structures within a memory function approach is given in Ref. 34.

For the high mobility AlGaAs-GaAs heterostructure system extrinsic
coupling is not so important and intrinsic effects on the 2D plasmon frequency can be studied[35]. Fig. 5a shows the relative change in transmission
$\Delta T/T$ for a GaAs heterostructure with a grating coupler of periodicity
a=880nm. The spectrum is normalized to a spectrum taken at B=8T. A magnetic field shifts the plasmon resonance, such that the effective conductivity at B=8T is low in the frequency range of Fig. 5a. This effect will be
discussed later. Two resonances are observed which are excited via the
first and second Fourier component of the grating and correspond to wavevectors $q_1=2\pi/a$ and $q_2= 2\cdot2\pi/a$. In the differential spectrum in Fig. 5b
(here $\Delta T/T$ is evaluated for two slightly different densities) three plasmon resonances are resolved. This allows us to study in detail the q-dependence of the dispersion, as shown in Fig. 5c. So far we have treated
the plasmons in a nonlocal and strictly 2D limit. In higher orders of q
several effects become important[20,30,31]. Two of these effects should be
mentioned. Evaluation of (6) in higher order of q gives[24,29] $\omega_p^2=$
$\omega_L^2+3/4(q\cdot v_F)^2$. Here v_F is the Fermi velocity and ω_L is the "local" plasmon frequency (22) considered so far. For high q the finite thickness of
the 2D system becomes also important[36-38]. The plasmon frequency may then
be written:

$$\omega_p^2 = \omega_L^2(1+ \frac{1}{g_v} \cdot \frac{3}{4}\ a^*\cdot q\ -\ F(q)\cdot d_e\cdot q)\ . \tag{25}$$

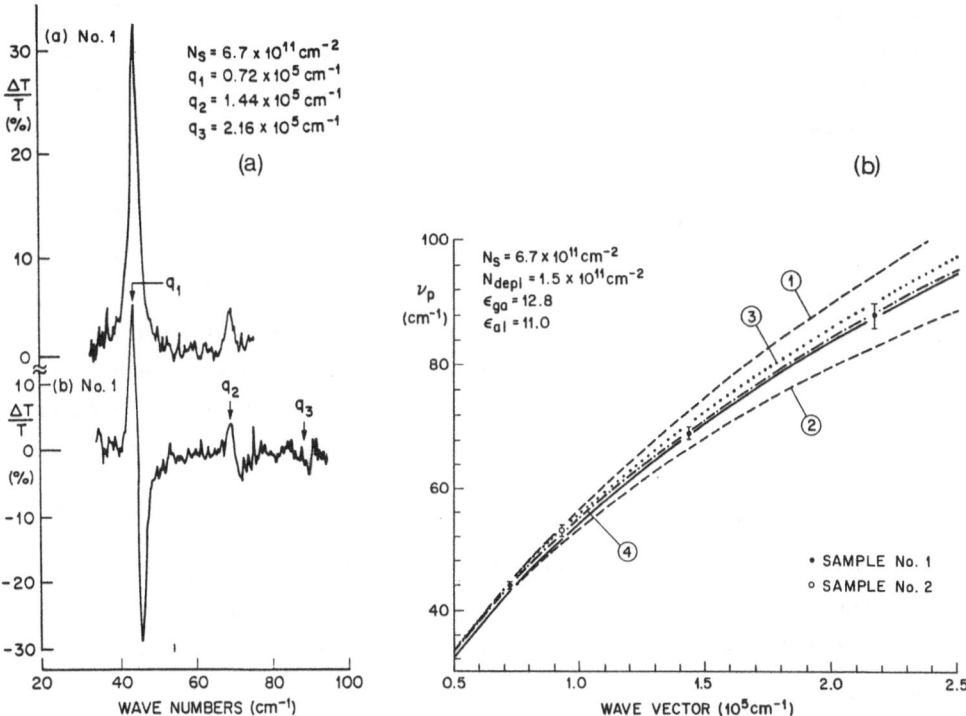

Fig. 5: Experimental plasmon resonances for a space-charge layer in GaAs with $N_S = 6.7 \times 10^{11} cm^{-2}$. The arrows mark plasmon resonance positions. For the upper trace (a) transmissions at B=0 T and B≳8 T have been ratioed to determine $\Delta T/T$. The lower trace (b), qualitatively the derivative of the upper trace, is obtained by changing N_S slightly via the persistent photo-effect. (c) Theoretical and experimental plasmon dispersion. The solid line is the classical local plasmon dispersions. The curves marked 1-4 are defined as follows: curve 1, plasmon dispersion including nonlocal correction; curve 2, plasmon dispersion including finite-thickness effect; curve 3, plasmon disperion including nonlocal and finite-thickness corrections combined; and curve 4, plasmon dispersion including all correction terms (from Ref. 35)

The first correction term, the so-called nonlocal term $3/4(q \cdot v_F)^2$, can be rewritten in terms of the effective Bohr radius $a^* = a_o \bar\epsilon m_o/m^*$ and the valley degeneracy g_v. Both $g_v = 1$ for GaAs (instead of $g_v = 2$ for Si(100)) and the larger effective Bohr radius of GaAs because of the small effective mass of GaAs ($a^*_{GaAs} \approx 10 nm$, $a^*_{Si(100)} \approx 2 nm$) make the nonlocal effect much more important for GaAs than for Si(100). The finite thickness becomes impor-tant if q approaches $1/d_e$, the inverse spatial extent of the electron wavefunction in the z-direction. The formfactor F(q) has a value of order 1 and has been expressed slightly differently by various authors[36-38]. For the conditions here F(q) is positive, meaning that the finite thickness effect lowers the plasmon frequency. In Fig. 5c we compare the experimental plasmon dispersion with the calculations. Taking into account only the nonlocal corrections we would expect a significantly stronger increase of ω_p with q as is found in the experiment. The finite thickness effect alone would decrease the plasmon frequency below the observed values. However both effects together nearly cancel each other in agreement with the experimental results. In the next paragraph we will see that the nonlocal effect can be directly observed in a resonant experiment. Thus the experimental results (Fig. 5c) implicitly demonstrate the presence of finite thickness effects.

In Refs. 39 and 40 the effect of nonlocality was directly observed for plasmon excitation in a magnetic field. In a classical approach a magnetic field B (assumed perpendicular the 2D plane) shifts the plasmon frequency[24]

$$\omega_{mp}^2 = \omega_p^2(B=0) + \omega_c^2 \tag{26}$$

where $\omega_c = e \cdot B/m^*$ is the cyclotron frequency. (This result can be calculated, e.g., by determining the zero of $D^o(\omega_{mp}, k=q)=0$ in eq. (14) with the Drude conductivity in a magnetic field $\sigma_{xx}(\omega, B)$ which will be given below (eq.(32)). Eq. (26) shows that the magnetoplasmon frequency approaches ω_c with increasing B. Experimental spectra measured in an AlGaAs-GaAs heterostructure with a grating coupler of a=1180 nm are shown in Fig. 6a. At low wavenumbers a CR is observed which decreases in resonance position with decreasing B. At higher frequencies, magnetoplasmon resonances are observed which also shift with decreasing B to lower frequencies roughly as expected from (26). For B=2.55T in Fig. 6a a second resonance occurs at $\omega_\ell = 70cm^{-1}$, which is not observed at higher magnetic fields. With decreasing B this resonance approaches ω_u and increases in intensity, whereas the resonance ω_u decreases in intensity. The resonance positions and relative amplitudes are depicted in Fig. 6b. Here it becomes evident that obviously there is an interaction of the magnetoplamson dispersion (26) with harmonics of the cyclotron resonance $2\omega_c$. In a homogeneous system cyclotron resonance excitation is only allowed for $\omega_c = eB/m^*$, corresponding to transitions between Landau levels with difference in index $\Delta n=1$. The dynamic spatial modulation of the plasmon oscillation induces an inhomogeneity which allows transitions for $\Delta n = 2, 3 \ldots$ This leads to a strong resonantly enhanced interaction of magnetoplasmons with harmonics of the CR and causes a splitting of the magnetoplasmon dispersion at the crossing with $2\omega_c$.

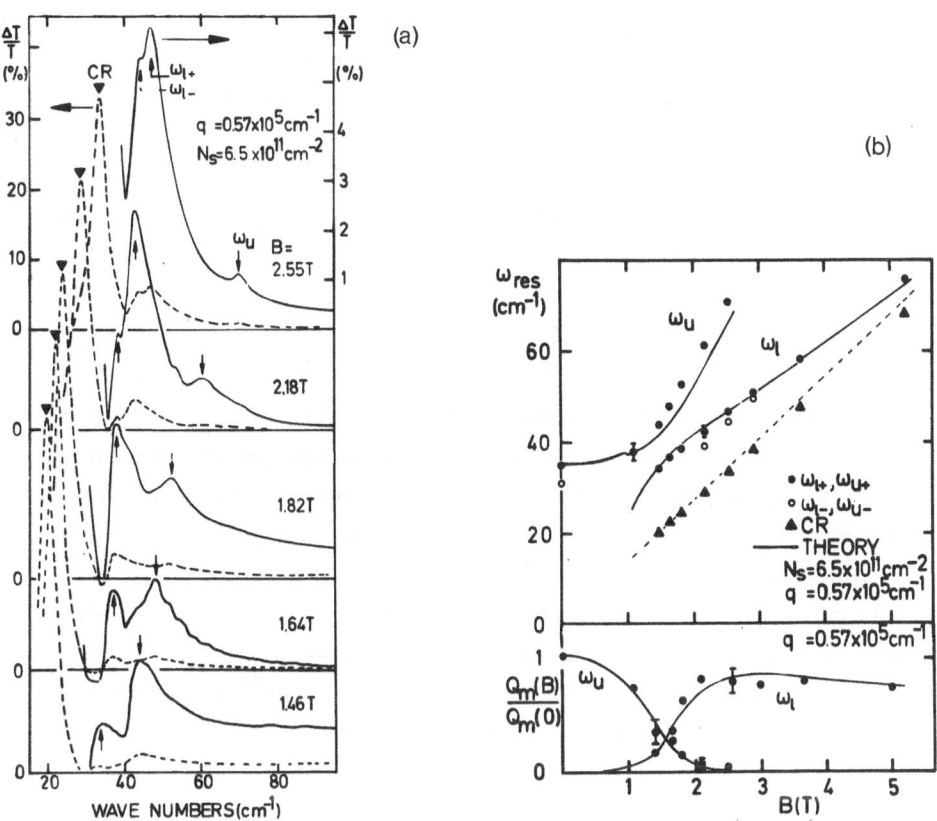

Fig. 6: (a) Cyclotron resonances (filled inverted triangles) and magnetoplasmon resonance for $N_S = 6.5 \times 10^{11} cm^{-2}$ and different magnetic fields B. The nonlocal interaction with the harmonic cyclotron frequency splits the magnetoplasmon dispersion into an upper (ω_u, downward-pointing arrow) and lower (ω_ℓ, upward-pointing arrow) branch. An additional splitting ($\omega_{\ell+}, \omega_{\ell-}$) due to the spatial charge density modulation is indicated. (Dashed lines, left scale; solid lines, expanded right scale). (b) Splitting of the magnetoplasmon dispersion due to the nonlocal interaction with the harmonic cyclotron frequency. Filled and open circles indicate experimental resonance position for the $\omega_{\ell+}, \omega_{u+}$ and $\omega_{\ell-}, \omega_{u-}$ resonances. The solid line is the theoretical dispersion (Ref. 42) calculated for the parameters of the experiment. In the lower part, experimental (filled circles) and theoretical (Ref. 42) (full lines) normalized excitation amplitudes are compared. $Q_m(B)$ is the peak absorption of the plasmon excitation in an ω sweep at fixed B (from Ref. 39)

The strength of the interaction, represented by the amount of the splitting, is governed by the parameter[40-42]

$$(qv_F/\omega_c)^2 = 3a^*q/g_v \quad .\tag{27}$$

Again for GaAs, due to the small mass and large a^*, this effect is very pronounced and can be observed very well in an experiment. As is shown in Fig. 6b the observed $\omega_\ell-\omega_u$ splitting and the experimental dependence of the amplitude on B is in excellent agreement with recent calculations[42] of magnetoplasmon excitations. These take into account nonlocal corrections in all orders of (qv_F/ω_c) and also finite values of the scattering time τ. The latter is important to describe the amplitude of the excitation. More details of magnetoplasmon-CR coupling, also for higher harmonics $3\omega_c$ and q, are described in Ref. 35.

In Fig. 6 we observe an additional small splitting $\omega_{\ell-}-\omega_{\ell+}$. This splitting is attributed to a static periodic charge density modulation. The charge density in the samples has been increased with short light pulses from a light emitting diode ($\lambda\approx820nm$) via the persistent photoeffect. Since the Ag stripes of the grating coupler are nontransparent the density of the ionized deep donors in the AlGaAs and thus the carrier density in the channel will be spatially modulated. We will discuss lateral superlattice effects on the 2D plasmon dispersion[43,44] in the next paragraph. This leads us to the rapidly expanding field of the physics of low dimensional electronic systems, where, starting from 2D systems, lateral structure of small sizes are fabricated which give rise to novel physical properties.

As an example we show in Fig. 7 a microstructured Si(100)-MOS system. These systems have been fabricated by first preparing a periodical surface profile in a photo resist layer on top of a homogeneous oxide using holographic lithography of two superimposed coherent laser beams[43,44]. This profile acts as a mask in a dry etching process. After etching the thickness of the oxide is d_1 in the region t_1 and is (except for the small region of the slopes) d_2 for the rest $t_2=a-t_1$ of the period a. A continuous layer of NiCr (3nm) is evaporated under varying angles onto the structured oxide. If a gate voltage V_g is applied between this gate and the substrate, an electron gas with a low charge density N_{S2} in the region t_2 and with high N_{S1} in the region t_1 is induced. d_1 and d_2 are, respectively, typically 20 and 50 nm.

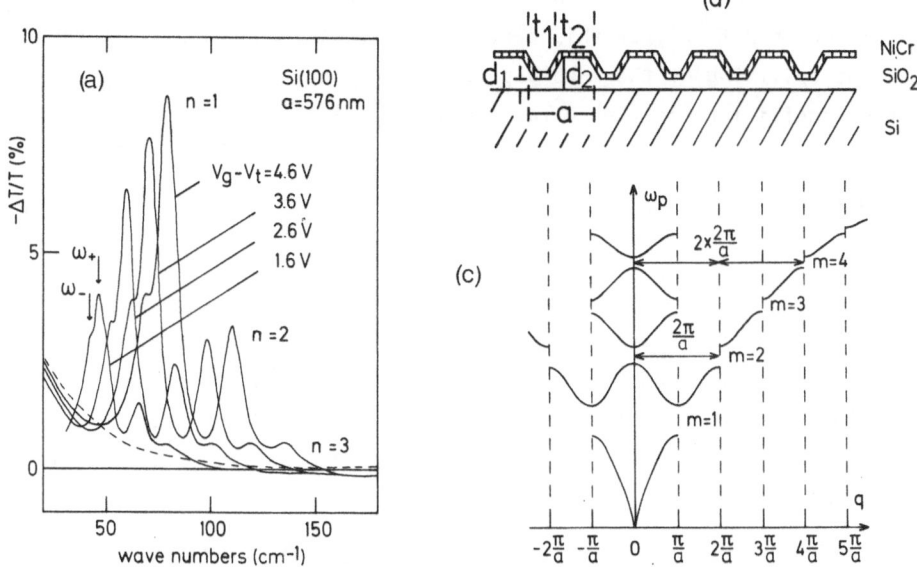

Fig. 7: Schematical geometry of a MOS sample with modulated oxide thickness (a), excitation of 2D plasmons with split resonances (b) due to the superlattice effect of the charge density modulation on the 2D plasmon dispersion (c) (from Refs. 43 and 44)

In Fig. 7b FIR transmission spectra are shown. Superimposed on the Drude background, well pronounced plasmon resonances are observed which shift with increasing gate voltage and corresponding charge densities to higher wavenumbers, roughly as is expected from the plasmon dispersion (22). Resonances ω_1 and ω_2 are indicated which are excited, respectively, via n=1 and n=2 reciprocal grating vectors $q_n = n$ $(2\pi/a)$. Characteristic for these charge density modulated systems is that the plasmon excitations are split into two resonances, ω_{n-} and ω_{n+}. To discuss the origin of this splitting, we show in Fig. 7c the plasmon dispersion in a charge density modulated system. The superlattice effect of the periodically modulated charge density creates Brillouin zones with boundaries $q = m\pi/a$, $m = \pm 1, \pm 2, \ldots$ If we fold the plasmon dispersion back into the first Brillouin zone, $-\pi/a < q < \pi/a$, then we expect a splitting of the plasmon dispersion at the zone boundaries and at the center of the Brillouin zone at q=0. The plasmon dispersion forms bands with minigaps at q=0 and $q=\pi/a$. Since we use the same grating that produces the Brillouin zones also for the coupling process $(q_n = n(2\pi/a))$, we can only observe the gaps at $m = 2, 4, \ldots$ The resonances observed in Fig. 7b are thus the lower and upper branches of this dispersion at q=0.

The plasmon dispersion in a charge density modulated system has been calculated in Ref. 45. In this perturbation theory approach it is found that the splitting of the m-th gap is proportional to N_m where N_m is the Fourier coefficient of the charge density Fourier series:

$$N_s(x) = \sum_{m=-\infty}^{+\infty} N_m \, e^{i2\pi mx/a} \tag{28}$$

To observe a large splitting of the plasmon resonance for $q=2\pi/a$, a system with a large second Fourier component N_2 in the charge density distribution has to be prepared. This explains a strong influence of the sample geometry on the amount of the experimentally observed splitting and, also, the fact that the splitting observed for higher gaps (m=4,6,...) is, in general smaller[43,44].

Another characteristic feature of plasmons in a periodically modulated system is that they have the character of standing waves, where both branches have different symmetry. Because of the symmetry, the upper branch ω_{1+} for the configuration in Fig. 7c has a radiative character, whereas the ω_{1-} branch is less radiative. Thus vice versa, FIR radiation can excite the ω_{1+} branch with higher efficiency. This is observed in Fig. 7b.

Physics in low dimensional electronic systems is a rapidly growing field (e.g.,Ref. 46,47). Recently structures, similar to those shown in Fig. 7a, have been prepared in GaAs-heterostructures where the electrons could be laterally confined and form 1D subbands[48]. Pioneering ideas and realizations in this field have been given by Sakaki and are discussed in his contribution of this volume[49].

Grating coupler induced intersubband resonances

Besides 2D plasmons, intersubband resonances (ISR) are characteristic dynamic excitations in heterostructures and quantum wells. ISR represent oscillations of the carriers perpendicular to the interface. Thus in highly symmetrical systems, e.g., electrons in Si(100) MOS systems or in GaAs-heterostructures, an E_z-component of the exciting electric field is necessary to excite these transitions. Strip-line and prism coupler arrangements have been used to study ISR on Si(100)[50,51]. Another very powerful method is the grating coupler technique[21,22]. As we have discussed above,

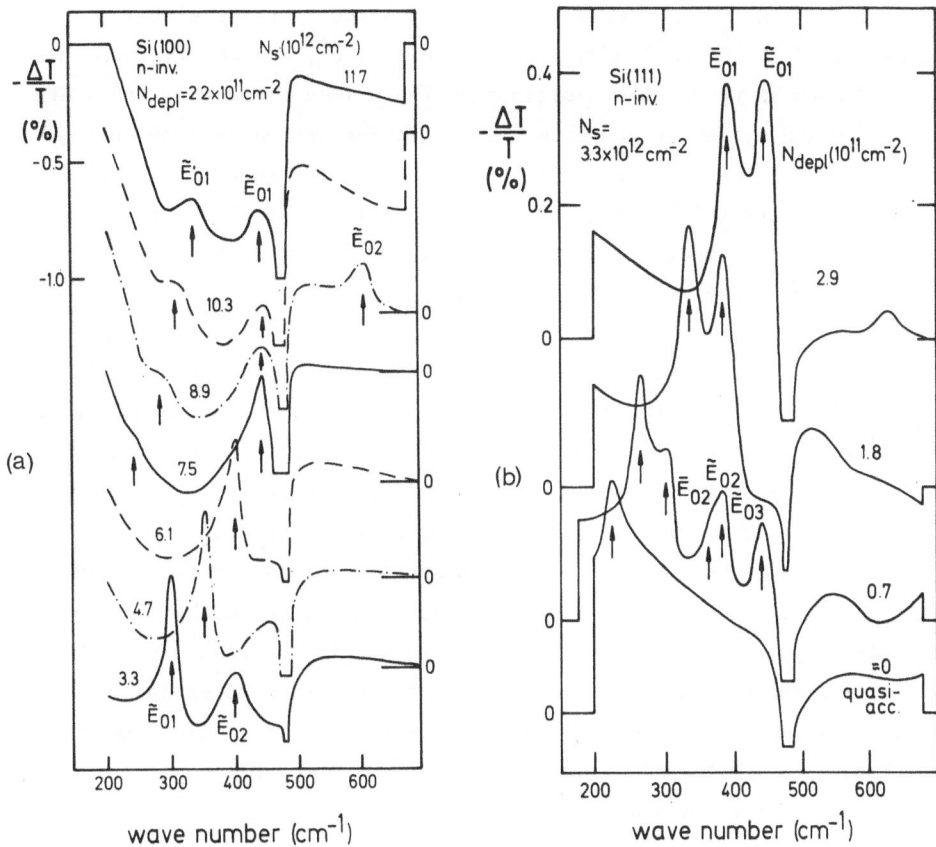

Fig. 8: (a) Grating coupler induced intersubband resonances for Si(100) at different charge densities N_S. Resonances $\tilde{E}_{01}$ and $\tilde{E}_{02}$ in the lower subband system and $\tilde{E}'_{01}$ in the second subband system are observed. In the regime of the optical phonon frequency of SiO_2 (about 480 cm^{-1}), a resonant coupling to polaritons is measured, which is not fully shown here for clarity. (b) Excitation of intersubband resonances on a Si(111) sample with a grating coupler for different N_{depl}. Directly parallel excited $\bar{E}_{oi}$ resonances and grating coupler induced depolarization shifted $\tilde{E}_{oi}$ resonances are present (from Ref. 22)

a grating coupler excites in the near field E_z components of the electric field (Fig. 4a). Experimental spectra measured on Si(100) samples with homogeneous charge density are shown in Fig. 8a. The grating periodicity is a=1800nm. For $N_S=3.3 \cdot 10^{12} cm^{-2}$ two resonances, $\tilde{E}_{01}$ and $\tilde{E}_{02}$, are observed which correspond, respectively, to resonant transitions from the lowest subband to the first and second excited subbands. With increasing N_S the resonances shift to higher frequencies, corresponding to a larger subband

337

separation in the steeper potential well at larger surface electric fields. For $N_S > 8 \cdot 10^{12} cm^{-2}$ additional resonances $\tilde{E}'_{01}$ are observed which can be attributed to resonant transitions in the primed subband system[1]. The primed subband system arises from the projection of four volume energy ellipsoids of Si onto the Si(100) surface and is separated in K-space by $0.86 \cdot 2\pi/A$ (A= crystal lattice constant) in [001] and equivalent directions. It is known that this subband system is occupied for $N_S > 7.5 \cdot 10^{12} cm^{-2}$ (Ref. 1).

The resonance energy measured in an ISR spectrum is not directly the subband spacing $E_{01} = E_1 - E_0$. Two effects shift the observed resonance with respect to the subband spacing. The first, the so-called exciton shift[52], results from the energy renormalisation when an electron is transferred to the first excited subband E_1 leaving a 'hole' in the E_0 subband. The exciton effect, characterized in a two-band model by β, shifts the resonance energy to smaller energies. A second effect, characterized by α, is the so-called depolarisation shift, which increases the resonance energy. It arises from the polarisation of the 2DEG in the finite potential well[53]. A strict calculation of this effect is included, if a full time dependent calculation of the Hartree potential is performed[54], instead of the static calculation of subband energies.

For Si(111) and Si(110), the surface bandstructure results from the projection of volume energy ellipsoids which are tilted with respect to the surface[1]. Due to the anisotropic energy contours, the j_x and j_z component are coupled and thus ISR can be excited with a parallel field component E_x.[12] Whereas parallel excited ISR (labelled $\bar{E}_{01}$) is not affected by the depolarisation shift

$$\bar{E}_{01} = E_{01} \cdot \sqrt{1-\beta'} \ . \tag{29}$$

the perpendicular excited resonance ($\tilde{E}_{01}$) is affected by both effects

$$\tilde{E}_{01} = E_{01} \cdot \sqrt{1+\alpha-\beta'} \ . \tag{30}$$

Thus if a grating coupler is used on Si(111) (Fig. 8b) both resonances, directly parallel excited resonances and perpendicular grating coupler excited (and thus depolarisation shifted) resonances are observed. We see from the spectra in Fig. 8b that the depolarisation shift slightly increases with the depletion field that is characterized by N_{depl}. It becomes much smaller for transitions to higher subbands (E_{02} and E_{03}). Very detailed investigations of grating coupler induced ISR for different surface orientation on Si are reported in Ref. 22.

For GaAs heterostructures only a few direct FIR spectroscopic investigations of ISR have been performed (see. e.g., Fig. 3 in Ref.35) so far. More often the method of tilted magnetic fields is applied[55-57]. If a magnetic field is tilted slightly by an angle Θ with respect to the surface normal, the Landau levels of different subbands couple, e.g. the 1-Landau level of the E_0 subband and the 0-Landau level of the E_1 subband. In resonance antilevel crossing occurs, leading to a splitting of the CR resonance at energy E_{01}. Thus, from the position of the splitting the subband separation can be found. The amount of the splitting is, for small Θ, $c_{01} \cdot \sin\Theta$. From the matrix element c_{01}, information on the wavefunctions and the shape of the potential can be extracted[56].

The CR-ISR coupling and the plasmon-CR coupling discussed above are examples of resonant coupling processes. As we have seen above very detailed information on the interaction can be extracted from resonant coupling processes. So far we have treated ISR and 2D plasmon resonance independently. The question arises whether it is possible to find conditions where both the fundamental parallel and perpendicular excitations can be matched in energy. At a first glance one would expect that the confinement of the 2D electrons perpendicular to the interface, and thus the ISR frequency, always should be larger than the parallel binding. In a multi-valley system with different subbands, however, the lowest subband can screen the higher subbands and thus lower the transition energies there. This can be realized in the Si(100) system by applying uniaxial stress in the [001] direction[17]. Uniaxial stress lowers the energy of two of the original four fold degenerated E' valleys so that they become occupied. Since the two subband systems are separated in K-space we have independent ISR in both systems. However, the 2D plasmon excitation is the collective oscillation of all carriers (for the optical branch of interest here all carriers oscillate in phase). The plasmon mass in this multivalley system is, as can be calculated from (6)

$$m_p^{-1} = (N_{So'} \cdot m_{po'}^{-1} + N_{So} \cdot m_{po}^{-1})/N_S \quad . \tag{31}$$

Here $N_{So'}/N_S$ and N_{So}/N_S are the relative occupations of the E_o', and the E_o subband, respectively. The plasmon mass in the E_o' subband is larger

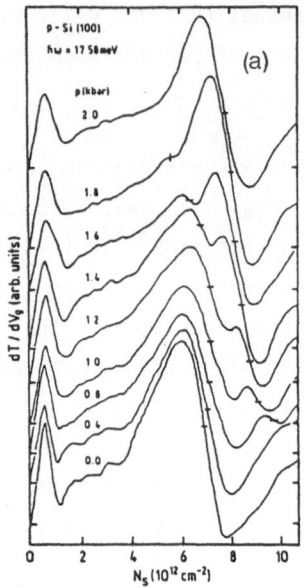

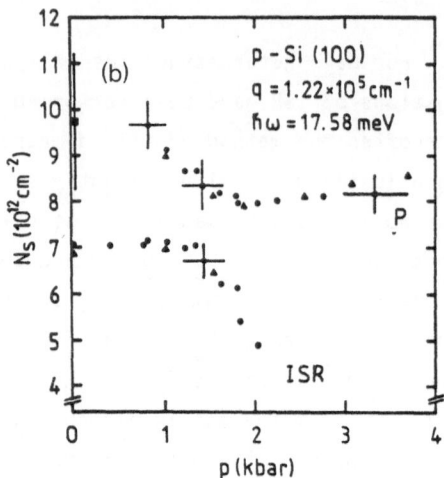

Fig. 9: (a) Derivative of the transmission T with respect to the gate voltage V_g vs. N_S for different values of the uniaxial stress at laser frequency 17.58 meV. (b) Experimental resonance positions from (a). The interaction of the plasmon (P) and ISR leads to a resonant anti-level crossing of the dispersions

($m_{po'} = 0.91 \cdot m_o$) than for the E_o subband ($m_{po} = 0.19\ m_o$) for q‖[001]. Thus, with increasing stress and occupation of E_o', m_p increases. Experimental spectra are shown in Fig. 9a. The plasmon resonance is detected in the derivative dT/dV_g of the transmission with respect to the gate voltage in a sweep of N_S at fixed laser frequency. The resonance position in N_S shifts with increasing stress, p, to higher N_S as is expected from (22) if, at fixed ω_p, m_p is increased via stress. At p≈0.8 kbar a second resonance occurs which shifts to lower N_S. This resonance is the E_{01}' intersubband resonance. From the resonance position in Fig. 9b we find that the interaction of ISR and plasmon leads to an antilevel crossing. The experimentally observed coupling strength[17] which governs the amount of the splitting, agrees with an estimated interaction strength using calculations in Ref. 58. It is also found in Fig. 9a that the ISR is only observed within the vicinity of the plasmon resonance. This reflects the fact, that the ISR is excited via the E_z components of the electromagnetic fields accompaning the 2D plasmon excitation. The resonance enhancement of $E_z(\omega,k)$ at ω_p and $q = k_p$ which we have discussed above leads to the enhanced excitation strength of the ISR observed in Fig. 9a.

340

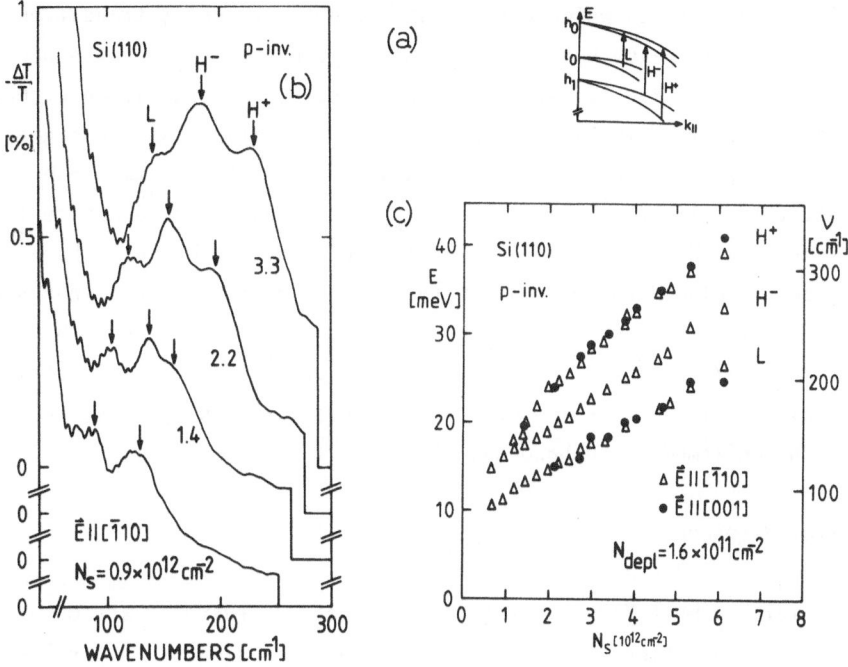

Fig. 10: Schematical surface bandstructure of holes with split spin states
(a), parallel excitation of intersubband resonances in hole inversion
layers of Si(110) (b), and experimental resonances positions vs charge
density N_S (c) (from Ref. 59)

So far we have considered 2D electron systems. In hole subband
systems interesting fundamental investigations can be performed too. I
will describe here experiments which are related to an inherent (i.e.,
without external magnetic field) lifting of the spin degeneracy in systems
without inversion symmetry[59]. This effect is extremely small in the bulk
of zincblende type crystals[60]. In MOS systems the surface electric field
induces an asymmetric potential well and it has been found from subband
calculations that in hole space-charge layers of Si, due to the strong
spin-orbit interactions of holes, this effect is important[61,62]. In Fig.
10 intersubband excitations in hole inversion layers of n-type Si(110)
(without grating coupler) are shown. The energy spectrum of 2D hole
space-charge layers consists of heavy and light hole type subbands. The
dispersion is strongly nonparabolic and for subband wavefunctions with

quasi momentum $K_\parallel \neq 0$ the spin degeneracy is lifted as is discussed above. Thus for intersubband spectroscopy on these systems one would expect smeared out broad resonances. Surprisingly the experimental resonances exhibit sharp features which can be attributed to resonant transitions into a spin split heavy hole subband, $h_0 \rightarrow h_1^+$ and h_1^-. These resonances can be resolved since, due to matrix element effects, the transition probability for parallel excitation is dominant for transitions at the highest occupied $K_\parallel$. Thus the resonance frequency measured at a certain density N_S is implicitly related to a certain $K_\parallel \approx K_{Fermi}$. In Fig. 10c we notice that with increasing N_S, which means implicitly with decreasing $K_\parallel$, the resonances labeled $H^-(h_0 \rightarrow h_1^-)$ and $H^+(h_0 \rightarrow h_1^+)$ approach the same energy of 12 meV at $N_S = 0$. The fact that the spin splitting goes to zero with $K_\parallel \rightarrow 0$, is a characterisitc feature of spin-orbit interaction in an asymmetric potential. The identification of the resonances as spin split states has also been confirmed by subband calculations[59].

Cyclotron Resonance

By far the most FIR investigations of heterostructures are devoted to the CR. Different aspects are covered in other articles of this volume which also include an extended list of references[8,9]. To demonstrate some characteristic features of CR, I would like to discuss here measurements on AlGaAs-GaAs heterostructure with front gates[63]. We have prepared heterostructures with a 50 nm Si-doped AlGaAs-layer and with 20 nm spacer layer. On top of this structure a semitransparent NiCr gate was prepared and contacts were made to the channel. Via a gate voltage that was applied between the gate and the channel we were able to tune N_S from $5 \cdot 10^{11} \text{cm}^{-2}$ to nearly zero. The charge density could be determined in situ under experimental conditions from magnetocapacitance measurements. Experimental CR-measurements are shown in Fig. 11 for different densities N_S. The CR excitation leads to a resonant decrease of the transmission at frequencies about 112 cm^{-1} and we see from Fig. 11 that not only the intensity of the signal changes with N_S but also the resonance position and halfwidth.

To extract from these CR curves information on the 2DES a Drude model is used very often to describe the 2D conductivity in a magnetic field:

$$\sigma_{xx}(\omega,B) = \frac{N_S e^2 \cdot \tau}{2m^*} ((1-i(\omega-\omega_c)\tau)^{-1} + (1-i(\omega+\omega_c)\tau)^{-1}) \, , \qquad (32)$$

$\omega_c = e \cdot B/m_c$ is the cyclotron frequency. For linearly polarised radiation σ_{xx} includes the active $(\omega - \omega_c)$ and the inactive mode $(\omega + \omega_c)$. Inserting (32) into the transmission formula (19) we find for the maximum change in transmission of the active mode at ω_c for circular polarised radiation

$$(-\Delta T/T)_{max} = 1 - (1 + b \cdot N_S \cdot \tau)^{-2} . \tag{33}$$

and for the halfwidth $\Delta\omega$ (HWHM)

$$\Delta\omega = \frac{1}{\tau} + b \cdot N_S . \tag{34}$$

with $b = e^2/(\varepsilon_0 cm^*(1+\sqrt{\varepsilon}))$. We see that for small N_S and τ the resonance amplitude increases linearly with N_S and τ $((-\Delta T/T)_{max} \approx 2 \cdot b \cdot N_S \cdot \tau)$ and the halfwidth of the resonance is directly $\Delta\omega = 1/\tau$. For large N_S and τ, however, the relative change of transmission $\Delta T/T$ cannot exceed 100% (50%) for circularly (linearly) polarised radiation. Then, with increasing N_S or τ, $(\Delta T/T)_{max}$ saturates and the resonance broadens with respect to $\Delta\omega = 1/\tau$. We can determine N_S by fitting the calculated transmission (eqs. (19) and (32)) to the experimental resonance profile in Fig. 11 and we find excel-

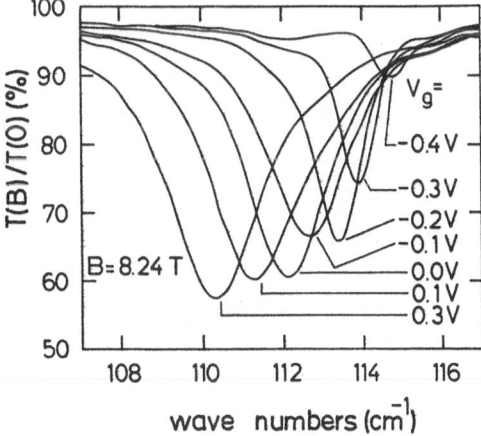

Fig. 11: Experimental CR excitation in frontgated AlGaAs-GaAs heterostructures. The transmission T(B) is measured at fixed B=8.24 T and is normalized to the transmission T(0) at B=0. From magnetocapacitance we find that the gate voltage V_g(V)= 0.3, 0.1, 0.0, -0.1, -0.2, -0.3 corresponds to densities N_S ($10^{11}cm^{-2}$)= 5.35, 3.94, 3.16, 2.37, 1.74, 0.80, respectively

lent agreement in the N_S regime where we can extract N_S reliably from magnetocapacitance. From the same fits we also can determine $m_c = e \cdot B / \omega_c$ and τ. Different aspect of what interaction determines m_c and τ are discussed in Ref. 8 and 9. A detailed analysis of the data in Fig. 11 shows[63] that the shifts of the resonance positions are not only caused by the non parabolic bandstructure of the GaAs system. In addition also effects due to filling factor dependent, combined and interacting CR of different Landau and spin levels and energy renormalisation effects at full Landau levels are important. Also the resonance halfwidth seems to be related in an interacting way to the resonance positions. Considering all experimental and theoretical results on CR in 2DESs, one has to confess that there are still many open questions left.

ACKNOWLEDGEMENT

In these lecture notes I have reported on investigations which have been performed at the Institut für Angewandte Physik in Hamburg and partly at the Max-Planck-Institut für Festkörperforschung in Stuttgart. I would like to thank very much all of my colleagues, as listed in the References, who have been working with me on these different subjects. I also acknowledge financial support from the DFG, Stiftung Volkswagenwerk and the BMFT.

REFERENCES

1   T. Ando, A.B. Fowler, and F. Stern, Rev. Mod. Phys. 54, 437 (1982)

2   J.F. Koch, Surf. Sci. 58, 104 (1976)

3   D.C. Tsui, S.J. Allen Jr., R.A. Logan, A. Kamgar, and S.N. Coppersmith, Surf. Sci. 73, 419 (1978)

4   J.P. Kotthaus, in: Springer Series in Solid State Sciences, Vol 53, Eds. G. Bauer, F. Kuchar, H. Heinrich, Springer Berlin (1984), p. 32

5   D. Heitmann, in: Festkörperprobleme (Advances in Solid State Physics), Vol. 25, Ed. P. Grosse, Vieweg Braunschweig (1985) p. 429

6   R.A. Höpfel and E. Gornik, Surf. Sci. 142, 412 (1984)

7   G. Abstreiter, this volume

8   J.C. Maan, this volume

9   E. Gornik, this volume

10 K. Muro, S. Mori, S. Narita, S. Hiyamizu, and K. Nanbu, Surf. Sci. 142, 394 (1984)

11 E. Batke and D. Heitmann, Infrared Phys. 24, 189 (1984)

12 T. Ando, T. Eda and M. Nakayama, Solid State Commun. 23, 751 (1977)

13 T. Ando, Z. Phys. B24, 33 (1976)

14 e.g. J.M. Ziman, 'Principles of the theory of solids', Cambridge, University press, Cambridge 1986

15 M.Ziesmann, D. Heitmann and L.L. Chang, Phys. Rev. B35, 4541 (1987)

16 for a review see, e.g., A.D. Broadman, Ed., 'Electromagnetic Surface Modes', Wiley, Chichester (1982)

17 S. Oelting, D. Heitmann, and J.P. Kotthaus, Phys. Rev. Lett. 56, 1846 (1986)

18 S.J. Allen Jr., D.C. Tsui, and R.A. Logan, Phys. Rev. Lett. 38, 980 (1977)

19 T.N. Theis, J.P. Kotthaus, and P.J. Stiles, Solid State Commun. 24, 273 (1977)

20 T.N. Theis, Surf. Sci. 98, 515 (1980)

21 D. Heitmann, J.P. Kotthaus, and E.G. Mohr, Solid State Commun. 44, 715 (1982)

22 D. Heitmann and U. Mackens, Phys. Rev. B33, 8269 (1986)

23 C.C. Grimes and G. Adams, Phys. Rev. Lett. 36, 145 (1976)

24 A.V. Chaplik, Soviet Phys. JETP 35, 395 (1972)

25 T.N. Theis, J.P. Kotthaus, and J.P. Stiles, Solid State Commun. 26, 603 (1978)

26 D.C. Tsui, E. Gornik, and R.A. Logan, Solid State Commun. 35, 875 (1980)

27 R.A. Höpfel, E. Vass, and E. Gornik, Phys. Rev. Lett. 49, 1667 (1982)

28 R.H. Ritchie, Phys. Rev. 106, 874 (1957)

29 F. Stern, Phys. Rev. Lett. 18, 546 (1967)

30 D. Heitmann, Surf. Sci. 170, 332 (1986)

31 A.V. Chaplik, Surf. Sci. Rep. 5, 289 (1985)

32 E. Batke and D.Heitmann, Solid State Commun. 47, 819 (1983)

33 E. Batke, D. Heitmann, A.D. Wieck, and J.P. Kotthaus, Solid State Commun. 46, 269 (1983)

34 A. Gold, Phys. Rev. B32, 4014 (1985)

35 E. Batke, D. Heitmann, and C.W. Tu, Phys. Rev. B34, 6951 (1986)

36 T.K. Lee, C.S. Ting, and J.J. Quinn, Solid State Commun. 16, 1309 (1975)

37 R.Z. Vitlina and A.V. Chaplik, Soviet Phys. JETP 54, 536 (1981)

38 J.I. Gerstens, Surf. Sci. 97, 206 (1980)

39 E. Batke, D. Heitmann, J.P. Kotthaus, and K. Ploog, Phys. Rev. Lett. 54, 2367 (1985)

40 T.K. Lee and J.J. Quinn, Phys. Rev. B11, 2144 (1975)

41 N.J.M. Horing and M.M. Yildiz, Ann. Physik 97, 216 (1976)

42 A.V. Chaplik and D. Heitmann, J. Phys. C.: Solid State Phys. 18, 3357 (1985)

43 U. Mackens, D. Heitmann, L. Prager, J.P. Kotthaus, and W. Beinvogl, Phys. Rev. Lett. 53, 1485 (1984)

44 D. Heitmann, J.P. Kotthaus, U. Mackens, and W. Beinvogl, J. Superlattices and Microstructures 1, 35 (1985)

45 M.V. Krasheninnikov and A.V. Chaplik, Sov. Phys. Semicond. 15, 19 (1981)

46 W.J. Skocpol, L.D. Jackel, E.L. Hu, R.E. Howard, and L.A. Fetter, Phys. Rev. Lett. 49, 951 (1982)

47 A.B. Fowler, A. Hartstein, and R.A. Webb, Phys. Rev. Lett. 48, 196 (1982)

48 W. Hansen, M. Horst, J.P. Kotthaus, U. Merkt, and Chr. Sikorski, to be published

49 H. Sakaki, this volume

50 P. Kneschaurek, A. Kamgar, and J.F. Koch, Phys. Rev. B14, 1610 (1976)

51 B.D. McCombe, R.T. Holm, and D.E. Schafer, Solid State Commun. 32, 603 (1979)

52 T. Ando, Z. Phys. B26, 263 (1977)

53 W.P. Chen, Y.J. Chen, and E. Burstein, Surf. Sci. 58, 263 (1976)

54 S.J. Allen, D.C. Tsui, and B. Vinter, Solid State Commun. 20, 425 (1976)

55 Z. Schlesinger, J.C.M. Hwang, and S.J. Allen, Phys. Rev. Lett. 50, 2098 (1983)

56 G.L.J.A. Rikken, H. Sigg, G.J.G.M. Langerak, H.W. Myron, and J.A.A.J. Perenboom, Phys. Rev. B34, 5590 (1986)

57 A.D. Wieck, J.C. Maan, U. Merkt, J.P. Kotthaus, K. Ploog, and G. Weimann, Phys. Rev. B35, 4145 (1987)

58 S. Das Sarma, Phys. Rev. B29, 2334 (1984)

59 A.D. Wieck, E. Batke, D. Heitmann, J.P. Kotthaus, and E. Bangert, Phys. Rev. Lett. 53, 493 (1984)

60 G. Dresselhaus, Phys. Rev. 100, 580 (1955)

61 E. Bangert, K.v.Klitzing, and G. Landwehr, Proc. 12. Int. Conf. on Physics of Semiconductors, Stuttgart, ed. M.H. Pilkuhn (Teubner, Stuttgart, 1974), p. 714

62 F.J. Ohkawa und Y. Uemura, Prog. Theor. Phys. Suppl. 57, 164 (1975)

63 K. Ensslin, D. Heitmann, H. Sigg, and K. Ploog, to be published

MAGNETO-OPTICAL PROPERTIES OF HETEROJUNCTIONS, QUANTUM WELLS AND SUPERLATTICES

J.C. Maan

Max-Planck-Institut für Festkörperforschung, Hochfeld Magnetlabor, 166X, F38042 Grenoble Cedex France

I. INTRODUCTION

In a magnetic field the continuous dispersion relations of the bandstructure are split into discrete Landau levels. The energy separation between these Landau levels can be measured optically, and this way information about the bandstructure is obtained. Since transitions are never infinitely sharp in real systems, additional information can be obtained from broadening. In the first part of this paper intraband absorption (cyclotron resonance) in heterojunctions and quantum wells in a magnetic field perpendicular to the layer, with emphasis on the consequences of non-parabolicity will be described. Furthermore several aspects which can contribute to the observed cyclotron linewidth will be mentioned. In the second part a discussion of interband (valence to conduction band) absorption will be given. In particular the effect of the complex valence bandstructure on the experimental results will be described. As in interband absorption both electrons and holes are involved, the effect of their interaction (exciton formation) on the results will be discussed. In the last part interband absorption in a superlattice with a magnetic field parallel to the layers will be discussed. Here a different effect of the magnetic field will be employed, namely that carriers in a magnetic field describe circular orbits in a plane perpendicular to the field, which have an orbit size that can be comparable to the superlattice periodicity. Therefore this type of experiments probes transport through the layers of the superlattice.

II. a <u>Intraband absorption (Cyclotron resonance)</u>

For simple parabolic bands the carrier motion is free electron like and the Hamiltonian for the motion in a magnetic field in the z direction in potential $V(z)$ is given by:

$$H = (\bar{p} - e\bar{A})^2 + V(z) \tag{1}$$

with the vector potential $\bar{A} = (0, xB, 0)$. For such simple cases the in plane motion (xy-plane) can be separated from the z-component by choosing for the wavefunction $g(\mathbf{x}, \mathbf{y}) \, f(z)$. The eigenvalue problem (1) becomes then equivalent to two separate equations. For the z-direction (1) reduces to:

$$\left[\left(\frac{1}{2m^*} \, p_z^2 + V(z) \right) \right] f(z) = E_n f(z) \tag{2}$$

with E_n the confinement energies in potential $V(z)$, m^* the effective mass and p_z the momentum operator. For an infinite square well of thickness d, E_n would be given by:

$$\frac{\hbar^2 \pi^2 (n+1)^2}{2m^* \, d^2},$$

For the in plane motion we have if $g(x,y) = e^{ik_y y} \varphi(x')$

$$\left[-\frac{\hbar^2}{2m^*} \frac{\partial^2}{\partial x'^2} + \frac{e^2 B^2 x'^2}{2m^*} + V(x) \right] \varphi(x') = E_\parallel \varphi(x') \tag{3}$$

with $E_\parallel$ the eigenvalues. In eq. (3) $x' = x + \hbar k/eB$, where $\hbar k_y/eB$ is the cyclotron orbit center coordinate. $V(x)$ is a one dimensional potential which for the moment is zero but the effect of which will be discussed later. Since the motion in the plane has translational symmetry, the eigenvalues have a degeneracy for all allowed k_y values, which is $2eB/h$ (neglecting spin splitting at present). The eigenvalues are the normal degenerate Landau levels given by $E_\parallel = (N + \frac{1}{2}) \frac{\hbar eB}{m^*}$ with $N = 0, 1, 2$ etc. the Landau level quantum number. The total energy E_t is simply given by: $E_t = E_\parallel + E_n$ (see Fig. 1). Cyclotron resonance measures the absorption from a Landau level N to N+1 and the resonance condition is just $\omega_c = eB/m^*$. It has been proven rigorously [1] that inclusion of electron-electron interaction in the Hamiltonian does not affect the resonance position, therefore any experimentally observed deviation from the simple linear relation between energy and magnetic field must be traced back to an energy dependent effective mass (non parabolicity) or

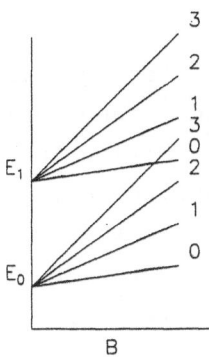

Fig. 1

Schematic energy level
structure of Landau levels
from two subbands.

interaction of carriers with impurities. In a real experiment
the resonance is measured as the absorption of light, and the
absorbed power (P) of linearly polarized light propagation
along the field direction (Faraday configuration) is given by:

$$P(\omega) = P_o \frac{1+(\omega^2+\omega_c^2)\tau^2}{|1+(\omega_c^2-\omega^2)\tau^2|+4\omega_c^2\tau^2} \qquad (4)$$

In this expression a Drude like description for the dynamical
conductivity is used where the level broadening is characterized
by a scattering time τ. [2]

Summarizing, for simple bands with cyclotron resonance
only the effective mass, the scattering time and, as the integrated absorption is proportional to the carrier density, the
carrier density can be measured. To a large extend the conduction band structure of GaAs and Si can be described by parabolic bands and since in this case the mass is the same as it
would be in bulk material, the interest of cyclotron resonance
is limited mainly to a study of the linewidth. This will be
discussed in II c. In many other semiconductors like InAs,
InSb, HgCdTe however there is an important band non-parabolicity, which has important consequences on CR in heterostructures.
In particular for non-parabolic bands it is possible to obtain
information about the subband structure and the Fermi energy
from the observed cyclotron mass.

II. b Effects of non-parabolicity

The main consequence of a non-parabolic (energy versus wavevector) dispersion relation, is that the second derivative of this dispersion relation, which is by definition the effective mass, is a function of the energy. Since the effective mass is dependent on the total energy $E_\perp + E$ and since the mass determines both the confinement energies E_n and the in plane energy $E_{//}$ the two directions cannot be separated anymore as before. The origin of the non-parabolicity is the interaction between different bands, and the simplest description of the effect is a two band model. Following Zawadzki [3] the hamiltonian for the conduction band in the two band model in the presence of a potential $V(z)$ is given by:

$$\left[\frac{1}{2m_o^*} \; P_z^2 \; - \; \frac{E_t - E_{//} - V(z)}{E_G} \quad (E_G + E_t + E_{//} - V(z)) \right] f(z) = 0 \qquad (5a)$$

where m_o^* is the bulk band edge effective mass. In a magnetic field:

$$E_{//} = -\frac{E_G}{2} + \left[\frac{E_G}{2}^2 + E_G D \right]^{1/2} \qquad (5b)$$

with

$$D = \frac{heB}{m_o^*} \quad (N + \frac{1}{2}) \qquad (5c)$$

For an infinite square well with thickness d, the energy levels can be calculated from Bohr Sommerfeld quantization:

$$\int P_z d_z = (n + \frac{1}{2}) \; \pi \hbar \qquad (6)$$

leading to an implicit quadratic equation for the energies.

$$\frac{(E_t - E_{//}) \; (E_G + E_t + E_{//})}{E_G} = \frac{\hbar^2 \pi^2 (n + \frac{1}{2})^2}{2 m_o^2 d^2} \qquad (7)$$

where all quantities have their usual meaning. Introducing E_n as the energy of subband n as $E_{//} = 0$ eq. 7 can be rewritten as [4]:

$$(N + \frac{1}{2}) \; \frac{\hbar eB}{m_o^*} = E_N (1 + \frac{E_N}{E_G}) - E_n (1 + \frac{E_n}{E_G}) \qquad (8)$$

The cyclotron effective mass is by definition $heB/(E_{N+1} - E_N)$, and as can be seen from eq. 8 depends on field, confinement energy E_n and Landau level quantum number N in a complex manner.

350

Fig. 2 shows the calculated Landau levels for a rectangular
InAs quantum well with a carrier density of $10^{12} cm^{-2}$ and 15nm
thickness. For this density and thickness two subbands are
occupied.

In a magnetic field each subband gives rise to a Landau
level fan which is non-linear due to non-parabolicity. Observable transitions take place between levels of the same subband

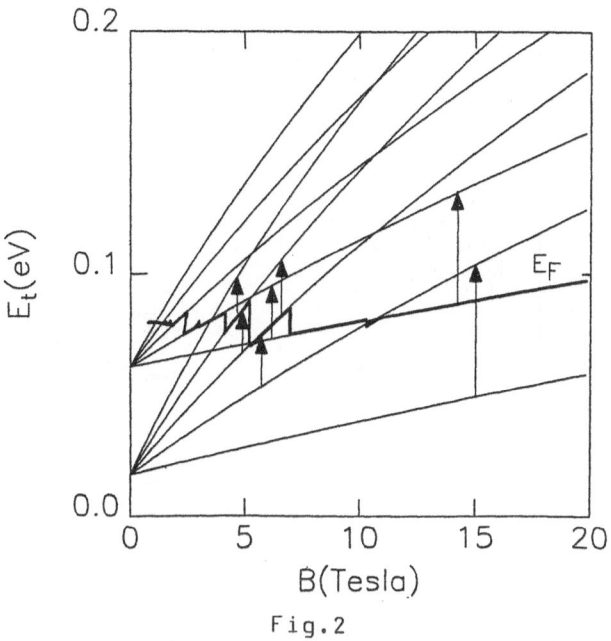

Fig.2

Calculated Landau levels for a 15nm InAs well, and position
of the Fermi energy for a carrier density of $10^{12} cm^{-2}$.
Allowed transitions at different magnetic fields are indicated
by the arrows.

below and above the Fermi energy. Since the degeneracy per
Landau level increases with magnetic field, the Fermi energy
jumps from Landau level to Landau level, and with increasing
field different transitions are involved in cyclotron resonance
as indicated in the figure. The cyclotron masses belonging
to the observable transitions are shown in Fig. 3. The masses
corresponding to the two different subbands show an oscillatory behaviour around the same average value, of $0.032m_o$ in
this example. Note that this value is substantially higher

than the bulk band-edge mass for InAs of $0.023m_0$. In principle
from the oscillatory behaviour of the mass and from its average
value important sample parameters like the location of the
Fermi energy and the subband energies with respect the bottom
of the band can be obtained.

For more complicated potential profiles the Landau levels
can be calculated in the same manner using Bohr-Sommerfeld
quantization [3] with appropriate boundary conditions. As an

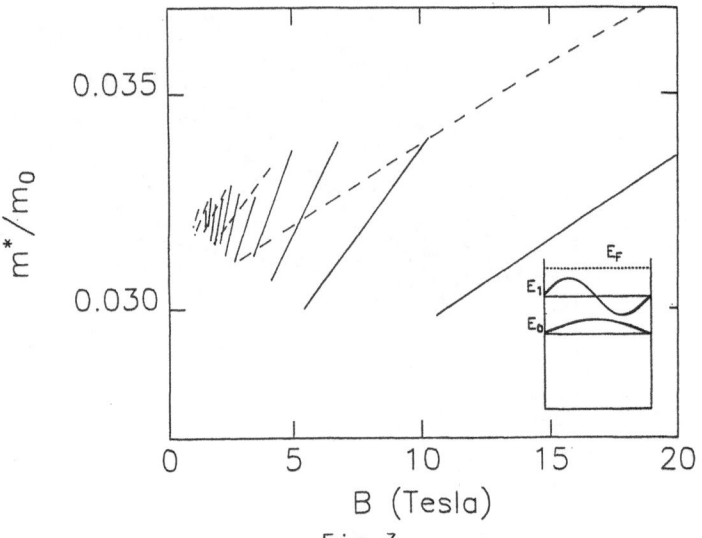

Fig.3

Cyclotron effective mass for a 15nm InAs well with $10^{12}cm^{-2}$
electrons. Transitions from the first subband are shown by the
drawn and those from the second by the dashed lines. The subbands
are 17meV resp. 61 meV above the bottom of the conduction band.

example for a triangular well with potential V=eFz with F the
electric field, the Landau levels can be obtained from the
following trancendental equation. [3]

$$(a+b)a^{1/2}b^{1/2}+(b-a)^{1/2}|n\left|\frac{b^{1/2}-a^{1/2}}{(b-a)^{1/2}}\right| = (\frac{E_G}{2})^{1/2} 2eFh(n+3/4) \quad (9)$$

where $a=E_t-E_{//}$ and $b=E_G+E_t+E_{//}$. The cyclotron effective masses
are calculated from this Landau levels for InAs with the same
carrier density as for the square well and an effective electric
field of $F=2.10^4V/cm$, and are shown in fig. 4 . As in the
square well the masses show an oscillatory behaviour due to

352

the location of the Fermi energy, however an important difference with the square well is that the two different subbands
oscillate around different average effective masses. This effect
is typical of a confining potential which is due to band bending [5,6]. The average effective mass is determined by the
average position of the Fermi energy with respect to the bottom
of the conducting band, weighted by the subband wavefunction

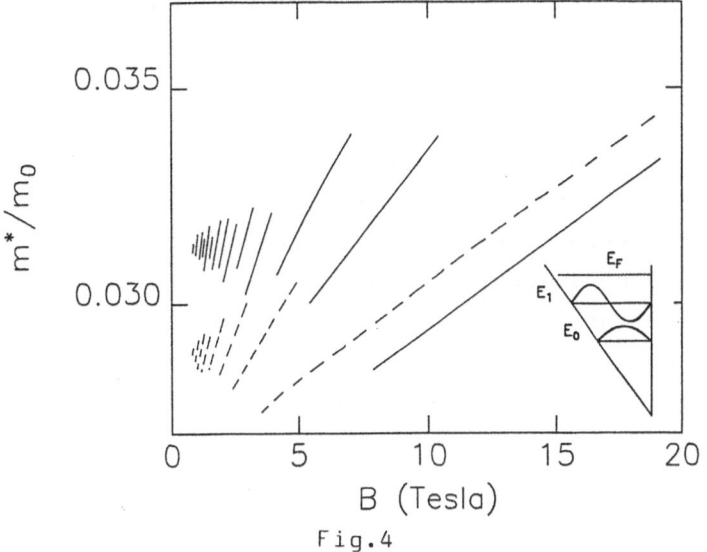

Fig.4

Cyclotron effective masses in a triangular well with a field
of 2.10^4 V/cm and 10^{12} cm^{-2} electrons. Transitions from the first
subband (43 meV) are indicated with the drawn and those from
the second (at 73 meV) by the dashed lines.

squared. For a rectangular well, carriers at the Fermi energy
in the two subbands are equally far from the bottom of the
conduction band and show the same mass. For the triangular
well carriers at the Fermi energy in the second subband are
on average closer to the conduction band than those in the
first subband. Therefore the two subbands show a different
mass, and the higher subbands have a lighter mass (see inset
in fig. 3 and 4). This different behaviour of the masses can
in principle be used to distinguish between wells of different
potential shape. [5,6]

II. c The cyclotron resonance linewidth

The problem of the cyclotron resonance linewidth in two
dimensional systems is neither theoretically nor experimentally clearly solved. Therefore here only a few important aspects
of the problem will be mentioned. The reason for this state
of matters is that the linewidth is a very complicated problem
and that depending on the experimental circumstances a different and even apparently contradictory behaviour can be observed, and indeed is observed. This situation is very similar
to that encountered for cyclotron resonance in bulk materials,
where also no general agreement exists between theory and experiments.

A first important remark is that electron-electron interaction alone can not influence the lineshape [1], for the same
reason that it cannot influence the resonance position, as
was mentioned before. Therefore the lineshape is always determined by the interaction of the electrons with the imperfections, be it impurities, interface roughness or others. The
effect of electron-electron interaction on the lineshape is
however present indirectly through the influence it has on
the way the electronic system interacts with impurities (Screening for instance). The second important point is that the
value of the magnetic field determines on one hand the cyclotron radius and on the other hand the filling factor of a Landau
level. Both these aspects affect the cyclotron resonance lineshape as a function of the magnetic field in a different manner. Lastly, it should be borne in mind that the experimentally
observed lineshape is not exclusively determined by the scattering time, which is related to the two previous points, but
also by simple non-parabolicity. [4]. This can easily be seen
from fig. 3 and 4, which shows that the resonance position
varies with Landau level occupancy, and if the separate peaks
are experimentally not well resolved this can easily give rise
to an apparently oscillatory linewidth.

For scatterers with a range a which is much less than
the magnetic length l Ando [7] has shown that the Landau level
width is given by:

$$\Gamma = \left(\frac{2}{\pi} \hbar \omega_c \frac{\hbar}{\tau} \right)^{1/2} \equiv \left(\frac{B}{\mu_o} \right)^{1/2} \tag{10}$$

which can therefore directly be related to the zero field
mobility and increases proportional to B. This behaviour is
quite generally observed in relatively speaking low mobility
samples (short range scatterers) like for instance Si-MOS
[8,9]. The broadening in this case may be seen as lifetime
broadening because the condition a/l<< 1 implies that each
carrier within one cyclotron orbit (length 2πl) may encounter
an impurity. (fig. 5b). It is important to note that since the
linewidth is related to the zero field dc mobility in this region, also its temperature dependence follows that of the mobility [9].

As the magnetic field is increased, l decreases and Ando
[7] has shown that as l≈a the level width depends on the Landau
level quantum number N. In a swept magnetic field experiment,
for a fixed number of carriers, N changes as the Fermi energy
jumps. Therefore different Landau levels with a different width
are involved in cyclotron resonance and this affects the cyclotron resonance lineshape. If this effect occurs within the width
of the line an oscillatory structure, which ressembles Shubnikov de Haas oscillations is observed superimposed on the absorption line. This effect too has been observed in several 2D

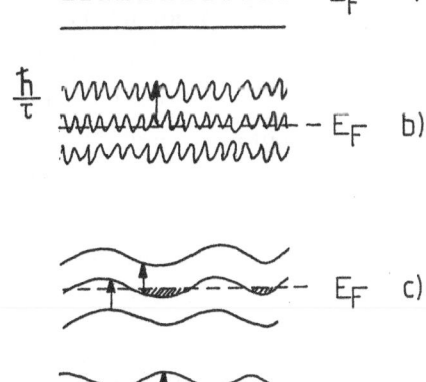

CYCLOTRON ORBIT CENTER

Fig.5
Schematic representation
of broadened Landau levels.
a) No broadening,(ideal case)
b) Short range scattering
l≫a. (Lifetime broadening)
c) Very long rang scatterers.
l≪a.
d) Medium range scatterers
l≈a.

systems with a moderate mobility, at least compared to the best
modulation doped heterostructures. [8,9.10]

As the size of the cyclotron orbit becomes even smaller
with respect to the range of the impurities (higher sample quality or higher fields) the lifetime broadening becomes less
relevant and rather an inhomogeneous broadening is important.
To understand this we add to the effective hamiltonian for the
in-plane motion (eq. 3) some fictitious one dimensional impurity potential $V(x)$. As before, $x'=x+hk_y eB$ is the position of
the cyclotron orbit center with respect to the potential. If
$V(x)=0$ all orbit centers have the same energy (the Landau level
degeneracy) and transitions from N to N+1 for the same k_y are
exactly at $h\omega_c$. [5a]. If $V(x)$ is weak and varying slowly on
the scale of the cyclotron orbit, it can be treated as a perturbation and the eigenenergies of (3) are the normal Landau levels
added to the potential; $E_N=(N+\frac{1}{2})h\omega_c + <V(x)>$. The effect of the
perturbing potential is that the Landau degeneracy is lifted.
In this case dc carrier transport is affected because this takes
place at the same energy within the same Landau level which
now has developed a dispersion (see fig. 5c). Transitions between Landau levels as before have the same energy difference
and the cyclotron resonance line would still be very sharp. If
the potential is stronger and varying rapidly on the scale of
the cyclotron orbit its effect will be to couple different
Landau levels and in general these will not be equidistant anymore which will affect the lineshape for transitions. (fig.5d).
In this latter case the width will depend of course on the
strength and the range of the perturbing potential (the nature
and the position of the impurities)[17], furthermore these in
turn will depend also on the way the electrons screen the impurity. Screening of an impurity means that electrons have to
rearrange themselves in real space as to make that at a certain
distance from the impurity its potential becomes negligible.
However in a magnetic field this means that empty orbit centers
($\hbar k_y/eB$) have to be available. The number of orbit centers per
unit area is just the Landau level degeneracy. Therefore if
a Landau level is full (integer filling factor) it cannot screen
the impurities, and a broad resonance line is expected. Screening will be most effective for half full Landau levels where
the linewidth is the narrowest. This behaviour has indeed been
observed. [12]. It is important to note that the linewidth

here is not related to the dc mobility anymore. This is demonstrated in particular by the very different temperature dependence of the two. For instance it has indeed been observed that
at a field where the linewidth is maximal (integer filling factor) the linewidth becomes narrower with increasing temperature
[12]. This is because screening becomes more effective as through
thermal excitation higher Landau levels become populated and
no completely full Landau levels exist anymore. At the same
time, the dc mobility always decreases with increasing temperature, which would correspond to a broadening of the line,
contrary to the observations.

III. Interband magneto-optics in quantum wells

III. a Excitonic effects

In an interband optical experiment the valence to conduction band absorption is measured as a function of energy and
magnetic field. For simple parabolic bands with conduction band
mass m_e and valence band mass m_n the transitions take place between valence and conduction band Landau levels with the same
index N and the energy difference between the two levels with
index N is given by:

$$\Delta E_N = (N + \frac{1}{2}) \frac{heB}{\mu} + E_G \qquad (11a)$$

where μ is the reduced mass given by:

$$\frac{1}{\mu} = (\frac{1}{m_e} + \frac{1}{m_h}) \qquad (11b)$$

However in an absorption experiment an electron is removed
from the full valence band, leaving a hole behind, and put into an empty conduction band. This negatively charged electron
and positively charged hole are bound together to form an exciton. For simple free carrier like bands the exciton is like
a hydrogen atom which has a binding energy (the Rydberg energy)
and a Bohr radius a_B which in a material with dielectric constant ϵ are given by:

$$Ry = \frac{\mu e^4}{2(h\epsilon)^2} \qquad a_B = \frac{h^2 \epsilon}{\mu e^2} \qquad (12)$$

The experimentally observed absorption is therefore not that
between free electron and hole states but rather between excitonic states. For GaAs the Rydberg energy of the exciton is
~4. meV and the Bohr radius ~140Å. This Bohr radius is of the
same order of magnitude as typical quantum wells, and there-

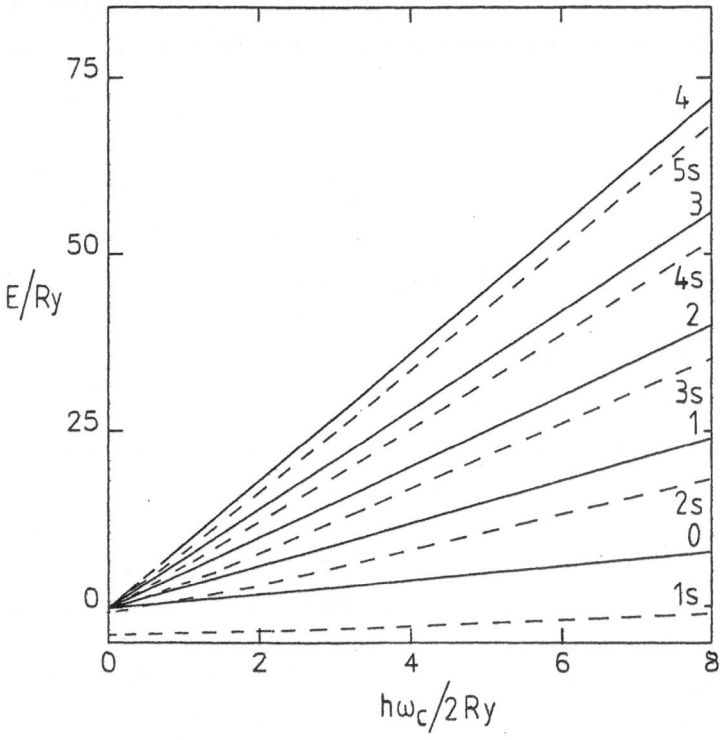

Fig.6
Magnetic field dependence of the s states of a hydrogenic two
dimensional exciton (dashed), and free interband electron hole
Landau levels. The energies are normalized to the three dimensional Rydberg energy.

fore the exciton binding energies are dependent on the well
thickness. For very thin wells an almost 2D like and for thick
wells a 3D like behaviour can be expected. The binding energies
of the s-states (which are the ones that are observed optically
in an interband experiment) in the two extreme cases (3D and
2D) at zero magnetic field are given by:

$$E_r = \frac{R_y}{(r+1)^2} \ (3D) \ \text{and} \ E_r = \frac{R_y}{(2r+\frac{1}{2})^2} \ (2D) \tag{13}$$

$$r = 0,1,2...$$

It can be seen that at zero field the purely 2D exciton has a four times higher binding energy than the 3D exciton, and that the higher excited states of the 2D exciton approach the continuum more quickly than in the 3D case. In real quantum wells the exciton is neither purely 2D nor 3D but has a binding energy somewhere between these extremal values depending on the thickness of the well compared to the exciton Bohr radius. The thickness dependence of the exciton binding energy has been extensively studied both experimentally and theoretically at zero magnetic field [13-16]. In order to explain the behaviour in a field the purely two dimensional exciton will be used as an example. In this case the Schrodinger equation for the 2D exciton in a magnetic field is given by:

$$\left[-\frac{\partial^2}{\partial \rho} - \frac{1}{\rho}\frac{\partial}{\partial \rho} - \frac{2}{\rho} + \frac{\gamma^2}{4}\rho^2 \right] \varphi_r(\rho) = E_r \varphi_r(\rho) \tag{14}$$

where ρ is the relative coordinates for the movement of the electron hole with respect to each other in units of the Bohr radius, γ is $\hbar\omega_c / 2R_y$, and E_r the eigenvalues given by(13) at B=o. Contrary to the three dimensional case where the problem can only be treated approximately under certain conditions the 2D case can be solved exactly, numerically. In fig. 6 the energy levels are shown and compared with the free reduced mass electron and hole Landau levels. It can be seen that at low fields ($\gamma<1$) the exciton character dominates while at higher fields ($\gamma>>1$) every hydrogenic level with quantum number r approaches the free Landau level with quantum numer N. Furthermore, the higher the quantum number the lower the field needed to recover the free carrier like behaviour. To be more concrete, for the case of GaAs $\gamma=1$ corresponds roughly to a field of 5T.

It is clear from fig. 6 that in principle the exciton binding energy can be obtained from interband magnetooptical experiments by careful analysis of the different behaviour as a function of magnetic field of the different transitions. In particular it is interesting to study samples with different well thicknesses compared to the Bohr radius. There are two types of experiments that address this question with this tech-

nique. In one case the field dependence of the ground state
(r=o) is studied (the diamagnetic shift). This field dependence
becomes weaker for thinner wells and with approximate theoretical
descriptions the binding energy can be evaluated from the observations. The accuracy of this technique depends strongly on
the theoretical model used to analyze the data. The other type
of analysis consists in studying the different extrapolation
to zero field of transitions with high r or N and that of the

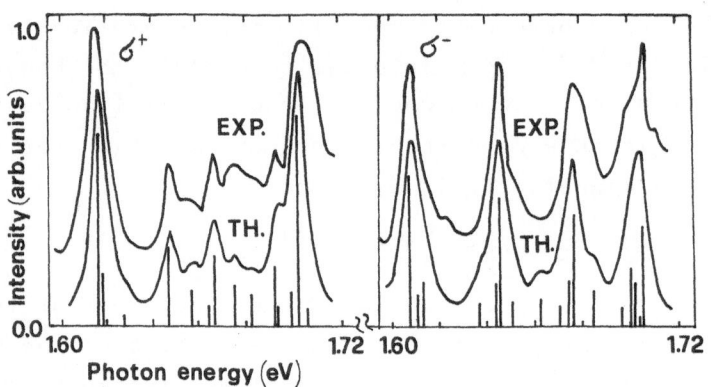

Fig.7

Measured (EXP) and calculated excitation spectra at 19T of a
8 nm GaAs quantum well and calculated transition intensities
(TH) for two circular polarizations. In order to obtain the
theoretical curves each of the indicated intensity bars has
been dressed with a Lorentzian profile 7meV. wide.
(After ref. 17)

ground state (See fig. 6). This type of study needs in principle
no theoretical description but its accuracy is strongly determined by the sample quality. Only in very good samples discrete
transitions can be observed at very low fields and this is of
course crucial for the precision.

III. b The complex valence band

     In the previous section the aspect of the excitonic nature
of interband transitions in a magnetic field has been discussed.
For the sake of simplicity a simple parabolic hole dispersion

relations has been used. However in real materials like GaAs the valence cannot realistically be described by a parabolic band. In quantum wells this approximation of a parabolic band is even less realistic than in bulk materials because the com-

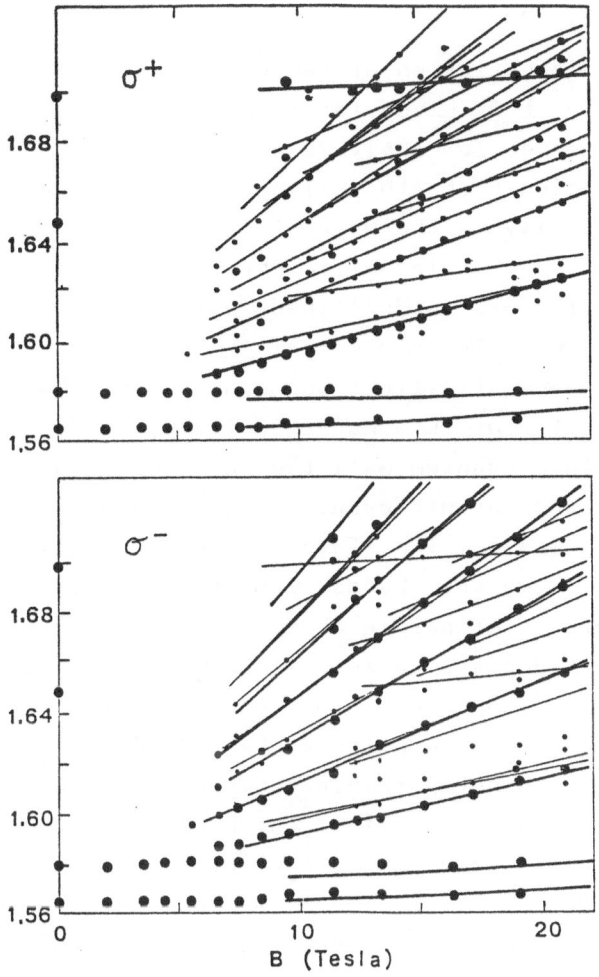

Fig. 8a and b.
Comparison between the calculated (lines) and the experimental transition energies (dots). Excitonic effects are included in a semiemperical way. (After ref. 17)

plexities of the valence band structure manifest themselves more strongly in these systems. This effect is clearly visible in the complex spectra of magnetooptical interband experiments. Therefore such experiments are also a good way to study this complex valence band, because many transitions can be observed yielding detailed data. This point is illustrated in fig.7 where two experimental spectra of interband transitions in a GaAs quantum well, for different circular polarizations are shown [17]. In addition a plot of transition energy versus magnetic field is shown in fig. 8a and 8b. It is evident that the spectra do not resemble the simple interband Landau level ladder shown in fig. 6.

A detailed description of the valence band structure and how to calculate it is given for example in ref. 18. Here recent results of a theoretical analysis of the complex spectra including the transition strength will be given. This part of this paper relies heavily on the work of Ancilotto et al. [17]. In the present context, it is sufficient to know that the bulk valence band structure is described in terms of the four $J=3/2$ valence bands, which in bulk materials correspond at zero field to the (spin degenerate) light and a heavy hole bands. At zero in-plane wavevector or zero magnetic field, these bands are decoupled and in a quantum well they give rise to heavy and light hole subbands at different energies as a consequence of their different masses. Therefore in a perpendicular magnetic field, there are two sets Landau levels for interband transitions; these are in principle observed as two excitonic transitions that extrapolate to zero field to two different energies However at finite magnetic field these two sets of Landau levels are coupled and a very complicated Landau level ladder results. In particular it is not possible anymore to associate a specific Landau level to a specific light or heavy hole subband, because all levels are of mixed character. In fact, the hamiltonian describing the valence and the conduction band is written as a 6x6 matrix acting on a basis of the six bands involved; namely the conduction band (spin up and spin down) and the four $J=3/2$ valence bands. This six component wavefunction in a magnetic field parallel to the growth direction can be written as:

$$F_N = \sum_{j=1}^{6} u_j c_j(z) \, \psi_N^j, \quad N = -2,-1,0\ldots \qquad (15a)$$

with:

$$u_1 = |S\uparrow\rangle$$

$$u_2 = |3/2, \ 3/2\rangle = |\ \tfrac{1}{\sqrt{2}} \ (X+iY)\uparrow\rangle$$

$$u_3 = |3/2, -1/2\rangle = |- \tfrac{1}{\sqrt{6}} \ (X-iY)\uparrow - \sqrt{\tfrac{2}{3}} \ Z\uparrow\rangle$$

$$u_4 = |S\downarrow\rangle \qquad\qquad (15b)$$

$$u_5 = |3/2, \ 1/2\rangle = |-\sqrt{\tfrac{2}{3}} \ Z\uparrow + \tfrac{1}{\sqrt{6}} \ (X+iY)\downarrow\rangle$$

$$u_6 = |3/2, -3/2\rangle = |- \tfrac{1}{\sqrt{2}} \ (X-iY)\downarrow\rangle$$

and ψ_N is the j-th component of the vector:

$$\psi_N = (\varphi_N, \ \varphi_{N-1}, \ \varphi_{N+1}, \ \varphi_{N+1}, \ \varphi_N^1 \ \varphi_{N+2})$$

u_j being the basis set of the periodic part of the wavefunction, φ_N harmonic oscillator wavefunctions, and c_j (z) are the envelope functions that are determined by the boundary conditions
requiring the continuity of the total wavefunction and the current operator at the interface. These coefficients depend on
the magnetic field and describe the degree of mixing between
the bands of different character. The intensity of an interband transition is proportional to the squared dipole matrix
element $M_{NN'}$:

$$|M_{NN'}|^2 = |\Sigma_{ij} \langle \mu_i | p \cdot \epsilon | \mu_j \rangle \int_{-\infty}^{+\infty} c_i(z) c_j(z) \int d^2\rho \, \varphi_N^i \, \varphi_{N'}^j |^2 \qquad (16)$$

By inspection it can be seen that the selection rule for interband transitions with circular polarized light is $\Delta N = \pm 1$ and
that the momentum matrix element $\langle u_j | p \cdot \epsilon | u_j \rangle$ involving the
$|J=3/2, M_J=1/2\rangle$ heavy hole states is $\sqrt{3}$ times stronger than the
one involving the light holes. Finally the strength of an observed transition is determined by the envelope functions of
the initial and final states, the intensity of which can only
be obtained after calculation. Therefore it is crucial that
not only the transition energies but also their intensities
are calculated as a function of the magnetic field in order
to allow a comparison with the experiments.

Fig. 9 shows an example of the calculated Landau levels
in the six band model to illustrate the mixing between the
different subbands and the resulting complex field dependence
of the holes and the simpler conduction band Landau levels.

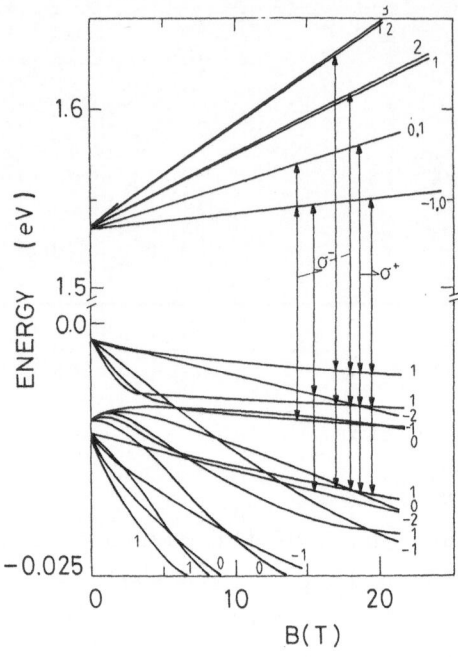

Fig.9

Calculated hole (bottom) and electron (top) Landau levels for a 12.5 nm GaAs quantum well. Indicated are allowed transitions for the two circular polarizations.

The $\Delta N = \pm 1$ allowed transitions for the different polarizations are also indicated. A comparison with the observed transitions which takes into account also the transition strength is shown in fig. 8a and b. In this comparison the effect of the exciton is included in a semiemperical way. In fig. 7 the calculated spectrum for two different polarizations is shown. To allow comparison with the experimental spectra the theoretical intensities are dressed with a Lorentzian broadening. This figure makes clear that a remarkable accurate description of the experimental results can be obtained, and is a clear demonstration of the usefulness of interband magneto-optical absorption in the analysis of the complex valence band in quantum wells and of the power of the envelope wavefunction theoretical calculation. It is important to note that it was essential to calculate the transition matrix elements as done by Ancilotto et al. [17] because many features of the spectra could not be explained without such a calculation. The $\Delta N = \pm 1$ selection rule does not distinguish between the parity of the different states and allows many more transitions that are actually very weak if the full calculation is performed. Furthermore as heavy and light hole like Landau levels cross (Fig. 9) the heavy hole like Landau level which gives the strongest intensity acquires a light hole like slope, although it still remains the strongest peak. This effect leads to a change of slope in the field dependence of the strongest transitions which could not be guessed from the calculated level structure alone, and which has lead to an overestimation of the exciton binding energy from the analysis of magnetooptical data [19].

It is worthwhile mentioning however that it is not yet possible to describe the results of magnetooptical experiments theoretically completely, because there is no theory which takes into account simultaneously the effect of the magnetic field, and that of the Coulomb interaction of electron and holes, with the real bandstructure.

IV. <u>Magnetic quantization in superlattices in a magnetic field parallel to the layers</u>

A superlattice is a layered periodic structure of alternating materials. For very thick or very high barriers, there is no coupling between successive wells and the system is just a set of isolated quantum wells, with properties of two dimensional systems as discussed in II and III. For thin barriers, the coupling between the wells leads to a dispersion relation of a finite width in the superlattice direction (minibands) separated by gaps (minigaps), as opposed to quantum wells which have flat subbands. To behave as a real superlattice the penetration depth of the wavefunction in the barriers must be much less than the periodicity. This length is characterized by $(2m^*V_b/h^2)^{1/2}$ with m^* effective mass and V_b the barrier height. For GaAs/GaAlAs superlattices it can be found that layer thicknesses <10nm are needed to see "real superlattice effects". For the GaAs/GaAlAs conduction band where the potential is simply rectangular, the bandstructure can be obtained from the solution of the Kronig Penney model. Fig. 10 shows how the original bandstructure is changed by the superlattice periodicity, and how the minibands and gaps are formed.

The central theme of this chapter will be the discussion of an interesting aspect of this new solid namely the quantization of its bandstructure by a magnetic field parallel to the layers. In a magnetic field, charged carriers describe circu-

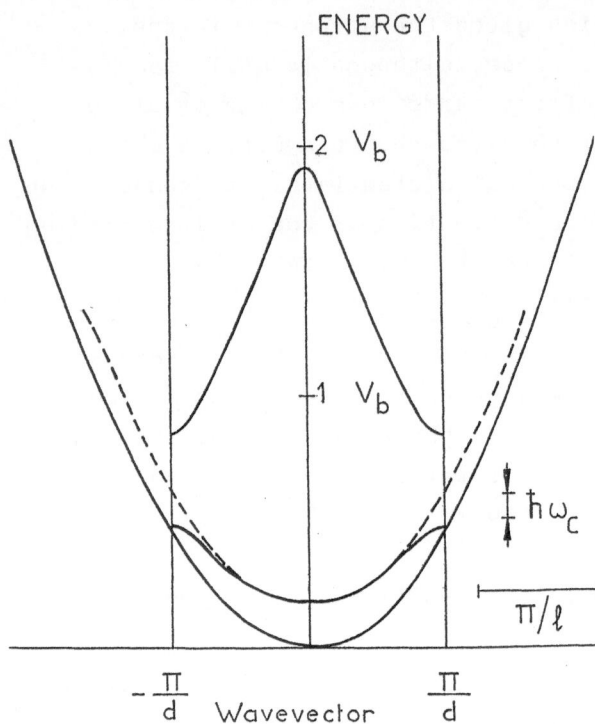

ENERGY

$-2\ V_b$

$-1\ V_b$

$\hbar\omega_c$

π/ℓ

$-\dfrac{\pi}{d}$ Wavevector $\dfrac{\pi}{d}$

Fig.10
Schematic diagram of
the minibands in a su-
perlattice and the ori-
ginal free electron
dispersion relation.

lar orbits with a cyclotron orbit radius R_N due to the Lorentz
force. This cyclotron orbit radius depends on the magnetic field
B and Landau quantum number N as:

$$R_N^2 = (2N+1)1^2 \qquad\qquad (17)$$

with 1 as before the magnetic length. At a field of 20T this
length is~5nm which is comparable to the superlattice periodicity and in a naive manner several type of orbits can be ima-
gined, as is shown schematically in fig.11. This is a rather
unique situation because for normal solids with a lattice period
of typically 0.5 nm such conditions could only be attained with
unrealistically high magnetic fields (of the order of 2000 T).
As can be guessed from fig. 12 experiments under these conditions are a way to study transport through the barriers (now
known as "vertical transport") which has become recently an
important research topic.

Taking the magnetic field in the z-direction and the superlattice potential to be V(x), (a rectangular, Kronig-Penney
type of potential) the hamiltonian of the system is given by
eq. 3 with V(x) added. This is then the same equation as that
discussed in the context of the cyclotron resonance linewidth.

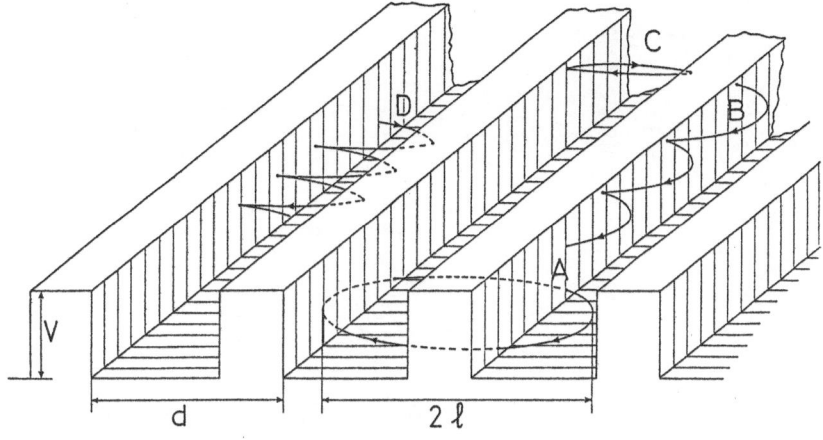

Fig.11

Schematic representation of different classical orbits in a superlattice with
a magnetic field parallel to the layers. Depending on the energy tunnelling
(A),skipping(B),lenslike(C) and doubly reflected orbits can be expected.

In the same way as discussed there, the effect of $V(x)$ is that
the energy becomes dependent on $\hbar k_y/eB$, the cyclotron orbit
center coordinate. Here $V(x)$ is a strong potential (the barrier
heigth) and its effect will be a mixing of the Landau levels
leading to a complicated level structure as a function of the
center coordinate. However since $V(x)$ is periodic, the solutions
will also be periodic in $\hbar k_y/eB$. To clarify this somewhat abstract discussion Fig. 12 shows the solutions of eq.3 in the
presence of a single potential step of height V_b. The results
can be understood in the following way: far away from the step,
the potential is constant and normal equidistant degenerate
Landau levels are obtained at energy E_N on the left side and
$E_N + \langle V_b \rangle$ on the right side (similar to fig. 5c). In the neigh-
bourhood of the interface the Landau levels develop a disper-
sion and are not equidistant anymore (similar to fig. 5d). It
is instructive to see that it is indeed at the length scale
of the cyclotron radius that deviations from the unperturbed
behaviour occur. The Larmor radius of a normal cyclotron orbit
with Landau level quantum number N being given by eq. 17, it
can be seen from the figure that indeed each Landau level N
develops a dispersion at a position $\sqrt{(2N+1)}.l$ from the barrier.
This observation corresponds to the intuitive classical image,
namely, that the energy levels will change "when the cyclotron

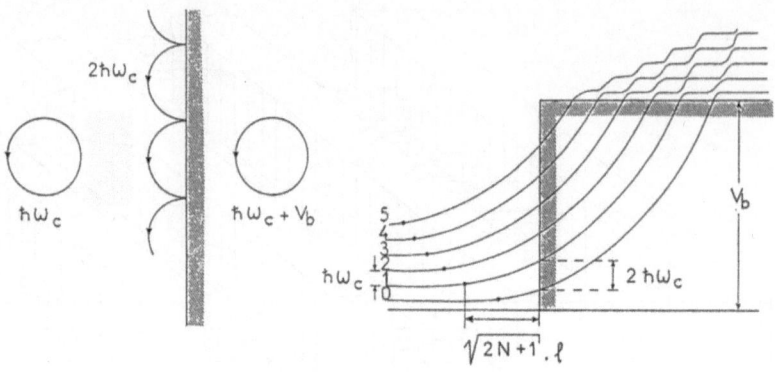

CYCLOTRON ORBIT CENTER
Fig.12
The magnetic energy levels at a potential step in a field of 20 T, and quasi-classical motion of an electron (left). Note that Landau level N develops at dispersion when its orbit center is at $(2N+1)^{1/2}\ell$ from the barrier.

orbit touches the barrier". One can carry this picture even
further as is shown in the left part of the figure. As long
as the cyclotron orbit radius is less than the distance between
the center coordinate and the barrier, normal circular orbits
can be expected while near the barrier carriers are reflected.
In the particular case that the center coordinate coincides
with the barrier, in one cyclotron period ω_c, the electron is
reflected twice to the barrier, and the spacing between Landau
levels is $2\hbar\omega_c$. The magnetic levels which "feel" the barrier,
are known in the literature as magnetic surface states and have
been both experimentally and theoretically studied. One final
aspect of fig. 12 should be mentioned which has no simple classical analogue, namely that the Landau levels "at the top of
the barriers" to the right, tend to intersect the continuation
of the surface states which come from the left. Where these
levels come close in energy they interact, leading to small
gaps between the two sets of Landau levels. I.e. the system
behaves as if two types of Landau levels were superimposed which
interact with each other at their intersections.

The superlattice Landau levels are now calculated as described before for several barrier thickness and the results
are shown in fig. 13. Going from the single interface to the
single well (fig. 13 top left and right) one observes that the

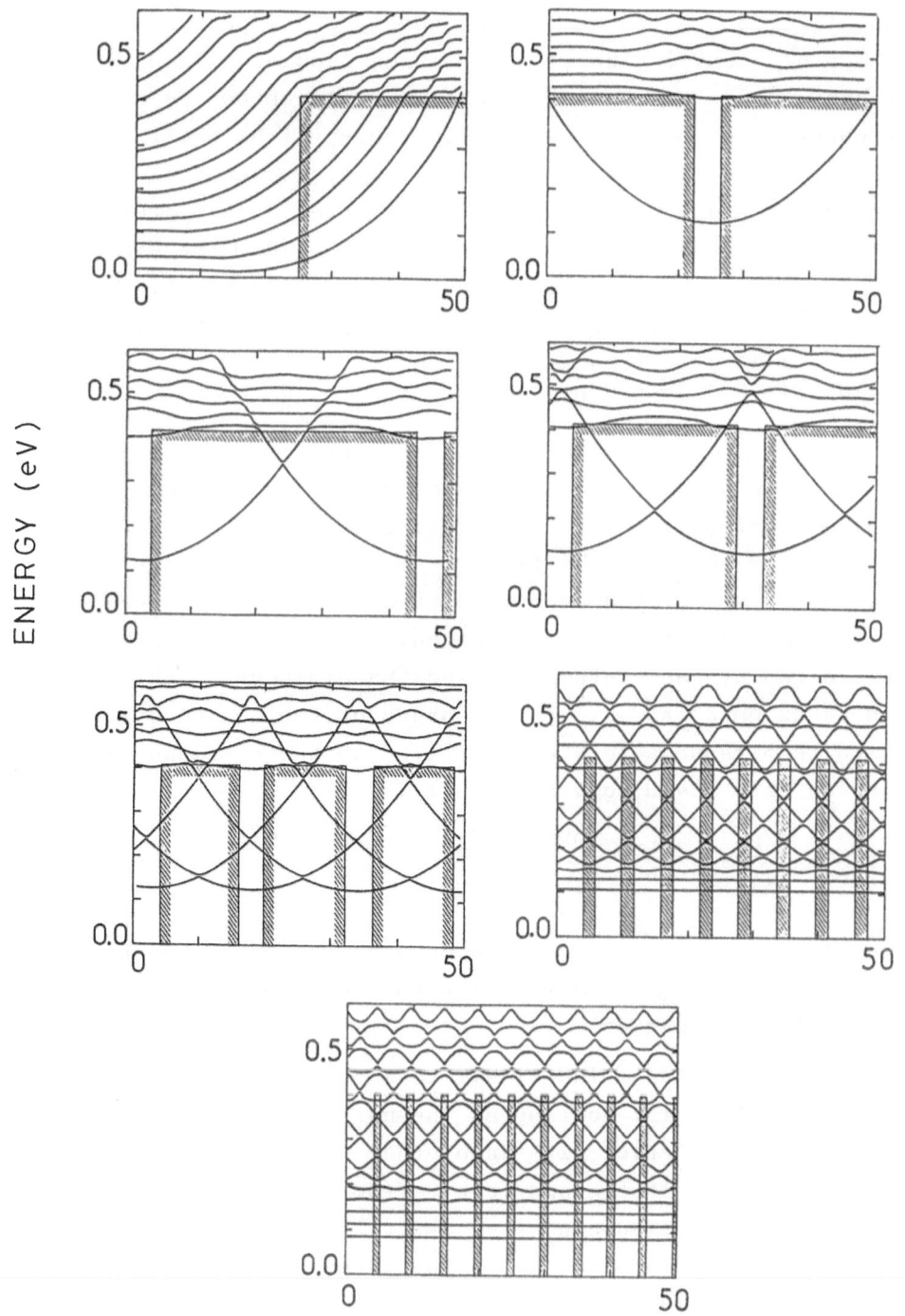

CYCLOTRON ORBIT CENTER (nm)

Fig.13

The magnetic energy levels at a field of 20T for various potentials as indicated in the figure. Note the development to flat Landau levels as the barriers become thinner.

size quantization shifts the lowest energy level upwards and
that the orbit center dispersion relation becomes parabolic-like.
The origin of this parabolic dispersion relation can be most
easily understood when the confinement energy (the shift of
the subband edge) is much larger than the cyclotron energy.
In this case the magnetic field can be treated as a perturbation. Following Ando [20] and Beinvogl et al. [21] the eigenvalues of eq. 3 with V(x) the potential of a single well can
be obtained in first order perturbation and are given by:

$$E = E_n + \frac{e^2 B^2}{2m^*} (\langle x^2 \rangle_n - \langle x \rangle_n^2) + \frac{1}{2m^*} (\hbar k_y + eB \langle x \rangle_n)^2 \tag{18}$$

where $\langle x \rangle$ and $\langle x^2 \rangle$ are the expectation values of x and x^2 for
the unperturbed wavefunctions of subband n. For a symmetric
well these wavefunctions are symmetric or antisymmetric, therefore $\langle x \rangle$ is always zero. The first term in eq. 18 is just the
confinement energy of the subband n at zero field, the second
term is a rigid shift (which in analogy to the field dependence
of the exciton ground state is called the diamagnetic shift)
which depends on B^2 and $\langle x^2 \rangle$, the spread of the wavefunction.
The third term reduces to $\hbar^2 k_y^2 / 2m^*$ i.e. to the k_y dispersion
relation at zero field, giving finally:

$$E = E_n + \frac{e^2 B^2}{2m^*} \langle z^2 \rangle_n + \frac{\hbar k_y}{eB}^2 \frac{e^2 B^2}{2m^*} \tag{19}$$

which explains the parabolic dependence of the energy on the
center coordinate $\hbar k_y / eB$ in fig. 13, the curvature of the para-
bola being field-dependent.

The evolution from single well to superlattice can now
be followed in fig. 13. As long as the barriers are thick and
the wells do not interact with each other every well has its
own set of magnetic levels and levels from different wells are
just superimposed. As soon as they start to interact, which
without magnetic field shows up as the development of a subband
in k_x with finite width, the parabolas from different wells
show an anticrossing at their intersections. This process however occurs more strongly at energies within the zero field
subband width and is much weaker in the superlattice minigaps.
For strongly interacting wells, this anticrossing is so important that the Landau levels become flat (dispersionless) within
the miniband width. As can be seen from eq. 19 the parabolas

are flatter at lower and steeper at higher magnetic fields. Therefore at lower fields more parabolas from more distant wells interact within the subband width increasing the number of flat Landau levels. With increasing magnetic field parabolas interact at a higher energy, until, at sufficiently high magnetic fields, this energy is outside the width of the miniband. Therefore the energy level separation increases with field and those levels which move out of the subband width become dispersive. The calculation shows that the flat Landau levels in fact behave almost like normal Landau levels in a parabolic band, i.e. as $(N+1/2)\hbar\omega_c$, while levels that for a given N and a given field approach the miniband edge develop a weaker field dependence and become dispersive [22].

Finally, it is interesting to note that this series of pictures can be summarized in the following way: Without potential variation degenerate flat Landau levels are found; when the potential varies on the scale of the CR orbit size the levels become very dispersive and finally, when the potential varies very rapidly and periodically, once again flat Landau levels are found. Somewhat intuitively one can say that, since in the absence of a potential variation Landau levels are flat due to translation symmetry, in the case of a very rapidly periodically varying potential in some form translation symmetry is recovered again.

These effects have indeed experimentally been observed [23]. For instance, in an interband absorption experiment transitions between flat Landau levels (at energies within the miniband width) can be observed as sharp peaks, whereas in the minigaps Landau levels are broadened and eventually are not observed anymore. For illustration fig. 14 shows spectra of interband absorption with a magnetic field perpendicular and parallel to the layers at 19 T; in the upper half of a superlattice and of a single quantum well, in the lower half. The main difference between these results is that in the case of the superlattice in both field configurations a clear, more or less regular, Landau level structure can be observed while for the quantum well this is only the case for the perpendicular field. A notable feature of the $B_{\parallel}$ spectra is that above a certain energy ($\backsim 1.8$eV in the present case) no transitions are observed anymore; as the field is increased transitions shift to higher energy and

the transitions close to this cut-off energy broaden and finally disappear. This cut-off energy is illustrated more clearly
in Fig. 15 where the high energy part of the superlattice spectra
in B$_\parallel$ is shown in more detail. Experimentally an almost linear
field dependence of the energy of the peaks has been found up
to energies close to the cut-off energy.

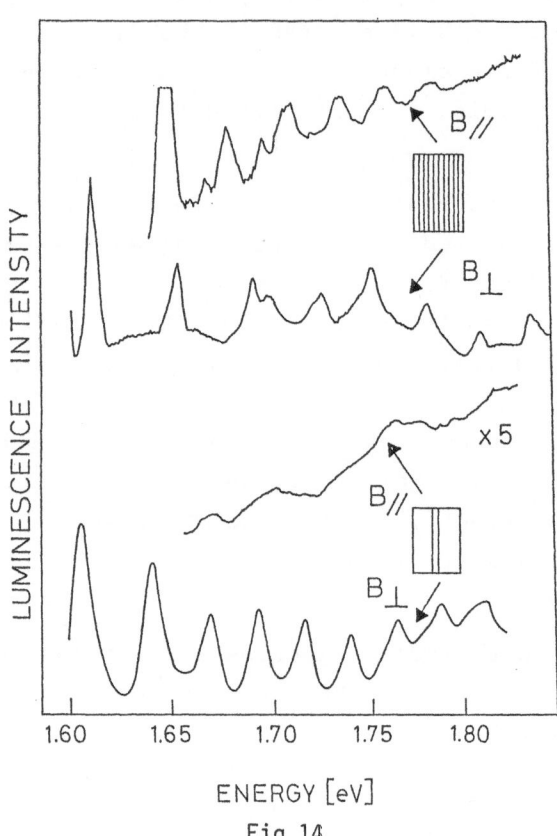

Fig.14

Excitation spectra of a superlattice (top) and a quantum well (bottom) at a fi
of 19T, applied parallel and perpendicular to the layers.

The general behaviour can easily be understood qualitatively from the description of the Landau levels given before. In
the perpendicular configuration for both the superlattice and
the quantum well electrons orbit in the plane of the layers
and the bands are splitted by a field like in a bulk material,

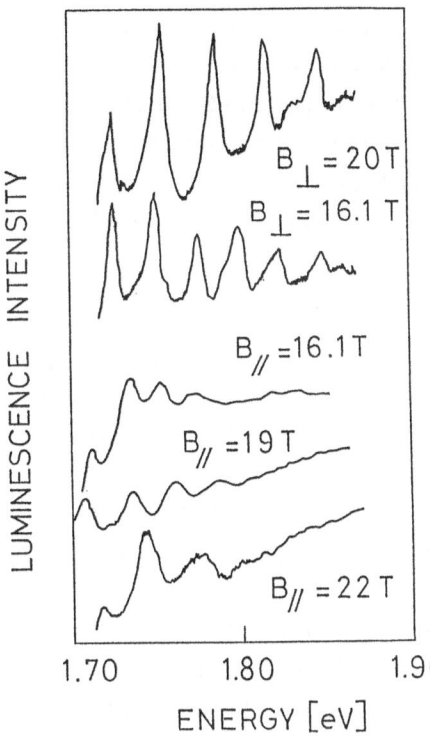

LUMINESCENCE INTENSITY

$B_\perp = 20\,T$

$B_\perp = 16.1\,T$

$B_{/\!/} = 16.1\,T$

$B_{/\!/} = 19\,T$

$B_{/\!/} = 22\,T$

1.70   1.80   1.90

ENERGY [eV]

Fig.15
Excitation spectra of a superlattice
with a field parallel to the layers.
Note the disappearance of observable
peaks as their energy is above 1.8 eV

as indeed is observed. In a parallel field for the quantum well
the orbit radius is larger than the well width, and therefore
no Landau levels can be observed. In fact no sharp Landau level
is formed but a band which experiences a small diamagnetic shift
(fig. 13) as discussed before. In the superlattice flat Landau
levels with a linear field dependence exist in the miniband
and are indeed observed. At the miniband edge these levels are
broadened which renders them unobservable as demonstrated experimentally in fig. 15. The complications arising from excitonic
effects and from the complex valence band structure should be
taken into account here as well as in a more detailed analysis.
At present, these effects are neglected because the main interest is not in these aspects. The results of interband magneto-
optics are in concordance with the cyclotron resonance measurements reported by Duffield et al. [24]. They observed cyclotron
resonance (transitions between the lowest two Landau levels
in their case) both in the perpendicular and in the parallel
field configuration. This behaviour is indeed expected from
the preceding discussion. As long as $\hbar w_c$ is less than the miniband width the Landau levels are flat and their separation is
$\hbar \omega_c$. Therefore cyclotron resonance can be observed as in a

normal bulk semiconductor. For higher magnetic fields this cyclotron resonance should be broadened as the upper level approaches the miniband edge. This limit was not reached in the experiments by Duffield et al. In this respect the technique of interband magnetooptics is more powerful because the full miniband can be scanned with the radiation energy.

References

1) W. Kohn, Phys.Rev.123,1242,(1961)

2) E.Gornik, in "Heterostructures and Semiconductor Superlattices", ed. G. Allan,G.Bastard,N.Boccara and M.Voos,Springer Berlin, (1986)

3) W.Zawadzki, in "Two-Dimensional systems, Heterostructures and Superlattices", ed. G.Bauer,F.Kuchar and H.Heinrich, Springer, Berlin, (1984) and
    W.Zawadaki, J.Phys.C. (Solid State Physics),229,(1983)

4) H.J.A. Bluyssen,J.C.Maan,P.Wyder,L.L.Chang and L.Esaki, Solid State Commun. 31,35,(1979) and ibid, Phys.Rev.B25, 5364,(1982)

5) U.Merkt,M.Horst,T.Evelbauer,J.P.Kotthaus, Phys.Rev.B34, 7234,(1986)

6) Y.Guldner,J.P.Vieren,P.Voisin,M.Voos,J.C.Maan, L.L.Chang and L.Esaki, Solid State Commun. 41,755,(1982)

7) T.Ando, J.Phys.Soc. Jpn.,36,959,(1974) and ibid. J.Phys. Soc. Jpn,38,989,(1975)

8) G.Abstreiter, P.Kneschaurek, J.P. Kotthaus and J.F. Koch, Phys.Rev.Lett.32, 104,(1974)

9) P.Voisin, Y.Guldner, J.P. Vieren, M.Voos, J.C. Maan, P.Delescluse and T.Linh, Physica 117B&1188,634,(1983)

10) D.Heitmann, M.Ziegmann and L.L.Chang, Phys.Rev.B34,7463, (1986)

11) R.Lassnig, in "Two-Dimensional systems, Heterostructures and Superlattices", ed. G.Bauer, F.Kuchar and H.Heinrich, Springer, Berlin, (1984)

12) Th.Englert, J.C.Maan, Ch. Uihlein, D.C.Tsui and A.C.Gossard, Solid State Commun.46,545,(1983)

13) R.C.Millar, D.A.Kleinmann, W.T.Tsang and A.C.Gossard, Phys. Rev.B 24,1134 (1981)

14) G.Bastard, E.E.Mendez, L.L.Chang and L.Esaki, Phys.Rev.B29, 1588, (1982)

15) P.Dawson, K.J.Moore, G.Duggan, H.I.Ralph and C.T.B.Foxon, Phys.Rev.B34,6007,(1986)

16) R.L.Greene and K.K.Bajaj, Solid State Commun. 45,831,(1983)

17) F.Ancilotto, A.Fasolino, and J.C.Maan, Proc. 2nd Int.Conf. on Superlattices Gotenborg, 1986, J. of Microstructures and Superlattices, to be published.

18) M.Altarelli, in "Heterostructures and Semiconductor Super-lattices", ed. G. Allan, G.Bastard, N.Boccara and M.Voos, Springer,Berlin,(1986)

19) J.C.Maan; G.Belle, A.Fasolino, M.Altarelli and K.Ploog, Phys.Rev.B30, 2253,(1984)

20) T.Ando,J.Phys.Soc.Jpn,39,411,(1975)

21) W.Beinvogl, A.Kamsar and J.F.Koch, Phys.Rev.B14,4274,(1986)

22) J.C.Maan, in "Two-Dimensional systems, Heterostructures and Superlattices", ed. G.Bauer, F.Kuchar and H.Heinrich, Springer, Berlin, (1984)

23) G.Belle, J.C.Maan and G.Weimann, Solid State Commun.56,65 (1985)

24) T.Duffield, R.Bhat, M.Koza, D.M.Hwang, P.Grabbe and S.J. Allen Jr.,Phys.Rev.Lett.56,2724,(1986)

BAND-GAP ENGINEERING FOR NEW PHOTONIC AND

ELECTRONIC DEVICES

Federico Capasso

AT&T Bell Laboratories
Murray Hill, New Jersey 07974

INTRODUCTION

In recent years there has been an intense research effort on semiconductor heterojunctions. This field is an excellent example of how basic science and technology interact and influence one another.

Early suggestions for using semiconductor heterojunctions to improve device performance are found in the celebrated transistor patent of Shockley (1). A few years later, Kroemer (2) proposed the concept of a compositionally graded semiconductor. By spatially varying the stoichiometry of a semiconductor, an energy band gap is produced that varies with position (graded band gap). Thus "quasi-electric forces," equal to the spatial gradient of the conduction and valence band edge, respectively, are exerted on the electrons and holes. This concept is one of the earliest and simplest examples of band-gap engineering, which is the spatial tailoring of the band gap or, more generally, the band structure of a semiconductor to achieve new material and device properties (3).

After the initial demonstrations of the homojunction semiconductor injection laser in the early 1960's, Kroemer suggested that carrier confinement in a low-gap region clad by wide-gap heterojunction barriers would make population inversion and laser action possible at much lower current densities (4). The demonstration of a continuous wave (CW) heterojunction laser at 300 K (5) was made possible by the growth of high-quality AlGaAs/GaAs heterojunctions by liquid phase epitaxy and paved the way to high-performance lasers for lightwave communications.

The next breakthrough was the invention of molecular beam epitaxy (MBE) at Bell Laboratories by Cho and Arthur (6). This epitaxial growth technique allows multilayer heterojunction structures to be grown with atomically abrupt interfaces and precisely controlled compositional and doping profiles over distances as short as a few tens of angstroms. Such structures include quantum wells, which are a key building block of band-gap engineering. These potential energy wells are formed by sandwiching an ultrathin lower gap layer (of thickness comparable or smaller than the carrier thermal de Broglie wavelength, which is $\approx 250\,\mathring{A}$ for electrons in GaAs at room temperature) between two wide-gap semiconductors (for example, AlGaAs). The spacing and position of the discrete energy levels in the well depend on the well thickness and depth (6). The well depth is the energy difference between the bottom of the conduction bands (or the valence bands in the case of holes) in the two materials and is also termed band discontinuity. Band discontinuities play a central role in the design of heterojunction devices such as quantum well lasers (7). These lasers have important applications in the area of optical recording.

If many quantum wells are grown on top of one another and the barriers are made so thin (typically <50 Å) that tunneling between the coupled wells becomes important, a superlattice is formed. This concept was first proposed by Esaki and Tsu at IBM in 1969 (8). It has since attracted a tremendous interest, which was spurred by the development and the unique capabilities of MBE and other growth techniques such as metallo-organic chemical vapor deposition (MOCVD) (9). Superlattices are new materials with novel optical and transport properties introduced by the artificial periodicity. Other superlattices with intriguing properties can be obtained by periodically alternating ultrathin $n-$ and p-type layers [nipi (n-type intrinsic p-type intrinsic) superlattices] (10), by periodically grading the composition (sawtooth superlattices) (11), or by alternating undoped layers with doped layers that have a wider band gap (modulation-doped superlattices) (12). Modulation doping in a single heterojunction is the key element of a new high electron mobility transistor (13). Strained layer superlattices have also shown interesting properties and have potential as components of new devices (14).

The rich variety of available combinations of band gaps, semiconductor alloys, and lattice constants is the main feature of band-gap engineering (15,16). A few years ago, I realized that the ability to engineer the band diagram of a semiconductor structure in an almost arbitrary and continuous way opened the door to exciting possibilities for optical detectors and other devices. These are reviewed here.

Solid-State Photomultipliers and Graded Gap Transistors

One device currently in widespread use for light detection is the avalanche photodiode (APD). In an APD, the photoinjected carriers (electrons or holes or both) gain energy from the electric field and undergo ionizing collisions (impact-ionization) with the atoms of the lattice. These collisions create more carriers across the forbidden band gap, which in turn impact-ionize other atoms and so on. The minimum energy required to create an electron-hole pair by an ionizing collision is called the ionization threshold energy. This energy is greater than the band gap for reasons of energy and momentum conservation.

Low-noise operation is important if an APD is to detect the low power levels of signals that emerge from long distances in optical fibers. Avalanche multiplication, however, generates extra noise, which adds to the shot noise of the incident photons. This so-called excess noise arises from the fluctuations of the avalanche gain. Avalanche multiplication occurs in two directions, since holes and electrons travel in opposite directions. If both electrons and holes undergo ionizing collisions at nearly the same rate, the avalanche is equally strong in both directions. Any slight fluctuation (a variation in the number of carriers created) in the electron-multiplication process is immediately fed back and amplified because of hole-initiated impact-ionization. This feedback strongly reinforces the original fluctuation so that the excess noise is very large. If only one type of carrier can ionize, however, the avalanche proceeds in only one direction. As a result, the avalanche-feedback process does not occur and excess noise is reduced.

To limit the excess noise caused by the avalanche, holes and electrons must ionize at vastly different rates. Ideally, one of the ionization rates should be zero, but no materials have that desirable characteristic. In silicon the ratio of these ionization rates (one rate for electrons, α, the other for holes, β) is greater than 20 to 1, so there is little avalanche noise. But silicon is not sensitive at wavelengths that fall in the low-loss, low-dispersion, long-wavelength region of optical fibers (1.3 to 1.6 μm). For this spectral region, no material currently can match the noise performance of silicon at shorter wavelengths; virtually all III-V materials (made from a group III and a group V element, for example, GaAs) have comparable ionization rates for electrons and holes.

Using band-gap engineering, one can artificially tailor the ratio of the ionization coefficients and therefore reduce excess noise. Figure 1 illustrates three structures designed to achieve this goal (17). The superlattice APD (Fig. 1a) alternates layers of high- and low-gap materials and restricts ionizing collisions to the low-gap regions. Carriers accelerate and gain energy but do not ionize in wide-gap regions. On entering the next well a free electron gains enough energy from the conduction-band discontinuity ΔE_c to ionize. However, the valence-band discontinuity ΔE_v is not large enough to supply a similar energy boost to free holes. Thus electrons enter the well with a higher kinetic energy than holes, so that electrons ionize more efficiently than holes if $\Delta E_c > \Delta E_v$. This concept, proposed at the University of Illinois (18), was demonstrated by Capasso and coworkers (19). They obtained α/β of 7 to 8 in an $Al_{0.45}Ga_{0.55}As$ (500 Å)/GaAs (500 Å) multiquantum well APD with 50 periods, which represents an improvement of a factor of 3 to 4 over α/β in GaAs (=2) (19).

A staircase APD (17), on the other hand, uses a superlattice of layers that are graded from low gap to high gap (Fig. 1b). Here, the band gap widens gradually in each layer but narrows abruptly at the layer interface. As in the earlier case, the band-gap difference is most visible in the conduction band, where the discontinuities form steps. As a reverse bias is applied, the sawtooth band diagram becomes a staircase. A free electron drifts toward the right in the graded layer but cannot ionize there, because the field is too low. However, the discontinuity furnishes all the energy an electron needs to ionize, so ionizing collisions occur only at the steps. Because the valence-band steps are of the wrong sign to enhance ionization and the electric field is too small to furnish the energy needed, holes do not ionize in this structure. The staircase APD is the solid-state analog of a photomultiplier, a type of vacuum tube detector with high internal amplification and negligible avalanche noise. As in a photomultiplier, the negligible avalanche noise of the staircase APD arises not only from the lack of feedback by ionizing holes but also from electrons ionizing at well-defined locations (the conduction band steps, which are equivalent to the photomultiplier dynodes) (17). This minimizes the intrinsic randomness of an avalanche. Such a structure can also operate at very low voltages ($\simeq 6$ V for

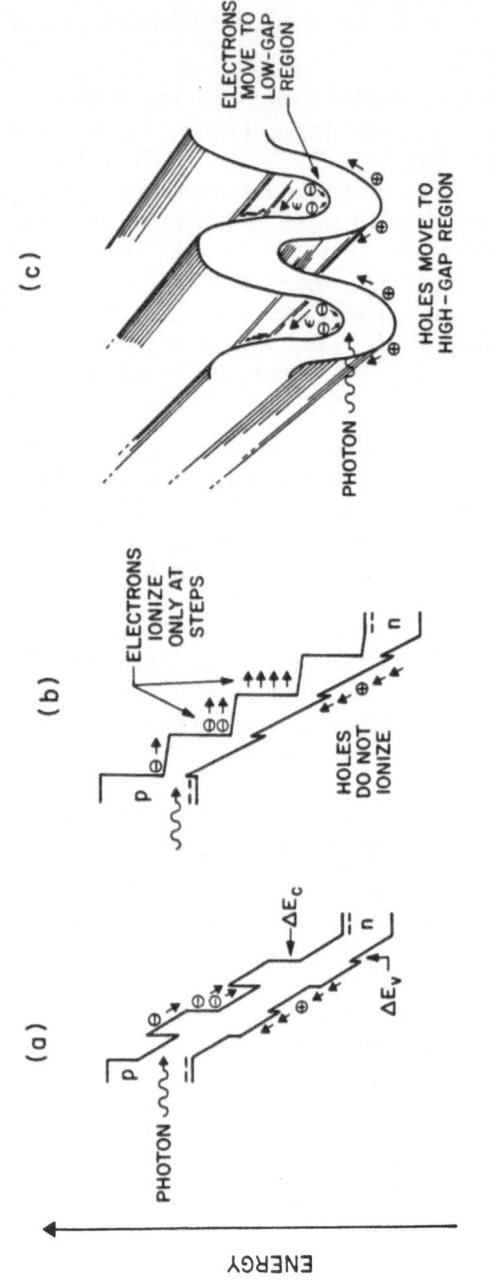

DISTANCE

ENERGY

Fig. 1.   Band-gap engineering applied to avalanche photodiodes (APD's). (a) Multiquantum well APD; (b) solid-state photomultiplier (staircase APD); and (c) channeling APD.

a five-stage device with a gain of $\simeq 32$ and a conduction band discontinuity of $\simeq 1$ eV), a feature not present in conventional avalanche photodiodes. This structure has not been fabricated and represents a challenge for the MBE crystal grower; promising materials are HgCdTe/CdTe and certain III-V alloys.

Another device, the channeling APD (Fig. 1c), consists of alternated p- and n-layers of different band gap, with lateral p- and n-contacts (20). The energy band diagram of this device is three dimensional. Here, the band boundaries resemble two gutters separated by a space, which is the band gap. Channels (formed by the pn junction) run parallel to the layers. A periodic, transverse electric field (perpendicular to the layers) results from the alternated $npnp$ layers, while an external field is applied parallel to the layers. The transverse field collects electrons in the low-gap n-layers, where they are channeled by the parallel electric field, so electrons impact-ionize in the low-gap n-layers. But the transverse field also causes holes created by electrons ionizing in the n-layers to transfer into the wide-gap p-layers before ionizing. Once they are there, holes cannot impact-ionize, because the band gap is too high to permit multiplication. The α/β ratio of this device is therefore very high and the avalanche noise very small. Channeling devices have novel applications in high-energy physics as solid-state drift chambers (21). Channeling devices are based on a new carrier depletion scheme which allows one to achieve total carrier depletion of a large volume of semiconductor material, regardless of the doping level as recently demonstrated (20), (21). Ultra-low capacitance detectors with large sensitive areas can be implemented in this way, (20), (21).

Another avalanche multiplication mechanism in superlattices suitable for a solid-state photomultiplier has been proposed (22,23) and demonstrated (23) (Fig. 2). Hot carriers in the barrier layers collide with carriers confined or dynamically stored in the wells and impact-ionize them out across the band-edge discontinuity. In this ionization process only one type of carrier is created, so that positive feedback of impact-ionizing holes is eliminated. Thus a quiet avalanche with small excess noise as in a photomultiplier is possible. This effect has some conceptual similarities with the impact-ionization of deep levels, in that the quantum well may be treated as an artificial trap.

One application of band-gap engineering to devices other than detectors may be in the development of a faster transistor. Invented at Bell Laboratories in the late 1940's, the bipolar transistor consists of three layers: an emitter, a base, and a collector. A current injected in the base layer controls the flow of electrons from the emitter to the collector layer. In conventional bipolar transistors, the base has a uniform band gap but no electric field. Therefore, electrons traversing the base travel relatively slowly by diffusion.

One way to speed-up electrons is to use a graded-gap material for the base. The gradual change in composition causes the conduction band in the base layer to "tilt" as if an electric field was present (Fig. 3). Electrons drift through the base much faster than in the conventional device. This concept was pioneered by Kroemer (2). Recently, AT&T Bell Laboratories demonstrated this concept in a phototransistor (24) and in a three-terminal bipolar transistor (25) grown by MBE. The graded-gap transistor had an AlGaAs emitter and a GaAs collector. Its p-type base, about 3000 Å thick, changed composition gradually, from the emitter at one end to the collector at the other. The grading gave the base an effective electric field of about 6 kV/cm. In initial tests, the device operated at frequencies up to 4 GHz with a dc current gain greater than 1000. Direct velocity measurements showed that electrons in the

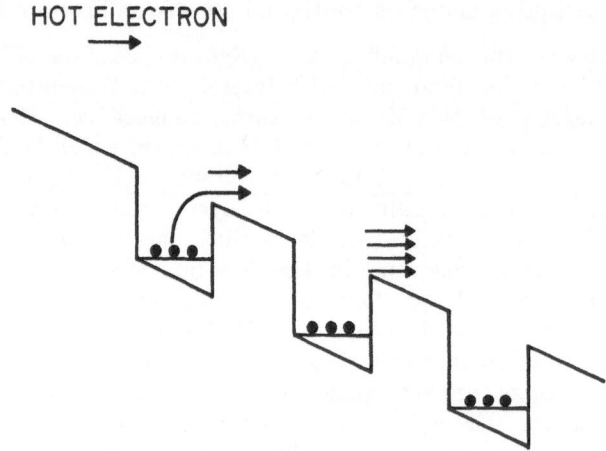

HOT ELECTRON

Fig. 2.   Band diagram of impact-ionization across the band discontinuity of electrons stored in the quantum wells by hot carriers in the barriers. This
phenomenon can also be used to implement a solid-state analog of a
photomultiplier.

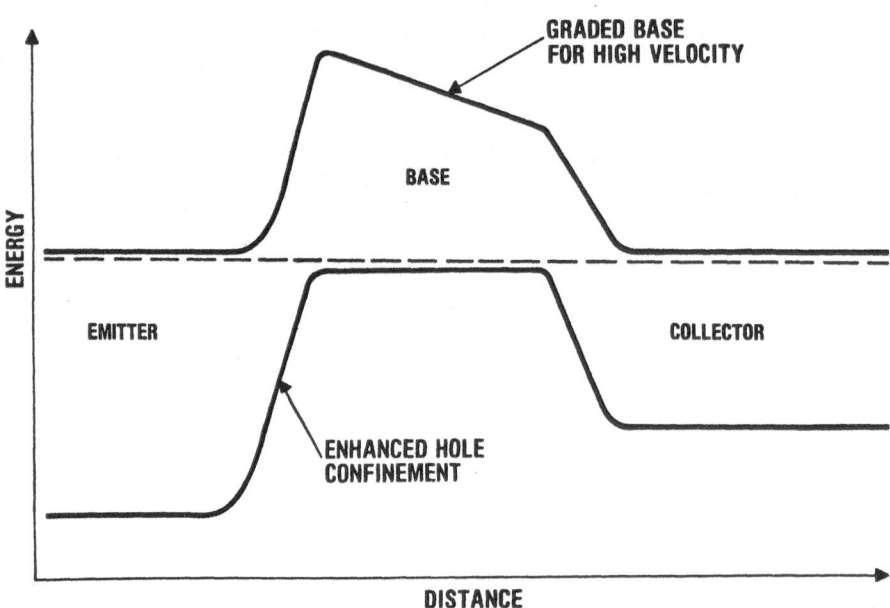

Fig. 3.   Band diagram in equilibrium of graded-gap base
bipolar transistor.

graded base can travel up to ten times as fas as electrons in the base of an ungraded transistor (3). Scientists expect operating frequencies above 50 GHz from these graded transistors. A maximum oscillation frequency of 45 GHz has been recently demonstrated (26).

Tunneling Devices

The formation of quantum resonances in narrow potential wells has opened the door to interesting transport phenomena and device applications. Consider an undoped double barrier (for example, AlAs/GaAs/AlAs grown by MBE) sandwiched between two heavily doped contact layers. Figure 4a shows how tunneling occurs with applied d-c bias. Electrons originate near the Fermi level to the left of the first barrier and tunnel through the well. Resonant tunneling occurs when the energy of the injected carriers becomes equal to the energy of one of the levels in the well. Maxima occur in the overall transmission through the double barrier and in the current-voltage curves (27). This negative differential resistance may be useful in ultrahigh-frequency device applications. Sollner et al. recently reported resonant tunneling double-barrier oscillators operating at frequencies up to 35 GHz (28). The resonant tunneling effect is essentially equivalent to the resonant enhancement of the transmission in an optical Fabry-Perot interferometer, provided that scattering in the double barrier is negligible (that is, transport must be coherent).

Resonant tunneling transistors have been recently proposed. These have an *npn* bipolar transistor structure with a quantum well in the base region rather than a graded layer (29). Figure 4, b and c, shows the band diagrams of two types of resonant tunneling bipolar transistors under operating conditions. The voltage between the base-emitter junction is such that electrons in the emitter resonantly tunnel through one of the energy levels of the well and are collected by the reversed-biased base-collector junction. Only when such resonance conditions are satisfied can electrons injected from the emitter reach the collector and give rise to transistor action. A plot of the collector current versus the base-emitter bias voltage displays a series of peaks corresponding to the quantum levels of the well. Resonant tunneling is accomplished not by applying an electric field to the double barrier but by varying the energy of the incident electrons. Thus the symmetry of the double barrier is maintained, and coherent resonant tunneling should be achieved with near unity transmission at all resonances, as in a Fabry-Perot resonator. In the first structure (Fig. 4b), electrons are ballistically launched into the quantum well by the conduction-band discontinuity at the base-emitter interface. In the second one (Fig. 4c), a parabolic quantum well is placed in the base. Such wells have been recently grown by MBE and exhibited the expected equal spacing of the energy levels (30).

The device in Fig. 4c is the electronic equivalent of a Fabry-Perot interferometer. By appropriately connecting the resonant tunneling transistor to a resistive load and a voltage supply, one can produce a device with N stable states, where N is the number of resonant peaks (29). In this configuration, the device serves as an N-state memory element (with N as high as $\simeq 8$), providing the possibility of extremely high density data storage. Memories of this sort and other circuits based on multiple negative resistance, such as counters, multipliers and dividers, partly bit generators, have been of interest for quite some time. However, since no physical device exhibiting multiple-valued negative differential resistance previously existed, such circuits were possible only with combinations of binary devices. This resulted in complex configurations with reduced density and speed. Because resonant tunneling transistors allow multiple-valued differential resistance in a single physical element, they have tremendous potential. Recently Capasso et al. demonstrated the first resonant tunneling bipolar transistor (31). These devices operate at room temperature and exhibit both negative transconductance and negative conductance in the common emitter configuration. The operating principle and overall structure of this device are similar to those of the resonant tunneling transistors previously proposed (Fig. 4b) (29). Nevertheless, the essential difference is that in the implemented resonant tunneling transistor (31) electrons are injected thermally into the quantum well, rather than quasi-ballistically. This makes the operation of the device much less critical, and allows operation at room temperature.

Tunneling in superlattices is also of considerable physical and practical interest. As pointed out by Capasso et al. (32), a superlattice can act as an effective mass filter (Fig. 5). Since the tunneling probability increases exponentially with decreasing effective mass, electrons are transported through a superlattice more readily than the heavy holes as long as the valence-band discontinuity is not negligible compared to the conduction-band discontinuity. Effective mass filtering is the basis of tunnel-photoconductivity, recently discovered at Bell Laboratories (32,33). In a classical photoconductor, photogenerated electrons and the electrons injected from the ohmic contacts can be viewed as moving through the semiconductor and around the circuit until they recombine with the slowly moving photogenerated holes. This produces a current gain whose value is given by the ratio of electron-hole pair lifetime to the electron transit time (34). In a superlattice, photogenerated heavy holes tend to remain localized in the wells as a result of the negligible tunneling probability. Electron states tend to be extended, because of the small electron effective mass ($\simeq 1/10$ of the heavy hole mass) and the ultrathin barriers. These extended states form a band referred to as a miniband. Thus, photoelectrons will be transported by band-type conduction. Since carrier mobility in the miniband depends exponentially on the superlattice barrier thickness (an effect caused by tunneling), the electron transit time, the photoconductive gain, and the gain-bandwidth product can be artificially tuned over a wide range. This offers a great versatility in device design, which is not available in standard photoconductors. High-performance infrared photoconductors that utilize effective mass filtering were recently demonstrated (32). The MBE-grown devices consisted of 100 periods of AlInAs (35 Å)/GaInAs (35 Å) between two contact layers. These devices responded to wavelengths in the 1.6- to 1.0- μm region and exhibited high current gain (up to 2×10^4) at low voltage (≤ 1 V) with response times in the range of 10^{-4} second. The low bias operation reduces device noise to only $\simeq 3 \times 10^{-14}$ W/Hz$^{1/2}$.

In superlattices with relatively thick barriers (≥ 100 Å) the electron states also become localized, and one can observe sequential resonant tunneling (Fig. 6) (35). Electrons tunnel from the ground state of a given well into the excited states of the adjacent well. This is followed by intrawell energy relaxation (mostly nonradiative) to the ground state. This process is repeated many times to produce a typical cascade process through the whole superlattice (Fig. 6a). The observation of this effect (35) has led to many exciting device possibilities. One is a solid-state infrared laser which emits radiation in the range from 5 to 250 μm (depending on the well thickness), as originally proposed by Kazarinov

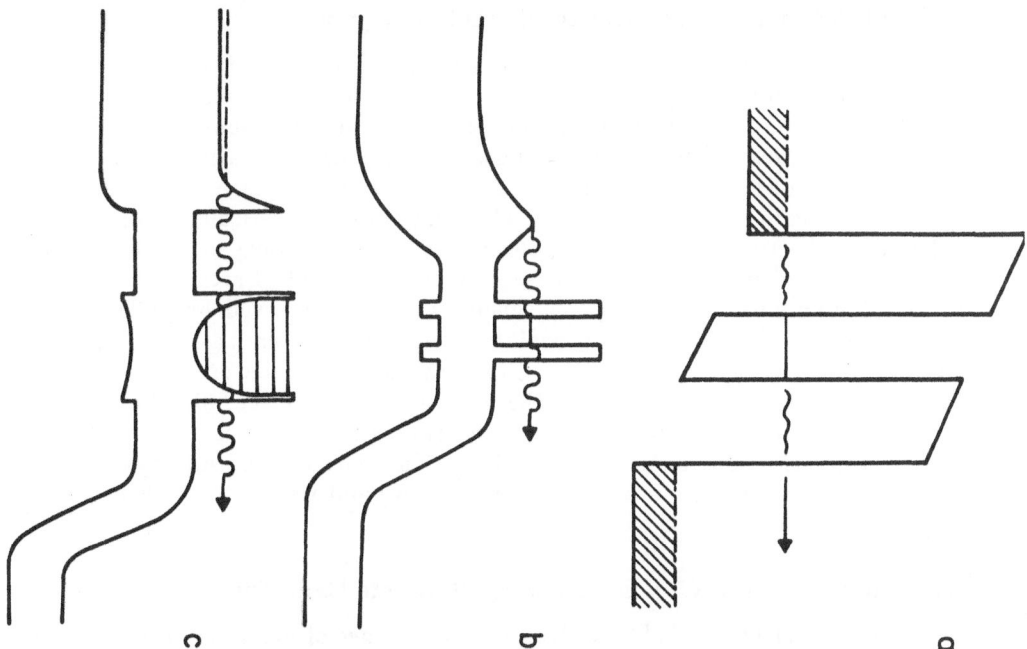

Fig. 4.   Resonant tunneling (RT) heterostructures. (a) RT through a double barrier; (b) RT bipolar transistor with near ballistic injection under operating
conditions; and (c) RT bipolar transistor with parabolic quantum well for
multiple-valued logic applications.

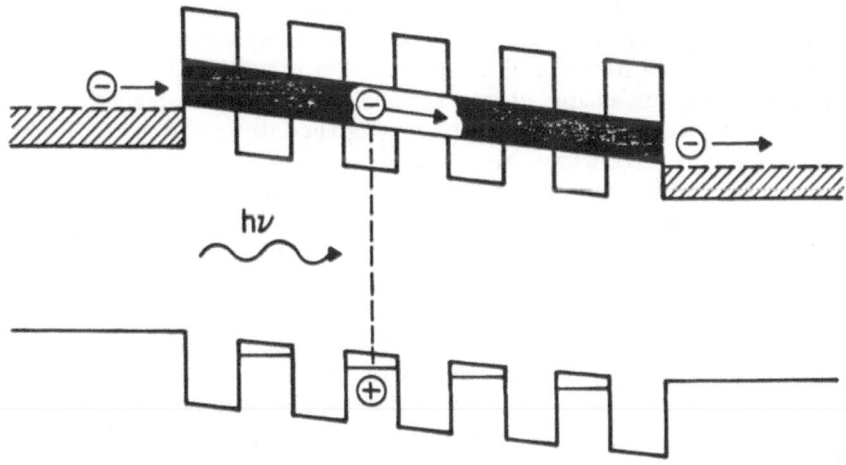

Fig. 5.   Band diagram of a superlattice photoconductor, which shows the effective mass filtering mechanism.

and Suris (36). The laser transition occurs between the third and second states of the quantum wells after a population inversion has been achieved between these levels (Fig. 6b). Another intriguing application is the generation of sub-Poisson light, that is, radiation with sub-shot noise properties (37).

Another important application of tunneling through superlattices deals with the creation of a spin-polarized electron source (38,39,40) and is illustrated in Fig. 6. It can be shown, based on selection rule arguments, that if one selectively excites with narrow bandwidth circularly polarized light the transition from the heavy hole subband to the ground state miniband in the conduction band the photoexcited electrons are theoretically 100% spin polarized. The total polarization is the result of the splitting of the light and heavy hole band degeneracy by the superlattice potential. Partial polarization in the photoluminescence using this scheme has been observed by Miller at al. (38). Of course the challenge is to maintain the electron polarization during the transport through the superlattice. Electrons could then be emitted into the vacuum with high yield by cesiating the p^+ GaAs surface layer (40), as shown in Fig. 7. To achieve that the tunneling time through the superlattice should be smaller than the spin depolarization time. This means that Al(Ga)As/GaAs superlattices with ultrathin barriers ($\simeq 15$ - 30 Å) and wells (15 - 30 Å) should be used.

The Ultimate Band-Gap Engineering: Tunable Band Discontinuities

The reader has probably realized the importance of band discontinuities in the design of band gap-engineered structures and quantum devices. The ability to artificially control such discontinuities would add a powerful degree of freedom in the design of materials and devices. Two methods have been successfully demonstrated independently by groups at Bell Laboratories (41) and the University of Wisconsin (42). The Bell Laboratories approach (Fig. 8) consists of the incorporation by MBE of ultrathin (≤ 50 Å) ionized donor and acceptor sheets within a few tens of angstroms from the heterojunction interface (planar doping). The electrostatic potential of this "doping interface dipole" is added to or subtracted from the dipole potential of the discontinuity. Since the separation between the charge sheets is on the order of or smaller than the carrier de Broglie wavelength, electrons crossing the interface "see" a new band discontinuity $\Delta E_c \pm e\Delta\phi$, where $\Delta\phi$ is the potential of the double layer. Using this technique, Capasso et al. demonstrated an artificial reduction of the conduction band discontinuity of the order of 0.1 eV in an $Al_{0.25}Ga_{0.75}As$/GaAs heterojunction (41). This lowering of the barrier produced an enhancement (by one order of magnitude) of the collection efficiency of photocarriers across the heterojunction as compared to identical heterojunctions without doping interface dipoles (41). On the other hand, Margaritondo and co-workers used ultrathin interlayers grown in the vicinity of the heterojunction (42). These interlayers also alter the band discontinuities. These techniques of modifying band offsets offer tremendous potential for heterojunction devices. The performance of many of these devices depends exponentially on band discontinuities. Thus artificial variations of these quantities by 1 or 2 kT (thermal energy, where k is Boltzmann's constant and T is temperature) can significantly alter that performance.

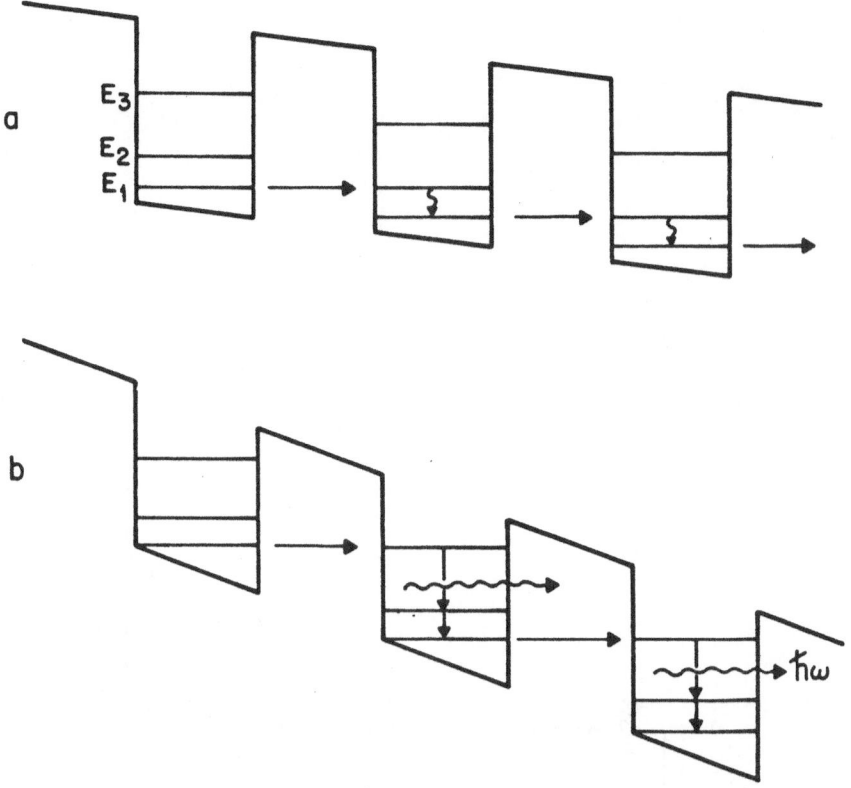

Fig. 6.   (a) Sequential resonant tunneling through a superlattice. Intrawell tran-
            sitions occur primarily by nonradiative relaxation. (b) Sequential resonant
            tunneling laser.

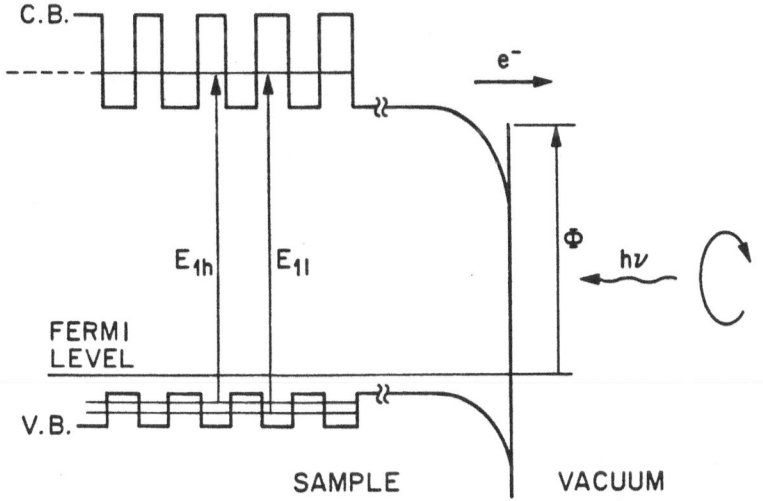

Fig. 7.   Schematics of band-diagram for spin-polarized electron source. Circularly
            polarized monochromatic photons of energy $hv = E_{1h}$ create spin polariz-
            ed electrons in the conduction band. These then tunnel through the
            superlattice and are emitted into the vacuum after crossing the negative
            electron affinity surface.

387

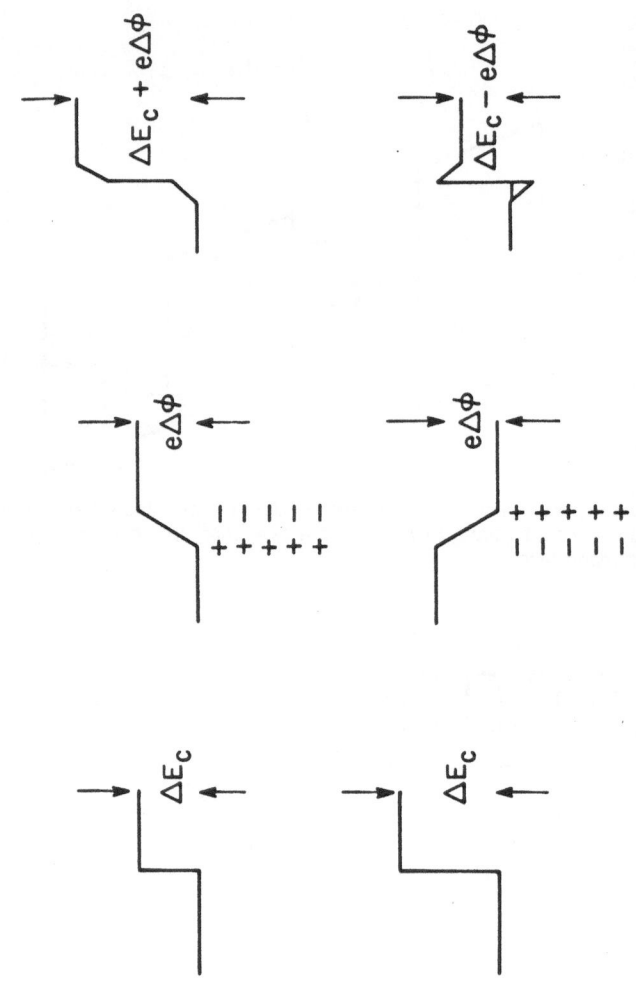

Fig. 8.   Tunable band discontinuities formed from doping-interface dipoles. (Top) The conduction band discontinuity is increased. (Bottom) Interchange of the acceptor and donor sheets reduces the band discontinuity. Tunneling through the spike and size quantization in the triangular well plays a key role in this reduction.

The key in this approach to designing microstructures and devices is the ability to model the energy band diagrams of semiconductor structures. With these models, scientists can visualize the behavior of electrons and holes in a device. Variable gap materials, superlattice, band discontinuities, and doping variations can be used alone or in combination to modify the energy bands almost arbitrarily and to tailor these bands for a specific application.

REFERENCES

1. W. Shockley, U.S. Patent 2 569, 347 (1951).

2. H. Kroemer, RCA Rev. *18*, 332 (1957).

3. For recent reviews on band-gap engineering see F. Capasso, in *Gallium Arsenide Technology*, D. K. Ferry, Ed. (Sams, Indianapolis, 1985), chap 8; F. Capasso, in *Picosecond Electronics and Optoelectronics*, G. Mourou, D. M. Bloom, C. H. Lee, Eds. (Springer-Verlag, Berlin, 1985), p. 112.

4. H. Kroemer, Proc. IEEE *51*, 1782 (1963).

5. Zh. I. Alferov et al., Sov. Phys. Semicond. *4*, 1573 (1971) [translated from Fiz. Tekh. Pouprovodn. *4*, 1826 (1970)]; I. Hayashi, M. B. Panish, P. W. Foy, S. Sumski, Appl. Phys. Lett. *17*, 109 (1970).

6. A. Y. Cho and J. R. Arthur, *Progress in Solid-State Chemistry*, J. O. McCaldin and G. Somorjai, Eds. (Pergamon, New York, 1975), vol. 10, p. 157; M. B. Panish, Science *208*, 916 (1980).

7. N. Holonyak, R. M. Kolbas, R. D. Dupuis, P. D. Dapkus, IEEE J. Quantum Electron. *QE-16*, 170 (1980).

8. L. Esaki and R. Tsu, IBM J. Res. Dev. *14*, 61 (1970).

9. H. M. Manasevit, Appl. Phys. Lett. *12*, 156 (1969); R. D. Dupuis, Science *226*, 623 (1984).

10. G. Döhler, J. Vac. Sci. Technol. *B1*, 278 (1983).

11. F. Capasso, S. Luryi, W. T. Tsang, C. G. Bethea, B. F. Levine, Phys. Rev. Lett. *51*, 2318 (1983).

12. R. Dingle, H. L. Störmer, A. C. Gossard, W. Wiegmann, Appl. Phys. Lett. *33*, 665 (1978).

13. H. L. Störmer, R. Dingle, A. C. Gossard, W. Wiegmann, M. D. Sturge, Solid State Commun. *29*, 705 (1979); R. Dingle, A. C. Gossard, H. L. Störmer, U.S. Patent 4 *194*, 935 (1980); T. Mimura, S. Hiyamizu, T. Fuji, K. Nanbu, Jpn. J. Appl. Phys. *19*, L125 (1980).

14. For strained layer superlattices in III-V materials, see G. C. Osbourn, J. Vac. Sci. Technol. *21*, 469 (1982); for Ge-Si strained layer superlattices, see J. C. Bean, Science *230*, 127 (1985).

15. J. M. Woodall, Science *208*, 908 (1980).

16. V. Narayanamurti, Phys. Today *37* (no. 6), 24 (1984).

17. F. Capasso, W. T. Tsang, G. F. Williams, IEEE Trans. Electron Devices *ED-30*, 381 (1983).

18. R. Chin, N. Holonyak, G. E. Stillman, J. Y. Tang, K. Hess, Electron. Lett. *40*, 38 (1982).

19. F. Capasso, W. T. Tsang, A. L. Hutchinson, G. F. Williams, Appl. Phys. Lett. *40*, 38 (1982).

20. F. Capasso, IEEE Trans. Electron Dev. *ED-29*, 1388 (1982); F. Capasso, R. A. Logan and W. T. Tsang, Electron. Lett. *18*, 760 (1982).

21. E. Gatti and P. Rehak, Nucl. Instrum. Methods *225*, 608 (1984).

22. S. L. Chuang and K. Hess, J. Appl. Phys. *59*, 2885 (1986).

23. F. Capasso et al., Appl. Phys. Lett. *48*, 1294 (1986). Near single-carrier type multiplication has been achieved using this scheme in a graded well structure. See J. Allam, F. Capasso, K. Alavi and A. Y. Cho, IEEE Electron. Dev. Lett. *EDL-8*, 4 (1987).

24. F. Capasso, W. T. Tsang, C. G. Bethea, A. L. Hutchinson, B. F. Levine, ibid. *42*, 93 (1983).

25. J. R. Hayes, F. Capasso, A. C. Gossard, R. J. Malik, W. Wiegmann, Electron. Lett. *19*, 818 (1983); R. J. Malik et al., Appl. Phys. Lett. *46*, 600 (1985).

26. G. J. Sullivan, P. M. Asbeck, M. F. Chang, D. L. Miller, K. C. Wang, paper III A-6, Technical Digest of the 44th Annual Device Research Conference, Amherst, MA, 23 to 25 June 1986.

27. L. L. Chang, L. Esaki, R. Tsu, Appl. Phys. Lett. *24*, 593 (1974). For a recent review on resonant tunneling in heterostructures, see F. Capasso, K. Mohammed, A. Y. Cho, IEEE J. Quantum Electron. *QE-22*, 1853 (1986).

28. T. C. L. G. Sollner, P. E. Tannewald, D. D. Peck, W. D. Goodhue, Appl. Phys. Lett. *45*, 1319 (1984).

29. F. Capasso and R. A. Kiehl, J. Appl. Phys. *58*, 1366 (1985). A unipolar resonant tunneling transistor with a quantum well in the emitter was demonstrated shortly hereafter by N. Yokoyama, K. Inamura, S. Muto, S. Hiyamizu, and H. Nishi [Jpn. J. Appl. Phys. *24*, L853 (1985)].

30. R. C. Miller, D. A. Kleinmann, A. C. Gossard, O. Munteanu, Phys. Rev. B *29*, 3740 (1984).

31. F. Capasso, S. Sen, A. C. Gossard, A. L. Hutchinson, J. H. English, IEEE Electron. Dev. Lett. *EDL7*, 573 (1986).

32. F. Capasso, K. Mohammed, A. Y. Cho, R. Hull, A. L. Hutchinson, Appl. Phys. Lett. *47*, 420 (1985).

33. _____, Phys, Rev. Lett. *55*, 1152 (1985).

34. A. Rose, *Concepts in Photoconductivity and Allied Problems* (Wiley, New York, 1963).

35. F. Capasso, K. Mohammed, A. Y. Cho, Appl. Phys. Lett. *48*, 478 (1986).

36. R. F. Kazarinov and R. A. Suris, Sov. Phys. Semicond. *5*, 707 (1971) [translated from Fiz. Tech. Pouprovodn. *5*, 797 (1971)].

37. F. Capasso and M. C. Teich, Phys. Rev. Lett. *57*, 1417 (1986).

38. C. K. Sinclair, Journal de Physique, Colloque C2, Suppl. to No 2, *46*, 669 (1985).

39. R. C. Miller, D. A. Kleinmann and A. C. Gossard, Inst. Phys. Conf. Ser. *43*, 1043 (1979).

40. R. Houdre´et al., Phys. Rev. Lett. *55*, 734 (1985).

41. F. Capasso, A. Y. Cho, K. Mohammed, P. W. Foy, Appl. Phys. Lett. *46*, 664 (1985); J. Vac. Sci. Technol. *B3*, 1245 (1985).

42. D. W. Niles, G. Margaritondo, P. Perfetti, C. Quaresima, M. Capozi, Appl. Phys. Lett. *47*, 1092 (1985); P. Perfetti et al., Phys. Rev. Lett. *57*, 2065 (1986).

HOT-ELECTRON SPECTROSCOPY AND TRANSISTOR DESIGN

M J Kelly, A P Long, P H Beton and T M Kerr

GEC Research Limited
Hirst Research Centre
East Lane
Wembley HA9 7PP
United Kingdom

ABSTRACT

The generation of hot electrons and their relaxation over short distances has been investigated in multilayer semiconductor structures. Scattering due to electron-electron interactions has been identified as the principal reason why hot-electron transistors have so far failed to give adequate current gain. Novel structures have been designed to circumvent this interaction in different ways, leading to the observation of ballistic transport and viable hot-electron transistors.

INTRODUCTION

The quest for a hot-electron transistor is twenty-five years old[1]. The generic structure is of three low resistance regions, emitter, base, and collector separated by two barriers, the emitter-base barrier taking the signal and the base-collector barrier the load (see Figure 1). If one imagines the signal driving a current I, of which a fraction α of the electrons cross the base and are collected by the barrier, the current gain of this device is $\beta = \alpha/(1-\alpha)$. If electrons were to cross ballistically then β would become very large. Furthermore, if the base is very thin, the transit time -~0.1 psec for 100 nm thickness -is short and a high speed device is also promised. (In fact it is now appreciated that the charging time of the emitter-base barrier is a major limitation on speed). Until recently only α~0.3 was achieved, and it is the progress over the last two years on hot-electron spectroscopy, made possible with specially designed structures, that is giving the hot-electron transistor a new lease of life. With a careful understanding of why previous generations of hot-electron transistor failed, it has been possible to achieve $\alpha > 0.9$ in new test structures grown by molecular beam epitaxy or metal-organic chemical vapour deposition. For the first time, a viable hot-electron transistor looks possible.

The recent spectroscopy work has been carried out in four laboratories, at AT&T Bell[2,3], IBM[4,5], Fujitsu[6,7] and GEC[8,9]. In figure 1

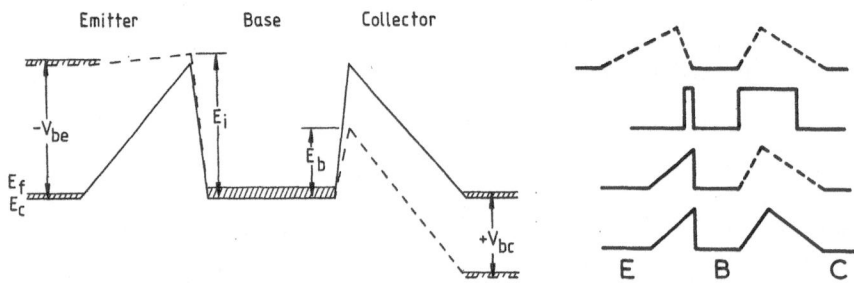

Fig. 1 The hot electron transistor (left) and conduction band profiles (right) of various hot-electron spectrometer structures, with the full (dashed) lines indicating composition (space charge) barriers.

we show the types of structure being considered. With the one exception described below in section 3, the base is a thin, n-doped layer (<100 nm, $10^{18}cm^{-3}$) of GaAs. The barriers are of two forms -either compositional with $Al_xGa_{1-x}As$, or doping, with the planar doped barrier (a thin p-doped region, surrounded by undoped GaAs, the former being depleted). The compositional barriers can be used either as graded layers as launching pads (at GEC), or as tunnelling injectors (at IBM and Fujitsu, the latter using resonant tunnelling). The AT&T work has concentrated on the planar-doped barrier (also part of the GEC spectrometer), which has the advantage of greater flexibility in designing the I-V characteristics: the height of the barrier comes from the total p-doping, and the lever arm from the position of the doping layer with respect to the base and collector. It is less capable of injecting or analysing a very narrow energy spread of carriers, because of the local potential fluctuations on the 10 nm scale that characterises the typical doping level. Indeed these fluctuations result, not in a knife edge discrimination of electron energies either in injection or in analysis but rather in a serated edge. Similarly the tunnelling injector admits a distribution equal in width to the Fermi energy in the emitter (i.e ~10's of meV), with a modest differential decrease in transmission of the lower energy electrons. The energy cutoff in the graded structure is likely to be sharper than that of the planar doped structure in that it is the Al distribution on the group III sublattice that determines the potential fluctuations. Even that allows tunnelling through the last 10 nm or so, an effect which could be reduced by clipping the top off the barrier. Practical advantages in terms of transistor performance of the graded AlGaAs emitter over that of the tunnel emitter are the inherently lower capacitance and higher current density.

In this chapter the research on and lessons from hot-electron spectroscopy are summarised in section 2, and the implications for transistor design are described in section 3.

2. HOT ELECTRON SPECTROSCOPY

To explain the low base transfer efficiency α, strong scattering of the hot electrons is taking place even with 100 nm. Furthermore, because optical experiments can be used to infer a mean-free-path due to optical phonon scattering in excess of 100 nm for undoped GaAs, the doping in the base is playing a role[10]. This doping level must be kept high for a low base resistance and a high speed device. If the base is reduced in thickness to increase the chance of ballistic transport, doping has to be increased to compensate. In the first spectroscopy work, the collector barrier was made larger under zero bias than the emitter barrier. During operation, the emitter barrier is forward biassed to inject electrons, and the collector barrier is reversed biassed and its height lowered through the leverage effect. As this reverse bias is increased, the increasing collector current can be analysed directly to spectrum analyse those electrons that have crossed the base. (The effective barrier height is linear with the base-collector bias)[3,9]. The results with base widths of 170, 120, 85 and 65 nm could all be accounted for with a mean-free path for inelastic scattering of order 40 nm.

The GEC results (of Figure 2) corroborate this estimate, and detailed simulations of both sets of data revealed further interesting physics. The Monte Carlo simulation of hot carriers traversing the base could produce a hot-electron spectrum in quantitative agreement with experiment only by invoking two types of scattering mechanism of equal importance. The first is the scattering from coupled optic phonon-plasmon

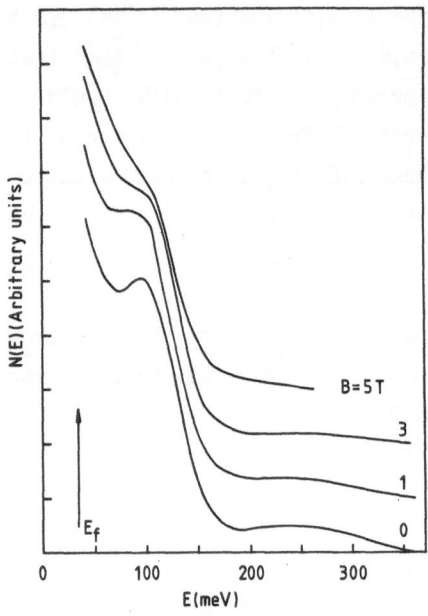

Fig. 2

Hot electron spectrum as a function of applied magnetic field (injection energy of 310 meV, base width 100 nm, doping $5.1-^{17}cm^{-3}$) showing the suppression of ballistic contribution with field.

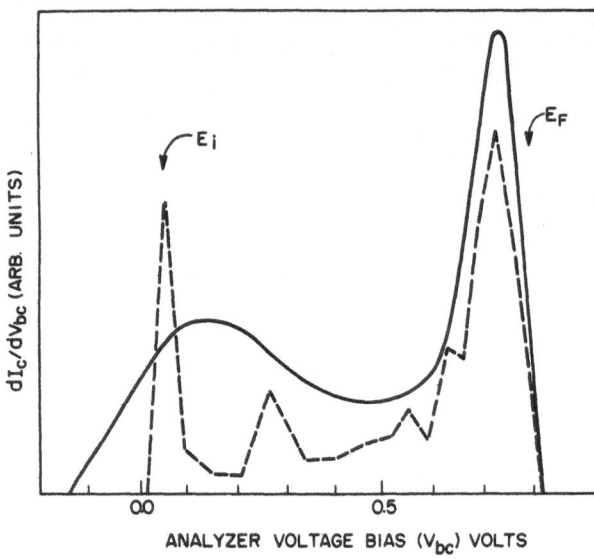

Fig. 3.

The GEC Monte Carlo simulations of the AT&T hot-electron spectrum showing (dashed-line) the fate of injected hot-electrons, and the remaining contribution from electrons injected from the Fermi sea in the base.

modes[9,11,12]. At carrier densities of $\sim 10^{18} cm^{-3}$, the plasma frequency is of order 30-60meV, comparable with the polar optic phonon energy, and so these collective modes couple. The second mechanism[9] is the single electron interaction which excites electrons from the Fermi sea in the base to a sufficient energy to cross the collector barrier when this latter is low. The Monte Carlo simulations show that this contribution accounts for most of the low energy peak (see Figure 3). The increase in scattering due to electron-electron interactions (both collective and single particle) explains the ~40 nm mean free path inferred from the decay of the ballistic peak in the hot electron spectrum.

The fact that we are observing ballistic electrons can be confirmed by applying a magnetic field at right angles to the hot electron current[13]. After crossing the emitter barrier, the excess forward kinetic energy collimates the electron current strongly in the forward direction. The Lorentz force will bend the electron trajectory, resulting in both an increased path length within the base, and in converting some forward momentum into the transverse direction so that these electrons may now no longer surmount the collector barrier. One observes the reduction of the ballistic peak amplitude, and the shift in position of the peak with increasing magentic field. We have recently completed a Monte Carlo simulation of the effect of the magnetic field[14]. Not only is the general picture confirmed, but the simulations can be used to deduce information regarding the angular and energy distribution of the injected electrons from an analysis of the precise form of the suppression and shift of the ballistic peak.

Several other aspects of injected hot electrons have been investigated. Under pressure the satellite valleys of GaAs are reduced in relative energy, and the IBM group have observed[15] the expected drop in ballistic contribution once intervalley transfer is added as a new scattering mechanism. In very thin base regions (< 30 nm) they have also seen a further set of oscillations in the collector current which are interpreted in terms of a form of enhanced transmission/reflection due to quantum size effects in the base[16]. At AT&T, the spectroscopy technique has been applied to the heterojunction bipolar transistor, and strong interactions with a p-doped base and a rapid thermalisation is observed[17].

The Fujitsu group have used a resonant tunnel structure for the injector. Since one can obtain a negative differential conductance at the emitter-base barrier, a richer variety of transistor characteristics follows, culminating in the demonstration of a one-transistor flip-flop circuit[18]. In principle this form of injector could produce a quasi-monochromatic current of electrons, and the sharpness of features in the collector current suggests a narrow energy range.

So far the quasi-ballistic peak is relatively wide in energy. As already suggested, the barriers are not perfectly sharp and introduce an energy spread. The one-scattering energy loss event is typified by an energy of ~50 meV, but the interaction of the coupled mode with the continuum results in a considerable energy dispersion, introducing a further spread. So far no one has resolved the single loss peak from the ballistic peak, and this represents a considerable challenge for the design of both the emitter and collector. (The Monte Carlo simulations clearly show the one-loss feature[9], but energy dispersion wipes out any distinctive features for electrons that have undergone two or more energy loss collisions).

3. HOT-ELECTRON TRANSISTOR DESIGN

The spectroscopy work on the GaAs/AlGaAs system explains just why poor hot-electron transistor performance was achieved. The doping/thickness combination for the base does not allow the simultaneous achievement of high speed and good current gain. Two immediate options are available[19]: new materials and a 2-dimensional electron gas base.

The limitation of the GaAs system is the L-valley approximately 0.3eV above the Γ minimum. Electrons with energy in excess of that can scatter very effectively into the satellite valleys where both their direction of motion is randomised, and their velocity reduced because of the higher effective mass. Other materials systems seem the answer. Two studies of hot-electrons in InGaAs/InAlAs have been reported[20,21]. In InGaAs, the satellite valleys are 0.55eV above the Γ minimum, which promises superior performance. The InAlAs barrier may be used to inject higher energy into InGaAs since the valley separation is large. Electrons may now suffer several inelastic scattering events and still surmount the collector barrier. The Fujitsu group achieved 94% transfer efficiency at 77K over 25 nm of $10^{18}cm^{-3}$ doped InGaAs with injected energy of 600 meV. We have undertaken detailed simulations[22] of this materials system and infer an effective collector barrier height of 210 meV when under the operating conditions of reverse bias. The same simulations suggest an optimum configuration of a 50 nm base doped at $10^{18} cm^{-3}$ with an injection energy less than or equal to 550 meV. This device could combine a viable gain (>10) with the potential for sub-picosecond switching speeds.

The other alternative to the GaAs doped base is the use of an inverted high-electron-mobility transistor structure. The base is undoped, and the collector modulation doped to provide carriers in the form of a two-dimensional electron gas in the (low resistance) base. The qualitatively different dispersion of the two-dimensional plasmon, and the phase-space for single-electron excitations, is such that rather less scattering is expected of hot-electrons passing at right angles to the electron gas. No results have been published, but our own preliminary studies indicate the predicted improvement in a test structure. A similar structure, using the two dimensional charge induced in a narrow quantum well[23] for the base, has achieved a high or value (~0.96) with a low base sheet resistance of 310Ω.

The potential of a hot-electron transistor for fast switching applications depends not only on the efficient base transfer but also on the charging time of the various capacitances in the structures. There are several avenues for device optimisation, the shape of the barriers (including a rounding of the collector barrier profile in order to reduce quantum mechanical reflections), the precise doping profile to reduce capacitance and resistance values, and the important technology issue, the precise revealing of thin buried layers and the fabrication of highly selective ohmic contacts, especially to the base.

The competition from other transistor families is fierce. The heterojunction bipolar transistor is a principal competitor, but the technology of a unipolar and majority carrier device should be marginally less demanding. The field-effect transistors can achieve comparable switching speeds only with extremely short gate lengths -the technology trade off here is that very fine lithography can be relaxed if the depth profile technology is mastered. In the coming years it is likely that some high speed prototype devices will become available, and the issues will shift to reliability, reproducibly, noise and other concerns associated with real circuits.

Some of this work is supported by the Marconi Electron Devices Ltd,

EEV, MOD, and EEC. We thank C E C Wood for discussions and P M Wood for processing our structures.

REFERENCES

1. For an introducton see: S Luryi, "Hot-Electron-Injection and Resonant-Tunnelling Heterojunction Devices", in "Heterojunctions: a Modern View of Band Discontinuities and Device Applications", (editors F. Capasso and G. Margaritondo) (North Holland 1987)

2. J R Hayes and A F J Levi, "Dynamics of Extreme Nonequilbrium Electron transport in GaAs", IEEE Journal of Quantum Electronics, QE-$\underline{22}$, 1744-52 (1986), and references therein

3. J R Hayes, A F J Levi and W Wiegmann, "Hot Electron Spectroscopy of GaAs' Phys. Rev. Letts $\underline{54}$, 1570-2 (1985)

4. M Heiblum, M I Nathan, D C Thomas and C M Knoedler "Direct Observation of Ballistic Electron Transport in GaAs", Phys. Rev. Lett $\underline{55}$ 2200-3 1985)

5. M Heiblum, I M Anderson and C M Knoedler, "d.c. Performance of Ballistic Tunnelling Hot-Electron Transfer Amplifiers", Appl. Phys. Lett, $\underline{49}$, 207-9 (1986)

6. S Muto, K Imamura, N Yokoyama, S Hiyamizu and H Nishi "Sub picosecond Base Transit Time Observed in a Hot Electron Transistor", Electronics Letts $\underline{21}$, 555-6 (1985)

7. N Yokoyama, K Imamura, S Muto, S Hiyamizu and H Nishi, "A New Functional Resonant Tunnelling Hot Electron Transistor (RHET)", Jap. J. Appl. Phys. $\underline{24}$, Part 2, L853-4 (1985)

8 A P Long, P H Beton, M J Kelly and T M Kerr, "Hot-electron Injection by Graded $Al_xGa_{1-x}As$" Electronics Letters, $\underline{22}$, 130-1, (1986)

9 A P Long, P H Beton and M J Kelly, "Hot Electron Transport in Heavily Doped GaAs", Semiconductor Science and Technology $\underline{1}$ 63-70, (1986)

10 L F Eastman "The Limits of Electron Ballistic Motion in Compound Semiconductor Transistors", in Gallium Arsenide and Related Compounds 1981 (Inst. Phys. Conf. Ser. No. 63) p245-50

11 M A Hollis, S C Palmateer, L F Eastman, N V Dundekar and P M Smith, "Importance of Electron Scattering with Coupled Plasmon-Optical Phonon Modes in GaAs Planar-Doped Barrier Transistors", IEEE Electron Device Letters, EDL-4, 440-3 (1987)

12 A F J Levi, J R Hayes, P M Platzman and W Wiegmann, "Injected Hot Electron Transport in GaAs", Phys. Rev. Lett, 55, 2071-3 (1985)

13 J R Hayes, A F J Levi and W Wiegmann "Magnetic Field Dependence of Hot Electron Transport in GaAs", Appl. Phys. Lett, 47, 964-6 (1985)

14 P H Beton, A P Long and M J Kelly, "Hot Electron Transport in GaAs in the Presence of a Magnetic field", to be published.

15 M Heiblum, E Calleja, I M Anderson, W P Dumke, C M Knoedler and L Osterling, "Evidence of Hot-Electron Transfer into an Upper Valley in GaAs", Phys. Rev. Lett 56, 2854-7, (1986)

16 M Heiblum, M V Fischetti, W P Dumke, D J Frank, I M Anderson, C M Knoedler and L Osterling, "Electron Interference Effects in Quantum Wells; Observation of Bound and Resonant States", Phys. Rev. Lett. 58 816-9 (1987).

17 J R Hayes, A F J Levi, A C Gossard and J H English "Base Transport Dynamics in a Heterojunction Bipolar Transistor", Applied Phys. Lett 49, 1481-3 (1986)

18 N Yokoyama and K Imamura, "Flip-Flop Circuit using a resonant-tunnelling Hot Electron Transistor (RHET)" Electronics Letters 22, 1228-9 (1986)

19 A F J Levi, J R Hayes and R Bhat " 'Ballistic' Injection Devices in Semiconductors", Applied Phys. Letts. 48, 1609-11 (1986)

20 U K Reddy, J Chen, C K Peng and H Morkoc, "InGaAs/InAlAs Hot-Electron Transistor", Appl. Phys. Letts. $\underline{68}$, 1799-801 (1986)

21 K Imamura, S Muto, T Fiujii, N Yokoyama, S Hiyamizu and A Shibatomi, "InGaAs/InAlGaAs Hot-Electron Transistors with Current Gain of 15" Electronics Letters $\underline{22}$, 1148-9 (1986)

22 A P Long, P H Beton and M J Kelly, "Hot Electron Transport in $In_{0.53}Ga_{0.47}As$", to be published

23 C Y Chang, W C Liu, M S Jame, Y M Wang, S Luryi and S M Sze, "Induced Base Transistor Fabricated by Molecular Beam Epitoxy" IEEE Electron Device Letters, EDL-7 497-499, (1986)

NOVEL TUNNELLING STRUCTURES: PHYSICS AND DEVICE IMPLICATIONS

M J Kelly, R A Davies, N R Couch, B Movaghar and T M Kerr

GEC Research Limited
Hirst Research Centre
East Lane
Wembley HA9 7PP
United Kingdom

ABSTRACT

The new crystal growth techniques (molecular beam epitaxy and metal-organic chemical vapour deposition) allow sufficient control over the layer thicknesses and integrity that resonant tunnelling phenomena can be explored with some precision. Beyond the double barrier diode and the uniform superlattice are a number of novel tunnelling structures, the physics of two of which, the superlattice tunnel diode and the short graded-parameter superlattice, will be discussed in some detail. A short discussion will also be given of the potential role of tunnelling in both two and three terminal devices.

INTRODUCTION

Because the control over the bandgap and doping profile is approaching the atomic scale in the direction of growth in both molecular beam epitaxy and metal-organic-chemical vapour deposition, one can design and fabricate tunnel structures with a precision never previously available. The emphasis within this chapter is on novel tunnelling structures, and some comments on their device implications are included. The physics of double barrier diodes is covered in the chapter by Mendez, and of regular superlattices in the chapter by Capasso. The work from GEC has concentrated on two novel structures -a superlattice tunnel diode[1-3] and a graded parameter superlattice[4-6], the former with microwave oscillator possibilities and the latter providing a photodetector with low dark currents. Some mention will also be made of structures devised by Nakagawa[7,8], Yokoyama[9,10] and others. While much of the work to date has been performed on two-terminal structures, recent reports describe the role of tunnel structures as the barriers[10], or indeed the base region[11,12], of transistor structures. Increased device functionality can be obtained from tailored current-voltage characteristics. One important point to emerge from all this work is that detailed and quantitative theories are still relatively primitive when it comes to contemplating a CAD of tunnelling devices with considerations of real-time behaviour, noise and transient behaviour, and circuit environment. Nevertheless, resonant tunnelling and hot-electron injection[13] do hold the key to new generations of devices.

2. TUNNELLING PHENOMENA

The textbook introduction to the tunnelling of an electron through a single barrier is deceptively simple, while in practice, the role of scattering processes (elastic or inelastic from several sources), space charge, high fields, and time-dependent phenomena complicate the detailed interpretation of current-voltage data from semiconductor tunnel barriers. In multiple barrier structures, further complications arise from distinctions between resonant and non-resonant tunnelling, and the sequential versus coherent nature of the tunnelling process[14]. In spite of these concerns, the subject of wide-spread research, we have adopted a pragmatic approach. We use relatively simple models for the design and modelling of multilayer tunnelling structures in a search for exploitable effects and pursue in parallel a theoretical programme aimed at a greater understanding of the more fundamental aspects of vertical transport[15].

While very high speed was the original rationale behind the study of tunnelling phenomena, which occur over very short length scales, limitations now are perceived to arise from the charging time of the active region's capacitance, which in turn depends on features such as the contact resistance.

3. THE SUPERLATTICE TUNNEL DIODE

In a regular superlattice, the quasi-bound levels within a single well interact via tunnelling with the levels in adjacent wells. The result is a series of minibands, which reflect high and low pass transmission bands for electrons as a function of their incident energy. In the first experimental study of vertical transport in a multilayer semiconductor structure, ESAKI and CHANG[16] reported negative differential resistance at low temperatures, and further oscillations in the current as the voltage was increased. Their interpretation was not in terms of the Stark regime[15,17] (with a voltage equal to the miniband width dropped across the superlattice), but rather of a high-field domain set up at the thickest barrier allowing electrons to tunnel between minibands. Although nominally uniform, the limitations of growth control will make one barrier thicker than the rest. Further high-field domains cause the further oscillations.

In the following decade, no further studies were reported. We commenced a study of tunnelling through 'non-ideal' superlattices, i.e. ones with non-equal barrier and well thicknesses, and have undercovered a much wider range of tunnelling phenomena, with both transport[1-4] and optical techniques[6,7,18,19].

The superlattice tunnel diode[1-3] is simply understood with respect to Figure 1. Two superlattices are separated by a thicker barrier. The electron states in the superlattices follow from a simple Kronig-Penney analysis. We dope the structure to produce a "metallic" system with the Fermi level in the lowest miniband. Our superlattice tunnel diode is now simply a conventional metal-insulator-metal tunnel barrier, but one in which the superlattice miniband structure is imposed on the initial and final electronic states for the tunnelling process. As the voltage across the structure, most of which is dropped across the thicker central barrier, is increased, the minibands of the superlattice move in and out of relative alignment giving rise to the negative differential conductance (n.d.c.) features. This simple picture is adequate for a qualitative underntanding of the data we have acquired, and as a design aid for increasing the n.d.c, but it does raise some interesting physics questions:

(i)   A long superlattice is not strictly required, only three or four periods suffice[3], provided only that a sufficient dynamic range is introduced between the high and low levels of transmission for the electrons.

(ii)   The structure is rather more tolerant of fluctuations in the layer thicknesses[20] and integrity than the double barrier diodes, while still providing a strong principal n.d.c feature.   Indeed, we can take this a stage further and perform a spectroscopy on the mini-bands. We have been able to correlate the appearance of further n.d.c. features with the regularity of the superlattice layers and the ability to tunnel from the first miniband of one superlattice to the third of the other[21]. Variations in layer thicknesses tend to blur out the regions of very high or very low transmission probability, and the gap between the second and third miniband is quite sensitive to thickness fluctuations.

(iii)   The transport is strictly three dimensional, and there are, in reality, no gaps in the density of states.   Indeed the strength of the n.d.c features relies on an absence of scattering, from whatever mechanism. Our own studies of magnetotransport[22] reveal a quenching of the n.d.c. in modest magnetic fields (~10T), with a zero-bias anomaly appearing at still higher fields. A simple semiclassical picture of the trajectory of a tunnelling electron, combined with the three-dimensional picture of the electron states, gives a qualitative explanation (Figure 2).   One is, however, in the regime where semiclassical assumptions are not strictly valid. When the characteristic magnetic energy, $\hbar\omega_c$, exceeds the miniband width, for example, the electron states are more properly hybrid electro-magnetic in character[23].

(iv)   The sequence of structures grown in pursuit of large and exploitable n.d.c. features, do reveal the importance to be attached to the integrity of the central barrier layer. An absence of scattering centres and fluctuations in either the alloy composition or barrier thickness, all contribute to an increased peak-to-valley ratio[24].   (Excess currents due to poor passivation or to surface leakage also reduce the relative strength of the tunnelling n.d.c.)

(vi)   Other experiments on these structures have included applications of high pressure, for which the data is consistent with bulk deformation potentials when used to interpret shifts in the I-V characteristics except for some minor effects at ~10 kbar, tentatively associated with 'DX' centres and with indirect-gap (X-symmetry) states becoming the low energy route through the barrier[25].

(vii) TEM studies have been performed on all the structures[26], the data from which is an important element in checking physics interpretations -after all minor variations in layer thickness can alter the interpretation.   Because of the high doping level in the contact region, no optical sensitivity has been observed.

The design principles by which improvements in the n.d.c. have been achieved are:

(i)   the increase of the tunnel component of the current with respect to the thermionic component. By increasing the mini-band gap with 43% Al concentration in the AlGaAs barriers, and by decreasing the GaAs well thicknesses, we can reduce the thermionic current. A decrease in the thickness of all barriers increases the tunnel current,

(ii)   a reduction in possible scattering mechanisms in the tunnel barrier that degrade the peak-to-valley ratio. The absence of doping, and a thinner barrier reduce the scattering. Binary material barriers eliminate alloy scattering, but see the discussion in Section 3 below on indirect-gap state tunnelling,

(iii) a doping profile that minimises the series resistance without introducing impurities at the tunnel barrier, but with suitable account taken of the space-charge implications from abrupt changes in doping. (This effect has only recently come to be recognised, with in some cases a re-interpretation of tunnelling data).

So far only rudimentary device-related work has been undertaken to confirm the superlattice tunnel diode as a microwave source[2]. More advanced work on the double barrier diode has led to oscillations 56 GHz, and circuit analysis predicts >100 GHz potential[27]. All these devices have characteristics similar to that of the Esaki (tunnel) diode[28]. The intrinsically high capacitance might be offset in several ways: (i)

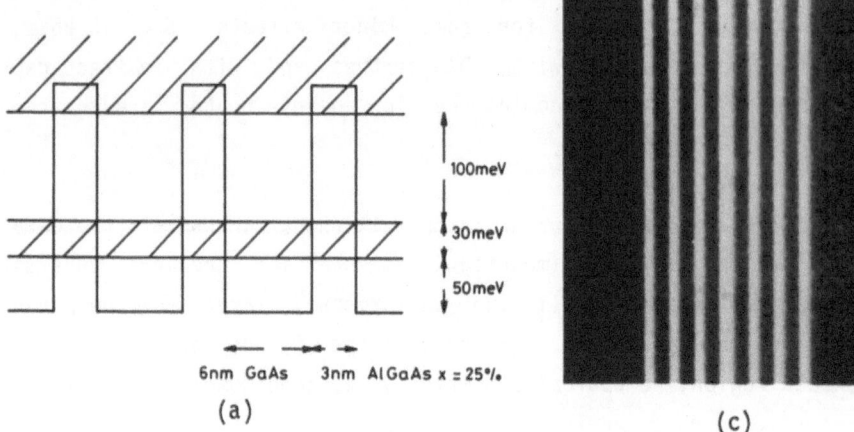

(a)

(c)

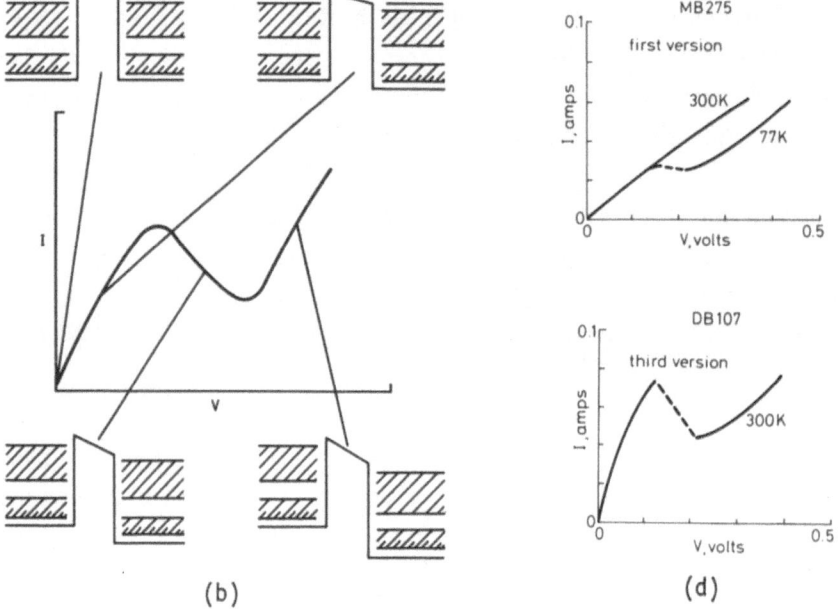

(b)

(d)

Figure 1:   The superlattice tunnel diode, showing (a) the design of the
superlattice bandstructures, (b) the principle of operation of the
superlattice tunnel diode, (c) a dark -field TEM image of a recent
structure with total active length of 55 nm (courtesy of E Britton and W M
Stobbs, University of Cambridge), and (d) negative differential resistance
at room temperature.

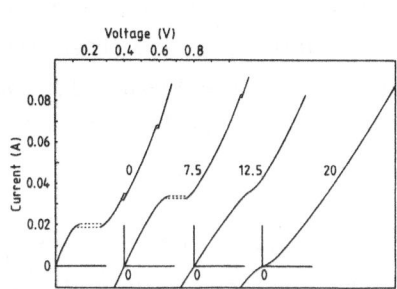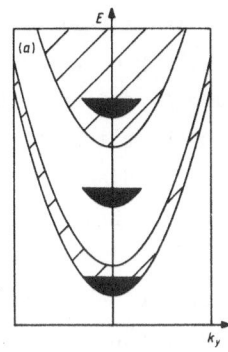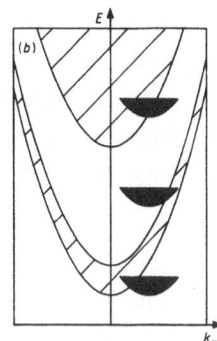

Figure 2: The supression of negative differential conductance with a magnetic field applied perpendicular to the current. The band diagrams (in the $k_{||}$ plane) show the effects on the occupied initial states with respect to the final states of an applied (a) electric and (b) electric and magnetic field. The n.d.c. feature is suppressed once overlap between initial and final states is achieved for all biases.

decreasing the area although this reduces power , (ii) vertically stacking several diodes (also a potential method for increasing the power handling) and (iii) fine-tuning of the doping profiles. The absence of doping in the high-field regime eliminates one important source of device failure, namely field assisted migration of the dopants.

4. GRADED PARAMETER SUPERLATTICE

Nakagawa et al proposed[7] and demonstrated[8] the possibility of creating the resonant tunneling condition in a superlattice only for a finite applied electric field by using a gradation of the superlattice period, but maintaining the mark-space ratio of well: barrier thicknesses. Electron states in progressively narrow wells are aligned only under finite bias to produce the resonant tunnel condition.

In our studies of hot-electron injection we considered both a linearly graded gap[29] (a layer of $Al_xGa_{1-x}As$ with x increasing from 0 to 0.3 over 50 nm) and its coarse digital analogue[4-6] (a short AlAs/GaAs multilayer structure) as possible launching pads. While the linear grading is more demanding on MBE growth, the effort is rewarded with superior launching properties at high current densities. The test structure for the digital analogue, a five 10 nm period structure, but with a graded barrier/well ratio, as the i-region of a p-i-n structure (Figure 3), has produced a wealth of interesting physics:

(i)   in the absence of the i-region, the structure is just a p-n diode
        but just inside the p-n junction turn-on, a negative differential
        conductance feature due to resonant tunnelling is observed[4],
        corresponding to the alignment of the Fermi energy in the GaAs and
        the bound levels in adjacent wells by the net field across the
        i-region near the flat band condition,

(ii)   the simulation of the I-V characteristics, given TEM confirmation
        of the AlAs layer thickness, can produce quantitative agreement of
        the bias and current level of the resonant tunnelling peak[4] only if
        the barrier height invoked is the offset between the Γ conduction
        band minimum of GaAs, and the X minimum of AlAs. (Note: more
        sophisticated simulations that include the Bloch functions in
        addition to the plane wave parts do show this X-mediated tunnelling
        to be possible[30]).

(iii)   as a function of increasing temperature, or magnetic field
        perpendicular to the current direction, or applied hydrostatic
        pressure, the peak-to-valley ratio of the n.d.c. feature falls
        linearly[31], and with the same slope if a common energy scale (kT
        with k the Boltzmann constant, $\hbar\omega_c$ with ω_c the cyclotron frequency,
        Ξp with Ξ the deformation potential and p the pressure (see Figure
        4). We have no convincing explanation of these facts.

(iv)   The structure is photosensitive, and the decay of current induced
        by a sub-picosecond optical pulse showed two characteristic times,
        one set by the RC constant of the device, but a longer time that
        was consistent with sequential tunnelling (in the Luryi scheme) of
        a few electrons through the thickest barrier if its height were the
        Γ GaAs to Γ AlAs offset[6]. The reverse-bias dark current of this
        structure was factor 100 lower than that expected if the
        superlattice were removed[5]. (v) The detailed electroluminescence[18]
        of this structure revealed peaks that could be associated with
        electron states confined by the X-conduction band minimum, not the
        Γ minimum, in AlAs. A similar conclusion was arrived at by
        magneto-optical studies where the photocurrent was monitored as a
        function of magnetic field and optical excitation energy[32]. In
        detail, the electroluminscence observes recombination of light
        holes and those electrons tunnelling through the Γ-X potential,
        while the photoconductivity shows features associated with light

and heavy holes, and, again, electron n=1 and n=2 quantum well states confined by the Γ-X potential.

(v) An equivalent structure was grown in which the AlAs barriers were replaced with $Al_{0.43}Ga_{0.57}As$ barriers of the same thickness, and no n.d.c. features were observed, where our simulations of the electron current predicted a weak feature. This may be because a hole contributions tends to cancel the effect, or that alloy fluctuations in the barriers (which TEM confirms are all in place) blur out the weaker n.d.c. feature. Optical investigations are in progress on this structure[18,32] latter, reflecting a better distribution of the Al and hence the more effective barrier it represents.

(vi) This structure is the simplest possible version of a quasi-graded gap structure of the kind proposed by Capasso[33] as an element in avalanche photodiode and other structures. The analysis of the dynamics of carriers in such structures led to the conclusion that only a few periods would be required[34]. Our structure bears out the general conclusions of that analysis, but still the speed is relatively low (0.3ns FWHM) because of the details of our contacting layers and their effect on the RC time constant.

5. OTHER TUNNELLING STRUCTURES

We have already cited the work of Nakagawa et al[7] on their CHIRP (coherent interfaces for reflection and penetration) superlattice. Their n.d.c feature was relatively weak[8], a result of the elaboration of the 60 layer structure and its sensitivity to layer fluctuations, and of the high level of doping throughout. The same group have studied the simplest version of the superlattice tunnel diode in a three (thin-thick-thin) barrier structure[35]. The striking n.d.c. seen, with large peak-to-valley ratio, is related now to tunnelling between the bound levels in the two wells under the condition of aligning the filled state in one well with an empty excited state of higher local energy in the other. To achieve the large value of the peak-to-valley ratio, the barriers are made relatively thick (5.1 nm and 10.2 nm) with the consequence of a low current density ($\sim 10^{-3}$ of that in Figure 1).

If we were to remove the central thicker barrier in Figure 1, leaving a

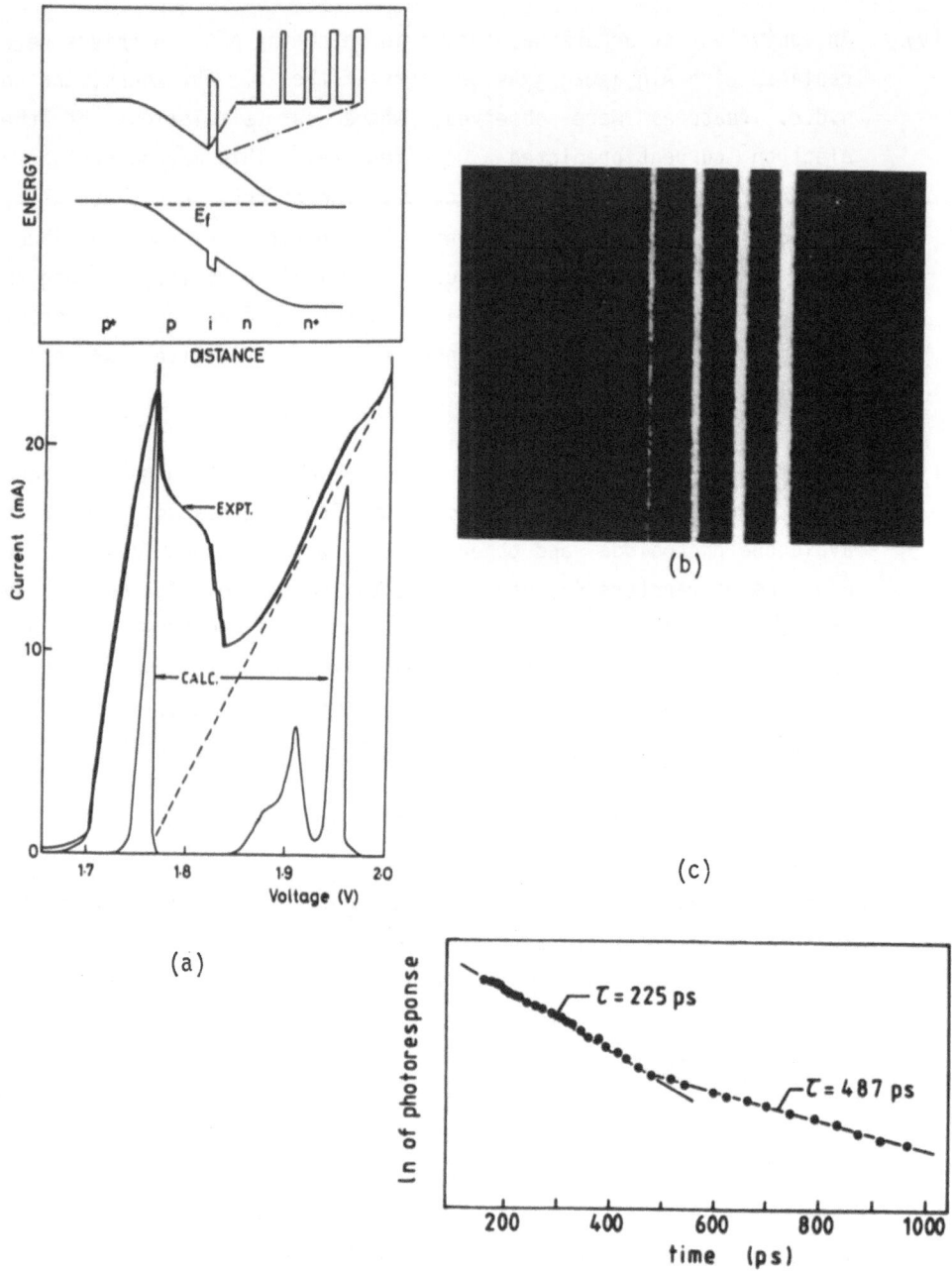

Figure 3: The graded-parameter superlattice showing (a) a schematic of the structure, and the I-V characteristics with a simple simulation (b) a TEM image (courtesy of E.M. Britton and W M Stobbs, University of Cambridge), and (c) the photoresponse showing two characteristic relaxation times.

wider well with two superlattices bounding either side we arrive at the structure of Reed et al[36]. (Their three AlAs and GaAs layers, on either side of the 4.5 nm wide well, were only 0.7 nm thick). The observed n.d.c. feature had a room-temperature peak-to-valley ratio of 1.8:1, and the I-V characteristics were very symmetrical compared with data from conventional double barrier diodes, suggesting that the effect of inverted growth is averaged out in this symmetrical structure. One interesting point to emerge is the relative low effect barrier height inferred from this data, suggestive of tunnelling via X-minimum states in AlAs.

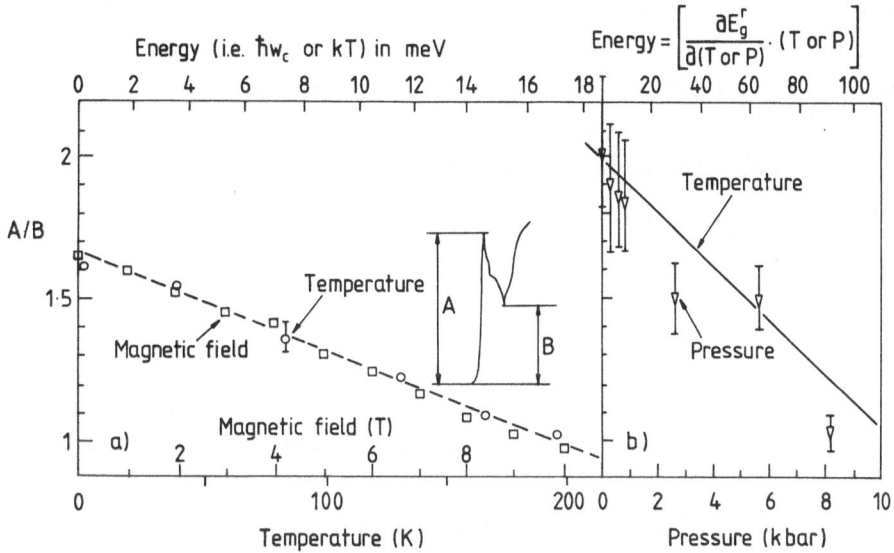

Figure 4: The peak-to-valley ratio of the negative-differential resistance peak in the graded parameter superlattice as a function of temperature, magentic field and pressure.

The structures described so far are all two-terminal, and the most of these tunnelling phenomena are likely to be exploited within three terminal structures. Several of these have been proposed recently and realised, but we concentrate here on the results of the Fujitsu group[9,10,37] as they have exploited resonant tunnelling in the emitter barrier of a hot electron transistor, and hot-electron phenomena is the theme of the other GEC chapter. Negative transconductance, which follows from the resonant tunnelling emitter base characteristics can be used to produce a one-transistor flip-flop, a basic element in memory and logic circuits where seven transistors are required in the conventional circuit. This is an early example of the greater functionality of devices that is made possible by the internal sophistication of tunnelling characteristics.

6. SIMULATIONS

Given a profile for the potential energy $V(z)$, that includes any applied fields or affects associated with space charge, in a multilayer tunnel structure, one can in principle solve the one-electron Schrodinger equation to calculate the transmission and reflection amplitudes as a function of the incident electron energy. (With constant potential, constant field or a few other special cases, the solution is analytical). With each eigenstate, one can calculate the particle flux (from $\hbar(\Psi^*\nabla\Psi - \Psi\nabla\Psi^*)/2im$), and then proceed to the tunnelling current by integrating over the (momentum) eigenstates:

$$I_{left \rightarrow right} = \frac{1}{4\pi^3} \frac{e\hbar A}{m^*} \iint_{k\perp > 0} 2\pi k_\perp dk_\perp k_z dk_z \ (1-|R|^2)f(E)(1-f(E+V))$$

where R is the reflected component of an incident wave, and the two last factors ensure that the initial states are occupied and the final states empty, for a bias V between the contacts[38]. The net current is $I_{left \rightarrow right}$ minus $I_{right \rightarrow left}$. The earliest work was in uniformly doped structures where space charge effects were neglected. Recent structures use tailored doping profiles to reduce scattering in the tunnel barriers, and the space charge effects must be incorporated -indeed the role of X or Γ state tunnelling mechanisms in AlAs has been clarified only by incorporating the space charge[39]. At present no simulations include scattering -from impurities, interface steps, alloy fluctuations or phonons. However it is possible to eliminate steps with atomic layer epitaxy, use binary material layers to exclude alloy scattering and work at low temperatures and low voltages to eliminate phonons to reduce the importance of these effects. At this stage, another key approximation, the use of the effective mass-cum-envelope function formalism which neglects the Bloch part of the wavefunction may in fact be able to be tested, and the connection rules of wavefunctions at interfaces be made

414

available experimentally. In the interim, further elaboration of the simple Kronig-Penney model is not warranted for the design and simulation of the more complex tunnelling structures.

7. THEORY

The simple models for electron states and transport invoked in the simulations needs justification or replacement. The high local electric fields encountered can induce gross changes in the electron states (the Stark regime) and severely modify the phonon and other scattering rates that are calculated or inferred from low field conditions. Several length and time scales become comparable and of competing importance. The Stark length (over which a potential is dropped equal to the miniband with), the localisation length (regarding well-width fluctuations as the origin of disorder), the elastic and inelastic transport scattering length and the superlattice period are all in the 5-50 nm range.

In an attempt to formulate a theory for high-field transport and magnetotransport in superlattice structures, the following conclusions have been reached[15] (with F = field, W = miniband width, d = superlattice period) (i) in the low field regime, the carrier drift velocity V_d, is simply $V_d = (eF/kT)D$ but the Diffusivity D may be renormalised from its bulk GaAs value by several factors -inter-Stark level, inter-miniband scattering etc. (ii) in higher fields where phonon emission is possible, and for an infinite and ordered system $V_d=(eF/kT)(W/Fd)^2$. The unperturbed "Stark formulation" leads to a low field 1/F divergence in electronic drift which is analogous to the "infinite conductivity" of the linear response regime in the absence of all scattering processes. Both are artifacts which are removed by "renormalisation" (allowing for elastic and inelastic scattering). If this latter is characterised by an inelastic scattering time τ, the renormalised result is $V_d = (eFd\tau/h)(wd/2h)/(1+(eFd\tau/h)^2)$ which gives an ohmic low field regime, but a F^{-1} limit in high fields. When the superlattice is finite, and comparable with the Stark or transport scattering lengths, a further renormalisation is required. (iii) with any applied magnetic field, a new length and time scale (cyclotron radius and period) are introduced. To first order, a magnetic field parallel to the direction of current through multilayers has no effect, although in the presence of disorder, any percolative motion is suppressed leading to a positive magnetoresistance. Weak magneto-Stark-phonon resonances can be expected in the I-V curves for high

B and F fields and long scattering times. For B $\perp$ current (i.e. 11 to layers), the electron states change, and the scattering rates are modified. Indeed the Stark-Landau system involves totally localised electron states in the plane $\perp$ B. The re-derivation of transport coefficients leads to a positive magnetoresistance and the Stark regime itself is manifest only at higher electric fields than in the case when B = 0. In the very high magnetic field, the inter-Landau level coupling can be introduced as a perturbation, but the drift velocity is more complex.

At present the restricted range of samples that so far have been examined experimentally do not approach the conditions where the simple limiting cases strictly apply, at least to the extent of extracting simple transport coefficients. Extensions of the theory to apply to the examples of section 3 and 4 above are being sought.

8.   DEVICE IMPLICATIONS

Although the short characteristic times and lengths associated with the active regions of tunnelling devices lead to predictions of very high speed devices, more mundane matters associated with capacitances and resistances place strong limitations on the frequency/power/efficiency/noise performances that are likely to be achieved. At first sight, the large device capacitance is intolerable, but recent analyses of double barrier diodes at modest frequencies (up to 1 GHz) reveal an effective capacitance set not by the tunnelling region length[40], but rather the distance between the high doping contact regions, once the tunnel structure itself is undoped in order to suppress any impurity-related transport channel as an alternative to the tunnel mechanism. To date the best figure is a 60 μW output at 56 GHz. Another factor of 10 in output is required. Our own structures are predicted to be capable of operating at frequencies >100 GHz, and our estimates of the noise performance of a highly efficient local oscillator are encouraging. (The noise was estimated by assuming it to be dominated by Shot noise, as in the Esaki diode). The most detailed general device analysis of the double barrier diode has been by Luryi[41], with prediction of >200 GHz upper frequency limits.

Many of the three-terminal superlattice devices that have so far been proposed and/or realised (see Luryi[41] for a discussion) are based on the incorporation of a superlattice as one element (emitter-base barrier, the base itself etc) of a vertical or field effect transitor. The device technology is thus a simple extension of that required for conventional devices, with only a recalibration for etching or implantation of superlattice/multilayer components. The drive towards greater functionality (c.f. the one-transistor flip-flop[10]) seems the most promising direction, leading to a reduction in transistor count in the basic circuits for logic, memory, or special purpose signal processing. It should be noted that maximum current densities in tunnel diodes are still relatively low, compared with graded composition barriers diodes, particularly if a large peak-to-valley ratio is required, so that the tunnel device in a circuit may be slower to operate. Furthermore thin, highly doped, intermediate layers for the base region of vertical devices must be designed with considerable care, if the resistance is to be kept acceptably low. The use of HEMT -technology in some cases is able to alleviate the base resistance problem, although contacting to the two-dimensional electron gas in a manner compatible with the rest of the structure is difficult[13]. While proof-of-principle exists in many cases for the novel devices, the technology for reliable circuits is still to be achieved.

The more imaginative device ideas are likely to emerge once the technology for the definition of features in the lateral direction approaches that now achieved, via growth, in the depth direction. We can expect a greater degree of the negative differential resistance effect in our superlattice tunnel oscillator if we could constrain the vertical motion of carriers to be nearly one-dimensional. The reduction of the lateral degrees of motion could be achieved either by etching or ion-implantation, but only when the remaining conducting channel has a diameter of order upto 0.05 μm will the effects be prominent. Many structures could be bonded in parallel to achieve useful current densities and low resistance, but will the retained integrity of the superlattice after processing be adequate?

The work has been supported in part by Marconi Electronic Device

Limited, the English Electric Valve Company, the Ministry of Defence, and the Commission of the European Communities. We thank C E C Wood for discussions.

REFERENCES

1   R A Davies, M J Kelly and T M Kerr, "Tunnelling between two strongly coupled superlattices", Phys. Rev. Lett. 55, 1114-6 (1985)

2   R A Davies, M J Kelly and T M Kerr, "Room Temperature Oscillation in a Superlattice Structure", Electronics Lett, 22, 121-3 (1986)

3   R A Davies, M J Kelly and T M Kerr "Tailoring the I-V Characteristics of a Superlattice Tunnel diode", Electronics Letts. 23, 90-2, (1987)

4   N R Couch, M J Kelly, T M Kerr, E Britton and W M Stobbs "Evidence for the Role of Indirect Gap Tunnelling through Thin AlAs Barriers", Semiconductor Science and Technology 2, 244-7 (1987)

5   N R Couch, D G Parker, M J Kelly and T M Kerr "Low Dark-current GaAs/AlAs Graded-Parameter Superlattice PIN Photodetector", Electronics Letts. 22, 636-7, (1986)

6   D G Parker, N R Couch, M J Kelly and T M Kerr "On the High-Speed Photoresponse of a Quasi-Graded Band-Gap Superlattice p-i-n Photodiode", Appl. Phys. Letts, 49, 939-41, (1986)

7   T Kakagawa, N J Kawai, K Ohta and N Kawashima "New Negative-Resistance Device by a CHIRP Superlattice", Electronics Letts 19, 822-3 (1983)

8   T Nakagawa, H Imamoto, T Sakamoto, T Kojima, K Ohta and N J Kawai, "Observation of Negative Differential Resistance in CHIRP Superlattices", Electronics Letts 21, 882-4 (1985)

9   N Yokoyama, K Imamura, S Muto, S Hiyamizu and H Nishi "A New Functional Resonant Tunnelling Hot Electron Transistor (RHET)" Jap. J. Appl. Phys., Part 2, 24 L853-4 (1985)

10 N Yokoyama and K Imamura "Flip-Flop Circuit Using a
    Resonant-Tunnelling Hot Electron Transistor (RHET)", Electronics
    Lett, 22, 1228-9 (1986)

11 F Capasso, S Sen, A C Gossard, A L Hutchinson and J H English,
    "Quantum-Well Resonant Tunnelling Bipolar Transistor Operating at
    Room Temperature", IEEE Electron Devices Lett. EDL 7, 573-6, (1986)

12 J F Palmier, C Minot, J L Lievin, F Alexandre, J C Hamand, J Dangla,
    C Dubon-Chevallier and D Ankri, "Observation of Bloch Conduction
    Perpendicular to Interfaces on a Superlattice Bipolar Transistor",
    Appl. Phys. Lett 49, 1260-2, (1986)

13 See accompanying chapter M J Kelly, A P Long and P H Beton, "Hot
    Electron Spectroscopy and Transistor Design",

14 F Capasso, K Mohammed and A Y Cho "Resonant Tunnelling through Double
    Barriers, Perpendicular Quantum Transport Phenomen in Superlattices,
    and their Device Applications", IEEE Quantum Electronics QE-22,
    1853-69, (1986)

15 B Movaghar "Theory of High-field Transport in Semiconductor
    Superlattice structures", Semiconductor Science and Technology 2
    185-206 (1987)

16 L Esaki and L L Chang, "New Transport Phenomena in a Semiconductor
    Superlattice", Phys. Rev. Lett., 33. 495-7 (1974)

17 M Saitoh "Stark Ladders in Solids" J. Phys. C5, 914-26 (1972).

18 R T Phillips, N R Couch and M J Kelly "Electroluminesence from a
    Short Asymmetric GaAs/AlAs Superlattice", submitted for Semiconductor
    Science and Technology

19 S R Andrews and R T Harley "Optically Detected Tunneling Between
    Quantum Wells", Proceedings of the SPIE meeting in "Quantum wells and
    Superlattices", Marriott Bay, Florida, March 1987.

20 R A Davies, M J Kelly, T M Kerr, C J Hetherington and C J Humphreys,
    "Geometrical and Electronic Structure of a Semiconductor
    Superlattice", Nature 317, 418-9, (1985)

21 M J Kelly, R A Davies, A P Long, N R Couch, P H Beton and T M Kerr,
    "Vertical Transport in Multilayer Semiconductor Structures",
    Superlattices and Microstructures 2, 313-7, (1986)

22 R A Davies, D J Newson, T G Powell, M J Kelly and H W Myron,
    "Magnetotransport in Semiconductor superlattice", Semiconductor
    Science and Technolopgy 2. 61-4 (1987)

23 J C Maan "Combined Electric and Magnetic Field Effects in
    Semiconductor Heterostructures", in "Two-Dimensional Systems,
    Heterostructures and Superlattices", editors G. Bauer, F Kucher and H
    Heinrich (Springer-Verlag 1984), p183-9

24 This is also a factor in the double barrier structures, e.g. M.
    Tsuchiya, H Sakaki and J Yoshino "Room Temperature Observation of
    Differential Negative Resistance in an AlAs/GaAs/AlAs Resonant
    Tunnelling diode", Jap. J. Applied Phys. 24, L466-8, (1985)

25 C Sotomayer-Torres, private communication and to be published

26 Courtesy of E Britton, W M Stobbs, C J Hetherington and C J Humphries

27 E R Brown, T C H G Sollner, W D Goodhue and C D Parker,
    "Millimeter-band Oscillations Based in Resonant Tunnelling in a
    Double-barrier Diode at Room Temperature", Appl. Phys. Lett, 50, 83-5
    (1987)

28 c.f. S M Sze "Physics of Semiconductor Devices", 2nd Electron (Wiley,
    New York, 1981)

29 A P Long, P H Beton and M J Kelly, "Hot Electron Transport in Heavily
    Doped GaAs", Semiconductor Science and Technology 1, 63-70, (1986)

30 N A Cade, S H Parmer, N R Couch and M J Kelly "Indirect Gap Resonant
    Tunneling in GaAs/AlAs" (to be published)

31 N R Couch, D G Parker, M J Kelly and T M Kerr "Direct and Indirect-Gap
    Tunnelling in a Graded-Parameter GaAs/AlAs Superlattice" Proceedings
    of the 18th International Conference on the Physics of Semiconductors,
    (Stockholm 1986), editor O. Engstrom, (World Science 1987) p 247-50
    (Pressure measurements courtesy of P Klipstein)

32   F Singleton, R J Nicholas, N J Pulsford, N R Couch, and M J Kelly "Quasi Bound states in an asymmetric GaAs/AlAs superlattice", to be published.

33   F Capasso, H M Cox, A L Hutchinson, N Olsson and S G Hummel "Pseudo-quaternany GaInAsP Semiconductors: A New $Ga_{0.47}In_{0.53}As/InP$ Graded Gap Superlattice and its Application to Avalanche Photodiodes", Appl. Phys. Lett. <u>45</u>, 1193-5, (1984)

34   T Weil and B Vinter "Calculation of Carrier Transport in Pseudo-Quaternary Alloys", Surf. Sci., <u>174</u>, 505-8 (1986)

35   T Nakagawa, H Imamoto, T Kojima and K Ohta "Observation of Resonant Tunnelling in AlGaAs/GaAs Triple Barrier Diodes", Appl. Phys. Lett. <u>49</u>, 73-5 (1986)

36   M A Reed, J W Lee and H. L. Tsai., "Resonant Tunnelling Through in Double GaAs/AlAs Superlattice Barrier, Single Quantum Well Heterostructure", Appl. Phys. Lett., <u>49</u>, 158-60 (1986)

37   S Muto, "Transport Characteristic in Heterostructure Devices" in the Proceedings of "High Speed Electronics" (Copenhagen), editors B Kallback and H Beneking (Springer-Verlag 1986)

38   R A Davies, "Simulations of the Current-Voltage Characteristics of Semiconductor Tunnel Structure", GEC Journal of Research, 1987, to appear

39   A R Bonnefoi, T C McGill, R D Burnham and G B Anderson., "Observation of Resonant Tunnelling through GaAs Quantum Well States Confined by AlAs X-point Barriers", Appl. Phys. Lett <u>50</u>, 344-6, (1987), and see reference 4 therein.

40   J M Gering, D A Crim, D G Morgan, P D Coleman, W Kopp and H Morkoc "A Small-Signal Equivalent-Circuit Model for $GaAs-Al_xGa_{1-x}As$ Resonant Tunneling Heterostructures and Microwave Frequencies", J. Appl. Phys, <u>61</u>, 271-6 (1987)

41   S Luryi "Hot-Electron-Injection and Resonant-Tunneling Heterojunction Devices" to appear in "Heterojunctions: A Modern View of Band Discontinuities and Device Applications" editors F Capasso and G Margaritondo (North Holland 1987, to appear)

OPTO-ELECTRONICS IN SEMICONDUCTORS QUANTUM WELLS STRUCTURES: PHYSICS AND APPLICATIONS

D.S. Chemla

AT&T Bell Laboratories
Holmdel, N.J. 07733

1- INTRODUCTION

In recent years the development of lightwave communication systems has driven an extensive research in the field of opto-electronics. The capability of optical fibers to carry extremely high bit rates, larger than those that electronic can handle, has raised a lot of attention into optical signal processing. Optics is well suited for parallel processing, hence permitting high throughout, and a number of optical processes in condensed matter have a very fast response time. Therefore it is predictable that optical switching elements are going to become important components in future information processing systems. These devices are based upon the possibility to induce large changes in the refractive index, n, or in the absorption coefficient, α , of some medium. The rational behind this statement goes as follow. When an optical field E_ω propagates through a length l of matter it experiences a phase change $\phi = (2\pi n/\lambda + l\alpha/2) \times l$, thus an external perturbation can be used for switching this field if it can causes a phase change of the order of $\mathrm{Re}(\Delta\phi) \approx \pi$ or $\mathrm{Im}(\Delta\phi)' \approx 1$. Ideally one would like to dispose of media in which large and fast changes Δn or $\Delta\alpha$ can be induced with low energy. Furthermore it is also desirable for these media to be robust, to have good optical quality and to be compatible with other electronic and/or optical technologies they must interface. All these requirements point obviously toward semiconductor based materials.

Investigations of bulk semiconductors has shown that they exhibit a number of nonlinear optical processes[1] that have been used for laboratory demonstrations[2]. Unfortunately the magnitude of these effects is not large enough to be used in practical systems. A promising approach, that originates from the recent progresses in crystal growth[3,4], is to fabricate "artificial" semiconductor nanostructures with enhanced nonlinear optical response. Semiconductors systems whose dimensions are of the order of the natural lengths (10-500Å) which governs the electrons quantum mechanic of optical response (Bohr radii) and electronic transport (de Broglie wavelengths), present properties that are not found in the parent bulk materials[5]. The semiconductor nanostructures the most extensively investigated are quantum wells and superlattices[6]. In these lectures we will review the physics of nonlinear optical properties of semiconductor quantum well under optical and electrical excitations and we will discuss some applications to prototype opto-electronic devices.

3- ABSORPTION IN SEMICONDUCTOR QUANTUM WELLS

The optical transitions in QW are governed by excitonic effects that are strongly enhanced in quasi two dimensional semiconductor structures[7-8]. This results from the reduction of the average distance between the electron and the hole that increases the binding energy. For the lowest lying states, the confinement in the QW essentially limits the interaction with the lattice to vibration modes of the QW material, hence the sensitivity of QW excitons to temperature is very similar to that of the bulk material. This combination results in one of the most remarkable excitonic property of III-V material QW structures: the persistence of well resolved exciton resonance at room temperature and even higher in the absorption spectrum[9-12].

Excitons at high temperature are unstable against collision with thermal phonons. For example in GaAs QW the LO-phonon energy (36 meV) is much larger that the exciton binding energy (9-10 meV), thus a single exciton LO-phonon collision can ionize the exciton and release a free e-h pairs with a substantial excess energy. It was found that the exciton linewidth varies as the sum of a constant inhomogeneous term Γ_o and a term proportional to the density of LO-phonons of the QW material; $\Gamma(T) = \Gamma_o + \Gamma_{ph}/[exp(\hbar\Omega_{LO}/kT)-1]$.

The parameter Γ_{ph}, that describes the averaged strength of the interaction with phonons in GaAs

QW ($\Gamma_{ph} = 5meV$) is smaller than but comparable to that in the bulk ($\Gamma_{ph} = 7meV$). This confirms that collisions with the QW LO-phonons is indeed the dominant broadening mechanism and that its efficiency is hardly changed as compare to the bulk[13]. The temperature broadening of the exciton resonance can be interpreted as a reduction of life time so that at room temperature excitons live only 0.4 ps in GaAs QW. Simple thermodynamic arguments show that the excitons have a very small probability to reform[13]. Consequently at room temperature a photon absorption process can be pictured as; $\hbar\omega \rightarrow X \rightarrow e\text{-}h$. As we shall see latter this sequence induces the most interesting dynamics in the nonlinear optical response of QW.

4-NONLINEAR OPTICAL EFFECTS IN QUANTUM WELLS

It is important to distinguish the physical mechanisms at the origin of the nonlinear optical response of semiconductors, these are very different whether real or virtual excited populations are generated. Furthermore in the former case effects of non-thermal populations are significantly distinct from those induced by thermalized populations.

We consider the experimental studies in which real and thermalized e-h plasma are responsible for the nonlinearities. The plasma can be generated directly above the gap or indirectly following exciton creation by resonant absorption or phonon assisted absorption[9,13-16]. For thermalization to be completed, observation must take place a few ps after excitation. The changes in the absorption spectrum depend only on the density of e-h pairs N_{eh}. As N_{eh} increases a very smooth evolution is seen, the two resonances weaken and are progressively replaced by a step-like edge which shape fits well the 2D free e-h absorption profile. These changes are interpreted as follow[17,18], when the plasma density increases the energy of the free-pairs states decreases (band gap renormalization) whereas that of the bound states only experiences a small change because of their charge neutrality. The binding energy of the bound states measured from the renormalized continuum decreases thus the resonances thus loose oscillator strength. At low density where the continuum is still far from the resonance only the loss of oscillator strength is apparent. However if the density continues to increase the resonance merge in the continuum which eventually remains the only spectral feature observed.

The corresponding nonlinearities are very large but they saturate rather quickly, in terms of an "effective" third order susceptibility they translate in a very large value $\chi^{(3)} \approx 6 \times 10^{-2} \, esu$. The impressive magnitude of the room temperature nonlinear effects in GaAs QW was demonstrated by observation degenerate four wave mixing using as sole light source a cw laser diode. A diffraction efficiency of 0.5×10^{-4} was observed in a 1.25 μm sample, using only a $17W/cm^2$ pump intensity[19]. Optical bistable devices using GaAs QW as active medium in nonlinear Fabry-Perot etalons have been operated and have raised much interest for application to all optical logic[20]. GaAs/AlGaAs QW have been used as saturable absorber to passively modelock GaAs diode lasers. Stable operation producing pulses as short as 1.6 ps has been demonstrated[21].

The most efficient mechanism governing the saturation of excitonic resonances in bulk semiconductors is the long range direct screening (DS) of the Coulomb interaction by a e-h plasma. Experiments on bulk GaAs have shown that the screening by a charged plasma is much more effective than by a neutral gas[22,23] because the neutral excitons can only screen by making energy consuming transitions to some excited state[17]. If the processes by which plasmas and excitons reduce excitonic absorption were the same in the bulk and in QW, resonant excitation at room temperature would produce a weak bleaching of the resonances that would increase substantially as the inefficient excitons are ionized in an efficient plasma. Experiments using femtosecond spectroscopic techniques[24,25] were designed to observe these effects. It was found that the dynamic of the hh-exciton peak under resonant fs-excitation is well described by the generation of an exciton gas which density follows the integral of the pump laser pulse and transforms in free e-h pairs in a ionization time $\tau_i \approx 300fs$ [26]. The time constant of the process is in excellent agreement with that evaluated from the thermal broadening of the resonances. However these experiments also lead to the surprising results that in 2D systems saturation of exciton resonance by a gas of exciton is more efficient than by a warm e-h plasma[26]. This demonstrate that it is incorrect to extend directly the hierarchy of the processes responsible for excitonic saturation in 3D to the case of 2D quantum wells.

In addition to the direct screening the absorption strength of exciton resonance can be diminished by the effects of Pauli exclusion principle: phase space filling (PSF) and exchange interaction (EI). These effects are extremely efficient but only at very short distance and only involve Fermions of the same spin. In the case of exciton they arise from the composite nature of the bound e-h pairs[17]. Excitons behave as Bosons only in the limit of very dilute gas and at high density it is absolutely necessary to account for their composite nature. In 3D the because of the efficiency of the the long range DS and the effects of the short range PSF and EI are seldom recognized although they are incorporated in the correct many body theories[17]. The unusual behavior of quantum wells has been recently explained by assuming that in 2D the strength of DS is strongly weakened as compared to PSF and EI[27]. In this case the effect of a plasma or of a gas of excitons on the exciton resonances is directly related to the overlap between the distribution of photoexcited e-h pairs (bound or unbound) and the single particle states out of which the exciton relative motion wavefunction in k-space U(k) is built up. At low temperature e-h plasma which fill up state close to k = 0, where U(k) is centered, is more efficient than the excitons in bleaching the resonance absorption, however as the temperature increases the thermal e-h distributions expand more and more toward high energy and the plasma efficiency decreases as E_{1S}/kT, here E_{1S} is the exciton bidding energy. The model also explains the effects seen at the lh-exciton. Finally the theoretical value of the nonlinear cross section is in very good agreement with the experimental one.

The assumption that DS is smaller than EI and PSF in quantum wells was tested directly by generation of a non-thermal carrier distribution in the continuum, between the first (n_z = 1) and the second subband (n_z = 2) transitions less than one LO-phonon energy above the gap and measuring the evolution of the absorption spectrum including the two resonances as a function of the pump-probe delay between[28]. At small delays the carriers do not occupy states at k = 0 in the n_z = 1 subband so that only the weak DS is effective. As they thermalize the carrier occupy states at the bottom of the subband and the effects of Pauli principle (EI+PSF) are turned on and strongly bleach the n=1 resonance. Meanwhile the n_z = 2 excitons, which are at higher energy and in any case made of single particle states nearly orthogonal to those occupied by the carriers, only feel the DS. The effects at the n_z = 2 excitons do not vary in time showing that DS remains essentially constant as the carriers thermalize. These experimental results put an upper limit on the magnitude of DS which is found to be at least six times smaller than the sum (EI + PSF). Yet another confirmation of the theory was obtained from the study of shift of the excitonic resonance at low temperature[29]. In 3D the exciton resonances remain remarkably at a constant energy in the presence of a e-h plasma even if the single particle states experiences strong red shift[17,22,23]. This is due to an almost perfect compensation of the red shift induced by the DS and a blue shift due to the effects of the exclusion principle. In this case where only exciton are present the blue shift can be interpreted as a short range hard-core repulsion. In the quasi 2D QW it is no more compensated by the weak exciton screening and thus should be measurable[29]. At low temperature the shift of the exciton is given by[27]; $\delta E_{1S} = -16 \pi N_x a_0^2$ (1-315 $\pi^2 / 2^{12}$) E_{1S} as a function of the exciton radius and density N_x. Experiments performed at low temperature on QW of various thicknesses have evidenced this blue shift when excitons are selectively generated or when they are formed after the cooling of a nonresonantly excited plasma. The shift is rater difficult to observe in thick layer but it is very clear in the narrowest QW where the dimensionality approaches the 2D limit[29,30]. The magnitude and density dependence have been found to be in excellent agreement with the theory.

5- ELECTROABSORPTION IN QWS.

Electroabsorption in QWS has been extensively studied and is now well documented, an exhaustive bibliography can be found in Ref. [31].

When an electrostatic field is applied parallel to the layer where the confinement potential is ineffective, the only force binding the electron and the hole is the Coulomb force and the effects are similar to those seen in bulk material[32]. The only difference with the case of massive samples is that the exciton binding energy and the in plane wave function are slightly modified[32]. The dominant process is a field ionization and theoretically a second order energy shift[33]. It is found that when a field is applied parallel to GaAs QW's the resonances broaden easily and disappear at moderate fields ($E_0 \approx$ 2-5×10$^4 V/cm$). No significant shift is observed and the evolution of the spectra are in qualitative agreement with the theory[32,33]. In this geometry the capacitance is very small so that it can be used to test how fast a change in the excitonic absorption change can be induced in a semiconductor. In recent the experiments[34] a potential was applied across the electrodes of a sample and the transmission was probe with a 120 fs broad continuum light pulse. Then the field in the QWS was shorten by a high density plasma across the inter-electrode gap was produced by a strong 50 fs red pulse. The 1/e recovery time of the absorption was measured to be 330 fs and seems to be still limited by electric field reflections in the electrodes. The absorption recovery time has to be compared to some characteristic times of the exciton dynamics at room temperature. For example a limit on how fast a resonance can recover is set by the mean time for exciton scattering by thermal LO-phonons (300 fs at room temperature). The change in excitonic absorption should follow the sudden application of a field in a time limited by how long it takes to the e-h pair to "feel" its Coulomb correlation (i.e the exciton classical orbit-time ~150 fs) or by how long it takes to the two particles to tunnel away one from another through the Coulomb well (i.e the mean time to field ionization $\approx 380 fs$ for a 8 kV/cm field[32]). All these values fall in the same range showing that the response measured in these experiments approaches the fundamental limits of the exciton dynamics. The intrinsic limits of the speed of electroabsorption are of course also of great technological interest for high speed sampling and modulation.

The electro-absorption for fields perpendicular to the layers presents features that are specific of QWS[32]. The electron and the hole charge distributions are pulled apart by the field and pushed against opposite well walls. If the thickness of the QW is large compared to the 3D Bohr radius the pair correlation is lost, but if the QW thickness is less than the 3D Bohr radius the particles are stopped from flying apart by a potential barrier and continue to interact with each another. Therefore as the QW thickness decreases the electroabsorption should show a continuous transition from the 3D behavior to a strongly quantized one. This has been analytically demonstrated recently for the Franz-Keldysh effect in infinitely deep wells[35]. The transition from the quantum confined to 3D behavior occurs over a narrow range of thickness for a given field and is accompanied by the growth of symmetry forbidden optical transitions from which the Franz-Keldysh oscillations originate. In the case of finite well depth, one can expect strong tunneling only for fields large enough to provide the carriers with an energy comparable to the potential barrier height by acceleration of the particles over the QW thickness. Indeed tunneling resonances calculations give very narrow the width of the resonances for moderate fields yet large enough to produce strong shifts of the absorption edge[32].

This behavior is actually observed in GaAs[32,36] and in high purity InGaAs/InP[37] QW When the static field is applied the whole absorption profile shifts toward lower energies, the height of the resonance somehow decreases and the peaks broaden. The magnitude of the effect is surprising, for example the maximum shift seen with yet resolved exciton peaks is about 40 mev at a field of $E_0 = 2.2 \times 10^5 V/cm$. This is 10 times the bulk exciton binding energy and correspond to 100 times the exciton classical ionization field. This effect can be described as an extreme quantization of the Stark effect hence its name; the "Quantum Confined Stark Effect" (QCSE)[38]. The overall behavior is now understood in details. The shifts of the resonances are well accounted for by summing the shifts of the single particle states to the change of the exciton binding energy due to the compression of the wavefunctions against the walls of the well[32]. The changes in height of the resonances and of the continua are well explained by several sum rules that have been demonstrated recently[31].

In order to apply large fields and yet avoid the flow of large currents, the samples are grown as p-i-n diodes with the QW's in the intrinsic region. The field across the QW's is varied by reverse biasing the diodes. Because of this geometry the sample also behaves as a photodetector, with an internal quantum efficiency close to unity[32]. This characteristic is important because the photocurrent give an electrical measure of the absorption that can be utilized in an external circuit to react on the bias voltage, thus providing a mechanism for feedback.

The most simple application of the QCSE is to high speed modulation of light. Modulation in the 100 ps range has been demonstrated both in the configuration where the light propagate perpendicular to the plane of the QW's and parallel to it in waveguides[36,39-41], in this geometry modulation depth as large as 10 dB has been measured. Recently the application to modulation has been extended to longer wavelengths using other III-V compounds[42-44]. The speed is presently limited by circuit consideration and can be further increased. The waveguide geometry is a natural configuration for integrated optics and the first steps toward optical integrated devices have reported[45,46].

As mentioned before the possibility to measure the absorption of p-i-n QWS by measuring the photo current leads to new devices. A simple example of such a device is a wavelength-selective voltage-tunable photodetector which operates on the following principle. Light, with photon energy less than that of the hh exciton peak at zero field, incident on the p-i-n QWS produces a photocurrent that is maximized for an applied field just large enough to shift the exciton to the photon energy. If the wavelength is changed then the photodetector response can be maximized again by changing the applied voltage, thus giving a mean to follow electrically the frequency of the light source[47,48]. This has potential applications in coherent light wave communications where it is often necessary to determine simultaneously the intensity and the wavelength of an optical beam.

The Self Electro-optic Devices (SEED) form an novel class of devices that utilize the p-i-n QWS capacity to operate as close coupled photodetector/modulator. To operate the p-i-n QWS is connected to an external circuit which imposes a relation between the applied voltage and the current passing through the device. These two quantities must also satisfy the response function imposed by the relation between absorption coefficient and field across the QWS, hence the possibility of electronic feedback on the optical behavior. The feedback can be positive or negative depending of the wavelength, both have been exploited in optical bistable gate, self-linearized modulator and optical level shifter and devices have been demonstrated in the propagation mode normal to the QWS as well as in waveguides[36,49-51].

The most interesting SEED operation may be that of low switching energy optical bistable gate[51]. The switching as well as the oscillations can be followed optically or electronically. The contrast between the two states is large because of the change in absorption coefficient is typically of the order of $5 \times 10^4 cm^{-1}$, 20/1 contrast ratios have been measured in waveguide SEED[36]. The importance of the SEED for optical switching resides in the extremely low operating energy of the device despite the absence of optical cavity, the tolerance of the SEED to fluctuation in the optical and electrical operating parameters and the natural structure for large planar arrays. Switching energy as low as 20 fJ (15 fJ electrical and 5 fJ optical) and 20 ns switching time have been demonstrated in large devices (100 um diameter). The limits for switching energy are expected to be in the few fJ range and switching time of a few ns can be expected from SEED's with diameter in the 10 um range. Recently operation of planar arrays of SEED including integrated photo-transistors for current load has been shown[52].

6- CONCLUSION

In these lectures we have reviewed the physics and application of QWS room exciton resonances in electro-optics and nonlinear optics. Some of these effects such as the QCSE are specific to QWS and the short lifetime of excitons at room temperature induced in the nonlinear optical response ultrafast transients that have no counterpart bulk semiconductors. These novel properties have been used to demonstrate interesting devices with potential applications to optical signal processing, optical logic and convenient source of ultrashort optical pulses. The general trends of the underlying physics are in the case of the QWS based on the III-V semiconductors.

Acknowledgements

Most of the work reviewed in these lectures has been performed in AT&T Bell Laboratories, in collaboration with D.A.B. Miller, A.C. Gossard, J.S. Weiner, W.H. Knox, T.C. Damen, J.E. Zucker, A. Pinczuk, S. Schmitt-Rink, C.V. Shank, R.L. Fork, J. Shah, T.H. Wood, and C.A. Burrus.

REFERENCES

[1]   D.S. Chemla, J. Jerphagnon, " Nonlinear Optical Properties of Semiconductors." Handbook of Semiconductors, Vol 2, Optical Properties of Solids, North-Holland Amsterdam (1980).

[2]   A. Miller, D.A.B. Miller, S.D. Smith Advances in Physics, 30, 697 (1981).

[3]   See for example, L.L. Chang and K. Ploog eds "Molecular Beam Epitaxy and Heterostructures" NATO Advanced Sciences Institute Series, Nijhoff, Dordrecht (1985).

[4]   See for example, J.B. Mullin, S.J.C. Irvine, R.H. Moss, P.N. Robson, D.R. Wight eds "Metal Organic Vapor Phase Epitaxy" North-Holland Amsterdam (1984).

[5]   For recent reviews see for example, F. Capasso and B.F. Levine J. Lumin. 30, 144 (1985) and D.S. Chemla J. Lumin. 30, 502 (1985).

[6]   See for example "The Physics and Fabrication of Microstructures and Microdevices" M.J. Kelly and C. Weisbuch eds Spinger-Verlag (1986).

[7]   R. Dingle, W. Wiegmann, C.H. Henry Phys. Rev. Lett. 33, 827 (1974).

[8]   R.C. Miller, D.A. Kleinman J. Lum. 30, 144 (1985).

[9]   D.A.B. Miller, D.S. Chemla, P.W. Smith, A.C. Gossard, W.T. Tsang, Appl. Phys. Lett. 41, 679 (1982).

[10] D.A.B. Miller, D.S. Chemla, P.W. Smith, A.C. Gossard, W. Wiegmann, Appl. Phys. B28, 96 (1982).

[11] J.S. Weiner, D.S. Chemla, D.A.B. Miller, T.H. Wood, D. Sivco, A.Y. Cho Appl. Phys. Lett. 46, 619 (1985).

[12] H. Temkin, M.B. Panish, P.M. Petroff, R.A. Hamm, J.M. Vandenberg, S. Sunski, Appl. Phys. Lett. 47, 394 (1985).

[13] D.S. Chemla, D.A.B. Miller, P.W. Smith, IEEE J. Quant. Electron. QE-20, 265 (1984).

[14] D.S. Chemla, D.A.B. Miller J. Opt. Soc. Am. B2, 1155 (1985) and reference therein.

[15] A. Von Lehnen, J. E. Zucker, J.P. Heritage, D.S. Chemla, Appl. Phys. Lett. 48, 1479 (1986).

[16] A. Von Lehnen, J. E. Zucker, J.P. Heritage, D.S. Chemla, Phys. Rev. B35, 6479 (1987).

[17] For a recent review of semiconductor optical nonlinearities close to the band gap see: H. Haug and S. Schmitt-Rink, Prog. Quant. Electron. 9, 3 (1984).

[18] S. Schmitt-Rink, C Ell, J. Lumin. 30, 585 (1985).

[19] D.A.B. Miller, D.S. Chemla, P.W. Smith, A.C. Gossard, W. Wiegmann, Opt. Lett. 4, 477 (1983).

[20] N. Peyghambarian, H. M. Gibbs, J. Opt. Soc. Am. B2, 1215 (1985).

[21] P.W. Smith, Y. Silberberg, D..B. Miller J. Opt. Soc. Am. B2, 1228 (1985).

[22] G.W. Fehrenbach, W. Schafer, J. Treusch, R.G. Ulbrich, Phys. Rev. Lett. 49, 1281 (1982).

[23] G.W. Fehrenbach, W. Schafer, R.G. Ulbrich, J. Lumin. 30, 154 (1985).

[24] R.L.Fork, B.I. Greene, C.V. Shank, Appl. Phys. Lett. 38, 671 (1981).

[25] W.H. Knox, M.C. Downer, R.L. Fork, C.V. Shank, Opt. Lett. 9, 552 (1984).

[26] W.H. Knox, R.L. Fork, M.C. Downer, D.A.B. Miller, D.S. Chemla, C.V. Shank, A.C. Gossard, W. Wiegmann, Phys. Rev. Lett. 54, 1306 (1985).

[27] S. Schmitt-Rink, D.S. Chemla, D.A.B. Miller, Phys. Rev. B32, 6601 (1985).

[28] W.H. Knox, C. Hirlimann, D.A.B. Miller, J. Shah, D.S. Chemla, C.V. Shank, Phys. Rev. Lett. 56, 1191 (1986).

[29] D. Hulin, A. Mysyrowicz, A. Antonetti, A. Migus, W.T. Masselink, H. Morkoc, H.M. Gibbs, N. Peyghambarian, Phys. Rev. B33, 4389 (1986).

[30] N. Peyghambarian, H. M. Gibbs, J. L, Jewell, A. Antonetti, D. Hulin, A. Mysyrowicz, Phys. Rev. Lett. 52, 2433 (1984).

[31] D.A.B. Miller, J.S. Weiner, D.S. Chemla, IEEE J. Quant. Electr., QE-22, 1816 (1986).

[32] D.A.B. Miller, D.S. Chemla, T.C. Damen, A.C. Gossard, W. Weigmann, T.H. Wood, C.A. Burrus Phys. Rev. B32, 1043 (1985).

[33] F.L. Lederman, J.D. Dow Phys. Rev. 13, 1633 (1976).

[34] W.H. Knox, D.A.B. Miller, T.C. Damen, D.S. Chemla, C.V. Shank, Appl. Phys. Lett. 48, 864 (1986).

[35] D.A.B. Miller, D.S. Chemla, S. Schmitt-Rink to be published in Phys. Rev. B (1986).

[36] J.S. Weiner, D.A.B. Miller, D.S. Chemla, T.C. Damem, C.A. Burrus, T.H. Wood, A.C. Gossard, W. Weigmann, Appl. Phys. Lett. 47, 1148 (1985).

[37] I. Bar Joseph, C. Klingshirn, D.A.B. Miller, D.S. Chemla, U. Koren, B.I. Miller, Appl. Phys. Lett. 50, 1010 (1987).

[38] D.A.B. Miller, D.S. Chemla, T.C. Damen, A.C. Gossard, W. Weigmann, T.H. Wood, C.A. Burrus, Phys. Rev. Lett. 53, 217 (1984).

[39] T.H. Wood, C.A. Burrus, D.A.B. Miller, D.S. Chemla, T.C. Damen, A.C. Gossard, W. Weigmann, Appl. Phys. Lett. 44, 16 (1984).

[40] T.H. Wood, C.A. Burrus, R.S. Tucker, J.S. Weiner, D.A.B. Miller, D.S. Chemla, T.C. Damen, A.C. Gossard, W. Weigmann, IEEE J. Quant. Electron. QE-21, 117 (1985).

[41] T.H. Wood, C.A. Burrus, R.S. Tucker, J.S. Weiner, D.A.B. Miller, D.S. Chemla, T.C. Damen, A.C. Gossard, W. Weigmann, Electronics Lett. 21, 693 (1985).

[42] K. Wakita, Y. Kawamura, Y. Yoshikuni, A. Asahi, Electronics Lett. 21, 338 (1985).

[43] K. Wakita, Y. Kawamura, Y. Yoshikuni, A. Asahi, Electronics Lett. 21, 574 (1985).

[44] T. Miyazawa, S. Tarucha, Y. Ohmori, H. Okamoto, Post deadline paper P1, Second International Conference on Modulated Semiconductor Structures, Kyoto, Japan, Sept. 1985.

[45] S. Tarucha, H Okamoto, Appl. Phys. Lett. 48, 1, (1986).

[46] Y. Arakawa, A. Larsson, J. Paslaski, A.Yariv, Appl. Phys. Lett. 48,561 (1986).

[47] T.H. Wood, C.A. Burrus, A.H. Gnauck, J.M. Wiesenfeld, D.A.B. Miller, D.S. Chemla, T.C. Damen, Appl. Phys. Lett. 47, 190 (1985).

[48] A. Larsson, A.Yariv, R.Tell, J. Maserjian, S.T. Eng, Appl. Phys. Lett.47, 866 (1985).

[49] D.A.B. Miller, D.S. Chemla, T.C. Damen, A.C. Gossard, W. Weigmann, T.H. Wood, C.A. Burrus, Appl. Phys. Lett. 45, 13 (1984).

[50] D.A.B. Miller, D.S. Chemla, T.C. Damen, A.C. Gossard, W. Weigmann, T.H. Wood, C.A. Burrus, Opt. Lett. 9, 567, (1984).

[51] D.A.B. Miller, D.S. Chemla, T.C. Damen, A.C. Gossard, W. Weigmann, T.H. Wood, C.A. Burrus, IEEE J. Quant. Electr. QE-21, 1462 (1985).

[52] D.A.B. Miller, J. Henry, A.C. Gossard, J.H. English, Appl. Phys. Lett. 49, 821 (1986).

INDEX

Absorption
  Drude, 320
  intraband, 321
  in quantum wells (see also
      Interband magneto-optics
      in quantum wells
    electroabsorption, 428-432
    photoluminescence excita-
        tion method, 269
    semiconductor, 424-425
    superlattice cutoff wave-
        length, 81
Absorption coefficients, 81,
    269
Absorption spectra
  excitons, 264
  techniques, 267-269
    dynamics of excitations,
        279
    transient/ultrafast, 272
Accommodation, mismatch, 105-
    113, 119
Accumulation layers, 221, 223,
    225
Acoustic-mode phonons
  deformation potential and
      piezoelectric scatter-
      ing, 140-143
  folded modes, 305, 306, 307
Aharonov-Bohm effect, 149
AlAs-GaAs (see also Molecular
    beam epitaxy, group III-
    V systems)
  barrier thickness, 57-63
  short-period superlattices,
      54-57
  tunneling structures, graded-
      parameter superlattice,
      410, 411
  ultrathin-layer, 63-66
AlAs-GaAs-AlAs
  bandgap-engineered devices,
      383
  tunneling via X-point states,
      182
AlGaAs, bandgap-engineered de-
    vices, 381

AlGaAs-GaAs heterostructures
  Coulomb scattering, 136-137
  cyclotron resonance, 342-344
  FIRS, 332
  hot-electron transistor, 394
  MBE process, 49, 51-52
  Raman spectroscopy, 303
  transport, 135
  tunneling structures, graded
      parameter SL, 409
  2D plasmon frequency, 330
Alloy scattering, 140
Anisotropic elasticity theory,
    106
Antimony, 122
Arsenic, MBE process, 50, 66
Asymmetry, 255 (see also
    Symmetry)
Auger coefficient, 288
Avalanche photodiode (APD),
    378-383

Back-scattering
  one-dimensional system, 135
  Raman, 302
Band diagram, real-space, 55,
    56
Band discontinuities, tunable,
    386-389
Bandedges (see also Conduct-
    ion bands: Valence
    bands)
  magnetic quantization, 373
  spatial variations, 2
Bandgap energies (see also
    Energy gaps)
  envelope function approxi-
      mation, 22
  renormalization
    in doped wells, n-type,
        34, 35
    optical effects in quan-
        tum wells, 425
  stress conditions and, 125
Bandgap engineering
  solid-state photomultipliers
      and graded gap trans-

Bandgap engineering (continued)
    istors, 378-383
  tunable band discontinuities,
    386-389
  tunneling devices, 383-386
Bandgaps (see also Energy
    gaps; Gaps)
  Brillouin zone folding, 128
  envelope function approx-
    imation, 22, 23
  indirect, 55
    Brillouin zone, 128
    tunneling, 179-182, 407
  periodically varying width,
    124
  short-period SLs, GaAs-AlAs,
    27
  SiGe, 122-124
  superlattice minigaps, 365,
    366, 373
  tunneling
    design of devices, 383-386
    through indirect-gap
      barriers, 179-182
Band mixing
  barrier thickness and, 60
  complex valence bands, 363
  heavy and light-hole, 266
  TM emission, 281
Band offsets, 5, 6, 309
Band ordering, in SiGe-Si,
  122-124
Band structure
  barrier thickness and, 58
  FIRS, 341
  hetero-interfaces, 6
  MBE process
    Hg-based structures, 72,
      85-90
    III-V short-period SLs,
      55-56
  plasmon dispersion, 330
  quasi-unidimensional hetero-
    structures, 38
  tunneling current, 161
Barrier potential, origin of,
  160
Barriers
  carrier confinement, 56
  Coulomb scattering, 136-137
  and electroabsorption, 429
  forbidden LO-phonon inten-
    sity, 309
  hot-electron transistors,
    393-394
  indirect-gap, 179-182
  superlattice
    magnetic quantization, 365
    368-370
  and optical properties, 57-63
  superlattice tunnel diode,

Barriers (continued)
    405, 406, 407
  and tunneling, 179-182, 185,
    411, 413
    in superlattices, 405, 406
    409
    transfer matrix method,
      167-170
    tunneling current, 161
    tunneling probability,
      163-165, 170
Beryllium, MBE process, 44-45,
  49
Binary semiconductors
  interface disorder, 278
  MBE, see Molecular beam
    epitaxy group, III-V
    systems
Binary-to-ternary layer growth,
  52
Binding energies
  cyclotron emission spectra,
    205-206
  exciton, 264-265, 276
  MBE process, 86, 89-90
  of s-states in interband
    experiment, 358-359
Bipolar transistors, 144
Bistable device, 426
Bloch-Gruneisen range, 143
Bloch states, 4, 23, 263, 264,
  266
Boltzmann constant, 410
Boltzmann equation, 133, 147-
  148
Born approximation
  Coulomb scattering rates, 136
  quantum Hall effect, 239-240
Boundary conditions, envelope
  function, 23
Bragg reflections, 53-54
Brillouin zone, 21
  FIRS, 335
  heterolayer electronic
    states, 27
  quasi-unidimensional hetero-
    structures, 37, 38
  Raman spectra, 305
  reflectivity modulation
    techniques, 270
  short-period SLs, GaAs-
    AlAs, 27
  strained-layer SLs, 128-129
  X-point tunneling, 184
Brooks-Herring predictions, 12
Buffer layer, strained-layer
  SLs
  design of, 116-118
  metastability of , 119-120
Burgers vector, 108, 109, 110
  111, 119

Capacitance, 197 (see also
    Magnetocapacitance)
  density of states, 2D elec-
    trons, 197-198
  electroabsorption, 428
  superlattice tunnel diode,
    407
Capture efficiency, 277
Carbon, MBE process, 48
Carrier concentration (see
    also Carrier density)
  in doped wells, n-type, 34,
    35
  quantum Hall effect, fract-
    ional, 253, 254
Carrier confinement, see Con-
    finement
Carrier density, 133 (see also
    Carrier concentration)
FIRS, 321
  Hall effect, 234
  lasers, 286, 288
  low-dimensional systems,
    290-291
  quantum Hall effect
    Landau quantization, 236
    Hall resistance, 231
  scattering
    Coulomb, 139
    interface, 140
  transport, 134, 135
Carrier distribution function,
    134
Carrier injection, short-
    period SLs, 55
Carrier leakage, lasers, 287
Carriers
  capture in quantum wells, 277
  electric field and, 145
  energy levels in doped het-
    erolayers, 31
  heating, 144-145, 146
  lasers, 286, 287, 288
  magnetotransport, 217
  warm and hot, 144-145
Carrier transport, see
    Transport
Cascade models, 273
Cathodoluminescence spectra,
    274
Cavity losses, 289
CdMnTe, 6
CdTe-CdMnTe, 15
Central barrier layer, SL
    tunnel diode, 406
Charge
  electroabsorption, 429
  energy levels in doped
    heterolayers, 33, 34
  fractional, 249
Charge accumulation, 171
  cyclotron resonance measure-

Charge density (continued)
    ments, 342, 343, 344
  FIRS, 335, 336, 341, 344
Charge density modulated sys-
    tem, plasmon disper-
    sion, 336
Charged impurity scattering,
    140
Chemical beam epitaxy (CBE),
    66
Chemical vapor deposition, 4
CHIRP superlattice, 411-412
Circularly polarised radiation
  bandgap-engineered devices,
    386
  FIRS, cyclotron resonance,
    343
  interband magneto-optics
    in quantum wells,
    360-361
Circular polarization spec-
    trum, 275, 276
CMOS-circuits, 104
Co-evaporation, MBE process,
    44-45
Coherent interfaces, 411
Coherent transport, see
    Transport
Commensurate growth, 113
Common anion rule, 96
Complex valence band, 360-365,
    373
Compression, strained-layer
    SLs, 124
Conductance
  coherent scattering and,
    150-151
  magnetic field and, 178
  negative differential, see
    Negative differential
    conductance
  tunneling structures, graded
    parameter superlattice,
    410
  Conduction
bandgap-engineered devices,
    386
  hopping, 151, 258
Conduction band, 266
  barrier thickness, 58, 59-63
  electronic states, 26-27
  envelope function approxi-
    mation, 21-24
  interband magneto-optics in
    quantum wells, exci-
    tonic effects, 357-360
  lasers, 282, 284-288
  strained-layer SLs, 123-124
  tunneling, 384
Conduction band offset
  short-period SLs, GaAs-
    AlAs, 27

Conduction band offset (continued)
strained-layer SLs, 123-124
Conductivity
Drude, 322
Drude-like description, 349
Hall device, 233
Shubnikov-de Haas effect,
218, 219
Confinement
acoustic phonons, 307
barrier thickness, 59-63
dielectric potential, 322
electrical and optical, 43
energy levels in doped heterolayers, 33
excitonic levels and radii,
265
heterointerface types, 5, 6
quantum wells, capture in,
277
magnetotransport, 217
Monte Carlo methods, 147-148
quantum Hall effect, fractional, 255
selection rules, 267
short-period SLs, 56
Confinement energy
and cyclotron effective
mass, 350
magnetic quantization, 370
non-parabolicity, 350
Confinement factor
lasers, 283, 284
low-dimensional systems,
290-291
Confinement potential, selection rules, 267
Continuous wave (CW) heterojunction laser, 377
Corbino geometry, 220, 232, 235
Coulomb interaction, 160, 251
of electron and holes, 365
electronic light scattering,
311-312
optical effects in quantum
wells
electroabsorption, 428, 429
nonlinear, 426
Coulomb scattering
mechanism of, 135-139
and phonon scattering, 141,
143
Coupling
barrier thickness, 58, 60
excitonic, 9, 10
plasmon, 330
Coupling constant, piezoelectric, 141
Critical thickness
vs. lattice mismatch, 105-106

Critical thickness (continued)
strained-layer SLs, 113, 114,
119-120, 309
Crystal lattice, see Strained layer superlattices
Cubic lattice cell, tetragonal
distortion, 106-108
Current
Hall 220
tunneling, 171-172
Current density, 321
Boltzmann equation, 134
and Hall plateau, 252
Hall resistance, 231
lasers, 287
tunneling, 171, 172
2D surface, 325
Current-voltage characteristics
rectangular barrier, 164
Si/SiGe MODFET, n-channel,
127
tunneling, 404
Cutoff energy, transitions in
superlattices, 372
Cutoff wavelength, HgTe-CdTe,
84-85
Cyclotron energy, 189, 236
Cyclotron frequency, 176, 410
Cyclotron mass, 236, 350-353
Cyclotron motion, 217, 236
Cyclotron orbit
superlattice, 366, 367-370
tunneling structures, 415
Cyclotron resonance
and background DOS, 190
FIRS, 342-344
high-mobility hetero-
structures, 326
magnetoplasmon dispersion,
332-334
plasmon coupling, 339-340
Landau levels, 208, 236
(see also Landau
levels)
magneto-optical properties,
189, 348-349
linewidth, 190, 354-357
with magnetic field para-
llel to layers, 365-374
magnetotransport, 217
and quantum well potential
shape, 350-353
tunneling structures, 176,
410
Cyclotron resonance spectroscopy
emission, 202-206
transmission, 206-209

Defects (see also Disorder)
  dislocations, misfit, 108-111
  electron interactions, 354-
    357
  and scattering
    Coulomb, 135-139
    deformation potential and
      piezoelectric, 140-143
Deformation potential scat-
    tering, 140-143
Degeneracy, 135
  Boltzmann equation, 134
  energy levels in doped
    heterolayers, 31, 32
  fractionally quantized
    states, 250
  Landau, 240, 356
  quasi-unidimensional hetero-
    structures, 36-37
  rectangular or centrosym-
    metric quantum wells, 30
  transverse optical phonons,
    302
Density of states
  comparison of, 3, 4
  Gaussian
    magnetic field and, 194-
      198
    specific heat measurement,
      190-194
    temperature-dependent re-
      sistivity, 198-199
  in-plane dispersion relation,
    28-31
  Landau levels, see Landau
    levels
  lasers, 284
  low-dimensional systems,
    290-291
  magnetocapacitance, 197-198
  modulation-doped materials,
    281
  optical techniques
    dielectric function, 262
    dynamics of excitations, 279
    lasers, 284, 285, 286
    reflectivity modulation
      techniques, 270
  quantum Hall effect (see also
    Quantum Hall effect)
    fractional, 256, 257
  quasi-unidimensional sub-
    bands, 38-40
  resonant tunneling via
    Landau levels, 177
  superlattice tunnel diode,
    406
  2D system, 218, 219
Depolarisation shift, 338
Device simulation, see Simu-
    lation programs

Diamagnetic shift, 360, 370
Dielectric constant
  dynamics of excitations, 279
  and reflectivity changes,
    270
Dielectric function
  interband optical phenomena,
    262
  of 2D electronic systems,
    321-329
Diffraction spectroscopy, MBE
    process, 53-54
Diffusion, MBE process, 49,
    76-79, 80
Dimensionality, reduced, 2, 3,
    4 (see also Low-
    dimensional electronic
    systems)
Dipole moment, s and p state,
    265, 266
Dipoles, interface, 96
Direct-indirect transitions,
    26-28
Direct screening, 426, 427
Discontinuity, heterointer-
    faces, 6
Dislocations, 14
  crystal lattice distortion,
    110
  MBE process, Hg-based struc-
    tures, 77
  misfit, 108-111, 120, 309
    equilibrium theory, 111-
      113
    general properties, 108-
      111
Disorder (see also Interface
    disorder)
  exciton peaks, 264
  dynamics of excitations, 279
  and low-dimensional systems,
    289-290
  MBE process, 53-54
  optical methods, 278
  quantum Hall effect, 239-240
    fractional, 252, 253, 254,
      255
Dispersion
  barrier thickness, 60
  magnetoplasmon, 332-334
  optical phonons, 307
  plasmon, 330-331
Dispersion relation (see
    also Nonparabolicity)
  barrier height and, 365, 370
  in-plane, 28-31
  non-parabolic, 350-353
Doped heterolayers
  energy levels in, 31-35
Doping, 2, 312 (see also
    N-type doping; P-type

Doping (continued)
       doping)
  Coulomb scattering, 136-137
  heterolayers, energy levels
     in, 31-35
  hot-electron transistor,
     394, 398, 399
  impurity-related lumine-
     scence, 280
  MBE process, 44-45, 49
    materials, 44
    short-period SLs, 56-57
  modulation, 11-12, 280-282
  by secondary implantation,
     122, 124-127
  superlattice tunnel diode,
     407
Double-barrier heterostruc-
     tures, 33
  current-voltage character-
     istics, 175
  lasers, 284-288
  rectangular, 160
  resonant tunneling, 7, 8
  transfer matrix method, 169
  tunneling probability, 167-
     170
Double-crystal X-ray diffrac-
     tion, 58
Drift-diffusion equation, 146
Drude model, 321, 342-343
DX centers, 55, 407

Effective barrier symmetry,
     168
Effective mass, cyclotron, 350-
     353
Effective mass theory, 1, 264
Effusion cells, 44-45, 46, 73,
     74
E-h plasma, 425, 426, 428
Elasticity, strained crystal,
     106-108
Elastic mean free path, 151
Elastic scattering
  Boltzmann equation, 134
  coherent, 151
  quantum Hall effect, 234-235
Elastic strains, SL, 114
Electrical transport, see
     Transport
Electric dipoles, quantum-
     mechanical, 265
Electric field, 133
  electron temperature as
     function of, 220
  along growth axis, 30
  Hall resistance, 231
  LO-phonon Raman scattering
     (EFIRS), 307-309

Electric field (continued)
  and tunneling probability,
     168
  warm and hot carriers, 144-
     145
Electric Quantum Limit, 29, 31
Electroabsorption in QWs, 428-
     432
Electroluminescence, 410
Electron diffraction, 72
Electron-electron interaction,
     and cyclotron resonance
     lineshape, 354-357
Electron gas, 2D, 126-127
    (see also Cyclotron
     resonance)
Electron-hole interactions, in
     lasers, 286
Electron-hole pairs
  optical effects in quantum
     wells
    absorption, 424-425
    electroabsorption, 429
    nonlinear, 425, 426-428
  transitions, uncorrelated,
     263-264
Electronic light scattering,
     309
Electronic states
  energy levels in doped
     heterolayers, 31-35
  energy levels in quasi-unidi-
     mensional structures,
     35-40
    density of states, 38-40
    energy spectrum, 35-38
  envelope function approxi-
     mation, 21-24
  examples
    GaAs-AlAs short-period SL,
     26-28
    GaAs-GaAlAs heterolayers,
     24-26
    in-plane dispersion rela-
     tion and density of
     states, 28-31
  tunneling structures
    graded-parameter SL, 409,
     410
    superlattice tunnel diode,
     406
    theory, 415, 416
Electron localization, short-
     period SLs, 27
Electron mobility, in modula-
     tion doped structures
  quantum wells, 33
  strained-layer SLs, 126
Electron-spin-resonance,
     quantum Hall effect,
     243

Electron temperature, electric
    field and, 220
Energy gaps (see also
    Bandgaps; Gaps)
  vs. lattice constants, 5, 6
  quantum Hall effect
    fractional, 250, 257-258
    Landau quantization, 235,
      236
  SPS barriers, 55
  variable, 14
Energy levels
  cyclotron resonance, 349
  in doped heterolayers, 31-35
  electronic light scattering,
    311-312
  envelope function, 21-24
  excitons, optical methods,
    276
  impurity, 280
  in quasi-unidimensional heter-
    ostructures, 35-40
  reflectivity modulation tech-
    niques, 270
  relaxation between Landau
    levels, 209-210
Energy spectrum
  in quasi-unidimensional heter-
    ostructures, 35-38
  of 2D hole, FIRS, 341
Energy splitting, see Split-
    ting
Envelope function approximation,
    21-24, 71, 263, 265
  tunneling current, 161
Epitaxy, 104
  layers, confinement by, 43
  MBE process, see Molecular
    beam epitaxy
  strained-layer, 105, 113
  and superlattice growth, 4-6
Equilibrium theory
  and MBE processes, 105-106
  misfit dislocations, 111-113
Esaki diode, 159, 407
Esaki-Tsu superlattice, 57
Etalons, Fabry-Perot, 426
Evaporation, MBE process, 45
Excitation spectra
  magneto-optics
    quantum wells, 360-361
    superlattice, with field
      parallel to layers,
      372-373
  photoluminescence, 269-270
Excitons, 52
  absorption, 268, 424-425
  AlAs-GaAs SLs, 65-66
  barrier thickness and, 57-58
  dynamics of excitations, 279

Excitons (continued)
  electroabsorption, 428, 429
  energy level determina-
    tion, 276
  FIRS subband resonances,
    grating coupler-induced,
    338
  free exciton luminescence,
    275
  interband magneto-optics,
    357-360
  interface disorder, 278
  nonlinear, 425, 426, 427,
    428
  photoluminescence studies,
    267, 269, 270
  in quasi-2D systems, 43
  selection rules, 266-267
  time-evolution, 270
  transition probability,
    264-265
  trapping in short-period
    SLs, 27
Fabry-Perot etalons, 426
Fabry-Perot resonator, 383
Faraday configuration, 349
Far-infrared spectroscopy
    (FIRS)
  dielectric response, 321-
    329
    grating coupler, 328-329
    nonradiative waves, 326-
    328
    transmission at normal in-
      cidence, 326
  experimental results, 329-
    344
    cyclotron resonance, 342-
    344
    plasmons, 2D, 329-336
  experimental techniques,
    318-321
Femtosecond spectroscopic
    techniques, 426
Fermi-Dirac distribution, 134
    162, 191
Fermi-Dirac occupancy factor,
    286
Fermi energy, 176, 178
  from cyclotron mass, 349
  k-space spectroscopy, 281
  multiple subband occupancy,
    220-224
  nonparabolicity, 351
  quantum Hall effect, 234-
    235, 239-240
  rectangular-potential model,
    163
  2D gas, 217
Fermi length, 133

Fermions, 426-427
Fermi's golden rule, 262
Fermi sphere, 184
Field effect transistors, 12, 16, 127, 230 (see also Metal-insulator-semiconductor systems)
Filling factor, 190
  cyclotron resonance spectroscopy, 202, 206
  Landau levels
    energy relaxation time between, 209-210
    and Landau level maxima, 200-201
  magnetic field and, 354-357
  multiple subband occupancy, 221
  quantum Hall effect, 243-244 250, 255
Film strain, 108, 113
Finite quantum wells, selection rules, 267
Finite-thickness effect, plasmon dispersion, 330-332
Flip-flop current, 417
Flux quantum, quantized Hall resistivity, 251
Flux transient, MBE, 47
Folded acoustic phonons, 305, 306, 307
Forbidden LO-phonon scattering, 307-309
Formfactor, plasmon frequency, 332
Fourier analysis, 221, 223
Fourier component
  charge density distribution, plasmon, 336
  grating couplers, 328-329
Fourier series, charge density, 336
Fourier transform spectroscopy, 319-320
Fowler-Nordheim tunneling, 163 165, 167
Fractional charge, quasi-particles, 249
Fractional filling, 12
Fractional Quantum Hall effect, see Quantum Hall effect
Franz-Keldysh effect, 429
Free carrier states, quantization of, 217
Free electrons
  absorption experiments, 357-358
  luminescence, 275

GaAlAs, tunneling, 172
  through indirect-gap barriers, 179-182
  X-point states, 183, 184
GaAlAs-GaAs-GaAlAs, 172
GaAs, 144
  FIRS, 334
  low-dimensional systems, 336
  MBE process, 49-50
  minority electrons, 144
  optical effects in quantum wells
    absorption, 424-425
    nonlinear, 426
  plasmon resonances, 331
  scattering
    Coulomb, 136-137, 139
    interface, 140
    optical phonon, 143-144
  transport, 135
GaAs-AlAs, energy states, 26-28
GaAs-AlGaAs, 398
  interface disorder, 278
  nonlinear optical effects in quantum wells, 426
  scattering, alloy, 140
GaAs-GaAlAs, electronic states, 24-26
GaAs-GaAlAs-GaAs, 165
GaAs-GaAsP SL, 4
Gain
  lasers, 287, 288, 289
  low-dimensional systems, 290-291
GaInAs-AlInAs
  multiple subband occupancy, 221, 222
  parallel field depopulation, 227, 228
GaInAs-InAlAs, 278
GaInAs-InP
  interface disorder, 278
  parallel field depopulation, 227, 228
GaInAsP-based single QW lasers, 288
Gain curves, 286
Gaps (see also Bandgaps; Energy gaps)
  energy and mobility, 257
  graded
    graded gap transistors, 378-383
    GRIN-SCH lasers, 409, 411
  superlattice, 365, 366
  tunneling through indirect-gap barriers, 179-182
Gas, MBE source materials, 66

Gas, 2D electron (see also
Cyclotron resonance)
  parallel field depopulation,
    226
  two-dimensional, 217
GaSb-InAs-GaSb, MBE-grown, 13
Gate voltage, 335, 340
Gauge, Landau, 250
Gauge transformation, 252
Gaussian density of states,
  (see Density of
  states, Gaususian
Ge content
  Raman spectroscopy, 303
  strained-layer superlattices,
    118, 123-124
Geometry
  and cyclotron effective
    masses, 350-353
  of dipoles in quantum wells,
    266
  Hall devices, 232
  Raman back-scattering, 302
  scattering, 311-312
  Shubnikov-de Haas effect,
    218, 220
G-factor, 220, 243
Gliding, 110
Graded gap structure, 409, 411
Graded gap transistors, 378-383
Graded-index separate confinement heterostructure
  (GRIN-SCH) laser
  56, 285, 287
Graded-parameter superlattice,
  409-411, 412, 413
GRIN-SCH lasers, 56, 285, 287
Ground state
  bandgap-engineered devices,
    386
  with decrease in dimension-
    ality, 4
  quantum Hall effect, frac-
    tional, 250
  quasi-particles, 249
Group II-VI systems, 6 (see
  also Molecular beam
  expitaxy, Hg-based
  structures)
  envelope function approxi-
    mation, 21
  LO-phonon oscillations, 269
  photoluminescence excitation
    spectra, 269
Group III elements MBE process
  45, 52, 63
Group III-V systems, 6, 43,
  172, 176
  avalanche multiplication, 379
  envelope function approxi-
    mation, 21

Group III-V systems (continued)
  hot-electron transistor, 394
  interatomic distances, 101
  MBE processes, see Molec-
    ular beam epitaxy,
    group III-V systems
  optical effects in quantum
    wells
  electroabsorption, 430
  photoluminescence exci-
    tation spectra, 269
  resonant tunneling, 174, 175
  tetrahedral covalent radii,
    102
  tunneling structures, 410
    413
Group IV elements, interatomic
  distances, 101
Group IV-VI systems, 6
Group V elements, MBE process,
  46-47, 50, 66
Growth axis, applied electric
  field, 30
Growth chamber, 50
Growth interruption, 53-54
Growth rate, 49
Growth techniques, 378
  epitaxy, 4-6
  interface disorder, 287
  MBE process
    group III-V semiconduc-
      tors, 44-54
    Hg-based structures, 73-76
  strained-layer epitaxy, 113

Hall coefficient, 220
Hall effect (see also Quan-
  tum Hall effect)
  modulation doped strained-
    layer superlattice,
    126
  multiple subband occupancy,
    221
  physics, 230-234
  Shubnikov-de Haas effect,
    218, 219
Hall mobility
  as function of doping phase
    angle, 125
  MBE process, 91, 92, 93
Hall plateau, 238, 243, 253
Hall resistance, 14
Hall voltage, 218, 221, 253
Hartree approximation, 31
Hartree potential, 338
Heat capacity, 191
Heavy holes, 9
  AlAs-GaAs SLs, 65-66
  bandgap-engineered devices,
    386

Heavy holes (continued)
  barrier thickness, 59-60
  envelope function approxi-
      mation, 24
  FIRS, 341
  in-plane dispersions, 29-30
  mixing, 30
  optical properties
    band mixing, 266
    barrier thickness and, 57,
      58
    circularly polarized exci-
      tations, 275, 276
    electroabsorption, 431
    interband magneto-optics,
      364-365
    nonlinear, 426
  and tunneling, 179
  valence band structure, 362,
      364-365
HEMT-technology, 417
Heteroepitaxy, 1, 4-6, 312
Heterostructures
  electronic states, see
      Electronic states
  lattice-mismatched, 14
  magneto-optical properties,
      see Magneto-optical
      properties
Hg-based microstructures,
      MBE, see Molecular
      beam epitaxy, Hg-based
      structures
HgCdTe
  cyclotron resonance, 349
  multiple subband occupancy,
      221, 223, 224
  parallel field depopulation,
      225, 226, 227, 228
Hg compounds, 6
High-angle X-ray diffraction,
      53-54
High electron mobility transis-
      tor, 57
Hole energy, tunneling proba-
      bility versus, 8
Hole mobility, 91-93, 144
Holes, 312 (see also Heavy
      holes; Light holes)
  absorption experiments,
      357-358
  band mixing, TM emission,
      281
  FIRS, 341
  magnetotunneling, 179
  MBE process, 91-93
  transport, 135
Hole splitting, AlAs-GaAs SLs,
      65-66
Hopping conduction, 151, 258

Hot carriers, 144-145
Hot-electron injection, 409
Hot-electron spectroscopy,
      395-398
Hot-electron transistor
  design, 393-395, 398-400
  tunneling, 413
Hot phonon effects, hot car-
      riers, 145
Hydrides, MBE process, 66
Hydrostatic pressure, short
      period SLs, 27

Imperfections, see Defects
Impurities, 33
  cyclotron emission spectra,
      204-206
  cyclotron resonance, 349
  electron interactions, 354-
      357
  and Landau levels, 189
  optical properties
    luminescence, 280
    quantum well quantum effi-
      ciency, 274-275
  quantum Hall effect, 239-240
  and scattering, Coulomb,
      135-139, 312
  supperlattice tunnel diode,
      407
InAlAs, hot-electron transis-
      tor, 398
InAs
  cyclotron resonance, 349
  nonparabolicity, 351
InAs-GaSb system, 13
Indirect bandgap, 55
  Brillouin zone, 128
  tunneling, 179-182, 407
Inelastic diffusion, 151
Inelastic mean free path, 2
Inelastic scattering, (see
      also Raman spectro-
      scopy)
  phonon scattering rate, 143
  and tunneling resonances,
      185
Inelastic tunneling, 183
Infinite square well, 267
Infrared laser, solid-state,
      384
Infrared photoconductors, 384
Infrared spectroscopy
  FIRS, see Far-infrared
      spectroscopy
InGaAs, hot-electron transis-
      tor, 398
InGaAs-InAlAs, 398
InGaAs-InP, virtual states,
      176

InP-GaInAs, energy levels in doped heterolayers, 32
InP-InGaAs-InP, alloy scattering, 140
In-plane dispersions, 28-31
InSb, cyclotron resonance, 349
Integer filling factor, 356
Integrated circuits, strained-layer SL, 104
Interband magneto-optics in quantum wells (see also Magnetotransport)
  complex valence band, 360-365
  excitonic effects, 357-360
  magnetic quantization, 371
Interband transitions
  complex valence bands, 363
  dielectric function, 262
Interdiffusion
  at elevated temperatures, 305
  MBE process, 76-79
Interface dipole, 96
Interface disorder (see also Disorder)
  electron interactions, 354-357
  in low-dimensional systems, 289-290
  MBE process, 53-54
  optical methods, 278-279
  short-period SLs, GaAs-AlAs, 27
Interfaces (see also Scattering)
  MBE process, Hg-based structures, 76, 78, 81
  optical properties, quantum well quantum efficiency, 274-275
  scattering
    Coulomb, 136-137
    mechanisms of, 139-140
  tunneling current, 161
Inter-Landau level coupling, 416
Interstitial motion, MBE process, 80
Intersubband resonances
  barrier thickness and, 58
  FIRS, 320, 321, 322, 329, 336-342
Intraband absorption, 321 (see also Cyclotron resonance)
Intrasubband excitations, Raman spectra 311-312
Inversion layers, 330
Inversion symmetry splitting, 24, 25
Ionic scattering, 31, 137-138
Ionization threshold energy, avalanche multiplication, 378-379

Kane model, 22
Knudsen cells, 45-46

Kramers degeneracy, 30, 31, 32
Kramers-Kronig relations, 262
Kronig Penney model, 59, 365, 366, 405, 415

Landau gauge, 217, 235
Landau levels
  Bohr-Sommerfeld quantization, 352
  cyclotron resonance, 202-209, 349, 354-355
  density of states for 2D electrons
    cyclotron resonance spectroscopy, 202-209
    energy relaxation between levels, 209-210
    magnetization, 194-197
    magnetocapacitance, 197-198
    from radiative recombination spectra, 199-201
    specific heat, 190-194
    temperature dependence of resistivity, 198-199
  FIRS
    cyclotron resonance, 344
    intersubband coupling, 339
  formation of, 218, 219
  fractionally quantized state, 252-253
  interband magneto-optics in quantum wells, 360
  low-dimensional systems, 291
  magnetic field and, 354-357
  magneto-optical properties, see Magneto-optical properties
  multiple subband occupancy, 220-224
  nonparabolicity, 351
  quantum Hall effect
    physics of, 234-238
    real conditions, 239-245
  resonant tunneling via, 176-179
Landau quantization, 234-238
Larmor radius, 367
Lasers
  infrared, 384
  photoluminescence studies, see Photoluminescence
  principles of action in quantum wells, 282-289
Lattice constants, energy gaps at 4.2K vs., 5, 6
Lattice image, 52
Lattice mismatch, 4, 14 (see also Strained-layer superlattices)
  critical thickness vs., 105-106
  forbidden LO-phonon intensity, 309
  MBE process, Hg-based structures, 77, 90

Lattice vibrations, scattering, 140-143
Layered system, dielectric function, 324
Layer thickness (see also Thickness)
  forbidden LO-phonon intensity, 309
  GaAs-GaAlAs superlattices, 365
  magnetic quantization, 365, 368-370
  superlattice tunnel diode, 405, 406, 407
  tunneling structures, graded parameter SLs, 410
LCAO, 85
Lifetimes
  Landau level, 208, 209-210
  resonant vs. sequential tunneling, 185
Light holes, 9, 10, 92
  AlAs-GaAs SLs, 65-66
  barrier thickness, 59-60
  envelope function approximation, 24
  FIRS, 341
  in-plane dispersions, 29-30
  mixing, 30
  optical properties
    band mixing, 266
    barrier thickness and, 57, 58
    circularly polarized excitations, 275, 276
    complex valence bands, 364-365
  nonlinear, 427
  resonant tunneling, 175
  and tunneling, 179
    valence band structure, 362
Light scattering, Raman spectroscopy, 309-312
Linearly graded gap, 409
Linewidth
  cyclotron resonance, 207-209, 354-357
  low-dimensional systems, 289-291
Localized states
  Landau levels, 238
  quantum Hall effect, 239-240, 243
  short-period SLs, GaAs-AlAs 27
  transition from weak to strong, 139
Longitudinal optical phonons, 302

LO-phonons
  forbidden, 307-309
  group II-VI QWs, 269
  optical effects in quantum wells
    absorption, 424-425
    electroabsorption, 429
Loss rate, MBE process, 48
Low-dimensional electronic systems, 336
  energy levels in quasi-one dimensional heterostructures, 35-40
Luminescence (see also Photoluminescence)
  barrier thickness and, 58
  impurity-related, 280
Luminescence spectrum (see also X-ray photoelectron spectroscopy)
  dynamics of excitations, 279
  free exciton luminescence, 275, 276
  modulation-doped GaAs MQW, 281
  transient, 273

Magnetic compounds, 6
Magnetic field (see also Landau levels; Quantum Hall effect)
  cyclotron resonance, 348-349
  envelope function, 23-24
  interband magneto-optics in quantum wells, 360-365
  low-dimensional systems, 290-291
  and negative differential conductance, 409
  parallel field component depopulation, 224-228
  quantum Hall effect, 255
  parallel to SL layers, 365-374
    classical orbits, 367-370
    tunneling structures, 415
  quantization of free carrier states, 217
  quantized Hall effect, 12, 255, 256
  tunneling current in, 177-178
  tunneling structures, graded parameter SL, 410
Magnetic-field-induced freeze-out, 139
Magnetic SLs, 15
Magnetization
  Landau level width at half maximum, 208

Magnetization (continued)
  2D electron gas, 194-197
Magnetocapacitance
  CR-measurements, 342, 343
  density of states, 2D elec-
    trons, 197-198
  FIRS, cyclotron resonance,
    344
Magneto-optical properties
  interband
    complex valence band, 360-
      365
    excitonic effects, 357-360
    magnetic quantization, 371
  intraband absorption, 348-357
    complex valence band, 360-
      365
    cyclotron resonance, 348-
      349
    cyclotron resonance line-
      width, 354-357
    excitonic effects, 357-360
    non-parabolicity, effects
      of, 350-353
  Landau levels, see Landau
    levels
  magnetic quantization in SLs
    in parallel magnetic
    field, 365-374
Magnetoplasmons, 329
Magnetoresistance, 14, 217, 226
Magneto-Stark-phonon resonances,
  415
Magnetotransport
  fractional quantum Hall
    effect, 249-258
    energy and mobility gaps,
      257-258
    experimental picture, 252-
      257
    theory, 250-252
  multiple subband occupancy,
    220-224
  parallel field depopulation,
    224-228
  superlattice tunnel diode,
    406
  tunneling structures, 415
Magnetotunneling, 179
Manufacturing techniques
  lasers, 288
  MBE, see Molecular beam
    epitaxy
Matrix elements, 148
  envelope function, 23
  · optical properties
    lasers, 286
  transition, 364
Matthews theory, 119
MBE,   see Molecular beam
    epitaxy
Mean free path, 3

Memory element, N-state, 383
Merwe theory, 119
Mesoscopic quantum regime, 2,
  3, 4
Metal-insulator-semiconductor
  (MOS) systems, 12
  cyclotron resonance line-
    width, 355
  FIRS, 317, 334-335, 341
  magnetocapacitance, 197
  Si-MOSFETs, 12, 201, 217
Metallo-organic chemical vapor
  deposition (MOCVD),
  4-6, 16, 378
  hot-electron transistor, 394
  interface disorder, 278
Metallo-organic molecular beam
  epitaxy (MOMBE), 4-6,
  66
Metrology, quantum Hall
  effect, 245-246
Minibands
  Brillouin zone folding, 128
  magnetic quantization, 373
  negative differential con-
    ductance, 405-409
  superlattice, 365, 366
  superlattice tunnel diode,
    406, 407
Minority carrier mobility, 144
Misfit dislocatioins
  forbidden LO-phonon inten-
    sity, 309
  mechanism for generation,
    120
Mismatch, see Lattice mis-
    match; Strained-layer
    superlattices
Mobility
  minority carrier, 144
  optical techniques, dynam-
    ics of excitations,
    279
  quantum Hall effect, fract-
    ional, 257-258
  scattering, by acoustic
    phonons, 141-143
MOCVD, see Metallo-organic
    chemical vapor deposi-
    tion
Modal gain, lasers, 287
Modulation-doped field-effect
    transistor, 12, 16, 127
Modulation doping, 11-12, 31,
    32 (see also Doping)
  electron mobility in quantum
    wells, 33
  hot-electron transistor, 399
  luminescence, 280-282
Molecular beam epitaxy (MBE),
  15, 377-378
  equilibrium theory vs., 133

Molecular beam epitaxy (MBE),
      (continued)
   GaSb-InAs-GaSb quantum wells,
      13
   Raman spectra measurement,
      304
   resonant tunneling, 174
   superlattice types, 14-15
Molecular beam epitaxy, group
      III-V systems, 4, 43-66
   engineering on atomic scale,
      54-66
      barrier thickness, 57-63
      short-period SLs, 54-57
      ultrathin layer and mono-
         layer SLs, 63-66
      vs. equilibrium theory, 105-
         106
   gaseous source materials, 66
   hot-electron transistor, 394
   silicon, 121-122
   technology, solid source ma-
         terials, 44-54
      materials, 46-50
      mointoring, 51-54
      sources, 45-46
Molecular beam epitaxy, Hg-
      based structures
   bandgaps, SL, 80-85
   growth, 73-76
   interdiffusion, 76-80
   valence-band discontin-
      uity, 85-90
Monolayer superlattices
   alloy-like behaviour, 55
   MBE process, 63-66
Monomers, MBE process, 45
Monte Carlo methods
   hot carriers, 395, 396,
      397, 398
   transport, 147-148
Multi-band tight binding model,
      90

Narrow band gap semiconductors,
   221-224
N-channel SiSiGe MODFET, 127
N-doped structures, see
      N-type doping
Negative differential conduc-
      tance
   CHIRP, 411
   graded-parameter superlat-
      tice, 411
      magnetic field and, 178
      superlattice tunnel diode,
         405-409
Negative transconductance, 413-
      414

Nonlinear optical effects in
      quantum wells, 425-428
Nonparabolicity
   barrier thickness and, 58,
      60
   and cyclotron resonance line-
      width, 354
   effects of, 350-353
   valence subbands, 30
Nonradiative recombination,
      274-275
Nonradiative waves, FIRS, 326-
      328
Non-resonant tunneling, see
      Tunneling
Non-zero spin orbit coupling,
      30
Non-phonon excitonic PL trans-
      ition, 65-66
N-state memory element, 383
N-type doping
   energy levels, 32, 34, 35
   FIRS, 321
   hot-electron transistor,
      394
   lasers, 288
   MBE process
      group III-V devices, 44-45,
         49, 54-55
      Hg-based structures, 91,
         92, 93
      secondary implantation, 122,
         124-127
      short-period superlattices,
         56

Ohmic conductance, see
      Transport
One-band Kronig-Penney model,
      59
One-dimensional Schrödinger
      equation, 161
One-dimensional subbands, 336
One-dimensional systems, 289-
      291
   density of states, 3, 4, 177
   transport in, 135
One-transistor flip-flop, 417
Optical absorption, 8-9, 10
Optical confinement factor,
      283
Optically pumped laser, 8
Optical mode, hot carriers,
      145
Optical phonons
   energy relaxation between
      Landau levels, 209-210
   Raman spectroscopy, see
      Raman spectroscopy

Optical phonons (continued)
scattering, 143-144
Optical properties (see also
Far-infrared spectro-
scopy; Magneto-optics;
Raman spectroscopy
applications, 273
carrier capture, 277
energy level determinat-
tions, 276
excitation dynamics, 279-
280
free exciton luminescence,
275-276
impurity-related lumi-
nescence, 280
interface disorder, 278-279
modulation-doped samples,
280-281
quantum efficiency of QWs,
273-275
barrier thickness, 57-63
laser action principles, 282
additional properties, 288-
289
background, 282-283
single quantum well oper-
ation, 284-288
one- and zero-dimensional,
289-291
opto-electronics in quantum
wells
absorption, 424-425
electroabsorption, 428-432
nonlinear effects, 425-428
optical transition probability, 262-267
exciton effects, 264-265
selection rules, 265-267
uncorrelated electron-hole
pair transitions, 263-
264
techniques of optical spectroscopy, 267-273
absorption spectra, 267-269
excitation spectra, 269-270
photoluminescence, 267, 268
reflectivity, 270-271
transient measurement, 271-
273
Optic phonon-plasmon modes, hot-
electron spectrum, 395,
396, 397
Orbits
cyclotron, 356
in superlattice with magnetic
field parallel to layers,
365-374
Oscillator strength, 425

Parabolic dispersion relation,
370-371
Parallel excited ISR, 228
Pauli exclusion principle,
426-427
P-channel SiGe-Si-MODFET's,
127
P-doped structures, see P-
type doping
Peak-to-valley ratio
superlattice tunnel diode,
406, 407
tunneling structure, 413
CHIRP, 411
graded-parameter SL, 410
Perturbation theory
plasmon dispersion, 336
quantum mechanical time-
dependent, 322-323
time-dependent, 262
Phase-coherent distance, 1, 3
Phonon replicas, short-period
SLs, 27
Phonons
hot-electron spectrum, 395
396, 397
and Landau levels, 189
optical effects in quantum
wells
absorption, 424-425
electroabsorption, 429
tunneling structures, 415
Phonon scattering (see also
Raman spectroscopy)
acoustic mode, 140-143
optical mode, 143-144
quasi-one-dimensional system, 135
Phosphorus, MBE process, 50,
66
Photoconductivity, tunnel,
384, 410
Photocurrent spectroscopy,
8-9, 10
Photoelastic coefficients, 305
Photoluminescence, 8-9, 10
(see also X-ray
photoelectron spectros-
copy)
bandgap-engineered devices,
386
barrier thickness, 58, 60-62
free exciton luminescence,
275
impurity related, 280
MBE process
group III-V structures,
63-66
Hg-based structures, 83,
84, 86

Photoluminescence (continued)
  MBE process (continued)
    modulation-doped GaAs
      MQW, 281
    quantum Hall effect, frac-
      tional, 252, 253, 254,
      255
  techniques, 267-273
Photoluminescence excitation
    (PLE) method, 269-270,
    281
Photomultipliers, graded-gap
    devices, 378-383
Photoreflectance experiments,
    270-271
Photosensitivity, graded-para-
    meter SL, 410
Piezoelectric scattering, 140-
    143
P-i-n structure, 409, 430-431
Plasma
  electroabsorption, 429
  optical effects in quantum
    wells, 425, 426, 427
Plasmons
  FIRS, 326-329
    experimental results, 329-
      336
    grating coupler, 328-329
    intersubband coupling, 339-
      340
    hot-electron spectrum, 395,
      396, 397
    layered dispersion of, 312
    surface, 326-328
Plastic accommodation, 105-113
P-like degeneracy, 29
Polaritons, surface plasmon,
    326
Polarization (see also Circu-
    larly polarized radia-
    tion)
  bandgap-engineered devices,
    386
  circular polarization spec-
    tra, 275-276
  of scattered light, 311
Polarized luminescence experi-
    ments, 266-267
Potential barrier, 429 (see
    also Barriers; Tun-
    neling)
Pressure
  short-period SLs, GaAs-AlAs,
    27
  and tunneling, 181
Pseudomorphic growth, 113
P-state, dipole moment, 265,
    266
P-type doping
  bandgap-engineered devices,

P-type doping (continued)
    381
  confinement layers, short-
    period superlattices,
    56
  energy levels, 32
  lasers, 288
  MBE process
    group III-V semiconduc-
      tors, 44-45, 49
    Hg-based structures, 91,
      92, 93, 96
    nipi superlattices, 378
    resonant tunneling in
      double-barrier struc-
      tures, 7, 8
Pump-probe absorption experi-
    ments, 272

Q-dependence plasmon disper-
    sion, 330
Quantum Confined Stark Effect
    (QCSE), 430
Quantum efficiency, quantum
    wells, 273-275
Quantum electrodynamic (QED)
    calculations, 246
Quantum Hall effect, 12, 218
  fractional
    density of states oscill-
      ations, 198
    energy and mobility gaps,
      257-258
    experimental picture, 252-
      257
    filling factor, 255
    theory, 250-252
  multiple subband occupancy,
    221, 222
  physics and applications,
    229
    applications in metrology,
      245-246
    Hall effect, 230-234
    Landau quantization,
      234-238
    under real conditions,
      239-244
Quantum liquid, Laughlin ap-
    proach, 249, 251
Quantum states
  bandgap-engineered devices,
    383
  of free carrier state, 217
  magnetic, in superlattices,
    365
  semiconductor laser, 282
  subband, 220
Quantum well lasers
  operation, 284-288
  properties, 288-289

Quantum wells
  coupling, barrier thickness
    and, 58
  doped, 33
  optics see Magneto-optical
    properties; Optical
    properties
  optoelectronics, see Opto-
    electronics in quantum
    wells
  quantum efficiency, 273-275
  single
    evolution to superlattice,
      368-371
  in-plane dispersions, 29-30
    lasers, 284-288
  in superlattices, magnetic
    quantization, 365
  tunneling, resonant, 6-8,
    167, 405, 407
Quantum well thickness
  energy levels, in doped
    heterolayers, 34
  low-dimensional systems,
    290-291
  and tunneling in SLs, 405,
    407
Quasi-bound states
  resonances, 175
  resonant tunneling, 168
  tunneling probability, 170
Quasi-graded gap structure,
  411
Quasi-particles
  condensation of, 249
  energy and mobility gaps,
    257
  Laughlin approach, 250-251
Quasi-2D systems, 2, 43
Quasi-unidimensional heterostructures
  energy levels, 35-40
transport in, 135

Radiative efficiency, lasers,
  287
Radiative recombination
  optical properties
    lasers, 286, 287
    quantum well quantum effi-
      ciency, 274
  2D density of states from,
    199-201
Raman spectroscopy, 9, 11
  allowed phonon scattering,
    302-304
  electronic light scattering,
    309-312
  forbidden LO-phonon, 307-309

Ramon spectroscopy (con-
    tinued)
  MBE process, Hg-based struc-
    tures, 76, 86
  superlattice phonons, 305-
    307
Random ternary alloy, 55
Rayleigh scattering, 279, 280
Real-space energy band dia-
    gram, 55, 56
Recombination
  in low-dimensional systems,
    289-290
  quantum well quantum effi-
    ciency, 274-275
  radiative, see Radiative
    recombination
Rectangular barriers, 164,
    166-170
Rectangular double barriers,
    160
Rectangular-potential model,
    163
Rectangular quantum wells
  carriers at Fermi energy,
    353
  degeneracy, 30
  nonparabolicity, 351
  quasi-unidimensional hetero-
    structures, 36-37
Rectangular quantum wires,
    36-37
Reduced dimensionality, 16
    (see also Low dimen-
    sional electronic sys-
    tems)
Reflection high energy elec-
    tron diffraction
    (RHEED), 51-53, 87
  interface disorder, 278
  MBE growth, 51-53, 63, 87
Reflectivity, ultrafast meas-
    urement, 272
Reflectivity modulation tech-
    niques, 270-271
Relaxation times
  graded-parameter superlat-
    tice, 412
  between Landau levels, 209-
    210
Resistivity (see also Quan-
    tum Hall effect)
  coherent scattering, 151
  Hall device, 233
  Hall resistance, 231
  Shubnikov-de Haas effect,
    218, 219
  2D electron, 198-199

Resonance
  optical effects in quantum
      wells, 425, 427, 428
  plasmon, 340
Resonance energy, FIRS, 338
Resonant absorption, 425
Resonant electron tunneling, 1
Resonant Rayleigh scattering
      intensity, 279
Resonant scattering, see
      Scattering
Resonant tunneling, see Tun-
      neling, resonant
Roughness (see also Disorder)
  electron interactions, 354-
      357
  and scattering, 139-140
Rydberg energy, 357, 358

Sample preparation chamber,
      MBE, 50
Saturation
  cyclotron resonance spec-
      troscopy, 209
  energy relaxation time be-
      tween Landau levels,
      209-210
Sb doping, 125, 127
Scattering, 133
  coherent, 149-150
  cyclotron resonance spec-
      troscopy, 207-209
  dynamics of excitations, 279
  energy levels in doped heter-
      olayers, 31
  hot-electron spectrum, 395
  and Landau levels, 189, 209-
      210
  mechanisms, 135-144
    alloy, 140
    Coulomb, 135-139
    deformation potential, 140-
      143
    hole mobility, 144
    interface, 139-140
    optical phonon, 143-144
    piezoelectric, 140-143
  one-dimensional system, 135
  Raman spectra, 311-312
  superlattice tunnel diode,
      407
  tunneling, 404
    theory, 415
    X-point, 184
  hot-electron spectrum, 395
Scattering time, 207-209, 220
Schottky barriers, 307
Schrödinger equation, 31
  for isolated well, 167

Schrödinger equation (contin-
      ued)
  one-dimensional, 161
  quantum Hall effect, 235
  in quasi-unidimensional
      heterostructures, 35-38
  tunneling current, 161
  tunneling probability, 163
Screening
  by mobile carriers, 135
  scattering
    Coulomb, 138
    interface, 139
Screw dislocations, 111
Secondary implantation, 122,
      124-127
Secondary ion mass spectro-
      scopy (SIMS), 76
Se-Ge strained-layer SLs, 307
Selection rules, 265-267
Self Electro-optic Devices
      (SEED), 431-432
Semiconductor heterostruc-
      tures, resonant tun-
      neling in, 172-176
Separate confinement hetero-
      structure (SCH), 283,
      284, 285
Sequential tunneling
  graded-parameter SL, 410
  resonant tunneling versus,
      185-186
Seraphin coefficients, 270
Shallow donor, 55
Short-period superlattices,
      15, 44
  energy states, 26-28
  envelope function approx-
      imation, 24
  MBE, 54-57
Short range scatterers, 355
Shubnikov-de Haas oscilla-
      tions, 12, 198, 217,
      355
  multiple subband occupancy,
      220-224
  strained-layer superlattice,
      126
Si
  inversion layer scattering
    by acoustic phonons, 141
    Coulomb, 138-139
    interface, 140
    optical phonon, 143-144
  MBE process, 44-45, 49,
      121-122
  Raman spectroscopy, 303
  substrates, 104

SiGe
  band ordering, 122-124
  Raman spectroscopy, 303-304
  superlattices
    growth and properties, 121-129
    strained-layer, 14, 118
  Si-MOS, see Metal-insulator-semiconductor systems
Simulation programs
  transport physics, 146-149
  tunneling structures, 414-415
Single flux quantum, 251
Slip plane, 109
Solid-state infrared laser, 384
Solid-state photomultipliers, 378-383
Space charge, and tunneling, 404-407
Spatial modulation, short-period SLs, 26
Specific heat
  DOS measurement, 190-194
  Landau level width at half maximum, 208
Spectroscopy, see Far-infrared spectroscopy; Raman spectroscopy; X-ray photoelectron spectroscopy
Specular tunneling, 161, 185
Spherical symmetry, Boltzmann equation, 134
Spin degeneracy
  Boltzmann equation, 134
  FIRS, 341
Spin memory experiments, 276
Spin-orbit interactions
  barrier thickness and, 61
  FIRS, 341
  in-plane dispersions, 29
  quantum Hall effect, 255
Spin-polarized electron source, bandgap-engineered devices, 386
Splitting, 218
  barrier thickness and, 57-58
  FIRS, 341
  heterolayer electronic states, 24, 25
  of Landau levels, 220
  plasmon resonance, 336
  quantum Hall effect, 255
S-state
  dipole moment, 265, 266
  magnetic field dependence of, 358
Staircase APD, 379
Stark effects, 430
Stark-Landau system, 416

Stark regime, 415
Stark shifts, 9, 10
Sticking coefficient, 45
Stimulated emission, 8-9, 10
Stoichiometry, MBE, 47
Stokes shift, 27, 65-66
Strain
  FIRS, intersubband coupling, 339-340
  Raman spectroscopy, ultra-thin films, 303-304, 309
Strained-layer superlattices, 4, 14
  combination of heterostructure devices with ICs, 104
  envelope function approximation, 21
  growth and properties of Si/SiGe SLs, 121
    band order in SiGeSi, 122-124
    Brillouin zone folding, 128-129
    MBE processes, 121-122
    modulation doping, 124-127
  mismatch accommodation by strain or misfit dislocations, 105-113
    dislocations, 108-111
    equilibrium theory, 111-113
    tetragonal distortion, 106-108
  Raman spectroscopy, 303-304, 305, 306
  stability of, 114
    buffer layer design, 116-118
    buffer layer metastability, 119-120
    strain symmetrization, 115-116
  substrate material, 104
Strained overlayers, forbidden LO-phonon intensity, 309
Strain symmetrization
  and band edges of, 125
  by buffer layer, 117
  strained-layer SLs, 115-116
Stripe-geometry lasers, 288
Subbands (see also Cyclotron resonance; Intersubband resonances; Optical properties)
  barrier thickness, 60
  Boltzmann equation, 134
  cyclotron mass and, 349

Subbands (continued)
  FIRS, 336-342
  in-plane dispersions, 29-30
  Monte Carlo methods, 147-148
  multiple occupancy, 220-224
  nonparabolicity, 30
  parallel field depopulation,
      224-228
  Raman spectra, 311-312
  transport, 135
Sublimation, MBE process, 45
Substrate, 104
  absorption measurement, 268
  effective Ge-content, 123
  MBE process
    group III-V structures,
        49-50
    Hg-based structures, 74,
        77
  superlattices
    strained-layer, 118
    short-period, 56
Superlattices
  bandgap engineering, see
      Bandgap engineering
  development, 1-2, 4-6
  dilute-magnetic, 15
  doping, 13-14
  graded-parameter, 409-411
      412, 413
  magnetic quantization, 365-
      374
  magneto-optical properties,
      see Magneto-optical
      properties
  MBE process (see also Mo-
      lecular beam epitaxy)
    group III-V structures, 63-
        66
    Hg-based structures, 80-85
  Raman spectroscopy, 305-307
  short-period, 15
  strained-layer, see Strain-
      ed-layer superlattices
Superlattices tunnel diode,
    405-409
Surface morphology, MBE pro-
    cess, 49
Surface plasmon polaritons,
    326-328
Surface space charge layers,
    228
Symmetrical doping, 34
Symmetry
  heterolayer electronic
      states, 24, 25
  quantum Hall effect
    fractional, 252
    Landau quantization, 235
  strained-layer superlat-
      tices, 115, 117

Symmetry (continued)
  superlattice tunnel diode,
      407
Symmetry-breaking transi-
    tions, selection
    rules, 267

TEM studies, 52
  substrate-GaAs interface,
      274
  tunneling structures
    graded-parameter super-
        lattices, 410, 411, 412
    superlattice tunnel diode,
        407, 408
Tension, strained-layer super-
    lattices, 124
Ternary alloys, 6
  interface disorder, 279
  MBE growth of, 48, 52, 55,
      66
  ultrathin-layer SLs and, 55
Tetragonal distortion, 106-108
Tetrahedral covalent radii,
    102
Tetrameric molecules, MBE, 45
Thermal phonons, 424-425
Thermionic current, superlat-
    tice tunnel diode, 407
Thermodynamics, Landau levels
  magnetization measurement,
      194-197
  specific heat, 190-194
Thin films, Raman spectros-
    copy, 303
Thomas Fermi approximation,
    29
Three-dimensional carrier
    density, Hall resis-
    tance, 231
Three-dimensional systems
  Boltzmann equation, 133
    density of states, 3, 4
    scattering, interface, 139
Three-terminal superlattice,
    416-417
Tight binding approximation,
    71
Tilted magnetic fields, 339
    (see also Magnetic
    field, angle of)
Time-dependent phenomena, and
    tunneling, 404
Time resolution, ultrafast
    processes, 272, 273
TM emissions, 281, 282, 286
Transconductance, negative,
    413-414
Transfer-matrix method, 166-
    170, 169
Transient optical measure-
    ments, 271-273

Transistors
    graded-gap, 379
    hot-electron, 393-395, 397,
        398-400, 413
    resonant tunneling, 383
Transition probability, 262-
        267
Transitions
    free exciton luminescence,
        275, 276
    interband magneto-optics in
        quantum wells, 357-358
        complex valence bands,
            360-356
        cyclotron orbit and, 356
    magneto-optical properties
        of quantum wells, 363
    optical properties, 262-267
        barrier thickness and, 58
        lasers, 286
    between subbands, 322
Transit time, 170-171
Transmission, FIRS, 342, 343-
        344
Transport
    Boltzmann equation, 134-135
    bandgap-engineered devices,
        383
    coherence effects, 149-151
    device simulation, 146-149
    excitons, 279
    MBE process, 72, 90-95
    quantum Hall effect, 239-
        240, 244
    scattering mechanisms, 135-
        144
        alloy, 140
        Coulomb, 135-139
        deformation potential,
            140-143
        hole mobility, 144
        interface, 139-140
        optical phonon, 143-144
        piezoelectric, 140-143
    through superlattice bar-
        riers, 366
    superlattice tunnel diode,
        406
    theory, 134-135
    tunneling structures, 415,
        416
    2D carrier gases, 127
    type III to type I transi-
        tion, 90-95
    warm and hot carriers, 144-
        145
Trapping, short-period SLs, 56

Triangular well, 353
Tunable band discontinuities,
        386-389
Tunneling, 383-386
    hot-elecon injection, 394,
        397-398
    and quantum wells, 6-8
Tunneling, resonant, 1, 167
    bandgap-engineered devices,
        383-386
    charge accumulation, 171-172
    coherent scattering, 151
    current, 160-162
    electroabsorption, 429
    hot electron transistor, 413
    via Landau levels, 176-179
    probability, see Tunneling
        probability
    and quantum wells, 6-8
    in semiconductor heterostruc-
        tures, 172-176
    vs. sequential tunneling,
        185-186
    structures
        device implications, 416-
            417
        graded-parameter superlat-
            tice, 409-411, 412, 413
        modeling, 404
        other structures, 411,
            413-414
        simulations, 414-415
        superlattice tunnel diode,
            405-409
        theory, 415-416
        tunneling probability, 168
    through indirect-gap bar-
        riers, 179-182
    transit time, 170-171
    via X-point states, 182-184
Tunneling current, 160-162
    in magnetic field, 177-178
    non-resonant, 182
    through rectangular barrier,
        164
    X-profile states, 183
Tunneling current density, 171
Tunneling probability, 163-
        165, 171
    through double-barrier pro-
        file, 167-170
    vs. incident energy, 169
    resonant vs. sequential tun-
        neling, 185
    transfer-matrix method, 166-
        170
    WKB method, 165-166

Two-band model, non-parabolicity, 350
Two-dimensional electron systems, 189
  density of states, 3, 4
  epitaxial layers and, 43
  FIRS, 320, 321
  Hall devices, 232
  hot carriers, 145, 339
  quantum Hall effect, 12, 229, 230, 237–238
  scattering, interface, 139
  hot carriers, 145
  transport in, 2–3, 135
2D plasmons, see Plasmons
Two-particle excitation, 276
Type III–Type I transition, 90–95

Ultrafast measurements, 271–273
Ultrashort-period SLs, Raman spectra, 306, 307
Ultrathin structures, 1
  barrier thickness and, 58
  films, Raman spectroscopy, 303–304
  MBE process, 63–66
  superlattices, 44, 55
Ultrashort-period SLs, Raman spectra, 306, 307

Vacancies, MBE process, 77, 80
Valence bandedges, 2, 21
Valence band offset, 22
Valence bands, 266 (see also Bandgaps; Energy gaps; Gaps)
  barrier thickness and, 58, 62
  degeneracy, 29–30
  discontinuity
    MBE process, Hg-based structures, 85–90, 96
    tunneling, 384
  heavy-hole 9
  interband magneto-optics in quantum wells
    complex valence band, 360–365
    excitonic effects, 357–360
  short-period SLs, GaAs–AlAs, 27
  strained-layer superlattices, 124
Valence to conduction band absorption, 347
Valence subbands, see Subbands
Vapor phase epitaxy, 49
Vapor pressure data, MBE, 48

Variable-range hopping, 139
Velocity-field characteristics, 145
Vertical transport, 366
  short-period SLs, 405
  and tunneling in SLs, 55
Voltage
  magnetic field and, 178
  X-profile states, 183

Warm carriers, 144–145
Waveguides
  graded-index, 56
  SEED, 431
Wavelength dispersive spectroscopy (WDS), 76
Wavelength-selective voltage-tunable photodetector, 431
Weakly coupled parallel wires, 38
Wentzel-Kramers-Brillouin (WKB) method, 164, 165–166, 170, 185
Wigner crystal ground state, 250

X-ray diffraction, MBE process
  III-V structures, 53–55, 58, 62, 63–66
  Hg-based structures, 74–76, 79
X-ray photoelectron spectroscopy
  band offset measurement, 22
  MBE process, 72, 87, 89–90, 96
X-states
  optical properties, 61, 62–63, 65
  tunneling, 182–184, 410, 413
X-symmetry, SL tunnel diode, 407
X valleys, 27
X-well, optical properties, 62

Zero-bias anomaly, 406
Zero-dimensional systems, 3, 4, 289–291
Zero field mobility, 208, 355
Zero-gap semiconductors, 6
Zincblende type crystals, 341
ZnSe-ZnMnSe, 15
Zone folding, 128 (see also Brillouin zone)